S0-AKC-157

$$\frac{ac}{bd} = \frac{a}{b} \cdot \frac{c}{d} = \frac{a}{d} \cdot \frac{c}{b} \tag{2.45}$$

$$\frac{a}{d} = a \cdot \frac{1}{d} = \frac{1}{d} \cdot a \quad \text{provided } d \neq 0 \tag{2.46}$$

$$\frac{a^m}{a^n} = a^{m-n} \text{ if } a \neq 0, \, m \text{ and } n \text{ are positive}$$
$$\text{integers, } m > n \tag{2.47}$$

$$a^0 = 1 \quad \text{provided } a \neq 0 \tag{2.48}$$

$$\frac{a+b+c}{d} = \frac{a}{d} + \frac{b}{d} + \frac{c}{d} \quad \text{provided } d \neq 0 \tag{2.49}$$

$$(ax + by)(cx + dy) = acx^2 + (ad + bc)xy$$
$$+ bdy^2 \tag{2.50}$$

$$(x + y)^2 = x^2 + 2xy + y^2 \tag{2.51}$$

$$(x - y)^2 = x^2 - 2xy + y^2 \tag{2.52}$$

$$(a + b)(a - b) = a^2 - b^2 \tag{2.53}$$

$$a^2 - b^2 = (a + b)(a - b) \tag{2.54}$$

$$acx^2 + (ad + bc)xy + bdy^2 =$$
$$(ax + by)(cx + dy) \tag{2.55}$$

$$a^3 + b^3 = (a + b)(a^2 - ab + b^2) \tag{2.56}$$

$$a^3 - b^3 = (a - b)(a^2 + ab + b^2) \tag{2.57}$$

$$\frac{p}{q} = \frac{n}{d} \quad \text{if } pd = nq, \, q \neq 0, \, d \neq 0 \tag{3.1}$$

$$\frac{a}{b} = \frac{an}{bn} = \frac{a \div m}{b \div m} \quad \text{provided } b \neq 0,$$
$$n \neq 0, \, m \neq 0 \tag{3.2}$$

$$\frac{n}{d} = \frac{-n}{-d} = -\frac{-n}{d} = -\frac{n}{-d} \quad \text{provided } d \neq 0 \tag{3.3}$$

$$\frac{n}{d} \div \frac{p}{q} = \frac{n}{d} \cdot \frac{q}{p} \tag{3.4}$$

If $f(x)$, $g(x)$, and $h(x)$ are polynomials, then $f(x) = g(x)$ and $f(x) + h(x) = g(x) + h(x)$ are equivalent equations (4.1)

If k is a nonzero constant, then $f(x) = g(x)$ and $k \cdot f(x) = k \cdot g(x)$ are equivalent equations (4.2)

If $k(x)$ is a polynomial, then each root of $f(x) = g(x)$ is also a root of $k(x) \cdot f(x) = k(x) \cdot g(x)$ (4.3)

$$f(x) + h(x) > g(x) + h(x) \text{ for } f(x) > g(x) \tag{4.4}$$

$$f(x)h(x) > g(x)h(x) \text{ for } f(x) > g(x) \text{ and}$$
$$x \in \{x | h(x) > 0\} \tag{4.5}$$

$$f(x)h(x) < g(x)h(x) \text{ for } f(x) > g(x) \text{ and}$$
$$x \in \{x | h(x) < 0\} \tag{4.6}$$

$$\left(\frac{a}{b}\right)^n = \frac{a^n}{b^n} \tag{5.1}$$

$$(a^n)^m = a^{mn} \tag{5.2}$$

$$a^{-t} = \frac{1}{a^t} \text{ for } a \neq 0 \tag{5.3}$$

$$a^{1/k} = \sqrt[k]{a} \tag{5.4}$$

$$b^{j/k} = (b^j)^{1/k} = (b^{1/k})^j = \sqrt[k]{b^j} = (\sqrt[k]{b})^j$$
$$\text{if } \sqrt[k]{b} \text{ is real} \tag{5.5}$$

$$\sqrt[k]{ab} = \sqrt[k]{a}\, \sqrt[k]{b} \tag{5.6}$$

$$\sqrt[k]{\frac{a}{b}} = \frac{\sqrt[k]{a}}{\sqrt[k]{b}} \tag{5.7}$$

$$\sqrt[k]{\sqrt[j]{a}} = \sqrt[kj]{a} \tag{5.8}$$

$$d = \sqrt{(x_2 - x_1)^2 + (y_2 - y_1)^2} \tag{7.1}$$

$$\pi \text{ radians} = 180° \tag{7.2}$$

$$s = r\theta \tag{7.5}$$

$$\omega = \frac{\theta}{t} \tag{7.6}$$

$$v = rw \tag{7.7}$$

$$x^2 + y^2 = r^2 \tag{7.8}$$

$$\sin \theta \csc \theta = 1 \tag{8.1}$$

$$\cos \theta \sec \theta = 1 \tag{8.2}$$

$$\tan \theta \cot \theta = 1 \tag{8.3}$$

$$\tan \theta = \frac{\sin \theta}{\cos \theta} \tag{8.4}$$

$$\cot \theta = \frac{\cos \theta}{\sin \theta} \tag{8.5}$$

$$\cos^2 \theta + \sin^2 \theta = 1 \tag{8.6}$$

$$1 + \tan^2 \theta = \sec^2 \theta \tag{8.7}$$

$$1 + \cot^2 \theta = \csc^2 \theta \tag{8.8}$$

$$\cos (A - B) = \cos A \cos B + \sin A \sin B \tag{10.1}$$

Algebra and Trigonometry

Paul K. Rees
Louisiana State University

Fred W. Sparks
Texas Tech University

Charles Sparks Rees
University of New Orleans

Third Edition

McGraw-Hill Book Company
New York   St. Louis   San Francisco   Auckland   Düsseldorf
Johannesburg   Kuala Lumpur   London   Mexico   Montreal   New Delhi
Panama   Paris   São Paulo   Singapore   Sydney   Tokyo   Toronto

ALGEBRA AND TRIGONOMETRY

Copyright © 1962, 1969, 1975 by McGraw-Hill, Inc. All rights reserved. Printed in the United States of America. No part of this publication may be reproduced, stored in a retrieval system, or transmitted, in any form or by any means, electronic, mechanical, photocopying, recording, or otherwise, without the prior written permission of the publisher.

3 4 5 6 7 8 9 0 M U R M 7 9 8 7 6

This book was set in Baskerville by Ruttle, Shaw & Wetherill, Inc. The editors were A. Anthony Arthur and Shelly Levine Langman; the designer was Merrill Haber; the production supervisor was Joe Campanella.
The printer was The Murray Printing Company; the binder, Rand McNally & Company.

Library of Congress Cataloging in Publication Data

Rees, Paul Klein, date
    Algebra and trigonometry.

    1. Algebra. 2. Trigonometry. I. Sparks, Fred Winchell, date   joint author. II. Rees, Charles S., joint author. III. Title.
QA154.2.R43 1975 512'.13   74-16234
ISBN 0-07-051723-1

Contents

Preface

The topics presented in this, the third, edition of *Algebra and Trigonometry* are essentially the same as those in the earlier editions, namely, the topics that are essential for the successful study of analytic geometry and calculus. All the material has been carefully reviewed and touched up as needed for the sake of clarity, proper degree of rigor, and the modern spirit. Some of it has been completely rewritten, and there have been a few additions of new material as well as new treatments of previously included material.

The new material includes four sections leading to and on linear programming at the end of the chapter on Systems of Equations and of Inequalities as well as a section on Intercepts and Symmetry in the chapter Introduction to Polar Coordinates.

The material that has been rewritten includes Secs. 4.6 on linear inequalities, 4.8 on nonlinear inequalities, 7.1 on the distance formula, 7.2 on angles, and 7.10 on the domain and range of the trigonometric functions. It also includes Secs. 8.3 on identities, 10.1 to 10.6 on the cosine of an angle and identities, 11.6 on equations in quadratic form, 11.11 on quadratic inequalities, and 14.3 on variation. Finally, most of Chap. 6 on Relations, Functions, and Graphs, Chaps. 13 on Matrices and Determinants, 16 on Complex Numbers, and 27 on Probability have been rewritten. Much of the rewritten material was taken from *College Algebra,* Sixth Edition.

As in our other books, the exercises are a lesson apart, and the problems are in groups of four. The average class can cover the material adequately by doing every fourth problem, but some may need more or less drill. Answers are given to three-fourths of the problems in the back of the book, and the others are available in the Instructor's Manual. Essentially all the problems are new. There is a review exercise at the end of each chapter except for some of the relatively short chapters. There are about 95 regular and 20 review exercises with a total of some 3,800 problems.

A list of essential definitions, axioms, and theorems is printed on the endpapers for the convenience of the reader.

The original authors want to take this opportunity to welcome Charles Sparks Rees as a third author. He is the son of one of us and the namesake of the other.

Paul K. Rees
Fred W. Sparks
Charles Sparks Rees

ix

The Number System of Algebra

Mathematics is concerned with the task of building logical structures. Such structures begin with undefined terms and concepts. Other terms and concepts are defined in terms of these, and then assumptions are made about these terms and concepts. These assumptions are called *axioms,* which are statements that we agree to accept without proof. The structure is developed by proving statements called *theorems,* with each step in the proof based on the axioms and previously proven theorems.

Axioms

Theorems

We shall deal with the development and properties of the real number system, the application of the processes of arithmetic to letters and other symbols that represent numbers, and algebraic symbols and equations.

1.1 SETS

One of the basic concepts of mathematics is denoted by the word *set.* It is used in such everyday phrases as "a set of rules," "a set of dishes," "a set of drawing instruments," and in other expressions that refer to a collection. We assume that the reader is familiar with the meaning of the word *collection,* and we define a set in terms of it.

Set

A *set* is a collection of well-defined elements. By well-defined we mean there exists a criterion that enables us to say that an element belongs to the set or that it does not belong to the set.

We shall give examples of several sets of well-defined objects. In each case, we shall designate the set by the letter *S* and shall state the criterion that defines an element of the set.

S is the *choir* of the All Saints Church. The criterion that determines the membership of the choir is the list of names selected by the choir director.

S is the herd of cattle each animal of which bears the brand *X.* Here the criterion requires that an element of *S* must belong to the bovine species and must bear the brand *X.* Consequently, since a horse is not of the bovine species, a horse is not a member of *S.* Furthermore, a steer that does not bear the brand *X* is not a member of the set. If, however, a steer is branded with *X,* it is an element of the set.

S is the set of *counting numbers* 1, 2, 3, 4, and so on, that are less than 20 and are divisible by 3. In this case, the elements of *S* are the numbers 3, 6, 9, 12, 15, and 18, since no other counting number divisible by 3 is less than 20.

The phrase "the three best looking girls in Jones High School" would not define a set, since the criterion is a matter of personal opinion.

We employ both the *listing method* and the *rule method* for describing a set. In the listing method we tabulate the elements of a set and enclose the tabulation in braces, { }. In the rule method we enclose a descriptive phrase in braces. Examples 1 and 2 below illustrate the listing method, and examples 3 and 4 illustrate the rule method.

1

1 If S is the first three counting numbers, then $S = \{1, 2, 3\}$.

2 If Jones, Brown, Long, and Small are the only members of a football squad that play end position, and if E is the set of ends on the squad, then $E = \{\text{Jones, Brown, Long, Small}\}$.

3 The counting numbers less than 5 are 1, 2, 3, and 4. If we designate this set by S and the phrase "is less than" by $<$, and if we require that x stand for a counting number, we can represent the set in this way: $S = \{x \text{ such that } x < 5\}$ or by $S = \{x | x < 5\}$. The vertical bar in this notation is read "such that."

4 If x stands for a counting number such that $x + 3 = 5$ and S is the set of such counting numbers, then $S = \{x | x + 3 = 5\} = \{2\}$. In this case, S contains only one element.

If a is an element of the set S, we say that *a belongs* to S, and we express this statement by the notation $a \in S$. The notation $a \notin S$ means that a *does not belong* to S.

Equality of
two sets

The set S is *equal* to the set T if each element of S belongs to T and each element of T belongs to S. This relation is expressed as $S = T$.

For example, $\{a, c, e, g\} = \{c, a, g, e\}$ since each element of either set is an element of the other. Note that the order of the elements in the two sets does not affect the relation of equality.

We shall now consider the sets $S = \{a, b, c, d, e\}$ and $T = \{a, c, e\}$. Here each element of T is an element of S. This situation illustrates the following definition:

Subset

If each element of a set T is also an element of a set S, then T is a *subset* of S. Furthermore, if there are elements belonging to S that do not belong to T, then T is a *proper* subset of S.

We use the notation $T \subseteq S$ to denote the fact that T is a subset of S, and $T \subset S$ to indicate that T is a proper subset of S.

Equality of two
sets in terms of
subsets

Now if $T \subseteq S$ and $S \subseteq T$, then each element in either set belongs to the other, and therefore the two sets are equal, and we write $T = S$.

It may happen that the same set of elements belongs to each of two sets. For example, if $S = \{1, 2, 3, 4, 5, 6\}$ and $T = \{2, 4, 6, 8, 10\}$, the set $\{2, 4, 6\}$ is a subset of both T and S, and every element that is common to S and T belongs to $\{2, 4, 6\}$. We call this set the intersection of S and T.

Intersection
of two sets

The set of elements that is common to the sets S and T is called the *intersection* of S and T and is designated† by $S \cap T$.

Examples of the intersection of two sets are:

1 $\{x | x$ is a counting number divisible by $2\} \cap \{x | x$ is a counting number less than $10\} = \{2, 4, 6, 8\}$.

---

† The notation $S \cap T$ is read "S cap T," the "intersection of S and T," or "S intersection T."

2 If $A = \{x|x$ is a councilman$\}$ and $B = \{x|x$ is a member of the Kiwanis Club$\}$, then $A \cap B = \{x|x$ is a councilman who is a Kiwanian$\}$.

If, in the second example above, no councilman is a Kiwanian, then $A \cap B$ contains no elements and is an example of the empty set, or the null set, defined as follows:

Empty or null set   The *empty set,* or the *null set,* designated by the symbol $\varnothing$, is a set that contains no elements.

Other examples of the empty set are:

1 $\{x|x$ is a woman who has been president of the United States$\}$.

2 $\{x|x$ is a two-digit counting number less than 10$\}$.

3 $\{x|x$ is a former governor of California$\} \cap \{x|x$ is a former governor of Texas$\}$.

Disjoint sets   If $S \cap T = \varnothing$, the sets S and T are called *disjoint* sets.

To see another concept associated with the theory of sets, consider the sets $A = \{x|x$ is a student in a given college$\}$ and $B = \{x|x$ is a member of the football squad of that college$\}$. The complement of B with respect to A is the set $C = \{x|x$ is a student of the college who is not on the football squad$\}$. This illustrates the following definition:

Complement of a set   The *complement* of the set B with respect to A is designated by $A - B$, and $A - B = \{x|x \in A$ and $x \notin B\}$.

As a second example, we shall discuss two sets that have common elements. If $T = \{x|x$ is a student in college $C\}$ and $S = \{x|x$ is a member of the senior class of college $C\}$, then $T - S = \{x|x$ is a student not classified as a senior$\}$.

We shall next discuss the *union* of two sets. As an example, consider $S = \{1, 2, 3, 4, 5, 6\}$ and $T = \{2, 4, 6, 8, 10\}$. The elements 1, 3, and 5 belong to S but not to T; the elements 8 and 10 belong to T but not to S; and the elements 2, 4, and 6 belong to both S and T. Hence the elements of $V = \{1, 2, 3, 4, 5, 6, 8, 10\}$ are in S or in T or in both. The set V, called the union of S and T, illustrates the following definition:

Union of two sets   The *union* of the sets S and T is the set whose elements are in S or T or in both S and T, and it is designated† by $S \cup T$.

The following examples illustrate the concepts of union and intersection when three sets are involved. In the examples we shall use the sets $A = \{a, b, c, d\}$, $B = \{a, c, e, g\}$, and $C = \{a, e, r, t\}$. The parentheses indicate the operation that is to be performed first.

---

† The notation $S \cup T$ is read "S cup T" or "the union of S and T."

EXAMPLE 1   Show that $(A \cup B) \cup C = A \cup (B \cup C)$.

Solution

$$
\begin{aligned}
A \cup B &= \{a, b, c, d\} \cup \{a, c, e, g\} \\
&= \{a, b, c, d, e, g\}
\end{aligned}
$$

Hence, $(A \cup B) \cup C = \{a, b, c, d, e, g\} \cup \{a, e, r, t\}$
                                   $= \{a, b, c, d, e, g, r, t\}$

Furthermore, $B \cup C = \{a, c, e, g\} \cup \{a, e, r, t\}$
                                 $= \{a, c, e, g, r, t\}$

Consequently, $A \cup (B \cup C) = \{a, b, c, d, e, g, r, t\}$

Therefore, $(A \cup B) \cup C = A \cup (B \cup C)$

EXAMPLE 2   Show that $A \cup (B \cap C) = (A \cup B) \cap (A \cup C)$.

Solution

$$B \cap C = \{a, e\}$$

Hence, $A \cup (B \cap C) = \{a, b, c, d\} \cup \{a, e\}$
                                   $= \{a, b, c, d, e\}$

Also, $A \cup B = \{a, b, c, d, e, g\}$

and $A \cup C = \{a, b, c, d, e, r, t\}$

Therefore, $(A \cup B) \cap (A \cup C) = \{a, b, c, d, e\}$

and $A \cup (B \cap C) = (A \cup B) \cap (A \cup C)$

EXAMPLE 3   Show that $A \cap (B \cup C) = (A \cap B) \cup (A \cap C)$.

Solution

$$
\begin{aligned}
B \cup C &= \{a, c, e, g, r, t\} \\
A \cap (B \cup C) &= \{a, b, c, d\} \cap \{a, c, e, g, r, t\} \\
&= \{a, c\} \\
A \cap B &= \{a, b, c, d\} \cap \{a, c, e, g\} = \{a, c\} \\
A \cap C &= \{a, b, c, d\} \cap \{a, e, r, t\} = \{a\} \\
(A \cap B) \cup (A \cap C) &= \{a, c\}
\end{aligned}
$$

Hence, $A \cap (B \cup C) = (A \cap B) \cup (A \cap C)$

Universal set

The totality of elements that are involved in any specific discussion or situation is called the *universal set* and is designated by the capital letter U. For example, the states in the United States are frequently classified into sets, such as the New England states, the Midwestern states, the Southern states, and so on. Each of these sets is a subset of the universal set, which, in this example, is composed of all the states of the United States. Each of the clubs, athletic teams, academic classes, and other groups whose members are students of a given college is a subset of the universal set composed of the entire student body of the college.

Finally, we shall define the *cartesian product* of two sets. This definition involves the concept of an *ordered pair* of elements. The pair of elements

(x, y) is an *ordered pair* if the position of each element in the pair associates a specific property to it. For example, if given the sets D and R, we require that the first element of x of (x, y) belong to D and the second element y belong to R, then (x, y) is an ordered pair.

Cartesian product

The *cartesian product* of two sets D and R is the set of all ordered pairs (x, y) that can be formed such that $x \in D$ and $y \in R$; it is indicated by $D \times R$.

As a first example of a cartesian product, we shall consider the sets $D = \{a, b, c\}$ and $R = \{d, e\}$. Then $D \times R = \{(a, d), (a, e), (b, d), (b, e), (c, d), (c, e)\}$. Similarly, if $D' = \{1, 2, 3\}$ and $R' = \{2, 4, 6\}$, then $D' \times R' = \{(1, 2), (1, 4), (1, 6), (2, 2), (2, 4), (2, 6), (3, 2), (3, 4), (3, 6)\}$. Finally, $R \times D = \{(d, a), (d, b), (d, c), (e, a), (e, b), (e, c)\}$. Note that $R \times D \neq†$ $D \times R$.

A method for picturing sets and certain relations between them was devised by an Englishman, John Venn (1834–1923). The fundamental idea is to represent a set by a simple plane figure. In order to illustrate the method, we shall use circles. We represent the universal set U by a circle C and define U as the set of all points within and on the circumference of C.

We shall represent the various subsets of U by circles wholly within the circle C. Figure 1.1 illustrates the device.

---

† The symbol $\neq$ means "not equal to."

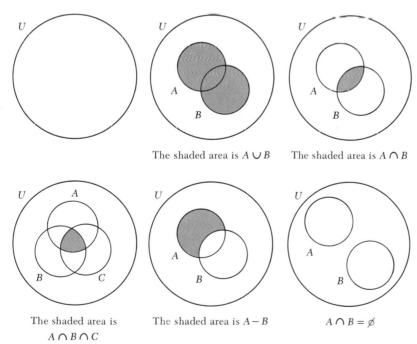

The shaded area is $A \cup B$ The shaded area is $A \cap B$

The shaded area is
$A \cap B \cap C$

The shaded area is $A - B$ $A \cap B = \emptyset$

Figure 1.1

Exercise 1.1 Operations on Sets

Use the listing method to describe the set in each of the problems 1 to 4.

1 $\{x|x$ is a counting number less than 27 and divisible by 7$\}$
2 $\{x|x$ is the name of a state of the U.S. that contains the word *New*$\}$
3 $\{x|x$ is a vowel in the word *consonant*$\}$
4 $\{x|x$ is a day of the week that contains u in its name$\}$

Use set-builder notation to designate each set in problems 5 to 8.

5 $\{3, 6, 9, 12\}$ **6** $\{1, 3, 5, 7, 9\}$
7 $\{a, e, i, o, u\}$ **8** $\{$Truman, Eisenhower, Kennedy, Johnson, Nixon, Ford$\}$

Find $A \cup B, A \cap B,$ and $A - B$ for the sets given in each of problems 9 to 16.

9 $A = \{2, 3, 5, 8, 12\}, B = \{2, 5, 7\}$
10 $A = \{1, 3, 5, 9, 14\}, B = \varnothing$
11 $A = \{2, 4, 6, 8, 10\}, B = \{3, 4, 5, 6\}$
12 $A = \{3, 7, 9, 10, 12\}, B = \{7, 8, 9, 10\}$
13 $A = \{x|x$ is one of the first five letters of the alphabet$\}$
   $B = \{x|x$ is a vowel$\}$
14 $A = \{x|x$ is an exact divisor of 15$\}, B = \{x|x$ is an exact divisor of 10$\}$
15 $A = \{x|x$ is an athlete at ABC College who lettered in basketball$\}$,
   $B = \{x|x$ is an athlete at ABC College who lettered in track$\}$
16 $A = \{x|x$ is a student at Beva College who is over 6 feet tall$\}$,
   $B = \{x|x$ is a student at Beva who is 6 feet tall or shorter$\}$

Perform the operations called for in problems 17 to 20 if
$A = \{x|x$ is a counting number less than 11$\}, B = \{x|x \in A$ and is divisible by 2$\}, C = \{x|x \in A$ and is divisible by 5$\},$ and $D = \{x|x \in A$ and is not divisible by 2 or 5$\}.$

17 $A \cap B \cap C \cap D, A \cup B \cup C, (A \cup C) - B$
18 $(A - B) \cup (C - D), (A \cup B) - (C \cup D), A - (B \cup C) - D$
19 $B \cap (C \cap D) = (B \cap C) \cap (B \cap D)$
20 $B \cup (C \cap D) = (B \cup C) \cap (B \cup D)$
21 If $A = \{a, b\}$ and $B = \{c, d, e\}$, find $A \times B$
22 If $A = \{a, b, c\}$ and $B = \{a, b, c\}$, find $A \times B$
23 If $A = \{2, 3, 5\}, B = \{2, 5, 7\},$ and $C = \{1, 3, 5\}$, find $(A \times B) \cap (A \times C)$
24 If $A = \{2, 4, 6, 8\}, B = \{3, 5, 7\},$ and $C = \{2, 3, 4\}$, find $(A \cap C) \times (B \cap C)$

If *A* is the set of points inside a rectangle and *B* is the set inside a triangle, construct Venn diagrams for problems 25 to 28 and shade the area that represents the set.

25 $A \cap B$ such that $A \cap B = \varnothing$
26 $A \cap B$ such that $A \cap B \neq \varnothing$
27 $A \cup B$ if $A \cap B \neq \varnothing$
28 $A - B$ if $A \cap B \neq \varnothing$

1.2 THE NATURAL NUMBERS

In Sec. 1.1 we used the term *counting number*. This brings up two questions: (1) What is a number? and (2) What is meant by counting? We shall not attempt to give a rigorous answer to these questions, but shall discuss a situation that illustrates the concept. First, consider the sets $A = \{a, b, c, d, e\}$ and $B = \{$Tom, Dick, Harry, Joe, Jim$\}$. Here we have two different sets since the elements of *A* are letters and the elements of *B* are names. There is, however, one aspect or property that is common to the two sets: the elements of either set can be paired with the elements of the other set; i.e., each element in *B* can be matched with an element in *A*, and each element in *A* can be matched with one in *B*. We call this matching a *one-to-one correspondence,* and the correspondence can be set up in many ways. As an illustration, we shall match the elements of the sets in the order in which they appear, proceeding from left to right:

We call the common property of the two sets the *number of elements* in each set. The Hindu-Arabic symbol for this number is 5, and the English name for it is *five*. We say that the sets *A* and *B* are equivalent, and they illustrate the following definition:

Equivalent sets   Two sets are *equivalent* if there exists a one-to-one correspondence between the elements of the two sets.

We can use this concept of equivalent sets to illustrate the meaning of the term *counting number*. For this purpose, we shall assume that the Hindu-Arabic number symbols and their order of succession are known, and we shall use these symbols as elements of sets. Now we say that the number *one* is the number associated with the totality of sets that are equivalent to the set $\{1\}$, the number *two* is the number associated with all sets equivalent to $\{1, 2\}$, and so on. Similarly, if *n* is a counting number expressed in the Hindu-Arabic notation, then the number *n* is the number associated with all sets equivalent to the set $\{1, 2, 3, \ldots, n\}$, where the dots indicate that the succession of number symbols is continued from 3 to *n*.

The process
of counting

The process of counting the elements of a set consists of establishing a one-to-one correspondence between the elements of the given set and the elements of $\{1, 2, 3 \ldots\}$, where the dots mean that the sequence of number symbols is continued until the correspondence is completed.

The numbers used in counting are called *natural numbers*.

Natural
numbers

The natural numbers are used for defining the numbers in the real number system, which will be discussed in the next section.

1.3 THE REAL NUMBER SYSTEM

Although the natural numbers may have sufficed for a primitive culture, the advance of civilization not only demanded but depended upon a progressive extension of the number system by the invention of other numbers. The extended system which we shall use in the first eight chapters of this book is called the *real number system*. In this section we shall define the various subsets of the set of real numbers and shall give a geometrical interpretation of each.

We shall use the straight line L and the unit length u in Fig. 1.2 for this purpose. We shall accept the terms "straight line," "point," "length," and "distance" as undefined terms. A *segment* of a straight line, or an *interval* on the line, is the portion of the line between two points on it. We shall say that two segments are equivalent or congruent if their lengths are the same. Furthermore, we shall use the fact that a geometrical method exists for dividing a line segment into any given integral number of equal parts, and we shall assume that any given length can be laid off on a straight line any desired natural number of times.

We choose the reference point P on the line L, a portion of which is shown in Fig. 1.2, and starting at P, we lay off successive intervals of length u to the right of P. Next, starting with the right end of the first interval to the right of P, we label the right ends of the intervals proceeding progressively to the right with the natural numbers $1, 2, 3, \ldots$, where the dots mean that the sequence of numbers is continued indefinitely. In this way, one and only one point on L is associated with each of the natural numbers. We shall use the notation (n) to refer to the point associated with the number n. For example, the point associated with 4 is denoted by (4).

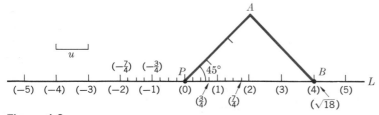

Figure 1.2

Geometric addition   One of the fundamental operations involving natural numbers is addition. We shall interpret this operation by use of the line L. The symbol indicating addition is $+$ and is read "plus." Thus, $a + b$ means that we are to add b to a. In order to obtain the point $(a + b)$, we start at the point a, move b unit intervals to the right, and thus reach the point $(a + b)$. For example, to get the point $(3 + 2)$, we start at (3) and move two units to the right and reach the point (5). Hence we say that $3 + 2 = 5$, and we call 5 the *sum* of 3 and 2.

We next call attention to the fact that if a is a natural number and if we start at the point P on L in Fig. 1.2 and move a intervals to the right, we arrive at the point (a). So far, we have not associated a number with the point P, and since the natural numbers are assigned to other points on L, we cannot use a natural number for this purpose. Therefore we shall introduce a new number, *zero,* denoted by 0, and assign 0 to P. Now if we start at P, or (0), and move a units to the right, we arrive at the point (a). Consequently, we define the number zero as follows:

Zero   The number *zero,* denoted by 0, is the number such that $0 + a = a$, where a stands for any given number. Hence it is called the *addition identity.*

We now define the operation $a + (-b)$ to mean that we start at (a) and move b units to the *left* of (a), as illustrated in Fig. 1.3. If (a) is to the right of the point (b) on L, then the point $a + (-b)$ is to the right of the point 0. If, however, (b) is to the right of (a), then the operation $a + (-b)$ brings us to a point that is to the left of (0), and so far we have associated no numbers with this portion of L. This will be our next task.

We first note that the operations $1 + (-1)$, $2 + (-2)$, and in general, $a + (-a)$ bring us to the point zero. Hence, $1 + (-1) = 2 + (-2) = a + (-a) = 0$. We now define the negative of the number a as follows:

Negative of a number   The *negative of the number a* is the number $-a$ such that $a + (-a) = 0$.

In order to associate a point on L with $-a$, we start at zero, lay off a units on L to the left of (0), and assign $-a$ to the left extremity of the ath interval. We call the point $(-a)$ the *reflection* of the point (a) on L with re-

Reflection of a point   spect to (0). Thus (-1) is the reflection of (1) with respect to zero, (-2) is the reflection of (2), and so on. In this way, we obtain the points in Fig. 1.2 that are labeled (-1), (-2), (-3), and so on.

Integers   We shall now define the set of *integers* to be the set of numbers composed of the natural numbers, zero, and the negatives of the natural numbers. Hereafter we shall call the natural numbers the *positive integers* and the negatives of the natural numbers the *negative integers.* The number zero is an integer that is neither positive nor negative. In the terminology

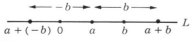

Figure 1.3

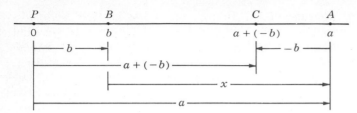

Figure 1.4

of sets, we have {integers} = {positive integers} ∪ {0} ∪ {negative integers}. Therefore, the set of positive integers is a proper subset of the set of integers, or {positive integers} ⊂ {integers}. Similarly, {0} ⊂ {integers} and {negative integers} ⊂ {integers}.

We shall agree that if p and q are integers, the statement $p > q$ means that (p) is to the right of (q) on L. Hence $5 > 3$, $2 > -6$ and $-3 > -7$. Similarly, $r < s$ means that (r) is to the left of (s) on L. Thus if a is a positive integer, then (a) is to the right of (0), so $a > 0$. Similarly, if a is a negative integer, $a < 0$.

Difference · Subtraction · We shall define the *difference* of a and b, designated by $a - b$, as the number x such that $b + x = a$. The procedure for finding x is called *subtraction,* and we shall show graphically by use of Fig. 1.4 that $x = a + (-b)$. In Fig. 1.4, the lengths of PB, PA, PC, BA, and AC are as indicated. Since $AC = -b$, $CA = b$, and it follows that $PB = CA = b$. Furthermore,

$$
\begin{aligned}
x &= BA \\
&= BC + CA \\
&= BC + PB \qquad \blacktriangleleft \text{ since } CA = PB \\
&= PC \\
&= a + (-b)
\end{aligned}
$$

Hence, $a - b = a + (-b)$.

EXAMPLE 1

Since $5 - 3 = 5 + (-3)$, then to get the difference $5 - 3$, we start at (5) and count three intervals to the left and arrive at (2). Hence, $5 - 3 = 2$.

EXAMPLE 2

To obtain the difference of 5 and 7, we have $5 - 7 = 5 + (-7)$; so we start at (5), count seven intervals to the left through (0), and arrive at (-2). Hence, $5 - 7 = -2$.

EXAMPLE 3

Similarly, the difference of -2 and 3 is $-2 - 3 = -5$.

The procedure for finding the difference of a and b when b is negative will be discussed in the next chapter.

The product of two integers a and b is indicated by $a \cdot b$, $a \times b$, $a(b)$,

10

Multiplication $(a)(b)$, or ab. The operation of obtaining the product is called *multiplication*. We shall first consider the case in which a is positive, and we shall illustrate the process with $a=3$ and $b=2$. We define the product $3 \cdot 2$ to be the sum of *three 2s*. Thus $3 \cdot 2 = 2 + 2 + 2 = 6$. Similarly, $2 \cdot 3 = 3 + 3 = 6$. Hence $3 \cdot 2 = 2 \cdot 3$. We readily can verify that $4 \cdot 5 = 5 \cdot 4$, $6 \cdot 9 = 9 \cdot 6$, and in general, $a \cdot b = b \cdot a$ if a and b are replaced by any two designated positive integers. However, at this point we cannot verify that $3(-2) = -2(3)$, since the latter product has not been defined. Nevertheless, we shall assume that if a and b are integers, then $a \cdot b = b \cdot a$. This is called the *commutative* property of multiplication and will be discussed more fully in the next chapter. By the commutative property, $-3 \cdot 2 = 2 \cdot -3$, and we define the latter product to be $-3 + (-3) = -6$. Hence $-3 \cdot 2 = -6$. This interpretation of multiplication does not suffice for $-2(-3)$. In the next chapter, however, we shall prove that if a and b are positive, then $-a(-b) = ab$. Hence $-2(-3) = 6$.

Since $3 \cdot a = a + a + a$ and $2 \cdot a = a + a$, it is logical to define $1 \cdot a$ as the number a; that is, $1 \cdot a = a$.

We shall next consider the question: If a and b are integers and $b \neq 0$, what number must be multiplied by b in order to obtain a? In other words, we seek the number x such that $bx = a$. The operation for deter-

Division mining the number x, if it exists, is called *division,* and this operation is the *inverse* of multiplication. The number x is called the *quotient* of a and b, and it is usually expressed in the form $\frac{a}{b}$ or a/b. A number expressed in either of these forms is called a *fraction.* Unless a is a multiple† of b, a/b is not an integer and hence is not associated with any point on L that represents an integer. We shall now show how to associate a point on L with a number of the type a/b, and we shall illustrate the method by use of $\frac{3}{4}$. We shall consider Fig. 1.5, which shows a portion of L on an enlarged scale, with the segment from 0 to 1 divided into four equal parts. Since the length of the segment from 0 to 1 is the unit u, the length of each of the four subdivisions of this segment is one-fourth of u. We now assign $\frac{1}{4}, \frac{2}{4}, \frac{3}{4}$, and $\frac{4}{4}$ to the points indicated in the figure. Hence, $(\frac{3}{4})$ is the right extremity of the third subdivision to the right of (0). Similarly, $(\frac{7}{4})$ is the right extremity of the seventh interval of length $\frac{1}{4}$ to the right of (0). We note that $\frac{4}{4}$ and 1 are associated with the same point, and so $\frac{4}{4} = 1$. We define the product $4 \times \frac{1}{4}$ as $\frac{1}{4} + \frac{1}{4} + \frac{1}{4} + \frac{1}{4}$, and by our graphical interpretation of addition, this sum is $\frac{4}{4}$, or 1. Hence we have $4 \times \frac{1}{4} = 1$.

---

† If a and b are nonzero integers, then a is a multiple of b if an integer n exists such that $a = nb$.

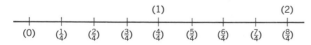

Figure 1.5

In general, to get the point associated with a/b, where a and b are positive integers, we first subdivide the interval from 0 to 1 on L into b equal parts. Each of these subintervals will be of length $1/b$. Then we lay off a of these intervals to the right of 0 and assign a/b to the right extremity of the ath interval. If $a < b$, then (a/b) is to the left of 1; if $a > b$, (a/b) is to the right of 1; and if $a = b$, (a/b) is the point 1.

Since (a/b) is the point that is the natural number a of the intervals of length $1/b$ to the right of (0), we call a/b a number, and since it is always a ratio of integers, we use the word *rational* to describe it. Since it is associated with a point to the right of (0), it is positive. The negative of a/b is $-(a/b)$, and the point on L associated with $-(a/b)$ is the reflection of (a/b) on L with respect to (0). The points $(\frac{3}{4})$, $(\frac{7}{4})$, $(-\frac{3}{4})$, and $(-\frac{7}{4})$ are shown in Fig. 1.2.

In the foregoing discussion, we assumed that a and b represented positive integers. We shall show in a later section that if a and b are both positive or both negative, then a/b is positive, and that if either of a or b is positive and the other is negative, then a/b is negative. We shall assume for the present that these statements are true, and we define a rational number as follows:

Rational
number

A *rational*† number is a number that can be expressed as the quotient of two integers.

Since $1 \cdot a = a$, we have, by the definition of a quotient, $a = a/1$. Hence any integer a can be expressed as the quotient $a/1$. Consequently, the integers are included in the rational numbers, or in the terminology of sets, {integers} $\subset$ {rational numbers}.

We shall now show that there is a point on L to the right of 0 that cannot be associated with a rational number. For this purpose, we shall first define the *square* of the number a, designated by a^2, as $a \cdot a$. Hence $3^2 = 3 \cdot 3 = 9$. We shall also define the positive *square root* of the number n, designated by $\sqrt{n}$, as the positive number whose square is n. For example, $\sqrt{16} = 4$ since $4 \cdot 4 = 16$. The pythagorean theorem states that the sum of the squares of the two legs of a right triangle is equal to the square of the hypotenuse. It is also proved in plane geometry that if one acute angle of a right triangle is 45°, the other acute angle is 45° and the sides which form the right angle are equal. We shall now construct a straight-line segment originating at point P in Fig. 1.2 that makes an angle of 45° with the positive direction of L. On this line we lay off three unit intervals starting at P, and we label the upper end of the third interval A. Then at A we construct a perpendicular to this line that intersects L at B. Since the acute angle at P is 45°, the acute angle at B is 45°, and therefore $PA = AB = 3$. Then, by the pythagorean theorem, the length of the line segment from P to B is

---

† The word *rational* is derived from the word *ratio*. The quotient a/b is also called the ratio of a to b. Hence a rational number is a number that expresses a ratio.

$\sqrt{3^2 + 3^2} = \sqrt{9 + 9} = \sqrt{18}$. Consequently, $\sqrt{18}$ is associated with the point B.

We shall now prove that $\sqrt{18}$ cannot be expressed as the quotient of two integers. For this purpose, we shall use the following definitions and assumptions, most of which will be discussed later.

1 An integer is *even* or *odd* according as it is or is not a multiple of 2.

2 If an even integer is multiplied by another integer, the product is an even integer. Conversely, if the product of two integers is even, then at least one of the integers is even.

3 The halves of equal numbers are equal.

4 The squares of two equal numbers are equal.

5 $(a \cdot b \cdot c)^2 = a^2 \cdot b^2 \cdot c^2$, and $\sqrt{a \cdot b \cdot c} = \sqrt{a} \cdot \sqrt{b} \cdot \sqrt{c}$.

6 If d is the greatest integer that is a divisor of a and b and if $a/d = q$ and $b/d = p$, then $a/b = q/p$.

7 $\dfrac{ab}{2} = \dfrac{a}{2} \cdot b = a \cdot \dfrac{b}{2}$

We shall assume that $\sqrt{18}$ can be expressed as the quotient of two integers. Hence, by assumption 6, there exist two integers q and p such that $\sqrt{18} = q/p$, where q and p have no common integral divisor greater than 1. Then, by the definition of a quotient, we have $p \cdot \sqrt{18} = q$, and we complete the proof by the following steps:

$$(p \cdot \sqrt{18})^2 = q^2 \qquad \blacktriangleleft \text{ by assumption 4}$$
$$p^2 \cdot (\sqrt{18})^2 = q^2 \qquad \blacktriangleleft \text{ by assumption 5}$$
$$p^2 \cdot (18) = q^2 \qquad \blacktriangleleft \text{ by the definition of } \sqrt{18}$$

$p^2 \cdot (18)$ is an even integer by assumption 2, since 18 is an even integer. Hence, q^2 is an even integer, and since $q^2 = q \cdot q$, q is an even integer by assumption 2. Consequently, $q = 2n$, where n is an integer. Now, replacing q by $2n$ in $p^2 \cdot (18) = q^2$, we have

$$p^2 \cdot 18 = (2n)^2$$
$$= 4n^2 \qquad \blacktriangleleft \text{ by assumption 4}$$
$$p^2 \cdot 9 = 2n^2 \qquad \blacktriangleleft \text{ by assumption 3}$$

Since $2n^2$ is an even integer, $p^2 \cdot 9$ is an even integer. Therefore, since 9 is odd, p^2 is even, by assumption 2. Furthermore, since $p^2 = p \cdot p$, it follows from assumption 2 that p is even. Hence, if $\sqrt{18} = q/p$, then p and q are both even, and this contradicts the assumption that q and p have no common divisor greater than 1. Consequently, $\sqrt{18}$ cannot be expressed as the quotient of two integers.

We shall call the number $\sqrt{18}$ an *irrational number* and shall define this type of number more precisely after the next paragraph.

It is proved in arithmetic that some rational numbers that are not integers can be expressed as terminating decimals and that others can be expressed as nonterminating periodic decimals.† For example, $\frac{1}{2} = 0.5$, $\frac{3}{4} = 0.75$, and $\frac{42}{37} = 1.135135135.\ldots$ By the statement $\frac{42}{37} = 1.135135135\ldots$, we mean that by annexing the cycle 135 a sufficient number of times, we obtain a decimal that differs from $\frac{42}{37}$ by a number that is less than any number that is chosen in advance. It is also proved in arithmetic that any terminating decimal can be expressed as the quotient of two integers, and in Chap. 15 we shall prove that any nonterminating periodic decimal also can be expressed as the quotient of two integers.

We now return to the discussion of $\sqrt{18}$. There is a process in arithmetic that enables us to express $\sqrt{18}$ approximately as a decimal with as many decimal places as is desired. The first six steps in this process yield 4, 4.2, 4.24, 4.242, 4.2426, and 4.24264. The process never terminates, and the decimal never becomes periodic, but each step yields a number whose square is nearer 18 than the square of the preceding number. Consequently, we say that $\sqrt{18} = 4.24264.\ldots$, where the decimal can be continued indefinitely by a repeated application of the square-root process.

Since every terminating decimal and every nonterminating periodic decimal can be expressed as the quotient of two integers and, furthermore, since the quotient of any two integers can be expressed as a terminating or a nonterminating periodic decimal, it seems reasonable to assume that a nonterminating nonperiodic decimal cannot be expressed as the quotient of two integers. It is proved in more advanced mathematics that this assumption is true. Therefore, such a number is not a rational number, and we call it an irrational number. This illustrates the following definition: An *irrational* number is a number whose decimal representation is nonterminating and nonperiodic.

Real number system

We shall now define the *real number system* as the set of numbers composed of the set of rational numbers and the set of irrational numbers.

In the terminology of sets, the above definition can be stated thus: {real numbers} = {rational numbers} $\cup$ {irrational numbers}. Furthermore, since no rational number is equal to an irrational number, {irrational numbers} $\cap$ {rational numbers} $= \varnothing$ (or the empty set).

It is proved in more advanced mathematics that each point on the line L is associated with one and only one real number. Furthermore, as we have stated previously, the numbers associated with the points to the right of (0) are positive, those associated with points to the left of (0) are nega-

---

† By *nonterminating periodic decimal* we mean that following the decimal point or following a certain digit at the right of the decimal point, the decimal consists of an indefinite number of repetitions of the same cycle of integers. For example, $0.325325325\ldots$, $0.125343434\ldots$, and $42.137245245245.\ldots$

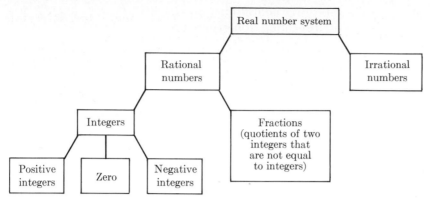

Figure 1.6

tive, and the number 0 is neither positive nor negative. Since a real number other than zero is either positive or negative, it is a *directed,* or *signed,* number. By the sign of the number a, we mean the direction of the point (a) from (0). For example, if the point (a) is to the right of (0), then a is positive, and if (a) is to the left of (0), a is negative. We shall agree that placing the negative sign, $-$, before a number changes the direction of the number. Thus the points (a) and $(-a)$ are on opposite sides of (0). If a is positive, $-a$ is negative; and if b is negative, $-b$ is positive.

In contrast to this interpretation of the negative sign, we shall agree that placing a positive sign, $+$, before a number does not affect the direction of the number. Consequently, $+n = n$ and $+(-n) = -n$.

Frequently we have occasion to refer to the absolute value of a number, and we shall define this concept as follows:

Absolute value   If a number n is positive, the *absolute value* of n, designated by $|n|$, is equal to n. If n is a negative number, then $|n| = -n$. If $n = 0$, then $|n| = 0$.

For example, $|5| = 5$, $|-3| = 3$, and $|0| = 0$.

Figure 1.6 shows the composition of the real number system in diagrammatic form.

Exercise 1.2 The Number System

Select the name from the real number system diagram that best describes the elements in the set in each of problems 1 to 12.

1 $A = \{1, 2, 5, 8\}$ **2** $A = \{-2, 0, 1, 3, 6\}$

3 $A = \{-3, 5, 8, -7\}$ **4** $A = \{-9, -8, -6, -3\}$

5 $A = \{\frac{1}{2}, \frac{2}{5}, \frac{4}{7}, \frac{5}{11}\}$ **6** $A = \{-\frac{7}{3}, \frac{2}{5}, 1, \frac{5}{9}\}$

7 $A = \{-3, 0, \frac{7}{2}, 1.2\}$ **8** $A = \{-1, 0, 1.7, \frac{3}{8}\}$

9 $A = \{x | x \text{ is the quotient of two integers}\}$

10 $A = \{x | x \text{ is a nonrepeating terminating decimal}\}$

11 $A = \{x|x \text{ is a repeating nonterminating decimal}\}$
12 $A = \{x|x \text{ is a nonrepeating nonterminating decimal}\}$

If $A = \{1, 4, \frac{2}{3}, \frac{5}{9}\}$, $B = \{2, 4, 5, 8\}$, and $C = \{1, \frac{3}{4}, \frac{5}{7}, \sqrt{2}, \sqrt{3}\}$, select the name from the real number system diagram that best describes the set D in each of problems 13 to 16.

13 $D = A \cap B$ **14** $D = A \cup B$
15 $D = B \cup C$ **16** $D = A \cap C$

In problems 17 to 20 insert $<$ or $>$ between each pair of numbers so that the resulting statement is true.

17	15	11	**18**	3	-5
	7	13		-8	-2
	42	$\frac{81}{2}$		-1	-3

19	$\frac{2}{5}$	$\frac{1}{3}$	**20**	$\sqrt{19}$	4
	$\frac{3}{7}$	$\frac{5}{9}$		$\sqrt{2}$	1.4
	$\frac{8}{17}$	$-\frac{9}{2}$		$\sqrt{2}$	1.42

Find the result of performing the operations indicated in each of problems 21 to 28.

21 $7 + (-2)$ **22** $(-4) + 6$
23 $(-9) + 4$ **24** $(-6) + (-3)$
25 $6 + |-2|$ **26** $|-4| + |-5|$
27 $|6 + (-2)|$ **28** $|(-4) + (-5)|$

Exercise 1.3   Review

1   Write out the elements of $\{x|x \text{ is a month of the year that contains}$ exactly 30 days$\}$.
2   Write out the elements of $\{x|x \text{ is a counting number less than 16 and}$ divisible by 3$\}$.
3   Designate $\{1, 3, 5, 7\}$ by use of set-builder notation.
4   Designate $\{$Truman, Barkley, Nixon, Johnson, Humphrey, Agnew, Ford$\}$ by use of set-builder notation.

Find $A \cup B$, $A \cap B$, and $A - B$ for the sets in problems 5 and 6.

5 $A = \{3, 5, 8, 12, 17\}$, $B = \{3, 8, 17\}$.
6 $A = \{x|x \text{ is one of the first six letters of the alphabet}\}$
    $B = \{x|x \text{ is a vowel}\}$.

7 If $A = \{1, 2, 3, 4, 5, 6\}$, $B = \{2, 4, 6\}$, and $C = \{3, 6\}$, find $A \cap B \cap C$, $A \cup B \cap C$, $(A \cup B) - C$.

8 If A, B, and C are as in problem 7, find $(A - B) \cap C$, $(A \cap B) \cup (B \cap C)$, $A - (B \cup C)$.

9 If $A = \{3, 7, 11\}$ and $B = \{2, 5\}$, find $A \times B$.

10 Find $A \times B$ if $A = \{x | x \text{ is less than 10 and divisible by 3}\}$ and $B = \{x | x \text{ is a one-digit number greater than 7}\}$.

11 What type of number is 3, -2, $\frac{11}{6}$, $-\frac{4}{9}$?

12 If $A = \{3, 2, 1.3, -\sqrt{2}\}$ and $B = \{5, 2, 3.1, \sqrt{5}\}$, what type of number is $A \cap B$?

13 Insert $<$ or $>$ between 3 and 2, $\frac{2}{9}$ and $\frac{1}{4}$, 1.42 and $\sqrt{2}$.

14 Find the value of $3 + (-5)$, $|3 + (-5)|$, $|(-2) + (-1)|$.

The Four
Fundamental
Operations

We defined the real number system and gave a brief discussion of the operations of addition, subtraction, multiplication, and division in Chap. 1, but left many questions unanswered. For example, what is the meaning of 0(3) and of $-4(-5)$? The use of the number line L does not enable us to answer these questions or many others that might occur to a thoughtful reader. We, however, shall develop the foundations for a logical structure in this chapter that will enable us to answer the above questions and that will serve as the basis for the fundamental operations and for solving equations.

2.1 A LOGICAL STRUCTURE

We could find by examining a plane geometry book that the book starts with some undefined terms and some definitions and that these are followed by statements that are called axioms. Each definition, axiom, and theorem depends on the undefined terms. The theorems are statements that are proved by use of the undefined terms, the axioms, and previously proved theorems. After a theorem is proved, however, it becomes a part of the foundation for the proof of other theorems. We shall use the same procedure in establishing the logical foundation for the algebra of numbers, and in the next section we shall define some of the terms that we shall employ.

2.2 DEFINITIONS

We shall use the letters of the alphabet to represent real numbers. A letter used in this way is called a variable, which we define as follows:

Variable   A *variable* is a symbol, usually a letter, that stands for, or that may be replaced by, a number from a specified set of numbers. The set of numbers is called the *replacement set*.

Replacement set

Constant   A *constant* is a symbol whose replacement set contains only one number.

EXAMPLE 1   An upright cylindrical water tank is 10 feet high and has a radius of 6 feet. It is supplied by a pipe with an automatic valve that closes when the tank is full and opens when the tank is drained to a depth of 1 foot. The volume of the water in the tank is given by the formula

$$V = \pi r^2 d$$

This formula contains the constants and variables tabulated on page 20.

19

Symbol	Replacement set	Classification
π	{3.1416 approximately}	Constant
r (radius)	{6}	Constant
d (depth)	$\{d\|1 \leq d \leq 10\}$	Variable
V (volume)	$\{V\|36\pi \leq V \leq 360\pi\}$	Variable

Fundamental operations   The four *fundamental operations* of algebra are addition, subtraction, multiplication, and division.

The sum of two numbers a and b is written $a + b$, the difference of a and b is expressed as $a - b$, the product of a and b is denoted by ab, and the quotient of a and b is expressed as a/b. We shall discuss these operations in more detail later.

Factor   In the product ab, the numbers a and b are called *factors* of ab.

The product $a \cdot a$ is expressed a^2. Similarly, $a \cdot a \cdot a = a^3$, $a \cdot a \cdot a \cdot a = a^4$, and in general

$$a \cdot a \cdot a \cdots a = a^n$$

$$n \text{ factors}$$

The numbers a^2, a^3, and a^n are called "a square," "a cube," and "the nth power of a," respectively.

Base and exponent   In the number a^n, a is called the *base* and n is the *exponent* of the base.

The result obtained by combining two or more numbers by means of one or more of the four fundamental operations of algebra is called an Expression   *expression*.

EXAMPLE 2 $a + b$, $3a + bc$, $\dfrac{a}{b}$, $\dfrac{3a + 2b}{a + b}$, and $\left(\dfrac{x^2 + 2y^2}{3 + 4y}\right)\left(\dfrac{x^3 - 3y^4}{2 - 5y}\right)$ are expressions.

Monomial   An expression that does not involve addition or subtraction is called a *monomial*.

EXAMPLE 3 a, $2ab$, and $\dfrac{2ab}{3bc}$ are monomials.

Multinomial   The sum of two or more monomials is a *multinomial*.

EXAMPLE 4　The expression $2a + \dfrac{3b^2}{2bc} - \dfrac{4ab}{2bc} + \dfrac{1}{2}\,a^3$ is a multinomial.

Term　Each monomial in a multinomial, together with the sign that precedes it, is called a *term* of the multinomial.

A multinomial consisting of exactly two terms is a *binomial,* and a multinomial consisting of three terms is a *trinomial.* If each term of a multinomial is an integral power of a number symbol or is the product of the integral powers of two or more number symbols, the multinomial is called a *polynomial.*

Polynomial

EXAMPLE 5　Examples of polynomials are $2a^4 + a^3 - 4a^2 + 5a + 3$ and $2x^4y + 2x^3y - 4x^2y^2 + 2xy^3 - 4y^2$.

Coefficient　If a monomial is expressed as the product of two or more symbols, each of the symbols is called the *coefficient* of the product of the others.

EXAMPLE 6　In the monomial $3ab$, 3 is the coefficient of ab, a is the coefficient of $3b$, b is the coefficient of $3a$, and $3a$ is the coefficient of b. We call the 3 in $3ab$ the numerical coefficient. Usually when we refer to the coefficient in a monomial, we mean the numerical coefficient.

Similar terms　Two monomials, or two terms, are called *similar* if they differ only in their numerical coefficients.

EXAMPLE 7　The monomials $3a^2b$ and $-2a^2b$ are similar, and the terms in $4(3a/5b) + 2(3a/5b)$ are similar.

2.3　THE RELATION OF EQUALITY

In Chap. 1 we defined the relation of equality as applied to two sets, and we stated that two numbers are equal if they represent the same point on the line L. We shall not attempt a general definition of this relation but shall state below several agreements, or axioms, dealing with it. These axioms define the properties of the relation of equality, although they do not actually define the relation itself. In the statements of the axioms, the letters used stand for real numbers.

$$a = a \qquad \blacktriangleleft \text{ reflexive axiom} \qquad (2.1)$$

$$\text{If } a = b, \text{ then } b = a \qquad \blacktriangleleft \text{ symmetric axiom} \qquad (2.2)$$

$$\text{If } a = b \text{ and } b = c, \text{ then } a = c \qquad \blacktriangleleft \text{ transitive axiom} \qquad (2.3)$$

By use of Eqs. (2.2) and (2.3), we can prove our first theorem, which is stated below.

THEOREM $$\text{If } a = b \text{ and } c = b, \text{ then } a = c \qquad (2.4)$$

Proof ▶   If $c = b$, then $b = c$ by Eq. (2.2). Hence, since $a = b$ and $b = c$, it follows that $a = c$ by Eq. (2.3).

$$\text{If } a = b, \text{ then } a + c = b + c \qquad \blacktriangleleft \text{ addition axiom} \qquad (2.5)$$

$$\text{If } a = b, \text{ then } ac = bc \qquad \blacktriangleleft \text{ multiplication axiom} \qquad (2.6)$$

If $a = b$, then a can be replaced by b in any statement involving algebraic expressions without affecting the truth or falsity of the statement $\blacktriangleleft$ replacement axiom   **(2.7)**

We shall state and prove a theorem that will be employed frequently in the remainder of this book.

THEOREM $$\text{If } a = b \text{ and } c = d, \text{ then } a + c = b + d \qquad (2.8)$$

Proof ▶

$$a = b \qquad \blacktriangleleft \text{ given}$$
$$a + c = b + c \qquad \blacktriangleleft \text{ by Eq. (2.5)}$$
$$= b + d \qquad \blacktriangleleft \text{ by Eq. (2.7), since } c = d$$

The following two examples illustrate the use of the above axioms and theorems.

EXAMPLE 1   Prove that if $a = b$ and $c = d$, then $a + d = b + c$.

Proof ▶

$$a = b \qquad \blacktriangleleft \text{ given}$$
$$a + d = b + d \qquad \blacktriangleleft \text{ by Eq. (2.5)}$$
$$= b + c \qquad \blacktriangleleft \text{ replacing } d \text{ by } c \text{ by Eq. (2.7)}$$

EXAMPLE 2   If $a = b$ and $c = d$, then $ar + cs = br + ds$.

Proof ▶   Since $a = b$ and $c = d$, then

$$ar = br \qquad \blacktriangleleft \text{ by Eq. (2.6)}$$
and
$$cs = ds \qquad \blacktriangleleft \text{ by Eq. (2.6)}$$
Hence,
$$ar + cs = br + ds \qquad \blacktriangleleft \text{ by Eq. (2.8)}$$

2.4 ADDITION

In Chap. 1 we demonstrated graphically that the sum of two integers is an integer, and we implied that the sum of two real numbers is a real number.

Closure
This property of a set of numbers is called *closure*. More precisely, we say: If the sum of any two numbers in a set of numbers is an element of the set, then the set is said to be *closed* under the operation of addition. We shall assume that this property holds for the operation of addition in the set of real numbers and state the assumption below.

If a and b are real numbers, there exists a real number c such that $a + b = c$ ◄ closure axiom for addition   **(2.9)**

The usual procedure for expressing $3a + 2b + 4a + 3b + a + 4b$ in the simplest form is to rearrange the terms as $3a + 4a + a + 2b + 3b + 4b$. Then we add the coefficients of the similar terms and get $8a + 9b$. This procedure brings up two questions: Why is the first expression equal to the expression with the terms rearranged? Why does $3a + 4a + a = 8a$? The answers to these questions depend upon the agreements or axioms dealing with addition, which we shall now state.

It is readily verified that $3 + 5 = 5 + 3$, that $4 + 7 = 7 + 4$, and that for any two numbers that we try, we obtain the same result regardless of the order in which the numbers are added. We shall assume that this is true for any two real numbers a and b and state the assumption below.

Commutative property of addition
$$a + b = b + a$$ ◄ commutative axiom for addition   **(2.10)**

As indicated by the phrase following the statement of the axiom, this property of addition is called the *commutative property* of addition.

Our next axiom deals with the sum of three numbers and is known as the *associative* axiom. We may easily verify that $(7 + 4) + 9 = 11 + 9 = 20$, and that $7 + (4 + 9) = 7 + 13 = 20$. We assume that this property is true for any three numbers, and thus we have the following axiom:

Associative property of addition
$$(a + b) + c = a + (b + c)$$ ◄ associative axiom for addition   **(2.11)**

This axiom, together with Eq. (2.10), enables us to say that when any two of three numbers $a + b + c$ are combined and the third added to this sum, the result is the same regardless of the way in which the first two numbers are chosen. We can prove this statement by writing all possible ways in which this operation can be performed and then showing that each of them is equal to some one combination, such as $(a + b) + c$. As an example, we shall prove that $(b + c) + a = (a + b) + c$ and that $c + (b + a) = (a + b) + c$.

First Proof ►
$$\begin{aligned}(b + c) + a &= a + (b + c) \qquad ◄ \text{ by Eq. (2.10)}\\ &= (a + b) + c \qquad ◄ \text{ by Eqs. (2.2)† and (2.11)}\end{aligned}$$

_____

† The axiom shown by Eq. (2.2) enables us to interchange the members in the associative axiom shown by Eq. (2.11) and have $a + (b + c) = (a + b) + c$.

Second Proof ▶

$$c + (b + a) = c + (a + b) \qquad \blacktriangleleft \text{ by Eq. (2.10)}$$
$$= (a + b) + c \qquad \blacktriangleleft \text{ by Eq. (2.10)}$$

The next axiom involves a combination of addition and multiplication and is illustrated by the following example:

$$(4 + 8)3 = 12 \times 3 \qquad \blacktriangleleft \text{ since } 4 + 8 = 12$$
$$= 36$$

Also $\qquad (4 \times 3) + (8 \times 3) = 12 + 24 = 36$

Hence we can obtain the result of the operation $(4 + 8)3$ either by adding the numbers in the parentheses and multiplying the sum by 3 or by multiplying 4 and 8 by 3 and adding the products. This property is known as the *distributive property* of multiplication with respect to addition. We shall assume that this property holds for any three numbers, and we shall state the assumption in this way:

Distributive property of multiplication with respect to addition

$$\boldsymbol{(a + b)c = ac + bc} \qquad \blacktriangleleft \text{ right-hand† distributive axiom} \qquad \textbf{(2.12)}$$

The axiom shown by Eq. (2.12) can be extended to cover situations in which the polynomial in the parentheses consists of more than two terms. As an example, we shall prove that

$$(a + b + d)c = ac + bc + dc$$

Proof ▶

By the statement following Eq. (2.11), $a + b + d = (a + b) + d$. Hence,

$$(a + b + d)c = [(a + b) + d]c$$
$$= (a + b)c + dc \qquad \blacktriangleleft \text{ by Eq. (2.12)}$$
$$= ac + bc + dc \qquad \blacktriangleleft \text{ by Eq. (2.12)}$$

The above axiom enables us to express the sum of two or more similar monomials as a monomial. For example,

$$3ab + 2ab + 4ab = (3 + 2 + 4)ab \qquad \blacktriangleleft \text{ by Eq. (2.12)}$$
$$= 9ab$$

In Chap. 1 we defined the number zero, or 0, and the negative of a number, and we interpreted the definitions graphically. We shall repeat these definitions below.

Zero, denoted by the symbol 0, is the number such that $a + 0 = a$. Consequently, since by Eq. (2.10) $a + 0 = 0 + a$, we have

$$\boldsymbol{a + 0 = 0 + a = a} \qquad \blacktriangleleft \text{ additive identity} \qquad \textbf{(2.13)}$$

The negative of the number a is the number $-a$ such that $a + (-a) = 0$, and since by Eq. (2.10) $a + (-a) = -a + a$, we have

---

† This statement is called the "right-hand" distributive axiom because the factor c is at the right of the binomial $a + b$. Later we shall prove that the left-hand distributive axiom, $c(a + b) = ca + cb$, is true.

$$a + (-a) = -a + a = 0 \qquad (2.14)$$

The negative of a is also called the *additive inverse* of a.

We shall next prove two very useful theorems involving addition. The first theorem is called the *cancellation theorem* for addition and is stated below:

If $a + b = a + c$, then $b = c$ ◄ cancellation theorem for addition **(2.15)**

Proof ►

$$a + b = a + c \qquad \text{◄ given}$$
$$a + b + (-a) = a + c + (-a) \qquad \text{◄ by Eq. (2.5)}$$

By Eq. (2.10), we have $a + b + (-a) = a + (-a) + b$ and $a + c + (-a) = a + (-a) + c$. Hence, by Eq. (2.4),

$$a + (-a) + b = a + (-a) + c$$

Hence, $\qquad 0 + b = 0 + c$ ◄ since $a + (-a) = 0$

Therefore, $\qquad b = c$ ◄ since by (2.13)

$$0 + b = b \text{ and}$$
$$0 + c = c$$

The second theorem follows:

If $a + b = d$ and $a + c = d$, then $b = c$ **(2.16)**

Proof ►

$$a + b = d \quad \text{and} \quad a + c = d \qquad \text{◄ given}$$
$$a + b = a + c \qquad \text{◄ by Eq. (2.4)}$$
$$b = c \qquad \text{◄ by Eq. (2.15)}$$

Difference of two numbers

The difference of the numbers a and b is expressed as $a - b$ and is defined to be the number x such that $a = b + x$. In other words,

If $a - b = x$, then $a = b + x$ **(2.17)**

We shall now prove that if $a = b + x$,† then

$$a + (-b) = a - b \qquad (2.18)$$

Proof ►

$$a = b + x \qquad \text{◄ given}$$
$$a + (-b) = b + x + (-b) \qquad \text{◄ by Eq. (2.5)}$$
$$= b + (-b) + x \qquad \text{◄ by Eq. (2.10), the commutative axiom}$$
$$= [b + (-b)] + x \qquad \text{◄ by Eq. (2.11), the associative axiom}$$
$$= 0 + x \qquad \text{◄ since by Eq. (2.14) } b + (-b) = 0$$
$$= x \qquad \text{◄ since by Eq. (2.13) } 0 + x = x$$

Hence, since $a - b = x$, we have

$$a + (-b) = a - b \qquad \text{◄ by Eq. (2.4)}$$

---

† This theorem was proved graphically in Sec. 1.3. Here we shall prove it by means of the axioms.

A theorem that is used in the removal of parentheses from an expression and in the insertion of parentheses in an expression is stated and proved below.

$$-(a + b) = -a - b \qquad\qquad (2.19)$$

Proof ▶ $(a + b) + [-(a + b)] = 0$ ◀ by Eq. (2.14)

Also

$(a + b) + (-a) + (-b) = -a + (a + b) + (-b)$
◀ by the commutative
axiom shown by Eq. (2.10)

$$= [(-a) + a] + b + (-b)$$
◀ by the associative
axiom shown by Eq. (2.11)

$$= 0 + b + (-b) \qquad ◀ \text{ since by Eq. (2.14)} -a + a = 0$$
$$= 0 \qquad\qquad ◀ \text{ since } b + (-b) = 0$$

Hence,

$$-(a + b) = (-a) + (-b) \qquad ◀ \text{ by Eq. (2.16)}$$
$$= -a - b \qquad\qquad ◀ \text{ by Eq. (2.18)}$$

This theorem can be extended to cover cases in which there are three or more terms in the parentheses. As an example, we shall prove that $-(a + b + c) = -a - b - c$.

Proof ▶ $$-(a + b + c) = -[(a + b) + c] \qquad ◀ \text{ by Eq. (2.11)}$$
$$= -(a + b) - c \qquad ◀ \text{ by Eq. (2.19)}$$
$$= -a - b - c \qquad ◀ \text{ by Eq. (2.19)}$$

As a final theorem in this section, we shall prove that

$$-(-a) = a \qquad\qquad (2.20)$$

Proof ▶ $$-a + [-(-a)] = 0 \qquad ◀ \text{ by Eq. (2.14)}$$
Also $$(-a) + a = 0 \qquad ◀ \text{ by Eq. (2.14)}$$
Hence, $$-(-a) = a \qquad ◀ \text{ by Eq. (2.15)}$$

2.5 THE ORDER RELATIONS

We shall employ the properties of the order relations in the discussion of addition and multiplication. In Chap. 1 we stated that "greater than" is denoted by $>$ and "less than" by $<$. We also stated that $a > b$ if the point (a) is to the right of (b) on the number line. This interpretation is not a definition of the relation, and we shall not attempt to formulate one. We shall, however, state four axioms that completely determine the properties of the relation. If we accept the relation "greater than" as undefined, we can define the relation "less than" as follows:

If a and b are real numbers, then $b < a$ if and only if $a > b$ **(2.21)**

We shall now state the four basic axioms dealing with the order relations. In the statement of the axioms, the letters used represent real numbers.

Exactly one of the statements $a = b$, $a > b$, $a < b$ is true

     ◀ trichotomy axiom   **(2.22)**

If $a > b$ and $b > c$, then $a > c$ ◀ transitivity axiom   **(2.23)**

If $a > b$, then $a + c > b + c$ ◀ additivity axiom   **(2.24)**

If $a > b$ and $c > 0$, then $ac > bc$; if $d < 0$, then $ad < bd$

     ◀ multiplicativity axiom   **(2.25)**

We shall now prove the following theorem:

If $a > 0$, then $-a < 0$ **(2.26)**

Proof ▶

$$a > 0 \qquad \text{◀ given}$$
$$a + (-a) > 0 + (-a) \qquad \text{◀ by Eq. (2.24)}$$
$$0 > -a \qquad \text{◀ since } a + (-a) = 0 \text{ and } 0 + (-a) = -a$$
$$-a < 0 \qquad \text{◀ by Eq. (2.21)}$$

Consequently, if a is a positive number, then $-a$ is a negative number.

Positive number   We now define a *positive number* as a number greater than zero and a *nega-*

Negative number   *tive number* as a number less than zero. Therefore:

If a is a positive number, then $a > 0$;
if b is a negative number, then $b < 0$ **(2.27)**

We shall now prove a theorem that enables us to tell whether or not a is greater than b.

$a > b$ if and only if $a - b > 0$ **(2.28)**

Proof ▶

$$a > b \qquad \text{◀ given}$$
$$a + (-b) > b + (-b) \qquad \text{◀ by Eq. (2.24)}$$
$$a - b > 0 \qquad \text{◀ since } a + (-b) = a - b \text{ and } b + (-b) = 0$$

Conversely, we shall assume that $a - b > 0$ and shall prove that $a > b$.

$$a - b > 0 \qquad \text{◀ assumed}$$
$$a + (-b) > 0 \qquad \text{◀ since by Eq. (2.18) } a + (-b) = a - b$$
$$a + (-b) + b > 0 + b \qquad \text{◀ by Eq. (2.24)}$$
$$a + [(-b) + b] > 0 + b \qquad \text{◀ by the associative axiom shown by Eq. (2.11)}$$
$$a + 0 > 0 + b \qquad \text{◀ since } (-b) + b = 0 \text{ by Eq. (2.14)}$$
$$a > b \qquad \text{◀ since by Eq. (2.13) } a + 0 = a \text{ and } b + 0 = b$$

2.6 LAW OF SIGNS FOR ADDITION

The extension of the number system to include negative numbers necessitates an extension of the notion of addition. We must define the meaning of the sum of two signed numbers and derive a law that enables us to decide whether the sum is positive or negative. The proof of this law (to be stated presently) involves the following theorem:

$$\text{If } r = s, \text{ then } -r = -s \tag{2.29}$$

Proof ▶

$-r + r = 0$	◀ by Eq. (2.14)
$-r + s = 0$	◀ replacing r by s
$-r + s + (-s) = 0 + (-s) = -s$	◀ by the addition axiom shown by Eq. (2.5)
$-r + [s + (-s)] = -s$	◀ by the associative axiom shown by Eq. (2.11)
$-r = -s$	◀ by Eq. (2.13), since $s + (-s) = 0$

The statement of the law involves the concept of absolute value, and the student is advised to review the definition near the end of Sec. 1.3. We shall now state and prove the following theorem:

If a and b are positive real numbers, then:

1 $a + b > 0$ and $|a + b| = |a| + |b|$

2 $-a - b < 0$ and $|-a - b| = |a| + |b|$

3 If $a > b$, then $a + (-b) > 0$ and $|a + (-b)| = |a| - |-b|$

4 If $a < b$, then $a + (-b) < 0$ and $|a + (-b)| = -|a| + |b|$

Proof of 1 ▶

$a > 0$	◀ since a is positive
$a + b > 0 + b$	◀ by Eq. (2.24)
$a + b > b$	◀ by Eq. (2.13), since $0 + b = b$
$a + b > 0$	◀ by Eq. (2.23), since $b > 0$

Also, since $a + b > 0$, then

$|a + b| = a + b$	◀ by the definition of absolute value
$= |a| + |b|$	◀ since $|a| = a$ and $|b| = b$

Proof of 2 ▶

$$-a - b = -(a + b) \qquad \text{◀ by Eq. (2.19)}$$

Hence, since $a + b > 0$, $-(a + b) < 0$ by Eq. (2.26). Consequently, $-a - b < 0$. Furthermore, $|-a - b| = |-(a + b)| = a + b$ by the definition of absolute value. Hence $|-a - b| = |a| + |b|$ since $a = |a|$ and $b = |b|$.

Proof of 3 ▶

$a > b$	◀ given
$a + (-b) > b + (-b)$	◀ by the additivity axiom shown by Eq. (2.24), with $c = -b$
$a + (-b) > 0$	◀ by Eq. (2.14), since $b + (-b) = 0$

Moreover, by the definition of absolute value,

$$|a + (-b)| = a + (-b) \qquad \blacktriangleleft \text{ since } a + (-b) \text{ is positive}$$
$$= a - b \qquad \blacktriangleleft \text{ by Eq. (2.18)}$$
$$= |a| - |-b| \qquad \blacktriangleleft \text{ since } a = |a| \text{ and } b = |-b|$$

Proof of 4 ▶ If $a < b$, then $b > a$ by Eq. (2.21). Hence,

$$b + (-b) > a + (-b) \qquad \blacktriangleleft \text{ by Eq. (2.24), with } c = -b$$
and $\qquad\qquad 0 > a + (-b) \qquad \blacktriangleleft \text{ since } b + (-b) = 0 \text{ by Eq. (2.14)}$
Consequently,

$$a + (-b) < 0 \qquad \blacktriangleleft \text{ by Eq. (2.21)}$$

Furthermore

$$|a + (-b)| = -[a + (-b)] \qquad \blacktriangleleft \text{ by the definition of absolute value}$$
$$= -a - (-b) \qquad \blacktriangleleft \text{ by Eq. (2.19)}$$
$$= -a + b \qquad \blacktriangleleft \text{ since } -(-b) = +b \text{ by Eq. (2.20)}$$
$$= -|a| + |b| \qquad \blacktriangleleft \text{ by the definition of absolute value}$$

If we state the above theorem in words, we have the following:

Law of signs for addition

The absolute value of the sum of two positive or two negative numbers is equal to the sum of their absolute values. The sum of the two numbers is positive if the two addends are positive, and the sum is negative if the two addends are negative. The absolute value of the sum of a positive number and a negative number is equal to the difference of their absolute values. The sum is positive if the number with the greater absolute value is positive and is negative if the number with the greater absolute value is negative.

The sign of a signed numeral is the sign that precedes it or is understood to precede it. For example, the sign of 3 is positive since a plus sign is understood to appear before it, and the sign of -3 is negative. Hence for two signed numerals, we may state the law of signs for addition as follows:

The sum of two signed numerals with the same sign is the sum of their absolute values preceded by the common sign of the addends. The sum of two signed numerals with different signs is the difference of their absolute values preceded by the sign of the addend with the greater absolute value.

EXAMPLE 1 $6 + 3 = 9$, since 6 and 3 are positive

EXAMPLE 2 $(-2) + (-4) = -6$, since -2 and -4 are negative

EXAMPLE 3 $-4 + 9 = 5$, since -4 and 9 have different signs and 9 has the greater absolute value

EXAMPLE 4 $3 + (-8) = -5$, since 3 and -8 have different signs and -8 has the greater absolute value

Binary operation   A *binary operation* in a set is a rule that assigns to each pair of elements of the set, taken in a definite order, a unique element of the set. The closure axiom shown by Eq. (2.9) states that if a and b are real numbers, there exists a real number c such that $a + b = c$. Hence addition is a binary operation in the set of real numbers. Furthermore, since the sum of two positive integers is a positive integer, addition is a binary operation in the set of positive integers.

Exercise 2.1

By use of the axioms and theorems dealing with addition, equality, and the order relations, prove that the statements in Problems 1 to 20 are true.

1 $a + (b + c) = b + (a + c)$
2   If $(b + c)a = ab + d$, then $ac = d$.
3   If $a + b = a$, then $b = 0$.
4   If $a + b = 0$, then $b = -a$.
5   If $(a + b) + c = a + (b + d)$, then $c = d$.
6   If $(a + b) + c = b + c$, then $a = 0$.
7   If $a + (b + c) = a$, then $b = -c$.
8   If $(a + b) + (c + d) = (a + e) + (c + f)$, then $b + d = e + f$.
9   If $a + (b + c) = d$ and $a - c = d$, then $b = -2c$.
10 If $(a + b) + (c + d) = 0$ and $(b + c) + (a - e) = 0$, then $d = -e$.
11 If $a + (b + c) = d$ and $a + b = d - e$, then $c = e$.
12 If $a + (b + c) = a$ and $d + (e + f) = d$, then $b - e = f - c$.
13 If $a > b$, $b > c$, and $d > 0$, then $ad > cd$.
14 If $a + b > a + c$, then $b > c$.
15 If $a + (b + c) > a + d$, then $b > d - c$.
16 If $a + b > c + d$, then $a - c > d - b$.
17 If $b > c$ and $ab > ac$, then $a > 0$.   HINT: Assume that $a = 0$ and that $a < 0$, and show that each assumption leads to a conclusion that contradicts the hypothesis that $ab > ac$.
18 If $a > 0$ and $ab > ac$, then $b > c$.
19 If $a > 0$, $ab > bc$, and $bc > ac$, then $b > c$.

20 If $a > 0$, $b > 0$, $d > 0$, and $(a-c)b > db$, then $a > c$. HINT: First prove that $a - c > d$.

21 Show that addition is a binary operation in the set of positive real numbers.

22 Give an example which shows that addition is not a binary operation in the set $\{x \mid 0 \leq x \leq 10\}$.

23 Give an example which shows that addition is not a binary operation in the set $\{x \mid x \text{ is an integer and } -6 \leq x \leq 1\}$.

24 Give an example which shows that the procedure for finding $a - b$ is not a binary operation in the set of positive integers.

2.7 ADDITION OF MONOMIALS AND POLYNOMIALS

The sum of two or more similar monomials is equal to a monomial that is obtained by use of the right-hand distributive axiom shown by Eq. (2.12) or its extension. For example,

$$12ab + 8ab + 5ab = (12 + 8 + 5)ab \quad \blacktriangleleft \text{ by Eq. (2.12)}$$
$$= 25ab$$

In this section we shall consider monomials that involve multiplication only, as illustrated by $3a$, $5x^2y$, and ab^3. We shall frequently encounter polynomials with all terms similar but in which some of the terms are preceded by the minus sign and others involve no numerical coefficient. In order to apply the distributive axiom to such polynomials, we shall assume† that the following statements are true:

$$a = 1 \cdot a$$
$$-(na) = -n \cdot a$$
$$0 \cdot a = 0$$

The procedure for adding two or more similar monomials and for combining by addition the similar terms in a polynomial is illustrated by the following examples.

EXAMPLE 1

$$-5b + b + 6b = -5 \cdot b + 1 \cdot b + 6 \cdot b$$
$$= (-5 + 1 + 6)b \quad \blacktriangleleft \text{ by the right-hand distributive axiom shown by Eq. (2.12)}$$
$$= (-5 + 7)b = 2b \quad \blacktriangleleft \text{ by the associative axiom shown by Eq. (2.11) and the law of signs for addition}$$

---

† We shall discuss these assumptions more fully in Sec. 2.9.

EXAMPLE 2

$$7x^2y - 3x^2y - 2xy^2 + 8xy^2 = 7 \cdot x^2y + (-3 \cdot x^2y) + (-2 \cdot xy^2) + 8 \cdot xy^2$$
$$= [7 + (-3)]x^2y + (-2 + 8)xy^2$$
$$= (7 - 3)x^2y + (-2 + 8)xy^2$$
$$= 4x^2y + 6xy^2$$

After some practice, the first and second steps in problems similar to Example 2 can be performed mentally, and thus we can proceed directly to the result in the third step. This is illustrated in Example 3.

EXAMPLE 3

$$2a - 3a + 6b + 4b - 7c + 9c = (2 - 3)a + (6 + 4)b + (-7 + 9)c$$
$$= -a + 10b + 2c$$

In the polynomial $4x^3 + 5x^2y - 6xy^2 + 2x^3 - 2x^2y + 2xy^2 - 3x^3 + 4xy^2$, we have three sets of monomials in which the elements of each set are similar. These sets are $\{4x^3, 2x^3, -3x^3\}$; $\{5x^2y, -2x^2y\}$; and $\{-6xy^2, 2xy^2, 4xy^2\}$. By repeated applications of the commutative axiom shown by Eq. (2.10), we can rearrange the terms in the polynomial so that the similar terms are to-gether, and then we can apply the distributive axiom. We must remember that each monomial, together with the sign that precedes it, is a term of the polynomial and that when a term is shifted from one position to another, the sign must be carried along with the remainder of the term. We shall now employ the above procedure to express the polynomial as the sum of two monomials.

Addition of polynomials

$$4x^3 + 5x^2y - 6xy^2 + 2x^3 - 2x^2y + 2xy^2 - 3x^3 + 4xy^2$$
$$= 4x^3 + 2x^3 - 3x^3 + 5x^2y - 2x^2y - 6xy^2 + 2xy^2 + 4xy^2$$
◄ by Eq. (2.10), the commutative axiom
$$= (4 + 2 - 3)x^3 + (5 - 2)x^2y + (-6 + 2 + 4)xy^2$$
◄ by Eq. (2.12), the right-hand distributive axiom
$$= 3x^3 + 3x^2y + 0xy^2 = 3x^3 + 3x^2y$$

The above procedure is called *combining the similar terms* in a polynomial.

The sum of two or more polynomials is obtained by repeated applica-tions of the commutative axiom and the distributive axiom. For example, to get the sum of $-4a + 2b - 5c$, $2a - 3c + 4b$, and $6c + 7a - 8b$, we proceed as follows: We first express the sum of the three polynomials in this way:

$$-4a + 2b - 5c + 2a - 3c + 4b + 6c + 7a - 8b$$

Now we arrange the terms in this polynomial so that similar terms are to-gether. We then apply the distributive law and get

$$-4a + 2b - 5c + 2a - 3c + 4b + 6c + 7a - 8b$$
$$= -4a + 2a + 7a + 2b + 4b - 8b - 5c - 3c + 6c \qquad \blacktriangleleft \text{ by Eq. (2.10)}$$
$$= (-4 + 2 + 7)a + (2 + 4 - 8)b + (-5 - 3 + 6)c \qquad \blacktriangleleft \text{ by Eq. (2.12)}$$
$$= 5a - 2b - 2c \qquad \blacktriangleleft \text{ by Eq. (2.11) and the law of signs for addition}$$

We usually employ a condensation of the above procedure, writing the given polynomials with each one after the first below the preceding and with the terms rearranged so that the similar terms are in the same column. Then we add the numerical coefficients in each column. In this way, we get

$$\begin{array}{l} -4a + 2b - 5c \\ 2a + 4b - 3c \\ \underline{7a - 8b + 6c} \\ 5a - 2b - 2c \qquad \blacktriangleleft \text{ sum} \end{array}$$

Frequently, we are required to add two or more polynomials when at least one of them contains one or more terms that are not similar to any term in at least one of the others. The method for dealing with such situations is illustrated by Example 4.

EXAMPLE 4

Solution

Add the polynomials $3x^2 + 4y^2 - 3xy + 7z^2$, $2x^2 + 4z^3$, and $4y^2 - 2z^2 - 2xy$.

We write the polynomials as shown below and perform the addition as indicated.

$$\begin{array}{l} 3x^2 + 4y^2 - 3xy + 7z^2 \\ 2x^2 \qquad\qquad\qquad\quad + 4z^3 \\ \underline{\qquad 4y^2 - 2xy - 2z^2 \qquad} \\ 5x^2 + 8y^2 - 5xy + 5z^2 + 4z^3 \end{array}$$

2.8 SUBTRACTION

The operation of subtracting the number b from the number a is the process of determining x such that $b + x = a$. The number a is called the *minuend,* b the *subtrahend,* and x the *difference* of a and b. We determine x by the method below.

Minuend
Subtrahend
Difference

$$\begin{array}{ll} b + x = a & \blacktriangleleft \text{ given} \\ b + x + (-b) = a + (-b) & \blacktriangleleft \text{ by Eq. (2.5)} \\ b + (-b) + x = a + (-b) & \blacktriangleleft \text{ by Eq. (2.10), the commutative axiom} \\ 0 + x = a + (-b) & \blacktriangleleft \text{ since } b + (-b) = 0 \text{ by Eq. (2.14)} \\ x = a + (-b) & \blacktriangleleft \text{ since } 0 + x = x \end{array}$$

Thus, to subtract one number from another, we add the negative of the subtrahend to the minuend. The statement "*a* minus *b*," or $a - b$, means that b is to be subtracted from a. The procedure is illustrated by the following examples.

EXAMPLE 1   Subtract 4 from 6.

Solution $6 \text{ minus } 4 = 6 + (-4) = 2$

EXAMPLE 2   Subtract $-16ab$ from $-20\ ab$.

Solution $-20ab \text{ minus } -16\ ab = -20ab + 16ab = -4ab$

EXAMPLE 3   Subtract $-18x^2yz$ from $-12x^2yz$.

Solution $-12x^2yz \text{ minus } -18x^2yz = -12x^2yz + 18x^2yz = 6x^2yz$

EXAMPLE 4   Subtract $3x - 2y - 9z$ from $5x + 3y - 6z$.

Solution
$$
\begin{aligned}
5x + 3y - 6z &\text{ minus } 3x - 2y - 9z \\
&= 5x + 3y - 6z - (3x - 2y - 9z) \\
&= 5x + 3y - 6z - 3x + 2y + 9z \qquad \blacktriangleleft \text{ by Eq. (2.19)} \\
&= 5x - 3x + 3y + 2y - 6z + 9z \qquad \blacktriangleleft \text{ by Eq. (2.10),} \\
&\qquad\qquad\qquad\qquad\qquad\qquad\quad \text{the commutative axiom} \\
&= (5 - 3)x + (3 + 2)y + (-6 + 9)z \qquad \blacktriangleleft \text{ by Eq. (2.12), the} \\
&\qquad\qquad\qquad\qquad\qquad\qquad\qquad \text{right-hand distributive axiom} \\
&= 2x + 5y + 3z
\end{aligned}
$$

The procedure in Example 4 is usually condensed as follows: We write the subtrahend below the minuend as indicated below, and then we mentally change the sign preceding each term of the subtrahend and proceed as in addition.

$$
\begin{array}{ll}
5x + 3y - 6z & \blacktriangleleft \text{ minuend} \\
3x - 2y - 9z & \blacktriangleleft \text{ subtrahend} \\
\hline
2x + 5y + 3z & \blacktriangleleft \text{ difference}
\end{array}
$$

Exercise 2.2   Addition and Subtraction

Perform the indicated operations in problems 1 to 8.

1	$7 - 3 + 2$	**2**	$11 - 5 - 1$
3	$-9 + 4 - 2$	**4**	$-8 - 7 + 12$

5 $|7| - |-2|$

6 $|-3| + |-4|$

7 $|8 - 13| - |-5|$

8 $|6 - 2| - |5 - 3|$

Add the expressions in problems 9 to 16.

9 $2x + 3y - 4z$
$-2x + y - z$
$3x - 4y + 4z$

10 $3a - 2b - 4c$
$2a + 4b + 7c$
$-5a + 6b - 3c$

11 $5p + 2q - 3r$
$4p + q - 4r$
$-7p - 3q + 6r$

12 $4r + 2s + t$
$-3r - 5s + 4t$
$6r + 3s - 3t$

13 $ab + cd - ac$
$3ab - 2cd - 5ac$
$2ab + 3cd + 6ac$

14 $4pq - 3pr - 5qr$
$-7pq + 4pr - 3qr$
$3pq - pr + 7qr$

15 $a^2b - 2ab - 4ab^2$
$3a^2b - 5ab + 6ab^2$
$4a^2b + 7ab - 3ab^2$

16 $2x^2y + 3xy^2 - 4x$
$-3x^2y + 5xy^2 + 7x$
$x^2y - 8xy^2 - 2x$

Combine similar terms in each of problems 17 to 24.

17 $3a + 2b - 4c - 2a - 5b - 4b + 5a - 6c + 10c$

18 $5x - 3y + 2z - 7x - 8y - 6z + 4x + 11y + 3z$

19 $4p + 3d - 6q + 2p - 5d - 5q - 6p + 4d + 10q$

20 $-5r + 7s - 4t + 8r - 5s - 2t - 3r + 6s + 7t$

21 $2a^2b + 3ab^2 - 4ab - 6ab^2 - 4a^2b + 5ab - 3a^2b - 2ab^2 + ab$

22 $3x^2y - 2xy^2 - 7xy + 4xy - 5xy^2 - 8x^2y + 5x^2y + 7xy^2 + 2xy$

23 $7p^3d - 4pd^2 - 2pd + 7pd^2 - 4pd + 3p^3d - 2pd^2 + 5pd - 4p^3d$

24 $2rs + 3rt + 4st - 5rt - 4rs + 3st - 7st - rs + 2rt$

In each of problems 25 to 32 subtract the second number or expression from the first.

25 32
17

26 41
-21

27 -36
15

28 -21
-37

29 $3a + 2b - 5c$
$2a - 3b - 6c$

30 $-2a - 3b + 4c$
$3a - 2b - c$

31 $5x - y + z$
$3x - 2y + 2z$

32 $-7x + 4y - 5z$
$-8x + 3y - 6z$

In problems 33 to 36, replace each variable by the indicated number, and then combine the result into a single number.

33 $x - y + |z|$; $x = 3$, $y = 2$, $z = -4$

34 $2|x| + 3y - |z|$; $x = -3$, $y = 2$, $z = -1$

35 $3x - |2y| - 2z; \; x = 2, \, y = -2, \, z = -3$

36 $|-x| - |-y| + 2z; \; x = 4, \, y = -3, \, z = 2$

2.9   AXIOMS AND THEOREMS OF MULTIPLICATION

In this section we shall state the axioms that determine the properties of multiplication and prove theorems that enable us to obtain the product of two signed numbers and to apply multiplication to monomials and polynomials.

Product
Multiplier
Multiplicand
Factor

      The operation of multiplying b by a is indicated by $a \times b$, $a \cdot b$, or $(a)(b)$, and the *product* is denoted by ab. The number a is the *multiplier*, and b is the *multiplicand*. The numbers a and b are also called the *factors* of ab.

      The first axiom given below is the closure axiom, and it states that the product of two real numbers is a real number.

Closure axiom for multiplication

If a and b are real numbers, there exists

a real number c such that $ab = c$ **(2.30)**

      We have seen the importance of the commutative and associative axioms in addition, and we shall presently state similar axioms for multiplication. It is readily verified that $3 \cdot 5 = 5 \cdot 3 = 15$ and that $3 \cdot 5 \cdot 7 = 3(35) = 5(21) = 7(15) = 105$. We assume that these properties hold for all real numbers and thus we may state the following axioms:

Commutative axiom for multiplication

$$ab = ba \qquad\qquad (2.31)$$

Associative axiom for multiplication

$$a(bc) = (ab)c \qquad\qquad (2.32)$$

These two axioms enable us to get the product $a \cdot b \cdot c$ by multiplying the product of two of the numbers by the third. Furthermore, this product is unique regardless of the choice of the two numbers for the first multiplication and the order in which the multiplication is performed. This means that

$$abc = a(bc) = (ab)c = a(cb) = (ac)b = c(ba) = (cb)a = \cdots$$

where the list of equalities can be extended to include all possible orders in which a, b, and c can be arranged and all choices of the two that are to be enclosed in parentheses. To prove this statement we must prove that each combination is equal to some of them. To illustrate the method we shall prove that $c(ba) = a(bc)$.

Proof ▶

$$
\begin{aligned}
c(ba) &= c(ab) &&\blacktriangleleft \text{ by the commutative axiom (2.31)}\\
&= (ab)c &&\blacktriangleleft \text{ by (2.29)}\\
&= a(bc) &&\blacktriangleleft \text{ by the associative axiom (2.32)}
\end{aligned}
$$

In Sec. 2.4 we stated the right-hand distributive axiom. We shall now prove that the left-hand distributive axiom

Left-hand distributive axiom

$$a(b + c) = ab + ac \qquad \text{(2.33)}$$

is true.

Proof ▶

$$
\begin{aligned}
a(b + c) &= (b + c)a && \blacktriangleleft \text{ by the commutative axiom (2.31)} \\
&= ba + ca && \blacktriangleleft \text{ by the right-hand distributive axiom (2.12)} \\
&= ab + ac && \blacktriangleleft \text{ by (2.31)}
\end{aligned}
$$

The numbers 1 and 0 play unique roles in multiplication. The role of 1 is established by definition. That is, we define $1 \cdot a$ to be a. Then, since $1 \cdot a = a \cdot 1$, we have

$$1 \cdot a = a \cdot 1 = a \qquad \text{(2.34)}$$

Multiplicative identity

For this reason 1 is called the *multiplicative identity* element. We shall now prove that $0 \cdot a = 0$.

Proof ▶

$$
\begin{aligned}
1 \cdot a + 0 \cdot a &= (1 + 0)a && \blacktriangleleft \text{ by the distributive axiom} \\
&= 1 \cdot a && \blacktriangleleft \text{ since } 1 + 0 = 1 \\
&= a && \blacktriangleleft \text{ by (2.34)}
\end{aligned}
$$

Also $\qquad 1 \cdot a + 0 = a \qquad \blacktriangleleft$ by (2.34) and (2.13)

Hence, $0 \cdot a$ and 0 must be equal, so $0 \cdot a = 0$. Therefore by the commutative axiom we have

$$0 \cdot a = a \cdot 0 = 0 \qquad \text{(2.35)}$$

The following axiom is essential for dealing with division and for computations involving fractions:

Reciprocal

If a is a real nonzero number, there exists a unique real number $1/a$, called the reciprocal of a, such that

$$a \cdot \frac{1}{a} = \frac{1}{a} \cdot a = 1 \qquad \text{(2.36)}$$

We shall employ this axiom to prove the very-useful *cancellation theorem* for multiplication.

$$\text{If } ab = ac \text{ and } a \neq 0, \text{ then } b = c \qquad \text{(2.37)}$$

Proof ▶

$$
\begin{aligned}
ab &= ac && \blacktriangleleft \text{ given} \\
\frac{1}{a} \cdot ab &= \frac{1}{a} \cdot ac && \blacktriangleleft \text{ by the multiplicativity axiom for equality (2.6)} \\
1 \cdot b &= 1 \cdot c && \blacktriangleleft \text{ by the associative axiom for multiplication (2.32) and (2.36)} \\
b &= c && \blacktriangleleft \text{ by (2.34)}
\end{aligned}
$$

2.10 LAW OF SIGNS FOR MULTIPLICATION

In this section we shall develop the law of signs for multiplication and explain methods for applying it. For this purpose, we shall first prove three theorems.

THEOREM 1

$$If\ a > 0\ and\ b > 0,\ then\ ab > 0$$

Proof ▶

$a > 0$	◀ given
$a \cdot b > 0 \cdot b$	◀ by (2.25)
$a \cdot b > 0$	◀ since $b \cdot 0 = 0$
$ab > 0$	◀ since $a \cdot b = ab$

THEOREM 2

$$If\ a > 0\ and\ b > 0,\ a(-b) = (-b)a = -ab$$

Proof ▶

$ab + a(-b) = a[b + (-b)]$	◀ by the left-hand distributive axiom
$= a(0)$	◀ since $b + (-b) = 0$ by (2.14)
$= 0$	◀ by (2.35)
Furthermore, $ab + (-ab) = 0$	◀ by (2.14)

Hence since $ab + a(-b) = 0$ and $ab + (-ab) = 0$, we have $a(-b) = -ab$ by (2.16). Moreover, by the commutative axiom $a(-b) = (-b)a$. This completes the proof.

THEOREM 3

$$If\ a > 0\ and\ b > 0,\ then\ (-a)(-b) = ab$$

Proof ▶

$a(-b) + (-a)(-b) = [a + (-a)](-b)$	◀ by the right-hand distributive axiom
$= 0(-b)$	◀ since $a + (-a) = 0$
$= 0$	◀ by (2.35)
Also, $a(-b) + ab = a(-b + b)$	◀ by the left-hand distributive axiom
$= a(0)$	◀ by (2.14)
$= 0$	◀ by (2.35)
Therefore, $(-a)(-b) = ab$	◀ by (2.16)

Since a and b are positive, we have by the definition of the absolute value of a number,

$$a = |a| = |-a| \qquad and \qquad b = |b| = |-b|$$

Furthermore, $|ab| = ab = |a|\,|b|$ ◀ since a, b, and ab are positive

Likewise, $\quad |(-a)(-b)| = |ab| = |a|\,|b| \quad$ ◀ since $(-a)(-b) = ab$

and $\qquad\qquad |a(-b)| = |(-b)a| = |-ab| = ab = |a|\,|b|$

Law of signs for multiplication

We summarize the above conclusions in the following law. *The absolute value of the product of two real numbers is equal to the product of their absolute values.* The product of two positive real numbers or of two negative real numbers is positive. The product of a positive real number and a negative real number or of a negative real number and a positive real number is negative.

EXAMPLE 1

$$4(7) = 28$$
$$6(-9) = -54$$
$$-5(8) = -40$$
$$-3(-6) = 18$$

It follows from the associative axiom (2.32) that the product of three or more positive numbers is positive. For example, $3 \cdot 5 \cdot 7 = (3 \cdot 5)7 = 15(7) = 105$, and $2 \cdot 6 \cdot 3 \cdot 8 = (2 \cdot 6)(3 \cdot 8) = 12(24) = 288$. Furthermore the product of an even number of negative numbers is positive, and the product of an odd number of negative numbers is negative. For example, $(-2)(-4)(-5)(-8) = [(-2)(-4)][(\ 5)(-8)] = 8(40) = 320$, and $(-4)(-6)(-3) = [(-4)(-6)](-3) = 24(-3) = -72$.

If we do not know whether certain factors in a product are positive or negative, we cannot express the product in terms of the absolute values of the factors. For example, the product of the absolute values of a and b is $|a|\,|b|$. However, if a is positive, then $ab = |a|\,|b|$ or $-|a|\,|b|$ according as b is positive or negative. If one or more of the factors are not clearly positive or clearly negative and if minus signs appear explicitly before some of the factors, the usual practice for expressing the product is as follows. First disregard the minus signs and write the product of the symbols that remain, then prefix a minus sign if the number of explicitly expressed minus signs is odd. If the number of explicitly expressed minus signs is even, we either prefix no sign before the product or prefix a plus sign.

EXAMPLE 2

$$(3x)(2y) = 6xy$$
$$(4x)(-3y)(5z) = -60xyz$$
$$(-7a)(8b)(-3c) = 168abc$$
$$(-2m)(-5n)(-7r) = -70\ mnr$$

2.11 LAWS OF EXPONENTS IN MULTIPLICATION

In Sec. 2.2 we stated that a^n, where n is a positive integer, is equal to $a \cdot a \cdot a \cdots$ to n factors. The number a is the *base*, and n is the exponent. We now consider the product $a^m a^n$. By definition this product is equal to the product of m a's and n a's, and this is the product of $m + n$ a's. Hence we have

$$a^m a^n = a^{m+n} \qquad \text{where } m \text{ and } n \text{ are positive integers} \qquad \textbf{(2.38)}$$

By the definition of the nth power of a number and the use of the commutative axiom, we can prove that

$$(ab)^n = a^n b^n \qquad \text{where } n \text{ is a positive integer} \qquad \textbf{(2.39)}$$

Proof ▶

$$(ab)^n = ab \cdot ab \cdot ab \cdots \text{ to } n \text{ factors}$$
$$= (a \cdot a \cdot a \cdots \text{ to } n \text{ factors})(b \cdot b \cdot b \cdots \text{ to } n \text{ factors})$$

◀ by the commutative axiom

$$= a^n b^n$$

Finally we prove that

Proof ▶

$$(a^n)^m = a^{mn} \qquad m \text{ and } n \text{ positive integers} \qquad \textbf{(2.40)}$$

$$(a^n)^m = a^n \cdot a^n \cdot a^n \cdots \text{ to } m \text{ factors}$$
$$= a^{n+n+n+\cdots \text{ to } m \text{ addends}} \qquad ◀ \text{ by (2.38)}$$
$$= a^{mn}$$

2.12 PRODUCTS OF MONOMIALS AND POLYNOMIALS

We employ the commutative, associative, and distributive axioms together with the law of signs and the law of exponents to obtain the product of two or more monomials, of a monomial and a polynomial, and of two polynomials. We shall illustrate the method with five examples.

EXAMPLE 1

$$3x^2 y \cdot 4xy^2 \cdot 6x^3 y^4 = 3 \cdot 4 \cdot 6 \cdot x^2 \cdot x \cdot x^3 \cdot y \cdot y^2 \cdot y^4 \qquad ◀ \text{ by the commutative axiom (2.31)}$$

$$= 72 x^{2+1+3} y^{1+2+4} \qquad ◀ \text{ by (2.38)}$$
$$= 72 x^6 y^7$$

EXAMPLE 2

$$-4ab^2 c^3 \cdot -2a^3 b^4 c \cdot 6a^2 bc^5 = -4 \cdot -2 \cdot 6 \cdot a \cdot a^3 \cdot a^2 \cdot b^2 \cdot b^4 \cdot b \cdot c^3 \cdot c \cdot c^5$$
$$= 48 a^6 b^7 c^9 \qquad ◀ \text{ by the law of signs and the law of exponents}$$

EXAMPLE 3 $3ab(2a - 4b + 7a^2b) = 3ab(2a) - 3ab(4b) + 3ab(7a^2b)$

$$\blacktriangleleft \text{ by the left-hand distributive axiom (2.33)}$$
$$= 6a^2b - 12ab^2 + 21a^3b^2$$

EXAMPLE 4 $(3x^2y - 6xy^2 - 8y^3)(-5x^3y^2) = 3x^2y(-5x^3y^2) + (-6xy^2)(-5x^3y^2)$
$$+ (-8y^3)(-5x^3y^2) = -15x^5y^3 + 30x^4y^4 + 40x^3y^5$$

$$\blacktriangleleft \text{ by the right-hand distributive axiom (2.12)}$$

The method for obtaining the product of two polynomials is illustrated in Example 5.

EXAMPLE 5 To obtain the product $(-5x^2 + 2xy + 3y^2)(3x^3 - 6x^2y + 2xy^2 - 4y^3)$ we first consider the second factor as a single number, apply the right-hand distributive axiom, and then complete the computation as indicated.

$(-5x^2 + 2xy + 3y^2)(3x^3 - 6x^2y + 2xy^2 - 4y^3)$
$= (-5x^2)(3x^3 - 6x^2y + 2xy^2 - 4y^3) + (2xy)(3x^3 - 6x^2y + 2xy^2 - 4y^3)$
$\quad + (3y^2)(3x^3 - 6x^2y + 2xy^2 - 4y^3)$ $\blacktriangleleft$ by the right-hand distributive axiom

$= -15x^5 + 30x^4y - 10x^3y^2 + 20x^2y^3 + 6x^4y - 12x^3y^2 + 4x^2y^3 - 8xy^4$
$\quad + 9x^3y^2 - 18x^2y^3 + 6xy^4 - 12y^5$ $\blacktriangleleft$ by the right-hand distributive axiom

$= -15x^5 + 30x^4y + 6x^4y - 10x^3y^2 - 12x^3y^2 + 9x^3y^2 + 20x^2y^3 + 4x^2y^3$
$\quad - 18x^2y^3 - 8xy^4 + 6xy^4 - 12y^5$ $\blacktriangleleft$ by the commutative axiom

$= -15x^5 + 36x^4y - 13x^3y^2 + 6x^2y^3 - 2xy^4 - 12y^5$

The above process is usually abbreviated to the method illustrated below.

$3x^3 - 6x^2y + 2xy^2 - 4y^3$	$\blacktriangleleft$ multiplicand
$- 5x^2 + 2xy + 3y^2$	$\blacktriangleleft$ multiplier
$-15x^5 + 30x^4y - 10x^3y^2 + 20x^2y^3$	$\blacktriangleleft$ multiplying by $-5x^2$
$+ 6x^4y - 12x^3y^2 + 4x^2y^3 - 8xy^4$	$\blacktriangleleft$ multiplying by $2xy$
$+ 9x^3y^2 - 18x^2y^3 + 6xy^4 - 12y^5$	$\blacktriangleleft$ multiplying by $3y^2$
$-15x^5 + 36x^4y - 13x^3y^2 + 6x^2y^3 - 2xy^4 - 12y^5$	$\blacktriangleleft$ adding coefficients

Exercise 2.3 Multiplication

Prove each statement in problems 1 to 8 by use of the axioms and theorems of this chapter.

1   If $a \neq 0$ and $ab = 0$, then $b = 0$.
2   If $a \neq 0$ and $a(b - c) = 0$, then $b = c$.
3   If $ab = cd$ and $cd = be$, then $a = e$.
4   If $a(b + c) = ab + d$, then $d = ac$.
5   If $a > b$ and $b > c$, then $a > c$.
6   If $a > b > 0$ and $c > d > 0$, then $ac > bd$.
7   If $a < b < 0$, then $a^2 > b^2$.
8   If $a > 0$, $a = b$, and $ac > bd$, then $c > d$.

Find the indicated product in each of problems 9 to 40.

9 $3x^2 \cdot 2x^3$	**10** $5x^4 \cdot 6x^3$	**11** $(4x^4)(5x^5)$
12 $(7x^3)(-3x^5)$	**13** $(2xy)(3x^2y^3)$	**14** $(-2x^2y)(-5xy^3)$
15 $(-4x^2y^4)(3x^3y^3)$	**16** $(5xy^6)(-2x^3y^2)$	**17** $(-2x^2y^3)^2$
18 $(-3xy^3)^3$	**19** $(4x^3y^4)^3$	**20** $(5x^2y^5)^2$
21 $2x^2(3x - 2y)$	**22** $3x^3y^2(2x - 3y)$	**23** $3x^2(2x^2 - 5xy)$

24 $4x^2y(3xy - 2y^2)$
25 $2ab^2c^3(3a^2b - 2bc^2 - 5a^2c)$
26 $3a^2bc^2(5abc - 3ab^2c - 2a^3b^2c)$
27 $-3ab^3c^4(-2abc + 3a^2bc^3 - 4ab^2c^2)$
28 $-2a^2b^2c(4abc^2 - 2a^2bc^3 - 5a^3b^2c^4)$

29 $(2x - 3y)(3x - 5y)$	**30** $(2x + 5y)(4x - 3y)$
31 $(3a - b)(2a + 3b)$	**32** $(5a - 7b)(a + 3b)$
33 $(2a - 3b)(3a^2 - 2ab + 4b^2)$	**34** $(3a + 2b)(2a^2 - 3ab - 2b^2)$
35 $(5a - b)(2a^2 + ab - 3b^2)$	**36** $(4a - 3b)(a^2 - ab + 2b^2)$
37 $(2x^2 - 3xy + y^2)(3x^2 + 2xy - 2y^2)$	**38** $(3x^2 - xy + 2y^2)(2x^2 + xy - 3y^2)$
39 $(5x^2 + 3xy - y^2)(3x^2 - xy + 2y^2)$	**40** $(2x^2 - 3xy + y^2)(x^2 + 3xy - y^2)$

2.13   DIVISION

As stated earlier, if $b \neq 0$, the quotient of a and b is expressed as $\dfrac{a}{b}$ or a/b.
We define this quotient as the unique number x such that $bx = a$.   That is,

$$\textit{If } b \neq 0, \textit{ then } \frac{a}{b} = x \textit{ if and only if } bx = a \qquad (2.41)$$

Dividend   The number a is the *dividend*, b is the *divisor*, and the procedure for com-
Divisor   puting x is called *division*.
Since $a \cdot 1 = a$, by (2.34), it follows that

$$\frac{a}{a} = 1 \qquad \text{provided } a \neq 0 \qquad (2.42)$$

The requirement $b \neq 0$ in (2.41) is specified for reasons that we next
discuss.   If $b = 0$, then $bx = 0 \cdot x = 0$ by (2.35).   Hence if $a \neq 0$, there is no

replacement in the set of real numbers for x such that $bx = a$, and so the quotient does not exist. If, however, $a = 0$ and $b = 0$, we have $0 \cdot x = 0$ for any replacement for x. Consequently $0/0$ does not exist as a unique number. Therefore if the divisor $b = 0$ in a/b, the quotient is not defined.

If in (2.41) $a = 0$ and $b \neq 0$, we have $0/b = 0$, since $b \cdot 0 = 0$. Hence

$$\frac{0}{b} = 0 \qquad \text{provided } b \neq 0 \tag{2.43}$$

The law of signs for division is derived from (2.41) and the law of signs for multiplication. If $bx = a$, then by the law of signs for multiplication, we have the following possibilities:

b	x	a
Positive	Positive	Positive
Negative	Positive	Negative
Negative	Negative	Positive
Positive	Negative	Negative

It follows from the above tabulation that if a and b are both positive or both negative, $x = a/b$ is positive. Furthermore, if either a or b is positive and the other is negative, $x = a/b$ is negative. Consequently, we have the following law of signs for division:

Law of signs for division

The quotient of two positive or of two negative numbers is positive. The quotient of a positive number and a negative number or of a negative number and a positive number is negative.

As examples, we have

$$\frac{12}{4} = 3 \qquad \frac{24}{-6} = -4 \qquad \frac{-18}{9} = -2 \qquad \frac{-9}{-3} = 3 \qquad \frac{-3}{4} = -\frac{3}{4} \qquad \frac{5}{-7} = -\frac{5}{7}$$

We shall now prove two theorems and a corollary which are essential for the simplification of quotients and for the multiplication and division of fractions. The first theorem follows directly from the definition of a quotient. Since $a \cdot 1 = a$, we have

$$\frac{a}{1} = a \tag{2.44}$$

The second theorem is

$$\frac{ac}{bd} = \frac{a}{b}\frac{c}{d} = \frac{a}{d}\frac{c}{b} \tag{2.45}$$

Proof ▶

In order to prove that $ac/bd = a/b \cdot c/d$ we shall let $a/b = x$ and $c/d = y$, and thus have $a/b \cdot c/d = xy$. Furthermore by the definition of a quotient (2.41), $a = bx$ and $c = dy$. Consequently, $ac = bx \cdot dy = bd \cdot xy$ by the commutative and associative axioms. Hence, by (2.41), $ac/bd = xy$. Therefore $ac/bd = a/b \cdot c/d$ since each is equal to xy.

Furthermore since $bd = db$, we have

$$\frac{ac}{bd} = \frac{ac}{db} = \frac{a}{d}\frac{c}{b}$$

This theorem can be extended to cover cases where there are more than two factors in the dividend and in the divisor. For example

$$\frac{ace}{bdf} = \frac{a(ce)}{b(df)} = \frac{a}{b}\frac{ce}{df} = \frac{a}{b}\frac{c}{d}\frac{e}{f}$$

If in (2.45) we let $b = c = 1$ and apply the commutative axiom, we have

$$\frac{a}{d} = a \cdot \frac{1}{d} = \frac{1}{d} \cdot a \qquad \text{provided } d \neq 0 \qquad \textbf{(2.46)}$$

Our next task is to prove the following law of exponents for division:

$$\begin{array}{ll} & \text{if } a \neq 0 \\ \frac{a^m}{a^n} = a^{m-n} & m \text{ and } n \text{ are positive integers} \\ & m > n \end{array} \qquad \textbf{(2.47)}$$

Proof ▶

By the law of exponents for multiplication (2.38) we have

$$a^n a^{m-n} = a^{n+(m-n)}$$
$$a^n a^{m-n} = a^m \qquad \text{◀ since } n + (m - n) = (n - n) + m = m$$

Consequently by the definition of a quotient (2.41) we have $a^m/a^n = a^{m-n}$.

If in (2.47), $m = n$, we have $a^n/a^n = a^{n-n} = a^0$, and a^0 has no meaning according to the previous definition of exponents. However, since $a \neq 0$, we have, by (2.42), $a^n/a^n = 1$. We therefore define

$$a^0 = 1 \qquad \text{provided } a \neq 0 \qquad \textbf{(2.48)}$$

2.14 MONOMIAL DIVISORS

We employ one or more of the theorems from (2.42) to (2.48) to obtain and simplify a quotient when the divisor is a monomial. The method is illustrated in the following examples.

EXAMPLE 1

$$\begin{aligned} 6x^8y^6z^3 \div 3x^4y^3z^2 &= \frac{6x^8y^6z^3}{3x^4y^3z^2} \\ &= \frac{6}{3}\frac{x^8}{x^4}\frac{y^6}{y^3}\frac{z^3}{z^2} \qquad \text{◀ by (2.45)} \\ &= 2x^{8-4}y^{6-3}z^{3-2} \qquad \text{◀ by (2.47)} \\ &= 2x^4y^3z \end{aligned}$$

EXAMPLE 2

$$24a^4b^{10}c^2 \div -3a^4b^7 = \frac{24a^4b^{10}c^2}{-3a^4b^7}$$

$$= \frac{24}{-3}\frac{a^4}{a^4}\frac{b^{10}}{b^7}\frac{c^2}{1}$$

$$= -8a^0b^3c^2$$

$$= -8b^3c^2 \qquad \blacktriangleleft \text{ since } a^0 = 1$$

If the dividend is a polynomial, we first employ the theorem stated and proved below.

$$\frac{a+b+c}{d} = \frac{a}{d} + \frac{b}{d} + \frac{c}{d} \qquad \text{provided } d \neq 0 \qquad \textbf{(2.49)}$$

Proof ▶

$$\frac{a+b+c}{d} = \frac{1}{d}(a+b+c) \qquad \blacktriangleleft \text{ by (2.46)}$$

$$= \frac{a}{d} + \frac{b}{d} + \frac{c}{d} \qquad \blacktriangleleft \text{ by the left-hand distributive axiom and (2.46)}$$

EXAMPLE 3

$$(6x^6y^5 + 4x^5y^4 - 3x^4y^3 - 2x^3y^2) \div 3x^3y^2 = \frac{6x^6y^5 + 4x^5y^4 - 3x^4y^3 - 2x^3y^2}{3x^3y^2}$$

$$= \frac{6x^6y^5}{3x^3y^2} + \frac{4x^5y^4}{3x^3y^2} - \frac{3x^4y^3}{3x^3y^2} - \frac{2x^3y^2}{3x^3y^2}$$

$$= 2x^3y^3 + \tfrac{4}{3}x^2y^2 - xy - \tfrac{2}{3}$$

2.15 QUOTIENT OF TWO POLYNOMIALS

Degree of a polynomial

A polynomial was defined in Sec. 2.2. The *degree* of a polynomial in any variable is the greatest exponent of that variable that appears in the polynomial. For example, $3x^4 + 2x^3y + 5x^2y^3 + xy^2$ is a polynomial of degree 4 in x and of degree 3 in y.

Before discussing the procedure for dividing one polynomial by another, we shall consider the definition of a quotient (2.41) with $a = 23$ and $b = 5$. For these replacements for a and b we have

$$5x = 23$$
$$x = \tfrac{23}{5} \qquad \blacktriangleleft \text{ dividing } 5x \text{ and } 23 \text{ by } 5$$
$$= 4\tfrac{3}{5} = 4 + \tfrac{3}{5}$$

Complete quotient
Partial quotient
Remainder

Hence if we divide 23 by 5, we obtain the integer 4 and the fraction $\frac{3}{5}$. We call the mixed number $4\frac{3}{5}$ the *complete quotient*, the integer 4 the *incomplete or partial quotient*, and the numerator of the fraction $\frac{3}{5}$ the *remainder*.

Likewise,

$$\frac{6x^2 + 4}{3x} = \frac{6x^2}{3x} + \frac{4}{3x} = 2x + \frac{4}{3x}$$

and we have the complete quotient $2x + 4/3x$, the partial quotient $2x$, and the remainder 4.

We may readily verify that in each of the above examples the following relation is satisfied:

Dividend = (divisor)(partial quotient) + remainder   ◀ since $23 = 5(4) + 3$
and $6x^2 + 4$
$= 3x(2x) + 4$

Hereafter in this section we shall use the word "quotient" to refer to the partial quotient.

In order to divide one polynomial by another, we first arrange the terms in each polynomial so that they are in the order of the descending powers of some variable that appears in each. Then we seek the quotient that is a polynomial and which satisfies the relation

Dividend = (divisor)(quotient) + remainder

where the degree of the remainder in the variable chosen as the basis of the arrangement of terms is less than the degree of the divisor in that variable. We shall illustrate the procedure by the following example.

EXAMPLE

Find the quotient obtained by dividing $6x^4 + 7x^3 + 6x^2 + 32x - 7$ by $3x^2 + 5x - 2$.

Solution

We first substitute the given polynomials in the relation

$$\begin{aligned} \text{Dividend} \quad &= (\text{divisor})(\text{quotient}) + \text{remainder} \\ 6x^4 + 7x^3 + 6x^2 + 32x - 7 &= (3x^2 + 5x - 2)(\text{quotient}) + \text{remainder} \quad (1) \end{aligned}$$

Since the degree of the dividend is 4, the degree of the divisor is 2, and the degree of the remainder is less than 2, it follows that the degree of the quotient is 2. Hence we write the quotient in the form $ax^2 + bx + c$ and determine the values of a, b, and c. For this purpose we substitute $ax^2 + bx + c$ for the quotient in (1) and get

$$\begin{aligned} 6x^4 + 7x^3 &+ 6x^2 + 32x - 7 \\ &= (3x^2 + 5x - 2)(ax^2 + bx + c) + \text{remainder} \quad (2) \end{aligned}$$

Next we apply (2.2) and (2.33) to (2) and get

$$\begin{aligned} (3x^2 + 5x - 2)ax^2 + (3x^2 + 5x - 2)bx &+ (3x^2 + 5x - 2)c + \text{remainder} \\ &= 6x^4 + 7x^3 + 6x^2 + 32x - 7 \quad (3) \end{aligned}$$

By inspection we see that the only terms in the left and right members of (3) that involve x^4 are $3ax^4$ and $6x^4$, respectively. Hence, $3ax^4 = 6x^4$ and it follows that $a = 2$. Now we substitute 2 for a in (3) and subtract $(3x^2 + 5x - 2)2x^2 = 6x^4 + 10x^3 - 4x^2$ from each member of (3) and get

$$(3x^2 + 5x - 2)bx + (3x^2 + 5x - 2)c + \text{remainder}$$
$$= -3x^3 + 10x^2 + 32x - 7 \qquad (4)$$

Again by inspection we see that $3bx^3 = -3x^3$. Hence we have $b = -1$, and we subtract $(3x^2 + 5x - 2)(-x) = -3x^3 - 5x^2 + 2x$ from each member of (4) and have

$$(3x^2 + 5x - 2)c + \text{remainder} = 15x^2 + 30x - 7 \qquad (5)$$

From (5) we see that $3cx^2 = 15x^2$; consequently, $c = 5$. Finally, by subtracting $(3x^2 + 5x - 2)5 = 15x^2 + 25x - 10$ from each member of (5), we obtain

$$\text{Remainder} = 5x + 3 \qquad (6)$$

Consequently, since $a = 2$, $b = -1$, and $c = 5$, the quotient is $2x^2 - x + 5$ and the remainder is $5x + 3$.

The usual form for performing the computation described in the preceding example is as follows:

divisor
$$
\begin{array}{r}
2x^2 - x + 5 \quad \text{quotient} \\
3x^2 + 5x - 2\,)\overline{6x^4 + 7x^3 + 6x^2 + 32x - 7} \qquad \blacktriangleleft \text{ dividend} \qquad (3) \\
6x^4 + 10x^3 - 4x^2 \qquad \blacktriangleleft (3x^2 + 5x - 2)2x^2 \\
-3x^3 + 10x^2 + 32x - 7 \qquad \blacktriangleleft \text{ subtracting} \qquad (4) \\
-3x^3 - 5x^2 + 2x \qquad \blacktriangleleft (3x^2 + 5x - 2)(-x) \\
15x^2 + 30x - 7 \qquad \blacktriangleleft \text{ subtracting} \qquad (5) \\
15x^2 + 25x - 10 \qquad \blacktriangleleft (3x^2 + 5x - 2)5 \\
5x + 3 \qquad \blacktriangleleft \text{ remainder} \qquad (6)
\end{array}
$$

The formal steps in the process of dividing one polynomial by another are the following:

Steps in dividing one polynomial by another

1   Arrange both dividend and divisor in ascending or descending powers of a letter that appears in both.

2   Divide the first term in the divisor into the first term in the dividend to get the first term in the quotient.

3   Multiply the divisor by the first term in the quotient and subtract the product from the dividend.

4   Treat the remainder obtained in step 3 as a new dividend and repeat steps 2 and 3.

5   Continue this process until the remainder obtained is such that the largest exponent of the letter chosen as the basis of arrangement in step 1 is less than the largest exponent of that letter in the divisor.

If, in the above example, the remainder is subtracted from the dividend, we get $6x^4 + 7x^3 + 6x^2 + 27x - 10$. It can be verified that this polynomial is the product of the dividend and the divisor, which illustrates the general principle

$$\text{Dividend} - \text{remainder} = (\text{divisor})(\text{quotient})$$
or $\qquad$ $$\text{Dividend} = (\text{divisor})(\text{quotient}) + \text{remainder}$$

which can be used as a check for division.

2.16 SYMBOLS OF GROUPING

Removing symbols of grouping

The use of parentheses has been explained in preceding sections. Symbols of grouping in addition to parentheses are frequently needed to make the meaning of certain expressions clear and to indicate the order in which operations are to be performed. In addition to parentheses, we use the brackets, [ ], and the braces, { }, for these purposes.

It is frequently desirable to remove the symbols of grouping from an expression, and we shall explain and illustrate the procedure with several examples. If an expression that is enclosed in parentheses is preceded or followed by a monomial factor, as in $x - 2y(3x - y + z)$, we apply the distributive law and replace $-2y(3x - y + z)$ by $-6xy + 2y^2 - 2yz$. Therefore, $x - 2y(3x - y + z) = x - 6xy + 2y^2 - 2yz$. Similarly, $a^2 + (a^3 - ab + b^2)2a = a^2 + 2a^4 - 2a^2b + 2ab^2$.

Since $-(n) = -1(n)$, the expression $x + y - (-2x^2 - y^2 + z) = x + y - 1(-2x^2 - y^2 + z^2) = x + y + 2x^2 + y^2 - z^2$ by use of the distributive law. Therefore, if an expression enclosed in parentheses is preceded by a minus sign, the parentheses can be removed if and only if the sign of each of the enclosed terms is changed. If an expression enclosed in parentheses is preceded by a plus sign and no monomial factor is indicated, as in $a + (b + c - d)$, it is understood that the factor $+1$ precedes the parentheses but is not expressed. Since multiplying a number by 1 yields a product equal to the number, the parentheses can be removed from $a + (b + c - d)$ with no further changes.

Usually when braces or brackets or both appear together with parentheses in an expression, one or more sets of grouping symbols will be enclosed in another set. When the symbols are removed from an expression of this type, it is advisable to remove the innermost symbols first. We shall illustrate the procedure with the following example.

EXAMPLE

$$3x^2 - \{2x^2 - xy - [x(x - y) - y(2x - y)] + 4xy\} - 3y^2$$

Solution

We shall start with the given expression and indicate the successive steps in removing the symbols of operations and explain the purpose of each step at the right.

$$3x^2 - \{2x^2 - xy - [x(x - y) - y(2x - y)] + 4xy\} - 3y^2$$
◄ given expression
$$= 3x^2 - \{2x^2 - xy - [x^2 - xy - 2xy + y^2] + 4xy\} - 3y^2$$
◄ applying the distributive law to the expression in the brackets
$$= 3x^2 - \{2x^2 - xy - [x^2 - 3xy + y^2] + 4xy\} - 3y^2$$
◄ adding similar terms in brackets
$$= 3x^2 - \{2x^2 - xy - x^2 + 3xy - y^2 + 4xy\} - 3y^2$$
◄ since $-[x^2 - 3xy + y^2] = -x^2 + 3xy - y^2$
$$= 3x^2 - \{x^2 + 6xy - y^2\} - 3y^2 \quad ◄ \text{ adding similar terms in braces}$$
$$= 3x^2 - x^2 - 6xy + y^2 - 3y^2 \quad ◄ \text{ since } -\{x^2 + 6xy - y^2\} = -x^2 - 6xy + y^2$$
$$= 2x^2 - 6xy - 2y^2 \quad\quad\quad ◄ \text{ adding similar terms}$$

Inserting symbols of grouping

If a pair of grouping symbols is inserted in an expression after a plus sign, no changes in signs are necessary. If the grouping symbols are inserted after a minus sign, the signs of all enclosed terms must be changed. For example,

$$x + y - z + w = x + (y - z + w)$$

and
$$a - b + c - d = a - (b - c + d)$$

Exercise 2.4   Division and Grouping Symbols

Perform the division indicated in each of problems 1 to 12.

1 $\dfrac{a^5}{a^2}$ **2** $\dfrac{b^7}{b^3}$ **3** $\dfrac{c^7}{c^5}$ **4** $\dfrac{d^6}{d}$

5 $\dfrac{a^3b^4}{a^2b^2}$ **6** $\dfrac{a^7b^5}{a^4b}$ **7** $\dfrac{8a^8b^4}{2a^4b^2}$ **8** $\dfrac{12a^7b^5}{4a^4b}$

9 $\dfrac{12a^5b^3c^4 - 8a^7b^4c^2 + 16a^4b^5c^3}{4a^4b^3c^2}$

10 $\dfrac{35a^9b^4c^7 + 21a^6b^7c^4 - 28a^4b^5c^5}{7a^3b^2c^4}$

11 $\dfrac{12a^4b^5c^6 - 8a^5b^6c^7 + 6a^6b^5c^4}{2a^2b^3c^4}$

12 $\dfrac{45a^8b^6c^4 + 27a^6b^8c^3 - 18a^7b^5c^7}{9a^5b^4c^2}$

In problems 13 to 20, divide the first expression by the second.

13 $2x^2 + 5xy - 3y^2, \; 2x - y$
14 $6x^2 + xy - 2y^2, \; 3x + 2y$
15 $6x^2 + 11xy - 10y^2, \; 3x - 2y$
16 $15x^2 + 19xy + 6y^2, \; 5x + 2y$
17 $2x^3 - x^2y + xy^2 + y^3, \; x^2 - xy + y^2$
18 $3x^3 + 5x^2y - 5xy^2 + y^3, \; 3x - y$
19 $4x^3 - 11x^2y + 10xy^2 - 3y^3, \; 4x - 3y$
20 $6x^3 + x^2y + 7xy^2 + 6y^3, \; 2x^2 - xy + 3y^2$

In problems 21 to 24, find the quotient and remainder if the first expression is divided by the second.

21 $a^3 + a^2 - 3a + 1, \; a^2 - a - 1$
22 $a^3 + a, \; a - 1$
23 $2a^3 - 5a^2 + 3a - 6, \; a - 2$
24 $3a^3 - 11a^2 + 7a - 3, \; 3a^2 - 2a + 1$

Remove the symbols of grouping and combine similar terms.

25 $2(2x - y) - 3(-3x + y)$
26 $3(2x + 3y) - 4(5x - 2y)$
27 $2a(a^2 - ab) - 3b(a^2 - ab - b^2)$
28 $-5a(a^2 - 2ab - b^2) + 3b(a^2 + 3ab - b^2)$
29 $2[x - 3(x - 3)] - 3[x - 2(x - 2)]$
30 $3[a - 2(a - 2)] + 2[3a - 2(3a - 2)]$
31 $a[a - 3(b - 2c) + 3] + 2a[2b - 2(a - c)]$
32 $a[-2a + 4(b + 3c) + a] - 2a[a - 2(2b - 2c) - 2a]$
33 $2\{2a - b[2a - c(2a - 1) + 2ac] - c\}$
34 $3a\{a^2 - a[2a - 2(a + 3) + 1] + a^2\}$
35 $2a - 3\{2a - 3[2a - 3(2a - 3) + 2a] - 3\}$
36 $3a - \{4a + 3[2a - 2(2a + 3b) - (a - b) - 6(a + b)] - (7a - 6b)\}$

2.17   SPECIAL PRODUCTS

Multiplication is an essential and sometimes tedious operation in algebraic computation.   Therefore, any short cuts in the procedure will increase one's speed and accuracy.   In this section we shall explain an efficient method for obtaining the product of two binomials and of two trinomials if the corresponding terms in the multiplicand and multiplier are similar.

THE PRODUCT OF TWO BINOMIALS The corresponding terms in $ax + by$ and $cx + dy$ are similar, and we obtain the product without explanation below.

$$(ax + by)(cx + dy) = ax(cx + dy) + by(cx + dy)$$
$$= acx^2 + adxy + bcxy + bdy^2$$
$$= acx^2 + (ad + bc)xy + bdy^2$$

Hence we have

$$(ax + by)(cx + dy) = acx^2 + (ad + bc)xy + bdy^2 \qquad (2.50)$$

By observing the product on the right, we see that to obtain the product of two binomials with similar terms, we perform the following operations:

Product of two
binomials with
similar terms

1 Multiply the first terms in the binomials to obtain the first term in the product.

2 Add the products obtained by multiplying the first term in each binomial by the second term of the other to get the second term in the product.

3 Multiply the two second terms in the binomials to get the third term in the product.

Symbolically we write the multiplication as follows:

$$(ax + by)(cx + dy) = acx^2 + (ad + bc)xy + bdy^2$$

Ordinarily, the computation required by these three steps can be done mentally and the result can be written with no intermediate steps.

EXAMPLE 1

Obtain the product of $2x - 3y$ and $7x + 5y$.

Solution

We write the indicated product as shown below, proceed as directed below the problem, and record the results in the positions indicated by the flow lines.

$$(2x - 3y)(7x + 5y) = 14x^2 - 11xy - 15y^2$$

Get these products mentally:

1 $2x \cdot 7x =$ ———————

2 $(2x \cdot 5y) + (-3y \cdot 7x) = 10xy - 21xy =$ ——

3 $-3y \cdot 5y =$ ———————

By suitably grouping the terms of two trinomials with similar terms we can use the above procedure to get the product of two trinomials. For example,

$(2a - 3b + 2c)(4a + 2b - 3c)$
$$= [(2a - 3b) + 2c][(4a + 2b) - 3c]$$
$$= (2a - 3b)(4a + 2b) + [2c(4a + 2b) - 3c(2a - 3b)] - 6c^2$$
$$= 8a^2 - 8ab - 6b^2 + 8ac + 4bc - 6ac + 9bc - 6c^2$$
$$= 8a^2 - 8ab - 6b^2 + 2ac + 13bc - 6c^2$$

THE SQUARE OF THE SUM OR THE DIFFERENCE OF TWO NUMBERS   If we apply formula (2.50) to $(x + y)^2$, we obtain

$$(x + y)^2 = (x + y)(x + y) = x^2 + 2xy + y^2 \qquad \textbf{(2.51)}$$

Similarly, $$(x - y)^2 = x^2 - 2xy + y^2 \qquad \textbf{(2.52)}$$

Square of sum or difference of two numbers

Therefore, *the square of the sum or the difference of two numbers is the square of the first term plus or minus twice the product of the first and the second terms plus the square of the second term.*

EXAMPLE 2   Use Eqs. (2.51) and (2.52) to obtain the square of $3x + 5y$, the square of $2a - 7b$, and the square of $x - 2y + 3z$.

Solution
$$(3x + 5y)^2 = 9x^2 + 30xy + 25y^2$$
$$(2a - 7b)^2 = 4a^2 - 28ab + 49b^2$$
$$(x - 2y + 3z)^2 = [(x - 2y) + 3z]^2$$
$$= (x - 2y)^2 + 6z(x - 2y) + 9z^2$$
$$= x^2 - 4xy + 4y^2 + 6xz - 12yz + 9z^2$$
$$= x^2 + 4y^2 + 9z^2 - 4xy + 6xz - 12yz$$

THE PRODUCT OF THE SUM AND THE DIFFERENCE OF THE SAME TWO NUMBERS   The product of the sum and the difference of the numbers a and b is expressed as $(a + b)(a - b)$. If we apply formula (2.50) to the product, we obtain $a^2 + ab - ab - b^2 = a^2 - b^2$. Hence we have

$$(a + b)(a - b) = a^2 - b^2 \qquad \textbf{(2.53)}$$

Product of sum and difference of two numbers

Therefore, *the product of the sum and the difference of the same two numbers is equal to the difference of their squares.*

EXAMPLE 3   By use of Eq. (2.53) find each of the following products:

$$(3x + 5y)(3x - 5y)$$
$$(x - 2y + 3z)(x - 2y - 3z)$$
$$(2a + 3b - c)(2a - 3b + c)$$

Solution

$$(3x + 5y)(3x - 5y) = 9x^2 - 25y^2$$
$$(x - 2y + 3z)(x - 2y - 3z) = [(x - 2y) + 3z][(x - 2y) - 3z]$$
$$= (x - 2y)^2 - 9z^2$$
$$= x^2 - 4xy + 4y^2 - 9z^2$$
$$(2a + 3b - c)(2a - 3b + c) = [2a + (3b - c)][2a - (3b - c)]$$
$$= 4a^2 - (3b - c)^2$$
$$= 4a^2 - 9b^2 + 6bc - c^2$$

Exercise 2.5   Special Products

By use of the appropriate method of Sec. 2.17, find the product in each of problems 1 to 36.

1	$(x + 3y)(x + 2y)$	**2**	$(x + 4y)(x - 2y)$
3	$(2x - y)(5x + y)$	**4**	$(3x - y)(4x - y)$
5	$(3b + 2c)(2b + 5c)$	**6**	$(2b + 5c)(3b + 4c)$
7	$(5b + 6c)(3b + 4c)$	**8**	$(7b + 3c)(4b + 5c)$
9	$(3x - 2y)(2x - 3y)$	**10**	$(4x - 3y)(3x - 2y)$
11	$(6x - 5y)(3x - 4y)$	**12**	$(5x - 2y)(3x - 5y)$
13	$(3x + 4y)(4x - 3y)$	**14**	$(6x + 5y)(2x - 7y)$
15	$(5x - 4y)(5x + 3y)$	**16**	$(2x - 9y)(3x + 4y)$
17	$(2x + y)^2$	**18**	$(3x + y)^2$
19	$(x - 3y)^2$	**20**	$(x - 4y)^2$
21	$(3a + 5b)^2$	**22**	$(2a + 3b)^2$
23	$(-2a - 5b)^2$	**24**	$(-3a - 4b)^2$
25	$(4a - 3b)^2$	**26**	$(5a - 2b)^2$
27	$(-3a + 5b)^2$	**28**	$(-4a + 7b)^2$
29	$(5x - y)(5x + y)$	**30**	$(3x - y)(3x + y)$
31	$(x + 4y)(x - 4y)$	**32**	$(x + 7y)(x - 7y)$
33	$(2x^2 + 3y)(2x^2 - 3y)$	**34**	$(5x - 2y^3)(5x + 2y^3)$
35	$(4x^3 - 5y^2)(4x^3 + 5y^2)$	**36**	$(3x^2 + 2y^4)(3x^2 - 2y^4)$

Express each product in problems 37 to 44 as the product of the sum and difference of two numbers and then find the product. For example, $(32)(28) = (30 + 2)(30 - 2) = 30^2 - 2^2 = 896$.

37	$(41)(39)$	**38**	$(53)(47)$
39	$(25)(15)$	**40**	$(26)(34)$
41	$(73)(67)$	**42**	$(37)(43)$
43	$(84)(96)$	**44**	$(63)(77)$

Express the trinomial in each of problems 45 to 52 as the sum and difference of two numbers and then find the square of the trinomial.

45	$(a + b + 2)^2$	**46**	$(a + 2b + 1)^2$
47	$(2a - b + 1)^2$	**48**	$(3a + b - 1)^2$
49	$(3x + y + 2)^2$	**50**	$(2x - 3y + 1)^2$
51	$(x - 3y - 4)^2$	**52**	$(x - 4y - 3)^2$

Express the trinomial in each of problems 53 to 60 as the sum and difference of the same two numbers and then find the product.

53	$(x - 3y + z)(x - 3y - z)$	**54**	$(2x - y + z)(2x + y - z)$
55	$(3x + 2y - z)(3x - 2y + z)$	**56**	$(2x + y + 5z)(2x - y - 5z)$
57	$(4x - 3y - 2z)(4x - 3y + 2z)$	**58**	$(4x - 3y - 2z)(4x + 3y + 2z)$
59	$(2x - 5y + 3z)(2x + 5y + 3z)$	**60**	$(3x + 2y + 5z)(3x - 2y + 5z)$

2.18 FACTORING

Prime number

A number is factored if it is expressed as the product of two or more other numbers. Several sets of factors may be possible. For example, $6 = 6 \cdot 1 = 3 \cdot 2 = 9 \cdot \frac{2}{3}$. In this section, however, we shall consider only *prime* factors. A *prime number* is an integer greater than 1 that has no integral factors except itself and 1. Since the only prime factors of 6 are 3 and 2, we say that 6 is factored if it is expressed as $3 \cdot 2$ or $2 \cdot 3$. In this section we shall consider only polynomials with integral numerical coefficients. We say that such a polynomial is factored if it is expressed as the product of two or more irreducible polynomials of the same type. A polynomial is

Irreducible polynomial

irreducible, or *prime,* if it cannot be expressed as the product of two or more polynomials of lower degree and if the coefficients have no common factor.

Skill in factoring is essential for computation involving fractions and for solving equations; it is useful in any field in which algebraic computation is necessary. The reader will find that several of the factoring methods depend on the special products studied in Sec. 2.17.

Factoring polynomials whose terms have a common factor

COMMON FACTORS The first step in factoring a polynomial in which each term is divisible by the same monomial is to express the polynomial as the product of the monomial and the quotient of the monomial and the polynomial. This procedure is justified by the distributive axiom. If the factors thus obtained are not prime, we continue factoring by applying the methods explained later. For example,

$$3x^3y^3 - 6x^2y^2 + 12xy = 3xy(x^2y^2 - 2xy + 4)$$
$$4a^5 - 12a^3 - 20a^2 = 4a^2(a^3 - 3a - 5)$$

This method can be extended to include polynomials in which the terms are products that have a common factor that is not a monomial. For example,

$$(a + b)(a - b) + 2(a + b) = (a + b)(a - b + 2)$$
$$(x - 1)(x + 2) - (x - 1)(2x - 3) = (x - 1)[(x + 2) - (2x - 3)]$$
$$= (x - 1)(x + 2 - 2x + 3)$$
$$= (x - 1)(-x + 5)$$

Factoring by grouping

Frequently the terms of a polynomial can be grouped in such a way that the groups have a common divisor, and then the above method can be applied.

EXAMPLE 1

Factor $2ab - ac + 2b^2 - bc$.

Solution

$$2ab - ac + 2b^2 - bc = a(2b - c) + b(2b - c)$$
$$= (2b - c)(a + b)$$

EXAMPLE 2

Factor $x^2 + 2xy - 2x - 4y$.

Solution

$$x^2 + 2xy - 2x - 4y = x(x + 2y) - 2(x + 2y)$$
$$= (x + 2y)(x - 2)$$

THE DIFFERENCE OF TWO SQUARES If we rewrite (2.53) in the form

$$a^2 - b^2 = (a + b)(a - b) \qquad (2.54)$$

Factors of the difference of two squares

we see that *the factors of the difference of the squares of two numbers are the sum and the difference of the two numbers.* For example,

$$x^2 - 9y^2 = x^2 - (3y)^2 = (x + 3y)(x - 3y)$$
$$9a^4 - 25 = (3a^2)^2 - 5^2 = (3a^2 + 5)(3a^2 - 5)$$
$$a^2 - (b + c)^2 = [a + (b + c)][a - (b + c)] = (a + b + c)(a - b - c)$$

TRINOMIALS THAT ARE PERFECT SQUARES By referring to (2.51) and (2.52), we see that a trinomial is a perfect square if two of the terms are perfect squares and the other term is plus or minus twice the product of the two square roots. If we arrange the terms in such trinomials so that the first and third terms are the perfect squares, then by (2.51) and (2.52) we conclude that *a trinomial that is a perfect square is the square of a binomial composed of the positive square roots of the first and third terms of the trinomial and connected by the sign of the second term.†* For example, we have

Factors of trinomials that are perfect squares

---

† The trinomial in this statement is also the square of a binomial composed of the negative square roots of the first and third terms of the trinomial connected by the sign of the second term. For example,

$$9x^2 + 6x + 1 = [(-3x) + (-1)]^2 = (-3x - 1)^2$$
$$16a^2 - 40ab + 25b^2 = [(-4a) - (-5b)]^2 = (-4a + 5b)^2$$

Ordinarily, however, we use the binomial in which the first term is positive.

$$9x^2 + 6x + 1 = (3x + 1)^2$$
$$4x^2 + 12xy + 9y^2 = (2x + 3y)^2$$
$$16a^2 - 40ab + 25b^2 = (4a - 5b)^2$$
$$(x + y)^2 + 2(x + y) + 1 = (x + y + 1)^2$$

QUADRATIC TRINOMIALS   A trinomial of the type $ax^2 + bx + c$, where a, b, and c are integers and $a \neq 0$, is a quadratic trinomial with integral coefficients.   Since we shall use (2.50) in factoring such a trinomial, we shall rewrite it here with the members interchanged.

$$acx^2 + (ad + bc)xy + bdy^2 = (ax + by)(cx + dy) \qquad (2.55)$$

Factors of quadratic trinomials   To use (2.55) in factoring $3x^2 - 10xy - 8y^2$, we must find four numbers, a, b, c, and d such that $ac = 3$, $bd = -8$, and $ad + bc = -10$.   The only possibilities for a and c are ± 3 and ± 1, and these numbers must have the same sign.   The possibilities for b and d are ± 4 and ∓ 2 or ± 8 and ∓ 1, where the double signs indicate that if one of the two numbers is positive, the other must be negative.   If we let $a = 3$ and $c = 1$, then $ad + bc = 3d + b = -10$, and this is true only when $d = -4$ and $b = 2$.   Therefore, $3x^2 - 10xy - 8y^2 = (3x + 2y)(x - 4y)$.

EXAMPLE 3   Factor $6x^2 + 47xy + 15y^2$.

Solution   To factor $6x^2 + 47xy + 15y^2$ into $(ax + by)(cx + dy)$, we first notice that we must have $ac = 6$, $bd = 15$, and $ad + bc = 47$.   The positive possibilities for a and c are 6, 1 and 3, 2; and those for bd are 15, 1 and 5, 3.   If we try $a = 6$, $c = 1$, then $ad + bc = 6d + b = 47$, and this is not satisfied by either of the possibilities for b and d.   However, if $a = 3$ and $c = 2$, we have $3d + 2b = 47$, and this is satisfied by $d = 15$, $b = 1$.   Hence

$$6x^2 + 47xy + 15y^2 = (3x + y)(2x + 15y)$$

EXAMPLE 4   Factor $12x^2 - 89xy + 60y^2$.

Solution   If the trinomial $12x^2 - 89xy + 60y^2 = (ax + by)(cx + dy)$, we must have $ac = 12$, $bd = 60$, and $ad + bc = -89$.   Since ac and bd are positive, a and c must have the same sign and b and d must have the same sign.   Also, since $ad + bc$ is negative, both members of one of these pairs must be negative. We shall assume that a and c are positive and try $a = 4$ and $c = 3$.   Then $ad + bc = 4d + 3b = -89$.   Obviously, the last relation and $bd = 60$ are satisfied if $d = -20$ and $b = -3$.   Therefore,

$$12x^2 - 89xy + 60y^2 = (4x - 3y)(3x - 20y)$$

THE SUM OR DIFFERENCE OF TWO CUBES If we divide $a^3 + b^3$ by $a + b$, we get $a^2 - ab + b^2$. Therefore,

$$a^3 + b^3 = (a + b)(a^2 - ab + b^2) \tag{2.56}$$

Similarly, $$a^3 - b^3 = (a - b)(a^2 + ab + b^2) \tag{2.57}$$

Consequently, we have the following rules:

One factor of the sum of the cubes of two numbers is the sum of the two numbers, and the other factor is the square of the first number minus the product of the first and second numbers plus the square of the second number.

Factors of the sum or the difference of two cubes

One factor of the difference of the cubes of two numbers is the difference of the numbers, and the other factor is the square of the first number plus the product of the first and second numbers plus the square of the second number.

As examples of these two rules we have

$$a^3 + 8b^3 = a^3 + (2b)^3 = (a + 2b)(a^2 - 2ab + 4b^2)$$
$$27x^3 - 64y^6 = (3x)^3 - (4y^2)^3 = (3x - 4y^2)(9x^2 + 12xy^2 + 16y^4)$$

MISCELLANEOUS EXAMPLES The following examples illustrate methods for converting certain binomials to the difference of two squares or to the sum or difference of two cubes. Note that success in factoring the expressions in Examples 5 to 8 depends on recognizing them as special products.

EXAMPLE 5

Factor $x^8 - y^8$.

Solution

$$\begin{aligned}
x^8 - y^8 &= (x^4)^2 - (y^4)^2 \\
&= (x^4 + y^4)(x^4 - y^4) &&\blacktriangleleft \text{ by (2.54)} \\
&= (x^4 + y^4)(x^2 + y^2)(x^2 - y^2) &&\blacktriangleleft \text{ by applying (2.54) to } x^4 - y^4 \\
&= (x^4 + y^4)(x^2 + y^2)(x + y)(x - y) &&\blacktriangleleft \text{ by applying (2.54) to } x^2 - y^2
\end{aligned}$$

EXAMPLE 6

Factor $64a^6 - 729b^6$.

Solution

$$\begin{aligned}
64a^6 - 729b^6 &= (8a^3)^2 - (27b^3)^2 \\
&= (8a^3 + 27b^3)(8a^3 - 27b^3) &&\blacktriangleleft \text{ by (2.54)} \\
&= [(2a)^3 + (3b)^3][(2a)^3 - (3b)^3] \\
&= (2a + 3b)(4a^2 - 6ab + 9b^2)(2a - 3b)(4a^2 + 6ab + 9b^2) \\
&&&\blacktriangleleft \text{ by (2.56) and (2.57)}
\end{aligned}$$

EXAMPLE 7

Factor $a^9 + 8b^6$.

Solution

$$\begin{aligned}
a^9 + 8b^6 &= (a^3)^3 + (2b^2)^3 \\
&= (a^3 + 2b^2)(a^6 - 2a^3b^2 + 4b^4) &&\blacktriangleleft \text{ by (2.56)}
\end{aligned}$$

Frequently, a trinomial can be converted to the difference of two squares by adding and subtracting a perfect square. The following example is an illustration of such situations.

EXAMPLE 8

Solution

Factor $4x^4 + 8x^2y^2 + 9y^4$.

In the trinomial $4x^4 + 8x^2y^2 + 9y^4$, $4x^4$ and $9y^4$ are perfect squares, and twice the product of their square roots is $12x^2y^2$. Hence if we add and subtract $4x^2y^2$ to the trinomial, we get

$$4x^4 + 8x^2y^2 + 9y^4 = 4x^4 + 12x^2y^2 + 9y^4 - 4x^2y^2 \qquad \blacktriangleleft \text{ adding and}$$
$$\text{subtracting } 4x^2y^2$$

$$= (2x^2 + 3y^2)^2 - (2xy)^2$$
$$= (2x^2 + 3y^2 + 2xy)(2x^2 + 3y^2 - 2xy) \qquad \blacktriangleleft \text{ by (2.54)}$$

Exercise 2.6 Factoring

Factor the expression in each of problems 1 to 68.

1 $2x^2 - 6x + 2$ **2** $6x - 9y + 3$

3 $3x^2y - 3x^3y - 6x^2$ **4** $5x^3y - 15x^2y^2 + 20xy^2$

5 $6x^2y^2 - 9x^2y^3 - 12xy$ **6** $10a^3b^3 - 15a^3b + 20a^2b$

7 $6a^3b^2 - 8a^2b^2 - 10a^2b^3$ **8** $15a^3b^4 - 9a^2b^4 + 3ab^3$

9 $(a + 2)(a - 1) + b(a + 2)$ **10** $x(x - y) + (x - y)^2$

11 $x(x - w) - (x - w)y$ **12** $(2x - y)(x + y) - 3(2x - y)$

13 $x^2 + xz + xy + yz$ **14** $xy - xz - y^2 + yz$

15 $2x^2 - 3xy - 2xw + 3yw$ **16** $2x^2 + 2xw + 3xy + 3yw$

17 $9x^2 - y^2$ **18** $4x^2 - 9y^2$

19 $9x^2 - 16y^2$ **20** $25x^2 - 49y^2$

21 $4x^2 - y^4$ **22** $25x^4 - y^2$

23 $9x^4 - 4y^6$ **24** $16x^6 - 9y^4$

25 $4a^2 + 20ab + 25b^2$ **26** $9a^2 + 12ab + 4b^2$

27 $25x^2 - 60xy + 36y^2$ **28** $49x^2 - 42xy + 9y^2$

29 $16x^2 - 24xy + 9y^2$ **30** $81x^2 - 72xy + 16y^2$

31 $25x^2 + 70xy + 49y^2$ **32** $49x^2 + 126xy + 81y^2$

33 $4a^2 - 4ab + b^2 - 16c^4$ **34** $9a^4 + 6a^2b + b^2 - 25c^6$

35 $9a^4 - 4b^6 - 20b^3c - 25c^2$ **36** $4a^6 - 20a^3b^2 + 25b^4 - 16c^2$

37 $4a^2 + 4ab + 4a + b^2 + 2b + 1$ **38** $9a^2 - 12ab + 6a + 4b^2 - 4b + 1$

39 $4 - 16a - 20b + 16a^2 + 40ab + 25b^2$

40 $81 + 36a - 54b + 4a^2 - 12ab + 9b^2$

41 $2a^2 + 5ab + 2b^2$ **42** $6a^2 + 7ab + 2b^2$

43 $10a^2 + 19ab + 6b^2$ **44** $21a^2 + 34ab + 8b^2$

45 $63x^2 - 46xy + 8y^2$	46 $12x^2 + 8xy - 15y^2$
47 $20x^2 + xy - 12y^2$	48 $30x^2 - 61xy + 30y^2$
49 $28x^2 - 65xy + 28y^2$	50 $63x^2 - 73xy + 20y^2$
51 $24x^2 - 23xy - 12y^2$	52 $15x^2 + 29xy - 14y^2$
53 $a^3 + 8b^3$	54 $27a^3 - b^3$
55 $8a^3 - 27b^3$	56 $27a^3 + 64b^3$
57 $a^9 - b^6$	58 $a^6 + b^9$
59 $a^6 - 27b^6$	60 $8a^3 + 27b^6$
61 $a^4 - 16$	62 $a^4 - 81$
63 $x^6 - 729$	64 $a^9 + b^9$
65 $a^4 - 2a^2 + 1$	66 $x^4 + 2x^2y^2 + 9y^4$
67 $x^4 - 21x^2y^2 + 4y^4$	68 $9x^4 + 3x^2y^2 + 4y^4$

2.19 FIELDS

In the preceding sections of this chapter we developed a logical structure that is the basis for the procedures involved in the four fundamental operations in the real number system. This structure consists of definitions, axioms, and theorems, which were stated as they were needed in the development. The set of axioms stated is not a minimal set since, as we shall show later, some of them can be proved from the others. We shall list below a set of fundamental axioms from which all other axioms and theorems of this chapter can be proved. The first group are called the *axioms of equality,* and the second, for reasons to be explained later, are the *axioms of a field.* In the statements, R stands for the set of real numbers; a, b, and c stand for elements of R; $+$ stands for the operation of addition; and $\times$ stands for multiplication. We shall show later that sets exist whose elements have the properties defined by these axioms under operations that differ from the usual addition and multiplication. The number in parentheses that follows the name of the axiom is the equation number attached to the same axiom earlier in this chapter.

AXIOMS OF EQUALITY

E.1 The reflexive axiom   (2.1)
$a = a$

E.2 The symmetric axiom   (2.2)
If $a = b$, then $b = a$.

E.3 The transitive axiom   (2.3)
If $a = b$ and $b = c$, then $a = c$.

E.4 The replacement axiom   (2.7)
If $a = b$, then a can be replaced by b in any mathematical statement without affecting the truth or falsity of the statement.

A.1   Closure axiom for addition   (2.9)
There exists a unique number $s \in R$ such that $a + b = s$.

A.2   Commutative axiom for addition   (2.10)
$a + b = b + a$

A.3   Associative axiom for addition   (2.11)
$a + (b + c) = (a + b) + c$

A.4   Identity element for addition   (2.13)
There exists an element $0 \in R$ such that $a + 0 = a$.

A.5   Inverse element for addition   (2.14)
There exists a unique element $a' \in R$ such that $a + a' = 0$. [Note that this statement is the same as that of Eq. (2.14) if a' is replaced by $-a$.]

M.1   Closure axiom for multiplication   (2.30)
There exists a unique element $p \in R$ such that $a \times b = p$.

M.2   Commutative axiom for multiplication   (2.31)
$a \times b = b \times a$.

M.3   Associative axiom for multiplication   (2.32)
$a \times (b \times c) = (a \times b) \times c$

M.4   Identity element for multiplication   (2.34)
There exists an element $1 \in R$ such that $a \times 1 = a$.

M.5   Inverse element for multiplication   (2.36)
If $a \neq 0$, there exists an element $a^{-1} \in R$ such that $a \times a^{-1} = 1$. [Note that in Eq. (2.36) a^{-1} is expressed as $1/a$.]

AM.1   Right-hand distributive axiom   (2.12)
$(a + b) \times c = (a \times c) + (b \times c)$
This axiom distributes the operation $\times$ over the operation $+$.

We employed the word *field* to describe the second group of axioms stated above. We now define a field as follows:

Field   A set S of elements is a field if there exist two binary operations $+$ and $\times$ (not necessarily the usual operations of addition and multiplication) such that all requirements of axioms A.1 to A.5, M.1 to M.5, and AM.1 are satisfied by the elements of S.

If $+$ and $\times$ denote the ordinary operations of addition and multiplication, then the elements of the set of real numbers R satisfy all the field axioms, and therefore R is a field. Furthermore, since a rational number is a real number, axioms A.2 to A.5, M.2 to M.5, and AM.1 hold if a, b, and c are rational numbers. Moreover, since

$$\frac{a}{b} + \frac{c}{d} = \frac{ad + bc}{ad} \qquad \text{and} \qquad \frac{a}{b} \times \frac{c}{d} = \frac{ac}{bd}$$

the sum and product of two rational numbers is rational.   Hence the closure axioms A.1 and M.1 hold, and the set of rational numbers is therefore a field.

If, however, a is an integer other than 1, there is no integer k such that $a \times k = 1$.   Hence the integer a has no inverse for $\times$ in the set of integers. Hence the set of integers is not a field.

We now discuss certain finite sets, called *residue sets,* and two operations, $\oplus$ and $\otimes$, that are not the usual operations of addition and multiplication.   As an example, we shall obtain the residue set of 7.   If we divide 26 by 7, we get the partial quotient 3 and the remainder 5, since $26 = (3 \times 7) + 5$.   The remainder 5 is called the *residue* of 26 modulo 7.   We use the word "modulo" for the phrase "with respect to the divisor."   The divisor is called the *modulus.*   Similarly, since $38 = (5 \times 7) + 3$, the residue of 38 modulo 7 is 3; and, since $5 = (0 \times 7) + 5$, the residue of 5 modulo 7 is 5. Finally, any integer can be expressed in the form $7n + r$, where n is an integer and r is an element of the set

Residue

$$S = \{0, 1, 2, 3, 4, 5, 6\}$$

Hence the residue of any specified integer modulo 7 is an element of the set S, and we call S the *complete residue* set modulo 7.   We now define the operations $\oplus$ and $\otimes$ as follows: If $a \in S$ and $b \in S$, then

Residue set

$$a \oplus b \text{ is the residue of } a + b \text{ modulo } 7$$
$$a \otimes b \text{ is the residue of } a \times b \text{ modulo } 7$$

EXAMPLES

1 $3 \oplus 6 = 2$ since $3 + 6 = 9 = (1 \times 7) + 2$
2 $1 \oplus 3 = 4$ since $1 + 3 = 4 = (0 \times 7) + 4$
3 $5 \oplus 2 = 0$ since $5 + 2 = 7 = (1 \times 7) + 0$
4 $6 \oplus 0 = 6$ since $6 + 0 = 6 = (0 \times 7) + 6$
5 $6 \otimes 5 = 2$ since $6 \times 5 = 30 = (4 \times 7) + 2$
6 $5 \otimes 3 = 1$ since $5 \times 3 = 15 = (2 \times 7) + 1$
7 $6 \otimes 0 = 0$ since $6 \times 0 = 0 = (0 \times 7) + 0$

We shall refer to $a \oplus b$ as the *residue sum* of a and b and $a \otimes b$ as the *residue product.*

By use of the above methods, we can construct the following tables of residue sums and residue products.   In the first table, the number in each square is $a \oplus b$, where a and b are, respectively, the number at the left end of the line and the number at the head of the column that contains the square.   Similarly, in the second table, the number in each square is $a \otimes b$.

Residue sums
modulo 7

$\oplus$	0	1	2	3	4	5	6
0	0	1	2	3	4	5	6
1	1	2	3	4	5	6	0
2	2	3	4	5	6	0	1
3	3	4	5	6	0	1	2
4	4	5	6	0	1	2	3
5	5	6	0	1	2	3	4
6	6	0	1	2	3	4	5

Residue products
modulo 7

$\otimes$	0	1	2	3	4	5	6
0	0	0	0	0	0	0	0
1	0	1	2	3	4	5	6
2	0	2	4	6	1	3	5
3	0	3	6	2	5	1	4
4	0	4	1	5	2	6	3
5	0	5	3	1	6	4	2
6	0	6	5	4	3	2	1

If we refer to the above tables, we see that 0 is the identity element for $\oplus$ and 1 is the identity element for $\otimes$, since if $a \in S$, then $a \oplus 0 = a$ and $a \otimes 1 = a$. Since 0 appears in each line of the first table, each element of S has an inverse under $\oplus$. Since 1 appears in each line of the second table except the line of zeros, each nonzero element of S has an inverse under $\otimes$. The closure axioms hold, since for each two elements in S there exist in S a unique residue sum and a unique residue product. By use of methods illustrated in the following examples, it can be verified that all the other field axioms hold, and therefore S is a field.

EXAMPLE 1   Show that $(2 \oplus 4) \oplus 6 = 2 \oplus (4 \oplus 6)$.

Solution
$$(2 \oplus 4) \oplus 6 = 6 \oplus 6 = 5$$
$$2 \oplus (4 \oplus 6) = 2 \oplus 3 \qquad \blacktriangleleft \text{ since } 4 \oplus 6 = 3$$
$$= 5$$

EXAMPLE 2   Show that $(2 \otimes 4) \otimes 6 = 2 \otimes (4 \otimes 6)$.

Solution
$$(2 \otimes 4) \otimes 6 = 1 \otimes 6 \qquad \blacktriangleleft \text{ since } 2 \otimes 4 = 1$$
$$= 6$$
$$2 \otimes (4 \otimes 6) = 2 \otimes 3 \qquad \blacktriangleleft \text{ since } 4 \otimes 6 = 3$$
$$= 6$$

EXAMPLE 3   Show that $(2 \oplus 4) \otimes 6 = (2 \otimes 6) \oplus (4 \otimes 6)$.

Solution
$$(2 \oplus 4) \otimes 6 = 6 \otimes 6$$
$$= 1$$
$$(2 \otimes 6) \oplus (4 \otimes 6) = 5 \oplus 3$$
$$= 1$$

It is proved in the theory of numbers that the complete set of residues modulo m is a field if and only if m is a prime number.

The complete set of residues modulo 4 is

$$T = \{0, 1, 2, 3\}$$

We show below the tables of residue sums and residue products in T.

$\oplus$	0	1	2	3
0	0	1	2	3
1	1	2	3	0
2	2	3	0	1
3	3	0	1	2

$\otimes$	0	1	2	3
0	0	0	0	0
1	0	1	2	3
2	0	2	0	2
3	0	3	2	1

By use of the above tables, it can be verified that all the field axioms hold in the set T for $\oplus$ and $\otimes$ except axiom M.5. The number 1 does not appear in the fourth line of the second table, and therefore 2 does not have an inverse under $\otimes$ Hence T is not a field.

Exercise 2.7 Properties of a Field

Which of the field axioms are never satisfied in the set of numbers in problems 1 to 12?

1   The set of positive integers
2   The set of nonnegative integers
3   $\{-4, -3, -2, -1, 0, 1, 2, 3, 4\}$
4   $\{x|x = 2n, n = \text{the set of nonnegative integers}\}$
5   $\{x|x = 2n + 1, n = \text{the set of integers}\}$
6   $\{x|x = 5n, n = \text{the set of integers}\}$
7   The set of positive prime numbers
8   $\{x|x = 3n + 1, n = \text{the set of nonnegative integers}\}$
9   $\{x|x = \dfrac{3n}{2}, n = \text{the set of integers}\}$
10 $\{x|x = n + \frac{1}{2}, n = \text{the set of integers}\}$
11 $\{x|x = \frac{1}{2}(n + 2), n = \text{the set of integers}\}$
12 $\{x|x = n^2, n = \text{the set of integers}\}$

Find the residues required in problems 13 to 16.

13   8 modulo 6   14   35 modulo 8
15   5 modulo 7   16   28 modulo 4
17   Write the complete set of residues modulo 3, and construct the tables of residue sums and residue products.
18   In view of the tables in problem 17, is the set $\{0, 1, 2\}$ a field under the operations $\oplus$ and $\otimes$?
19   Construct a table of residue products for the residue set $\{0, 1, 2, 3, 4, 5\}$ modulo 6.   Which of the axioms M.1 to M.5 do not hold for this set?
20   If the operations $a + {}^o b$ and $a \times {}^o b$ are defined as below, find $b + {}^o a$ in (a) and (b) and $b \times {}^o a$ in (c) and (d).   In each case, state whether or not the commutative axiom holds.
     (a) $a + {}^o b = a + b + \sqrt{ab}$ (b) $a + {}^o b = a + \sqrt{b} + ab$
     (c) $a \times {}^o b = ab(a + b)$ (d) $a \times {}^o b = ab^2(a^2 + b)$

Exercise 2.8 Review

1   Prove that if $a + (b + c) = d$ and $a - 2c = d$, then $b = -3c$.
2   Prove that if $a + (b + c) = d$ and $a + b = d - 2e$, then $c = 2e$.
3   Prove that if $a > c$ and $d < 0$, then $ad < cd$.

4 Prove that if $(a + b) + c < a + d$, then $b < d - c$.

5 Prove that $|-a| = |a|$.

6 Prove that if $a > 0$, $b < 0$, and $|a| < |b|$, then $a + b < 0$.

7 Evaluate: $|-6| + |-2|$

8 Evaluate: $|7 - 3| - |-5 + 1|$

9 Add $2x - 3y - 5z$, $-4x - 8y + 7z$, and $3x + 7y - 2z$.

10 Combine: $5x + 4y - 6z - 2y + 5z - 3x + 2z - x - 3y$

11 Subtract $3x - 2y + 4z$ from $7x + 3y + 6z$.

12 Prove that if $a < b$ and $b < c$, then $a < c$.

13 Prove that if $a < 0$ and $|a| > b > 0$, then $a^2 > b^2$.

Perform the indicated operations in each of problems 14 to 20.

14 $(-3x^3y^2z^0)(4x^2y^4z^5)$

15 $(2x^2y^3z)^3$

16 $2ab^2c^3(3abc - 4a^2bc^{-1} + 5a^4b^{-2}c^2)$

17 $(2x - 5y)(3x + 4y)$

18 $(3x^2 - 2xy - y^2)(2x^2 + 3xy + 4y^2)$

19 $15x^4y^3z^7/3xy^2z^3$

20 $(6x^2 + 11xy - 10y^2) \div (2x + 5y)$

21 Find the quotient and remainder if $2a^3 + a^2 + a + 16$ is divided by $2a + 3$.

22 Remove the symbols of grouping and combine similar terms in $a[2a - 3(b + 2c) + 2] - 2a[b - 3(a + c)]$.

Perform the indicated multiplications in problems 23 to 28.

23 $(2a + b)(3a + 2b)$

24 $(3a - 2b)^2$

25 $(4a + 3b)(4a - 3b)$

26 $(43)(37)$

27 $(2a + 3b - 1)^2$

28 $(3x - y + 2z)(3x + y - 2z)$

Factor the expression in each of problems 29 to 34.

29 $3x^2 - 5x + 2$

30 $6a^3b^2 + 15a^2b^3 - 9ab^4$

31 $25x^2 - 16y^2$

32 $4a^2 - 9b^2 + 6bc - c^2$

33 $4a^2 - 4ab + 12a + b^2 - 6b + 9$

34 $a^6 - b^6$

Fractions

In Chap. 1, we defined a rational fraction as the quotient of two integers. We shall extend the definition in this chapter by defining a fraction as the indicated quotient of two real numbers a and b. Skill in operations that involve fractions is essential for progress in algebra and for success in fields in which algebra is used; hence we shall explain the basic operations that are needed in dealing with fractions.

3.1 DEFINITIONS AND FUNDAMENTAL PRINCIPLE

In Chap. 2, we defined the number a/b as the quotient of a and b. We shall now define a/b as a fraction:

Fraction
Numerator
Denominator

A *fraction* is a number of the type a/b, where $b \neq 0$. The number a is the *numerator* of the fraction, and b is the *denominator*.

We shall frequently refer to the numerator and denominator as *members* of the fraction.

The letters a and b in the above definition may stand for integers, monomials, or other algebraic expressions.

Hence,

$$\frac{4a}{7} \qquad \frac{3x^2}{2xy} \qquad \frac{c^2 - d^2}{c^2 + d^2} \qquad \frac{(x + y)(x^2 + xy + y^4)}{(x - y)(x^2 + x^2y + y^2)}$$

are fractions.

Our next task will be to prove a theorem which states the condition under which two fractions are equal.

$$\frac{p}{q} = \frac{n}{d} \qquad \textit{if } pd = nq, \, q \neq 0, \, d \neq 0 \qquad \textbf{(3.1)}$$

and conversely,

$$\textit{If } \frac{p}{q} = \frac{n}{d}, \textit{ then } pd = nq$$

Proof of first statement ▶

$pd = nq$	◀ given
$pd \dfrac{1}{q}\dfrac{1}{d} = nq \dfrac{1}{q}\dfrac{1}{d}$	◀ multiplicativity axiom with $a = pd$, $b = nq$, and $c = (1/q)(1/d)$
$\left(d\dfrac{1}{d}\right)\left(p\dfrac{1}{q}\right) = \left(q\dfrac{1}{q}\right)\left(n\dfrac{1}{d}\right)$	◀ commutative and associative axioms
$p\dfrac{1}{q} = n\dfrac{1}{d}$	◀ since $d(1/d) = q(1/q) = 1$
$\dfrac{p}{q} = \dfrac{n}{d}$	◀ by (2.46)

Proof of second
statement ▶

$$\frac{p}{q} = \frac{n}{d} \qquad \text{◀ given}$$

$$\frac{p}{q} qd = \frac{n}{d} qd \qquad \text{◀ multiplicativity axiom}$$

$$pd \frac{q}{q} = nq \frac{d}{d} \qquad \text{◀ commutative and associative axioms and (2.46)}$$

$$pd = nq \qquad \text{◀ since } q/q = d/d = 1$$

Fundamental
principle of
fractions

We shall frequently have occasion to employ the *fundamental principle* of fractions:

If the numerator and denominator of a fraction are multiplied or divided by the same nonzero number, then the resulting fraction is equal to the original one.

In symbols, this principle is stated as

$$\frac{a}{b} = \frac{an}{bn} = \frac{a \div m}{b \div m} \qquad \textbf{provided } b \neq 0,\, n \neq 0,\, m \neq 0 \qquad (3.2)$$

Proof ▶

$$a(bn) = b(an) \qquad \text{◀ by the commutative and associative axioms}$$

Hence, $\quad \dfrac{a}{b} = \dfrac{an}{bn} \qquad \text{◀ by (3.1)}$

Now if we replace n by $1/m$, we have

$$\frac{a}{b} = \frac{a(1/m)}{b(1/m)} = \frac{a/m}{b/m} \qquad \text{◀ by (2.46)}$$

$$= \frac{a \div m}{b \div m}$$

EXAMPLE 1

$$\frac{3x}{4y} = \frac{3x \cdot 2x^2y^3}{4y \cdot 2x^2y^3} = \frac{6x^3y^3}{8x^2y^4}$$

EXAMPLE 2

$$\frac{15x^3y^7}{25x^5y^2} = \frac{15x^3y^7/5x^3y^2}{25x^5y^2/5x^3y^2} = \frac{3x^0y^5}{5x^2y^0} \qquad \text{◀ by (2.47)}$$

$$= \frac{3y^5}{5x^2} \qquad\qquad\qquad \text{◀ since } x^0 = y^0 = 1$$

There are three signs associated with a fraction: the sign preceding the fraction, the sign preceding the numerator, and the sign preceding the denominator. For example,

$$+\frac{+n}{+d} \qquad -\frac{+n}{-d} \qquad +\frac{-n}{-d}$$

We shall prove that if two of these signs in a given fraction are changed, the resulting fraction is equal to the given fraction.

Proof ▶

$$\frac{n}{d} = \frac{-1 \cdot n}{-1 \cdot d} = \frac{-n}{-d} \qquad \text{◀ by (3.2)}$$

Furthermore,

$$\frac{n}{d} = -(-1)\,\frac{n}{d} \qquad \blacktriangleleft \text{ since } -(-1) = 1$$

$$= -\left(\frac{-1}{1} \cdot \frac{n}{d}\right) \qquad \blacktriangleleft \text{ since } -1/1 = -1 \text{ by (2.44)}$$

$$= -\frac{-n}{d} \qquad \blacktriangleleft \text{ by (2.45)}$$

Similarly it can be proved that $n/d = -n/(-d)$. Consequently we have

$$\frac{n}{d} = \frac{-n}{-d} = -\frac{-n}{d} = -\frac{n}{-d} \qquad \textbf{provided } d \neq 0 \qquad (3.3)$$

EXAMPLE 3

$$\frac{-x}{y-x} = \frac{-(-x)}{-(y-x)} = \frac{x}{x-y}$$

3.2 CONVERSION OF FRACTIONS

Lowest terms of a fraction

Usually, fractions that are obtained as the result of operations are converted to *lowest terms*. A fraction is said to be in lowest terms if the members have no common prime factor. Also, in many operations we are required to convert a fraction into an equal fraction having a denominator or numerator in a specified form. We accomplish these conversions by use of the fundamental principle of fractions.

EXAMPLE 1

Convert $(a + b)/(a - b)$ to an equal fraction with denominator $a^2 - b^2$.

Solution

To convert $(a + b)/(a - b)$ to a fraction with $a^2 - b^2$ as a denominator, we multiply each member by $a + b$ and obtain

$$\frac{a+b}{a-b} = \frac{(a+b)(a+b)}{(a-b)(a+b)} = \frac{a^2 + 2ab + b^2}{a^2 - b^2}$$

EXAMPLE 2

Convert the set of fractions $\{a/xy,\ 3b/x^2y,\ 2c/xy^2\}$ to an equal set with x^2y^2 as a denominator of each element.

Solution

To convert $\{a/xy,\ 3b/x^2y,\ 2c/xy^2\}$ to an equal set with x^2y^2 as the denominator of each of the elements, we multiply the members of the first fraction by xy, of the second by y, and of the third by x and get

$$\left\{\frac{axy}{x^2y^2},\ \frac{3by}{x^2y^2},\ \frac{2cx}{x^2y^2}\right\}$$

To reduce a fraction to lowest terms, we divide both members by every factor that is common to each. If the members of the fraction are polynomials, it is advisable to factor each as a first step in the reduction.

EXAMPLE 3

Reduce $27a^2bc^4/36a^3b^2c^3$ to lowest terms.

Solution

We divide each member by the common factor with the greatest possible numerical coefficient and with the exponent of each variable as large as possible. This factor is $9a^2bc^3$. Thus we get

$$\frac{27a^2bc^4}{36a^3b^2c^3} = \frac{27a^2bc^4/9a^2bc^3}{36a^3b^2c^3/9a^2bc^3} = \frac{3c}{4ab}$$

EXAMPLE 4

Reduce $(x^3 - 2x^2y + xy^2)/(x^3 - xy^2)$ to lowest terms.

Solution

$$\frac{x^3 - 2x^2y + xy^2}{x^3 - xy^2} = \frac{x(x-y)(x-y)}{x(x-y)(x+y)}$$ ◀ factoring numerator and denominator

$$= \frac{x-y}{x+y}$$ ◀ dividing each member by $x(x-y)$

3.3 THE LOWEST COMMON DENOMINATOR

Lowest common multiple

The *lowest common denominator,* or lcd, of a set of fractions is the *lowest common multiple*, or lcm,† of the denominators of the fractions in the set. Usually, the lcd is obtained in factored form. It must be divisible by every denominator in the set, and it must have no more factors than are necessary to satisfy this requirement.

To get the lcd of the fractions in a set, we first factor each denominator. Then the lcd is the product of the different factors of the denominators each with an exponent that is equal to the greatest exponent of that factor in any denominator. For example, suppose the denominators of a set of fractions are

$$(x-2)^4(x+1)$$
$$(x-2)(x+1)^3(x-1)$$
$$(x-2)^2(x-1)^2$$

The different factors are $x-2$, $x+1$, and $x-1$. The greatest exponents of these factors are 4, 3, and 2, respectively. Therefore the lcd is $(x-2)^4(x+1)^3(x-1)^2$.

† The lcm of a set of polynomials is the polynomial of lowest degree in each variable, with numerical coefficients that have no common prime factor, that is a multiple of each polynomial in the set.

EXAMPLE

Convert each fraction in the set

$$\left\{ \frac{x-1}{(x-2)^2}, \frac{x+4}{(x-2)(x+1)}, \frac{x-3}{(x+1)^3} \right\}$$

to an equal fraction having the lowest common denominator.

Solution

We first note that the lcd is $(x-2)^2(x+1)^3$. Therefore, we must multiply each member of the first fraction by $(x+1)^3$, each member of the second by $(x-2)(x+1)^2$, and each member of the third by $(x-2)^2$. We then have

$$\frac{x-1}{(x-2)^2} = \frac{(x-1)(x+1)^3}{(x-2)^2(x+1)^3}$$

$$\frac{x+4}{(x-2)(x+1)} = \frac{(x+4)(x-2)(x+1)^2}{(x-2)^2(x+1)^3}$$

$$\frac{x-3}{(x+1)^3} = \frac{(x-3)(x-2)^2}{(x+1)^3(x-2)^2}$$

Exercise 3.1 Conversion of Fractions

Convert the fraction in each of problems 1 to 12 into an equal fraction with the expression to the right of the comma as denominator.

1 $\dfrac{8}{x-2}, \; 2-x$

2 $\dfrac{7}{4-x}, \; x-4$

3 $\dfrac{3a-2b}{a^2-b^2}, \; b^2-a^2$

4 $\dfrac{2a-5b}{b^3-a^3}, \; a^3-b^3$

5 $\dfrac{3x}{5y}, \; 15y^2$

6 $\dfrac{5y}{8x}, \; 16x^2$

7 $\dfrac{2ab^2}{7a^3b}, \; 21a^4b^3$

8 $\dfrac{3b^2c^3}{5b^2c}, \; 20b^4c^2$

9 $\dfrac{2x-3y}{x-y}, \; x^2-y^2$

10 $\dfrac{3x+5y}{2x+y}, \; -4x^2+y^2$

11 $\dfrac{x-y}{x^2+xy+y^2}, \; -x^3+y^3$

12 $\dfrac{x+2y}{x-y}, \; x^2-3xy+y^2$

Reduce the fraction in each of problems 13 to 24 to lowest terms.

13 $\dfrac{6x^2y}{9xy^2}$

14 $\dfrac{5x^3y^4}{10x^4y^2}$

15 $\dfrac{18a^3b^4c}{24a^2bc^2}$

16 $\dfrac{15ab^2c^3}{21a^3b^2c}$

17 $\dfrac{a^2+a-6}{a^2+5a+6}$

18 $\dfrac{a^2+3a+2}{2a^2+a-6}$

19 $\dfrac{6a^2 - 5a - 6}{9a^2 - 4}$ **20** $\dfrac{4a^2 + 16a + 15}{6a^2 + 7a - 3}$

21 $\dfrac{6a^2 + 7a - 5}{10a^2 - 11a + 3}$ **22** $\dfrac{10a^2 + 9a - 7}{5a^2 + 7a}$

23 $\dfrac{15a^2 + 7a - 2}{10a^2 - 17a + 3}$ **24** $\dfrac{12a^2 - 11a - 5}{6a^2 + 17a + 5}$

Convert each of the following sets of fractions to an equal set with a common denominator.

25 $\left\{\dfrac{2}{x}, \dfrac{3}{xy}, \dfrac{5}{y^2}\right\}$ **26** $\left\{\dfrac{4}{x^2y}, \dfrac{3}{xy^3}, \dfrac{2}{x^3y^2}\right\}$

27 $\left\{\dfrac{-2}{x^3y^4}, \dfrac{3}{xy^5}, \dfrac{2}{x^5y}\right\}$ **28** $\left\{\dfrac{1}{xy^3}, \dfrac{5}{x^4y}, \dfrac{3}{x^2y^4}\right\}$

29 $\left\{\dfrac{x + 2y}{x - y}, \dfrac{2x - y}{x + y}, \dfrac{x^2 - 3y^2}{x^2 - y^2}\right\}$ **30** $\left\{\dfrac{x + 3y}{x - 3y}, \dfrac{x - 2y}{x + 2y}, \dfrac{x^2 - 6y^2}{x^2 - xy - 6y^2}\right\}$

31 $\left\{\dfrac{x - 2y}{(x + y)(x - 3y)}, \dfrac{x - y}{(x - 3y)(2x - y)}, \dfrac{x + 3y}{(x + y)(2x - y)}\right\}$

32 $\left\{\dfrac{x - 2y}{(x + 2y)(2x + y)}, \dfrac{3x - y}{(x + 2y)(2x - y)}, \dfrac{x + 2y}{(2x + y)(2x - y)}\right\}$

3.4 ADDITION OF FRACTIONS

Theorem (2.46) states that $a/d = a \cdot 1/d$. Now if we employ this theorem and the left-hand distributive axiom (2.33), we have

$$\frac{a}{d} + \frac{b}{d} + \frac{c}{d} + \cdots = \frac{1}{d}(a + b + c + \cdots) = \frac{a + b + c + \cdots}{d}$$

Sum of two or more fractions   Hence the sum of two or more fractions with identical denominators is the fraction that has the sum of the given numerators as the numerator and the common denominator as the denominator. For example,

$$\frac{3a}{2xy} + \frac{5a}{2xy} - \frac{c}{2xy} = \frac{3a + 5a - c}{2xy} = \frac{8a - c}{2xy}$$

$$\frac{x + y}{x + 3y} + \frac{x - y}{x + 3y} - \frac{2x + y}{x + 3y} = \frac{(x + y) + (x - y) - (2x + y)}{x + 3y}$$

$$= \frac{x + y + x - y - 2x - y}{x + 3y}$$

$$= \frac{-y}{x + 3y}$$

If the denominators of the fractions to be added are different, we convert each fraction to an equal fraction with the lcd as the denominator and proceed as above.

EXAMPLE 1   Combine $\dfrac{1}{6x} + \dfrac{1}{3y} - \dfrac{3x+2y}{12xy}$ into a single fraction.

Solution   The lcd of the given fractions is $12xy$. To convert the given fractions to equal fractions having $12xy$ as a denominator, we multiply each member of the first fraction by $2y$ and each member of the second by $4x$. We thereby obtain

$$\frac{1}{6x} + \frac{1}{3y} - \frac{3x+2y}{12xy} = \frac{2y}{12xy} + \frac{4x}{12xy} - \frac{3x+2y}{12xy}$$

$$= \frac{2y + 4x - (3x+2y)}{12xy}$$

$$= \frac{2y + 4x - 3x - 2y}{12xy} \qquad \blacktriangleleft \text{ removing parentheses}$$

$$= \frac{x}{12xy} = \frac{1}{12y} \qquad \blacktriangleleft \text{ combining similar terms and} \\ \text{dividing both members by } x$$

EXAMPLE 2   Combine $\dfrac{9x^2 - 3y^2}{(9x^2 - y^2)y} - \dfrac{3}{3x+y} - \dfrac{2}{y-3x}$ into a single fraction.

Solution   The factored form of the first denominator is $(3x+y)(3x-y)y$. Now we note that the factor $3x+y$ is equal to the second denominator and the factor $3x-y$ is the negative of the third denominator. Consequently we employ (3.3), change both signs in the denominator of the third fraction, and also change the sign before the fraction. The problem then becomes

$$\frac{9x^2 - 3y^2}{(9x^2 - y^2)y} - \frac{3}{3x+y} + \frac{2}{3x-y}$$

The lcd is clearly $(3x+y)(3x-y)y$, and we complete the addition as follows:

$$\frac{9x^2 - 3y^2}{(3x+y)(3x-y)y} - \frac{3(3x-y)y}{(3x+y)(3x-y)y} + \frac{2(3x+y)y}{(3x-y)(3x+y)y}$$

$$= \frac{9x^2 - 3y^2 - 9xy + 3y^2 + 6xy + 2y^2}{(3x+y)(3x-y)y} = \frac{9x^2 - 3xy + 2y^2}{(3x+y)(3x-y)y}$$

Exercise 3.2   Addition of Fractions

Perform the addition indicated in each of problems 1 to 40.

1 $\dfrac{x+2}{3} + \dfrac{x-1}{2} + \dfrac{2x-1}{9}$ **2** $\dfrac{3x+2}{5} + \dfrac{2x-3}{3} + \dfrac{5x-1}{2}$

3 $\dfrac{4x+3}{7} - \dfrac{3x-5}{2} + \dfrac{5x-3}{14}$

4 $\dfrac{5x+1}{2} - \dfrac{2x-5}{4} + \dfrac{3x-7}{12}$

5 $\dfrac{2x}{9yz} - \dfrac{3y}{2xz} + \dfrac{z}{3xy}$

6 $\dfrac{5z}{3xy} - \dfrac{2y}{xz} - \dfrac{3x}{2yz}$

7 $\dfrac{3y}{2xz} + \dfrac{2z}{5xy} - \dfrac{7x}{10yz}$

8 $\dfrac{4x}{15yz} - \dfrac{2y}{5xz} + \dfrac{5z}{3xy}$

9 $\dfrac{2x-y}{3x+y} + \dfrac{5x}{2y}$

10 $\dfrac{3x+4y}{2x+y} - \dfrac{3y}{x}$

11 $\dfrac{3y}{2x} - \dfrac{x-5y}{2x+3y}$

12 $\dfrac{5x}{3y} - \dfrac{2x+5y}{x-2y}$

13 $\dfrac{x-y}{2x+y} + \dfrac{x+2y}{x+y}$

14 $\dfrac{3x-2y}{2x-y} - \dfrac{2x+y}{3x-y}$

15 $\dfrac{2x-3y}{x \mid 2y} - \dfrac{2x-y}{x+3y}$

16 $\dfrac{5x-3y}{3x-5y} + \dfrac{2x+5y}{2x-3y}$

17 $\dfrac{2x-y}{3x+5y} - \dfrac{2x^2-y^2}{(3x+5y)(x+y)}$

18 $\dfrac{x^2+18y^2}{(2x+3y)(3x-y)} + \dfrac{4x-6y}{3x-y}$

19 $\dfrac{x+3y}{2x+y} - \dfrac{3x^2+8xy}{6x^2+xy-y^2}$

20 $\dfrac{9xy-2y^2}{2x^2+5xy-3y^2} + \dfrac{5x-2y}{x+3y}$

21 $\dfrac{2}{x-y} + \dfrac{3}{x+y} - \dfrac{5x-y}{x^2-y^2}$

22 $\dfrac{3x}{2x+y} - \dfrac{2y}{x-2y} - \dfrac{3x^2-10xy-2y^2}{2x^2-3xy-2y^2}$

23 $\dfrac{2x-3y}{3x-2y} - \dfrac{3x+2y}{2x+3y} + \dfrac{5x^2+6y^2}{6x^2+5xy-6y^2}$

24 $\dfrac{x+4y}{3x-5y} - \dfrac{2x+y}{6x-y} - \dfrac{10xy+y^2}{18x^2-33xy+5y^2}$

25 $\dfrac{2}{x+2y} - \dfrac{3}{2x-y} + \dfrac{1}{x-y}$

26 $\dfrac{5}{3x+y} - \dfrac{2}{2x-3y} + \dfrac{4}{x-y}$

27 $\dfrac{7}{3x+2y} + \dfrac{2}{3x-y} - \dfrac{3}{x-2y}$

28 $\dfrac{1}{2x+5y} - \dfrac{3}{5x-2y} - \dfrac{2}{x+y}$

29 $\dfrac{x+2y}{x-y} - \dfrac{2x-y}{x+y} + \dfrac{x^2-6xy}{x^2-y^2}$

30 $\dfrac{x+3y}{2x-y} + \dfrac{3x+y}{2x+y} - \dfrac{8x^2+2y^2}{4x^2-y^2}$

31 $\dfrac{2x+5y}{3x-2y} - \dfrac{3x+y}{3x+2y} - \dfrac{22xy+12y^2}{9x^2-4y^2}$

32 $\dfrac{x+2y}{4x-3y} - \dfrac{3x-y}{4x+3y} + \dfrac{8x^2-24xy}{16x^2-9y^2}$

33 $\dfrac{x}{2x^2+xy-3y^2} + \dfrac{y}{2x^2-3xy+y^2} - \dfrac{x+y}{4x^2+4xy-3y^2}$

34 $\dfrac{2x}{x^2-2xy-3y^2} - \dfrac{y}{3x^2+4xy+y^2} + \dfrac{2x-y}{3x^2-8xy-3y^2}$

35 $\dfrac{3x+2y}{2x^2+3xy-5y^2} - \dfrac{x}{2x^2+7xy+5y^2} - \dfrac{2y}{x^2-y^2}$

36 $\dfrac{3x}{2x^2+3xy-2y^2} - \dfrac{2x+y}{2x^2-5xy+2y^2} + \dfrac{y}{x^2-4y^2}$

37 $\dfrac{2x-y}{x+3y} - \dfrac{x-3y}{2x+y} + \dfrac{3x+y}{3x-y}$

38 $\dfrac{5x+2y}{2x+5y} - \dfrac{3x+y}{x+3y} - \dfrac{2x-y}{x-2y}$

39 $\dfrac{3x-y}{2x+y} + \dfrac{2x-y}{3x+4y} - \dfrac{3x-4y}{3x+y}$

40 $\dfrac{4x+3y}{3x+2y} - \dfrac{2x+3y}{4x-3y} - \dfrac{3x-2y}{2x-3y}$

3.5 MULTIPLICATION OF FRACTIONS

In Sec. 2.13 we proved that $a/d \cdot c/b = ac/db$. By use of the associative axiom we can extend this theorem in this way:

Product of
two or more
fractions

and

$$\frac{a}{d}\frac{c}{b}\frac{q}{p} = \left(\frac{a}{d}\frac{c}{b}\right)\frac{q}{p} = \frac{ac}{db}\frac{q}{p} = \frac{acq}{dbp}$$

$$\frac{a}{d}\frac{c}{b}\frac{q}{p}\frac{r}{s} = \left(\frac{a}{d}\frac{c}{b}\frac{q}{p}\right)\frac{r}{s} = \frac{acq}{dbp}\frac{r}{s} = \frac{acqr}{dbps}$$

Therefore the product of two or more fractions is a fraction in which the numerator is the product of the numerators of the factors and the denominator is the product of the denominators.

EXAMPLE

$$\frac{3a^2}{2b}\frac{4b^2(a-b)}{9(a+b)}\frac{a^2-b^2}{a^2b^2} = \frac{3a^2(4b^2)(a-b)(a^2-b^2)}{9(2b)(a+b)(a^2b^2)}$$

◄ multiplying the numerators and denominators

$$= \frac{12a^2b^2(a-b)(a^2-b^2)}{18a^2b^3(a+b)}$$

$$= \frac{2(a-b)(a-b)}{3b}$$

◄ dividing numerator and denominator by $6a^2b^2(a+b)$

$$= \frac{2(a-b)^2}{3b}$$

3.6 DIVISION OF FRACTIONS

In Sec. 2.13 we defined $a \div b$, or a/b as the number x such that $bx = a$. Now if a and b are fractions so that $a = n/d$ and $b = p/q$, then $n/d \div p/q$ is the number x such that

$$\frac{p}{q} \cdot x = \frac{n}{d}$$

We shall determine x by use of the multiplicativity axiom, which states that if $a = b$, then $ac = bc$. Now letting $a = (p/q) \cdot x$, $b = n/d$, and $c = q/p$, we have

$$\frac{p}{q} \cdot x \cdot \frac{q}{p} = \frac{n}{d}\frac{q}{p}$$

or

$$\left(\frac{p}{q}\frac{q}{p}\right) \cdot x = \frac{n}{d}\frac{q}{p}$$

◄ commutative and associative axioms

Hence, $x = \dfrac{n}{d}\dfrac{q}{p}$ ◄ since $(p/q)(q/p) = 1$

Therefore, $\dfrac{n}{d} \div \dfrac{p}{q} = \dfrac{n}{d}\dfrac{q}{p}$ **(3.4)**

Quotient of
two fractions

Consequently, *to obtain the quotient of two fractions, we invert the divisor and multiply by the dividend.*

EXAMPLE 1

Divide $3x^2/4a$ by $6x^3/5a^2$.

Solution

$$\frac{3x^2}{4a} \div \frac{6x^3}{5a^2} = \frac{3x^2}{4a}\frac{5a^2}{6x^3} = \frac{15a^2x^2}{24ax^3} = \frac{5a}{8x}$$

EXAMPLE 2

Divide $\dfrac{x^2 - y^2}{x + 3y}$ by $\dfrac{x - y}{x^2 + 3xy}$.

Solution

$$\frac{x^2 - y^2}{x + 3y} \div \frac{x - y}{x^2 + 3xy} = \frac{x^2 - y^2}{x + 3y}\frac{x^2 + 3xy}{x - y}$$

$$= \frac{(x - y)(x + y)}{x + 3y}\frac{x(x + 3y)}{x - y}$$ ◄ factoring both members

$$= x^2 + xy$$ ◄ dividing both members
by $(x + 3y)(x - y)$

Exercise 3.3   Multiplication and Division of Fractions

Perform the indicated multiplications and divisions.

1 $\dfrac{xy^2}{z^3w}\dfrac{z^2w^3}{x^2y^3}\dfrac{x^3}{w^2}$

2 $\dfrac{x^3y^2}{w^4z}\dfrac{xz^2}{wy^3}\dfrac{w^3}{x^2}$

3 $\dfrac{15x^3y^4}{7w^5z^5}\dfrac{21x^2z^3}{5y^2w}\dfrac{2w^4}{9x^4}$

4 $\dfrac{18y^7w^3}{5x^4z^3}\dfrac{15x^3w^2}{6y^2z}\dfrac{2z^5}{3w^2}$

5 $\dfrac{11x^4y}{7w^3z^2} \div \dfrac{22xy^3}{21wz^3}$

6 $\dfrac{8x^3y^4}{6w^2z} \div \dfrac{35w^4z^3}{10xy^2}$

7 $\dfrac{30x^5w^4}{2y^3z^5} \div \dfrac{15x^2w^3}{6y^2z^2}$

8 $\dfrac{28x^3y^4}{15z^2w^2} \div \dfrac{21zw^5}{20x^2y^3}$

9 $\dfrac{5x^2}{6z^3}\dfrac{12y^2z}{15x} \div \dfrac{8x^2y}{10z^2}$

10 $\dfrac{10x^4}{9y^3}\dfrac{18y^4}{5z^4} \div \dfrac{4x^6}{7z}$

11 $\dfrac{7x^3z}{9y^2}\dfrac{18wy}{21x^2} \div \dfrac{2z^3}{9y^4}$

12 $\dfrac{14x^5}{35y^4}\dfrac{15y}{6z^2} \div \dfrac{z^3}{3y^2x}$

13 $\dfrac{2xy + xz}{6x - 18y}\dfrac{6x^2 - 18xy}{6y + 3z}$

14 $\dfrac{2x^2 + 3xy}{9x - 3y}\dfrac{3xz - yz}{4x + 6y}$

15 $\dfrac{x^2 - 2xy}{2xy + y^2} \dfrac{4x + 2y}{2x^2 - xy}$

16 $\dfrac{x^2 - 9y^2}{2xy + 5y^2} \dfrac{2x^2 + 5xy}{x^2 - xy - 6y^2}$

17 $\dfrac{xy + 2xz}{y^2 - 2yz} \div \dfrac{y^2 + 3yz + 2z^2}{y^2 - 4z^2}$

18 $\dfrac{x^2 - 4y^2}{x^2 + xy - 2y^2} \div \dfrac{x^2 - 2xy}{x^2 - y^2}$

19 $\dfrac{2x^2 + 3xy}{3xy - 2y^2} \div \dfrac{4x^2 - 9y^2}{3x^2 - 2xy}$

20 $\dfrac{2x^2 + xy - y^2}{x^2 + xy - 2y^2} \div \dfrac{2x^2 - xy}{x^2 + 3xy + 2y^2}$

21 $\dfrac{x^2 - y^2}{x^2 y} \dfrac{xy - 2y^2}{x - y} \dfrac{xy^3}{x^2 - xy - 2y^2}$

22 $\dfrac{xy - xz - y^2 + yz}{yz} \dfrac{yz - 2z^2}{xz - xy} \dfrac{x^2 y}{2xy - xz - 2y^2 + yz}$

23 $\dfrac{xy + 2xz + y^2 + 2yz}{xy + 2xz} \dfrac{xy + yz}{x^2 + xy + xz + yz} \dfrac{x^2 - xy - xz + yz}{xy - xz - yz + z^2}$

24 $\dfrac{2xy - xz + 4y^2 - 2yz}{y - 2x} \dfrac{2xy - y^2}{x^2 + xy - 2y^2} \dfrac{x^2 - xy + xz - yz}{z^3 - 2yz^2}$

25 $\dfrac{2xy - xz}{xy^2 + 2y^3} \dfrac{x^2 + 2xy}{z^2 - 2yz} \div \dfrac{2y^3 - y^2 z}{x^3 z}$

26 $\dfrac{3x^4 + 6x^2 y^2}{x^2 y - 9y^3} \dfrac{xy + 3y^2}{5x^5 + 10x^3 y^2} \div \dfrac{y^2}{x^2 - 3xy}$

27 $\dfrac{6x^2 y - 24y^3}{18x^2 - 36xy} \dfrac{18xy - 9y^2}{5xy - 10y^2} \div \dfrac{30x^2 - 75xy + 30y^2}{3xy - 6y^2}$

28 $\dfrac{x^2 - 2xy - 3y^2}{2x^2 + xy - y^2} \dfrac{2x^2 - 3xy + y^2}{x^2 + xy - 2y^2} \div \dfrac{x^2 - 9y^2}{x^2 - 4y^2}$

29 $\dfrac{x - y}{6xw} \dfrac{xw + yw - 3x - 3y}{2xw - 8w} \div \dfrac{x^2 - y^2}{4x - 16}$

30 $\dfrac{3x^2 - 2xy - y^2}{2xy + 4x^2} \dfrac{2x^2 - 3xy - 2y^2}{3x^2 + 4xy + y^2} \div \dfrac{2x^2 - xy - y^2}{xy^2 + y^3}$

31 $\dfrac{3x^2 + 6xy}{2x^2 y - 8xy^2 + 8y^3} \dfrac{6x^2 y - 12xy^2}{x + y} \div \dfrac{3xy^2 + 6y^3}{x - 2y}$

32 $\dfrac{3xy^2 + 9x^2 y + 6x^3}{x^2 y + xy^2 - 2y^3} \dfrac{2y^4 + 7xy^3 + 3x^2 y^2}{12x^3 - 18x^2 y - 12xy^2} \div \dfrac{6x^2 + 8xy + 2y^2}{x^5 - 2x^4 y + x^3 y^2}$

33 $\dfrac{(x - 2)x - 3}{(x - 4)x + 3} \dfrac{(x + 2)x - 3}{(x^2 - 4) - 3x}$

34 $\dfrac{x(x - 4) + 3}{x(x - 1) - 2} \dfrac{x(x - 4) - 5}{x(x - 1) - 6}$

35 $\dfrac{x(x - 3) - (x - 4)}{(x - 3)x - 4} \dfrac{(x - 2)x - (x + 4)}{x^2 - 4}$

36 $\dfrac{x(x + 3) + 4(x + 3)}{x(x + 3) + 2(x + 2)} \dfrac{x^2 - (2x + 3)}{x^2 - 9}$

37 $\dfrac{(x - 2)x + 1}{(x - 1)x - 2} \dfrac{(x - 2)x + (x - 2)}{x^2 - 1} \div \dfrac{x + 6}{x + 1}$

38 $\dfrac{x(x + 2) + 2(2x + 4)}{2x(x + 2) + (x + 3)} \dfrac{x^2 - (x + 2)}{x^2 - 4} \div \dfrac{x + 4}{x}$

39 $\dfrac{(x - 3)x - 4}{(x - 9)x + 20} \dfrac{(x + 3)x - 40}{(x^2 - 9) - 8x} \div \dfrac{x + 8}{x - 9}$

40 $\dfrac{(x + 1)x - 2}{(x + 1) - 4} \dfrac{(x - 2)x + 1(x - 2)}{(x - 2)(x - 1)} \div \dfrac{x + 2}{x - 3}$

3.7 COMPLEX FRACTIONS

A complex fraction is a fraction in which the numerator, the denominator, or both contain fractions. For example,

$$\frac{1 + \frac{1}{3}}{1 - \frac{1}{3}} \qquad \frac{\frac{2x - y}{y}}{x + \frac{2}{y}} \qquad \frac{\frac{a + b}{a - b} - \frac{a}{2b}}{2 - \frac{1}{a^2 b^2}}$$

are complex fractions.

We simplify a complex fraction by converting it to an equal fraction in lowest terms in which neither the numerator nor the denominator contains a fraction. A complex fraction can be simplified by first simplifying the numerator and the denominator and then finding their quotient. Usually, however, the most efficient method consists of the following steps:

Simplification of a complex fraction

1 Find the lowest common multiple of the denominators of the fractions that appear in the complex fraction.

2 Multiply each of the members of the complex fraction by the lowest common multiple found in step 1.

3 Simplify the result obtained in step 2.

For example, to simplify

$$\frac{4 - \dfrac{1}{x}}{16 - \dfrac{1}{x^2}}$$

we notice, as step 1, that the lowest common multiple of the denominators is x^2 and then proceed as follows:

$$\frac{4 - \dfrac{1}{x}}{16 - \dfrac{1}{x^2}} = \frac{x^2\left(4 - \dfrac{1}{x}\right)}{x^2\left(16 - \dfrac{1}{x^2}\right)} = \frac{4x^2 - x}{16x^2 - 1} \qquad \blacktriangleleft \text{ step 2, multiplying numerator and denominator by } x^2 \text{ and performing indicated operations}$$

$$= \frac{x(4x - 1)}{(4x + 1)(4x - 1)} \qquad \blacktriangleleft \text{ step 3, factoring the members}$$

$$= \frac{x}{4x + 1} \qquad \blacktriangleleft \text{ dividing each member by } 4x - 1$$

It is sometimes advisable to simplify the numerator and denominator of a complex fraction before multiplying by the lowest common multiple of the denominators. For example,

$$\frac{2-\dfrac{3}{a+2}}{\dfrac{1}{a-1}+\dfrac{1}{a+2}} = \frac{\dfrac{2a+4-3}{a+2}}{\dfrac{a+2+a-1}{(a-1)(a+2)}}$$

◄ adding the fractions in the numerator and denominator

$$=\frac{\dfrac{2a+1}{a+2}}{\dfrac{2a+1}{(a-1)(a+2)}}$$

◄ combining similar terms

$$=\frac{(a-1)(a+2)\dfrac{2a+1}{a+2}}{(a-1)(a+2)\dfrac{2a+1}{(a-1)(a+2)}}$$

◄ multiplying each member by the lcm of the denominators

$$=\frac{(a-1)(2a+1)}{2a+1}$$

◄ simplifying

$$=a-1$$

◄ dividing numerator and denominator by $2a+1$

Exercise 3.4   Complex Fractions

Reduce the following to simple fractions.

1 $\dfrac{1}{1-\frac{2}{3}}$ **2** $\dfrac{3}{1+\frac{5}{6}}$ **3** $\dfrac{3}{1-\frac{2}{5}}$ **4** $\dfrac{2}{1+\frac{3}{7}}$ **5** $\dfrac{2+\frac{1}{2}}{3-\frac{6}{7}}$

6 $\dfrac{5-\frac{5}{6}}{4+\frac{4}{9}}$ **7** $\dfrac{\frac{1}{2}+\frac{1}{3}}{\frac{3}{4}-\frac{7}{6}}$ **8** $\dfrac{\frac{2}{3}+\frac{1}{2}}{\frac{1}{2}+\frac{5}{6}}$ **9** $\dfrac{a-\dfrac{1}{a}}{1-\dfrac{1}{a^2}}$ **10** $\dfrac{a+\dfrac{1}{a^2}}{a-1+\dfrac{1}{a}}$

11 $\dfrac{x+\dfrac{8}{x^2}}{x-2+\dfrac{4}{x}}$ **12** $\dfrac{x^2-\dfrac{16}{x^2}}{x+\dfrac{4}{x}}$ **13** $\dfrac{\dfrac{x}{3}-\dfrac{3}{x}}{\dfrac{1}{x}+\dfrac{2}{3x}}$ **14** $\dfrac{\dfrac{x}{6}-\dfrac{1}{3}}{\dfrac{1}{2}-\dfrac{1}{x}}$

15 $\dfrac{\dfrac{1}{2}-\dfrac{4}{a}}{\dfrac{1}{a}+\dfrac{3}{2a}}$ **16** $\dfrac{\dfrac{x}{2y}-\dfrac{1}{2}}{\dfrac{x}{3y}-\dfrac{1}{3}}$ **17** $\dfrac{2-\dfrac{y}{x+2y}}{2+\dfrac{5y}{x-y}}$ **18** $\dfrac{1+\dfrac{y}{x+y}}{1+\dfrac{3y}{x-y}}$

19 $\dfrac{2+\dfrac{3y}{x-y}}{2-\dfrac{3y}{x+2y}}$ **20** $\dfrac{1+\dfrac{2y}{x+y}}{1+\dfrac{x}{x+y}}$ **21** $\dfrac{3x+\dfrac{x-5}{x-1}}{x-\dfrac{5}{3x-2}}$ **22** $\dfrac{x-\dfrac{2x+6}{2x+1}}{2x-\dfrac{7x+6}{x+3}}$

23 $\dfrac{2x-\dfrac{3x+4}{x-2}}{x-\dfrac{10x+4}{2x+3}}$ **24** $\dfrac{x-\dfrac{x}{x+2}}{x+\dfrac{1}{x+2}}$ **25** $\dfrac{\dfrac{a-2}{a+3}-\dfrac{a}{a-1}}{\dfrac{20}{a+3}-6}$ **26** $\dfrac{\dfrac{a+2}{a-2}-\dfrac{a}{a+2}}{3-\dfrac{4}{a+2}}$

27 $\dfrac{\dfrac{2}{2x-1}+\dfrac{1}{x+2}}{\dfrac{5}{2x-1}+2}$

28 $\dfrac{\dfrac{x+1}{x-1}-\dfrac{x}{x+1}}{\dfrac{4}{x-1}+3}$

29 $\dfrac{\dfrac{x}{x+1}-\dfrac{x^2}{x^2-1}}{\dfrac{1}{x-1}+1}$

30 $\dfrac{\dfrac{2x+3}{x+2}-\dfrac{2x}{x+1}}{\dfrac{x}{x+2}-1}$

31 $\dfrac{\dfrac{3}{2x-3}-\dfrac{2}{x+1}}{1-\dfrac{x+6}{2x-3}}$

32 $\dfrac{\dfrac{5}{2-x}-\dfrac{3}{1+2x}}{1-\dfrac{1+12x}{2-x}}$

33 $\dfrac{\dfrac{3}{2a+3}-\dfrac{5}{a+2}}{\dfrac{3a+1}{a+2}+4}$

34 $\dfrac{\dfrac{x}{2}-\dfrac{4x+12}{x+6}}{\dfrac{x}{2}+\dfrac{x-12}{x-4}}$

35 $\dfrac{\dfrac{2}{3a+1}-\dfrac{2}{a-1}}{\dfrac{7a+5}{3a+1}-1}$

36 $\dfrac{\dfrac{1}{x+1}+\dfrac{2}{x^2-1}}{\dfrac{2}{x-1}+1}$

37 $\dfrac{x-\dfrac{x}{2-\dfrac{z}{x}}}{y-\dfrac{y}{\dfrac{2x}{z}-1}}$

38 $\dfrac{\dfrac{1}{x}}{1-\dfrac{1}{1-\dfrac{3y}{x}}}$

39 $\dfrac{\dfrac{1}{x}}{1-\dfrac{1}{1+\dfrac{2x}{y}}}$

40 $\dfrac{1-\dfrac{3}{1-\dfrac{x}{y}}}{1+\dfrac{1}{1-\dfrac{x}{y}}}$

Exercise 3.5   Review

Convert the fraction in each of problems 1 to 4 into an equal fraction, with the expression to the right of the comma as denominator.

1 $\dfrac{7}{3-x}$, $x-3$

2 $\dfrac{4x^2y^3}{3x^3y}$, $15xy^2z$

3 $\dfrac{3x+4y}{x+y}$, x^2-y^2

4 $\dfrac{2x+5y}{x+2y}$, $x^2+5xy+6y^2$

Reduce the fraction in each of problems 5 to 8 to lowest terms.

5 $\dfrac{12x^3y^2z^4}{3xy^0z^3}$

6 $\dfrac{a^2-7a+10}{a^2-2a-15}$

7 $\dfrac{6a^2-5a-4}{9a^2-16}$

8 $\dfrac{3a^2+8a-3}{a^2+5a+6}$

Convert the set of fractions in each of problems 9 and 10 to an equal set with a common denominator.

9 $\left\{\dfrac{3}{x^2y^3}, \dfrac{-2}{xy^4}, \dfrac{5}{x^3y^2}\right\}$ **10** $\left\{\dfrac{2x+y}{x+2y}, \dfrac{x-y}{2x-y}, \dfrac{3x+2y}{2x^2+3xy-2y^2}\right\}$

Perform the indicated operations in each of problems 11 to 24.

11 $\dfrac{3x+5}{4} + \dfrac{2x-1}{3} + \dfrac{x-4}{6}$ **12** $\dfrac{3x}{4yz} + \dfrac{2y}{5xz} - \dfrac{z}{2xy}$

13 $\dfrac{2x+3y}{3x+2y} + \dfrac{3x}{5y}$ **14** $\dfrac{3x-1}{x-3} + \dfrac{x+3}{3x+1}$

15 $\dfrac{3}{x-3y} - \dfrac{2}{x+3y} + \dfrac{-x-14y}{x^2-9y^2}$ **16** $\dfrac{2}{2x+3y} + \dfrac{3}{3x-2y} - \dfrac{2}{x+2y}$

17 $\dfrac{2x-3y}{3x^2-7xy+2y^2} + \dfrac{x+2y}{6x^2+7xy-3y^2} - \dfrac{x+3y}{2x^2-xy-6y^2}$

18 $\dfrac{18x^3y^2}{7w^4} \dfrac{28x^0w^2}{9y^3z^3} \dfrac{3x^2z^2}{4yw^4}$ **19** $\dfrac{15x^2y^5}{14z^2w} \dfrac{7xw^2}{5y^3z} \div \dfrac{3x^2z^3}{4y^0w^3}$

20 $\dfrac{x^2-4y^2}{xy^3} \dfrac{x^2-3xy}{x+2y} \dfrac{x^2y}{x^2-5xy+6y^2}$

21 $\dfrac{x^2+2xz-xy-2yz}{xz} \dfrac{xy-yz}{xy-y^2} \dfrac{x^2z^0}{x^2+xz-2z^2}$

22 $\dfrac{2x^2+3xy-2y^2}{2x^2-xy-y^2} \dfrac{2x^2+3xy+y^2}{2x^2+xy-y^2} \div \dfrac{x^2+5xy+6y^2}{x^2+2xy-3y^2}$

23 $\dfrac{2x^2+4xy}{2x^2y+xy^2-y^3} \dfrac{6x^2y-3xy^2}{x^2-3xy} \div \dfrac{2x^2y+4xy^2}{x^2-2xy-3y^2}$

24 $\dfrac{x(x+2)-(x-3)}{(x+2)x-3} \dfrac{(x+1)x+(x-3)}{x^2-1}$

Reduce the following to simple fractions.

25 $\dfrac{4}{1+\frac{3}{5}}$ **26** $\dfrac{\frac{3}{7}+\frac{1}{3}}{\frac{5}{6}-\frac{9}{14}}$

27 $\dfrac{\frac{x}{2}-\frac{2}{x}}{\frac{1}{2}+\frac{1}{x}}$ **28** $\dfrac{1-\dfrac{x}{x-2y}}{2+\dfrac{x-2y}{x+y}}$

29 $\dfrac{\dfrac{x}{x-1}-\dfrac{x-2}{x-3}}{\dfrac{12}{x-3}+3}$ **30** $\dfrac{\dfrac{x}{x+2}-\dfrac{x^2-2}{x^2-4}}{\dfrac{1}{x-2}+1}$

31 $\dfrac{x-\dfrac{x}{1+\frac{y}{x}}}{z-\dfrac{z}{\frac{x}{y}+1}}$ **32** $\dfrac{1+\dfrac{1}{x}}{1-\dfrac{1}{1+\dfrac{2}{x-4}}}$

Linear and Fractional Equations and Inequalities

Heretofore we have been concerned with formal operations that followed prescribed rules of procedure. In this chapter we shall study the conditions under which two algebraic expressions are equal. For example, we shall explain methods for finding the replacements for x so that a statement such as $3x^2 - x + 1 = 2x + 3$ is true. A statement of this type is called an *equation*. The equation is a powerful tool in mathematics and is essential in the development and understanding of the physical sciences and engineering.

4.1 OPEN SENTENCES

The colors in a rainbow are violet, indigo, blue, green, yellow, orange, and red. Therefore the sentence

$$\text{Red is a color in a rainbow}$$

is true, while the sentence

$$\text{Brown is a color in a rainbow}$$

is false. Now we shall consider four aspects of the sentence

$$x \text{ is a color in a rainbow} \tag{1}$$

where x may be replaced by the name of any one of the existing colors. First, the sentence (1) contains the variable x, which holds a place for the name of some color. Second, the sentence is neither true nor false as it stands. Third, the sentence is a true statement if x is replaced by the name of one of the colors in the rainbow. Fourth, the sentence is false if x is replaced by the name of any other color.

Sentences of the type (1) illustrate the following definition:

Open sentence An *open sentence* is a statement containing a variable which is neither true nor false but which becomes true or false when the variable is replaced by an element chosen from an appropriate set.

In this chapter we shall consider open sentences of an algebraic nature. For example,

$$x + 2 = 5 \tag{2}$$

is an open sentence that states, "the sum of x and 2 is 5." The sentence is true when x is replaced by 3, but it is false if x is replaced by any other numeral.

The word problems that appear in texts in algebra can usually be translated into algebraic open sentences by using a letter to stand for the answer, or one of the answers, and then using the symbols of operations and the equality sign. For example, consider this problem.

Tom and Dick earned $500 during the summer, and Tom earned $10 more than Dick. Find the amount earned by each.

83

If we let x stand for the number of dollars earned by Dick, then $x + 10$ stands for the number of dollars earned by Tom. Since the total amount earned was $500, we can restate the first sentence in the problem as

$$x + (x + 10) = 500 \qquad (3)$$

The second sentence directs us to find the replacement for x that will make the open sentence (3) true. A little experimentation will lead to the discovery that the sentence (3) is true if x is replaced by 245. Then $x + 10 = 255$, and we see that Dick earned \$245 and Tom earned \$255.

Open sentences of types (2) and (3) are called equations, and the procedure for finding the replacement for the variable that makes such a sentence true is called *solving the equation*. In this chapter we shall discuss methods for solving certain types of equations; in the next section we define some of the terms we shall be using.

4.2 SOME DEFINITIONS

Each of the open sentences (2) and (3) is an example of an equation. We shall now define an equation. An open sentence which states that two expressions, at least one of which contains one or more variables, are equal is
Equation   called an *equation*.

Therefore,

$$2x - 5 = 4x + 1 \qquad (1)$$

$$\sqrt{\frac{x+1}{3}} = 2 \qquad (2)$$

and
$$\frac{1}{x-1} + \frac{2}{x+2} = \frac{3x}{(x-1)(x+2)} \qquad (3)$$

are equations.

Member   The expression on each side of the equality sign is called a *member* of the equation.

Each replacement for the variable which makes the equation a true
Root   statement is called a *root* of the equation. The set of all roots of an equation
Solution set   is called the *solution set* of the equation. The set of permissible replace-
Domain   ments for the variable is called the *domain* for the equation. The solution set is a subset of the domain.

Equations are classified in several ways, and we shall now consider two classes.

Conditional   An equation that is satisfied by some numbers in its domain but not by
equation   others is called a *conditional equation* as is an equation that is not satisfied by any number in the domain. An equation that is satisfied by each number
Identity   in the domain is called an *identity*.

EXAMPLE 1   Equation (1) of this section is true if $x = -3$; hence, -3 is a root. Equation (1) is not a true statement if x is replaced by any number other than -3; consequently, (1) is a conditional equation.

EXAMPLE 2   The right member of (2) is positive; consequently, the left member must also be positive. Therefore, the domain is all numbers greater than -1. In set language, the domain of (2) is $\{x | x > -1\}$.

EXAMPLE 3   Neither member of equation (3) is a number if x is replaced by 1 or -2; hence, the domain cannot include these numbers. If, however, we combine the fractions in the left member, we obtain the right member. Therefore, the equation is a true statement if x is replaced by any number other than 1 or -2. Consequently, (3) is an identity.

EXAMPLE 4   In the equation

$$x - 5 = |x + 1|$$

the right-hand member is an absolute value and is therefore zero or positive. Consequently, the left member must also be zero or positive. Hence, the domain is $\{x | x > 5\}$, and we have a conditional equation.

EXAMPLE 5   The equation

$$\frac{x - 7}{x + 3} = 1$$

is not true for any replacement for x since $x - 7$ and $x + 3$ are never equal. Therefore, the equation is a conditional equation with $\varnothing$ as its solution set.

The process of finding the solution set of an equation is called solving the equation.

We shall now introduce the symbol $f(x)$, read "f of x," which is very important in mathematics. In this chapter we shall use $f(x)$† to stand for an algebraic expression in x. This symbol will be discussed more fully in Chap. 6, where it will be given a broader interpretation.

Thus, if in a given discussion, we let $f(x) = 3x^2 - 2x + 1$, then throughout the discussion, $f(x)$ stands for the trinomial $3x^2 - 2x + 1$. The x in $f(x)$ may be replaced by a specific number, but then the x in the expression

---

† We call attention to the fact that $f(x)$ does not mean "f times x."

denoted by $f(x)$ must be replaced by the same number. Thus, if $f(x) = 3x^2 - 2x + 1$, then $f(2) = 3(2^2) - 2(2) + 1 = 9$, and $f(y) = 3y^2 - 2y + 1$.

The notations $h(x)$, $g(x)$, and $F(x)$ or any letter followed by a second letter enclosed in parentheses are also used for algebraic expressions. In this sense then, each of the following expressions is an equation:

$$f(x) = c \qquad \blacktriangleleft \text{ where } c \text{ is a constant} \qquad (4)$$
$$f(x) = g(x) \qquad\qquad\qquad\qquad (5)$$
$$f(x) + g(x) = k(x) \qquad\qquad\quad (6)$$

If r, r', and r'' are the respective roots of these equations, then we have

$$f(r) = c \qquad \blacktriangleleft \text{ from (4)}$$
$$f(r') = g(r') \qquad \blacktriangleleft \text{ from (5)}$$
$$f(r'') + g(r'') = k(r'') \qquad \blacktriangleleft \text{ from (6)}$$

Hereafter in this chapter we shall deal only with conditional equations and methods for solving them. The variable in an equation is often called the *unknown,* and we shall frequently refer to it in this way.

Unknown

4.3 EQUIVALENT EQUATIONS

The objective in solving an equation is to find a replacement for the variable that satisfies the equation. The simpler the equation is in form, the easier it is to solve. For example, consider the equations

$$7x - 45 = 5x - 43 \qquad\qquad (1)$$
and
$$2x = 2 \qquad\qquad\qquad\qquad (2)$$

At this stage, the only way that we can find a root of (1) is to guess at a number, substitute it for x, and see if it satisfies the equation. In (2), however, it is obvious that the root is 1. Now if we substitute 1 for x in (1) and combine terms, we get $-38 = -38$. Therefore 1 is also a root of Eq. (1). We call attention to the fact that Eqs. (1) and (2) are different statements. Each statement, however, is true if x is replaced by 1. Two equations of this type are said to be equivalent.

Equivalent equations

Two equations are *equivalent* if every root of each of them is also a root of the other.

In terms of the solution set, we can state the above definition in this way:

Two equations are *equivalent* if the solution set of each of them is the solution set of the other.

We make extensive use of the concept of equivalent equations in the process of solving an equation. We shall therefore consider the operations that can be performed on the members of a given equation so as to yield a simpler equation that is equivalent to the given one. We shall first state, illustrate, and prove the following theorem:

Operations that
yield equivalent
equations

If $f(x)$, $g(x)$, and $h(x)$ are polynomials,† then the two equations $f(x) = g(x)$ and $f(x) + h(x) = g(x) + h(x)$ are equivalent equations   (4.1)

To illustrate this theorem, let $f(x) = 3x + 1$, $g(x) = 2x + 5$, and $h(x) = -2x + 3$; then the equation $f(x) = g(x)$ is

$$3x + 1 = 2x + 5$$

Furthermore, $f(x) + h(x) = g(x) + h(x)$ becomes

$$3x + 1 + (-2x + 3) = 2x + 5 + (-2x + 3)$$

or $$x + 4 = 8$$

Now the theorem states that $3x + 1 = 2x + 5$ and $x + 4 = 8$ are equivalent. It can be readily verified that 4 is a root of each equation.

We shall now prove the theorem.

Proof ▶

If the number r is a root of $f(x) = g(x)$, then

$$f(r) = g(r)$$

Furthermore, $h(r)$ is a constant, and it follows from (2.5) that

$$f(r) + h(r) = g(r) + h(r)$$

Hence, r is a root of $f(x) + h(x) = g(x) + h(x)$.

Conversely, if r is a root of $f(x) + h(x) = g(x) + h(x)$, then

$$f(r) + h(r) = g(r) + h(r)$$

Consequently, by (2.5) we have

$$f(r) + h(r) - h(r) = g(r) + h(r) - h(r)$$

Therefore $f(r) = g(r)$ since $h(r) - h(r) = 0$, and it follows that r is a root of $f(x) = g(x)$.

EXAMPLE 1

Solve the equation

$$6x - 3 = 7 + 5x \tag{3}$$

Solution

In this equation $f(x) = 6x - 3$ and $g(x) = 7 + 5x$. The first step in the solution is to choose $h(x)$ so that $6x - 3 + h(x)$ contains only terms involving x and $7 + 5x + h(x)$ contains only constant terms. These conditions are satisfied if one term in $h(x)$ is 3 and the other term is $-5x$. Hence we let $h(x) = 3 - 5x$. Consequently the equation

$$6x - 3 + 3 - 5x = 7 + 5x + 3 - 5x \tag{4}$$

is equivalent to (3) by theorem (4.1). Furthermore, if we combine similar

---

† In most cases, the theorem is true if $f(x)$, $g(x)$, and $h(x)$ are not polynomials. The theorem as stated, however, suffices for our purposes.

terms in each member of (4), we obtain the very simple equation

$$x = 10$$

Since the members of (4) reduce to x and 10, respectively, 10 is a root of (4), and therefore, since (4) and (3) are equivalent, 10 is a root of (3). Hence the solution set of (3) is $\{10\}$.

In order to verify that 10 is a root of (3), we substitute 10 for x in the equation and find that each member is equal to 57.

The purpose of adding $h(x)$ to each member of $f(x) = g(x)$ is to obtain an equivalent equation in which each term in one member involves x and each term in the other member is a constant. This purpose is accomplished if $h(x)$ is the polynomial whose terms are the negatives of the terms involving x in $g(x)$ and the negatives of the constant terms in $f(x)$. For example, in the equation

$$6x - 2 + 4x + \tfrac{1}{2} = 3x + 5 - \tfrac{1}{2}x \tag{5}$$

the negatives of the terms involving x in the right member, or $g(x)$, are $-3x$ and $\tfrac{1}{2}x$, and the negatives of the constant terms in the left member, or $f(x)$, are 2 and $-\tfrac{1}{2}$. Hence we let $h(x) = -3x + \tfrac{1}{2}x + 2 - \tfrac{1}{2}$. Then, if we add $h(x)$ to each member of (5), we get

$$6x - 2 + 4x + \tfrac{1}{2} - 3x + \tfrac{1}{2}x + 2 - \tfrac{1}{2} = 3x + 5 - \tfrac{1}{2}x - 3x + \tfrac{1}{2}x + 2 - \tfrac{1}{2} \tag{6}$$

Combining the constant terms in the left member of (6) and the terms involving x in the right member, we obtain

$$6x + 4x - 3x + \tfrac{1}{2}x = 5 + 2 - \tfrac{1}{2}$$
$$7\tfrac{1}{2}x = 6\tfrac{1}{2}$$

or

We shall now state an additional theorem that is useful in solving equations. The proof of this theorem follows directly from the multiplicativity axiom and will be left as an exercise for the student.

Equivalent
equations

If k is a nonzero constant, then the equations $f(x) = g(x)$ and $k \cdot f(x) = k \cdot g(x)$ are equivalent equations $\qquad$ **(4.2)**

EXAMPLE 2   Solve the equation

$$\tfrac{1}{2}x + \tfrac{2}{3} = \tfrac{1}{4}x - \tfrac{1}{6} \tag{7}$$

Solution   We apply (4.2) and (4.1) to obtain a sequence of equations, each simpler in form than the preceding and each equivalent to (7), until we ultimately reach an equation in which the root is evident. We first note that the lcm of the denominators in (7) is 12. We therefore proceed as follows:

$$(\tfrac{1}{2}x + \tfrac{2}{3})12 = (\tfrac{1}{4}x - \tfrac{1}{6})12 \qquad \blacktriangleleft \text{ by (4.2) with } k = 12$$
$$6x + 8 = 3x - 2 \qquad \blacktriangleleft \text{ multiplying}$$
$$6x + 8 - 3x - 8 = 3x - 2 - 3x - 8 \qquad \blacktriangleleft \text{ by (4.1) with } h(x) = -3x - 8$$
$$3x = -10 \qquad \blacktriangleleft \text{ combining similar terms}$$
$$x = -\tfrac{10}{3} \qquad \blacktriangleleft \text{ by (4.2) with } k = \tfrac{1}{3}$$

Hence $-\tfrac{10}{3}$ is a root of (7). We verify this fact by replacing x in (7) by $-\tfrac{10}{3}$ and finding that each member is equal to -1.

4.4 LINEAR EQUATIONS IN ONE VARIABLE

We define a linear equation in one unknown as follows:

Linear equation in one variable
An equation $f(x) = g(x)$ is a *linear equation in one variable* if $f(x)$ and $g(x)$ are of degree 1 in x or if one of them is of degree 1 and the other is a constant.

For example, $ax = b$, $3x + 4 = 7$, and $2x - 7 = 8 - 4x$ are linear equations in one variable.

The procedure for solving a linear equation consists of the following steps:

1 If one or more of the coefficients or constant terms in the equation $f(x) = g(x)$ are fractions, multiply each member by the lcm of the denominators and equate their products. If the lcm of the denominators is k, then the equation becomes $k \cdot f(x) = k \cdot g(x)$.

2 Next formulate the polynomial $h(x)$ so that each term of $k \cdot f(x) + h(x)$ involves x and each term of $k \cdot g(x) + h(x)$ is a constant. Then write the equation $k \cdot f(x) + h(x) = k \cdot g(x) + h(x)$. This equation is equivalent to $f(x) = g(x)$.

3 Combine similar terms in each member of the equation obtained in step 2 and thus obtain an equation of the type $ax = b$.

4 If $a \neq 0$, we multiply each member by $1/a$ and thus obtain an equation of the type $x = b/a$. The number b/a is the root of the given equation since the equation obtained in each step is equivalent to the given equation.

5 Finally, replace x in the given equation by the number obtained in step 4 in order to verify the fact that it is a root.

EXAMPLE

Solution

Find the set indicated by $\{x \mid \tfrac{1}{2}x - \tfrac{2}{3} = \tfrac{3}{4}x + \tfrac{1}{12}\}$.

To obtain the required set, we must solve the equation

$$\tfrac{1}{2}x - \tfrac{2}{3} = \tfrac{3}{4}x + \tfrac{1}{12} \qquad (1)$$

We shall employ the five-step process for this purpose.

$$12(\tfrac{1}{2}x - \tfrac{2}{3}) = 12(\tfrac{3}{4}x + \tfrac{1}{12}) \qquad \blacktriangleleft \text{ by step 1 with lcm} = 12$$
$$6x - 8 = 9x + 1 \qquad \blacktriangleleft \text{ by the distributive axiom}$$
$$6x - 8 - 9x + 8 = 9x + 1 - 9x + 8 \qquad \blacktriangleleft \text{ by step 2 with } h(x) = -9x + 8$$
$$-3x = 9 \qquad \blacktriangleleft \text{ step 3, combining similar terms in each member}$$
$$x = -3 \qquad \blacktriangleleft \text{ step 4, multiplying each member by } -\tfrac{1}{3}$$

Step 5 is a verification. If we replace x by -3 in Eq. (1), we get

$$-\frac{3}{2} - \frac{2}{3} = \frac{-9 - 4}{6} = \frac{-13}{6}$$

from the left member, and

$$-\frac{9}{4} + \frac{1}{12} = \frac{-27 + 1}{12} = -\frac{26}{12} = -\frac{13}{6}$$

from the right member. Consequently

$$\{x | \tfrac{1}{2}x - \tfrac{2}{3} = \tfrac{3}{4}x + \tfrac{1}{12}\} = \{-3\}$$

Exercise 4.1

Find the solution set of the equation in each of the problems 1 to 20.

1 $5x = 3x + 6$

2 $2x = 6x - 8$

3 $7x + 2 = 5x + 2$

4 $8x - 2 = 2x + 4$

5 $3(3x - 2) + 4(x + 2) = 54$

6 $3(x + 10) - 2(2x + 7) = 21$

7 $5(2x + 3) - 2(3x + 5) = -3$

8 $7(2x - 9) + 3(3x - 15) = 30$

9 $\frac{3}{4}x - \frac{3}{4} = \frac{2}{3}x + \frac{1}{4}$

10 $\frac{3}{4}x - \frac{1}{2} = \frac{3}{5}x + 1$

11 $\frac{2}{5}x + \frac{3}{2} = \frac{3}{5}x - \frac{3}{2}$

12 $1 - \frac{2}{3}x = \frac{1}{2} - \frac{3}{4}x$

13 $\frac{5}{6}x - \frac{2}{9} - \frac{2}{9}x = \frac{5}{9}x + \frac{7}{9}$

14 $\frac{7}{9}x - \frac{5}{9} - \frac{1}{6}x = \frac{7}{12}x + \frac{4}{9}$

15 $\frac{2}{3}x + \frac{1}{2}x + \frac{1}{2} = \frac{5}{6}x + 1$

16 $\frac{1}{2}x + \frac{1}{3}x + \frac{1}{4} = \frac{3}{4}x + \frac{3}{4}$

17 $\frac{1}{3}\left(\frac{x}{4} + 9\right) = \frac{1}{4}\left(\frac{2x}{3} - 4\right)$

18 $\frac{1}{6}\left(\frac{x}{3} + 12\right) = \frac{1}{9}\left(\frac{x}{2} - 18\right)$

19 $\frac{1}{6}\left(\frac{1}{4}x + \frac{9}{2}\right) = \frac{1}{8}\left(\frac{1}{2}x + 2\right)$

20 $\frac{1}{5}\left(\frac{1}{2}x + 20\right) = \frac{1}{2}\left(\frac{1}{2}x + 2\right)$

21 Find $\left\{x \left| \dfrac{3x + 1}{4} = 9 - x \right.\right\}$.

22 Find $\left\{x \left| \dfrac{3x - 2}{2} = 2x - 3 \right.\right\}$.

23 Find $\left\{x \left| \dfrac{3x + 4}{5} + 2 = \dfrac{x + 7}{2} \right.\right\}$.

24 Find $\left\{x \left| \dfrac{2x - 4}{5} - \dfrac{x + 3}{4} = -\dfrac{1}{5} \right.\right\}$.

25 Find $\left\{x \left| \dfrac{5x+2}{6} + \dfrac{3x+2}{4} = x+2 \right.\right\}$.

26 Find $\left\{x \left| \dfrac{2x-3}{2} + \dfrac{3x-1}{7} = \dfrac{2x+1}{2} \right.\right\}$.

27 Find $\left\{x \left| \dfrac{3x+3}{4} + \dfrac{2x+1}{3} = \dfrac{9x+4}{6} \right.\right\}$.

28 Find $\left\{x \left| \dfrac{5x-8}{3} = \dfrac{3x-1}{2} - 1 \right.\right\}$.

Find the value of x in each of problems 29 to 36.

29 $\dfrac{ax}{b} - \dfrac{bx}{a} = \dfrac{(a+b)^2}{ab}$

30 $\dfrac{ax}{b} - \dfrac{9bx}{a} = a + 3b$

31 $\dfrac{x-2b}{a} - \dfrac{3}{2} = \dfrac{a-x}{2b} + \dfrac{1}{2}$

32 $\dfrac{a^2x-a}{b} - \dfrac{b+b^2x}{a} = 1$

33 $b(x+a) = b(a-x) + 2a + 2b$

34 $a(x+2) + b(x-2) = 3a - b$

35 $a(x+1) + b(x+1) = x(a+2b)$

36 $b(x+2) + a(x+3) = 2a + b$

4.5 FRACTIONAL EQUATIONS

Fractional equation If at least one algebraic fraction with the variable in the denominator appears in an equation, then the equation is a *fractional equation*. For example,

$$\frac{x}{x+1} + \frac{5}{8} = \frac{5}{2(x+1)} - \frac{3}{4}$$

is a fractional equation, but

$$\frac{x+1}{2} - 5x = 7$$

is not.

We employ the following theorem in solving a fractional equation:

THEOREM *If $k(x)$ is a polynomial, then each root of $f(x) = g(x)$ is also a root of*
$$k(x) \cdot f(x) = k(x) \cdot g(x) \tag{4.3}$$

Proof ▶ If r is a root of $f(x) = g(x)$, then $f(r) = g(r)$. Furthermore $k(r)$ is a constant. Hence by the multiplicativity axiom for equalities, we have $k(r) \cdot f(r) = k(r) \cdot g(r)$. Hence r is a root of $k(x) \cdot f(x) = k(x) \cdot g(x)$.

As an example, consider $f(x) = 3x - 2$, $g(x) = 5x + 8$, and $h(x) = x - 3$. Then $f(x) = g(x)$ becomes

$$3x - 2 = 5x + 8 \tag{1}$$

and $h(x) \cdot f(x) = h(x) \cdot g(x)$ is

$$(x - 3)(3x - 2) = (x - 3)(5x + 8) \tag{2}$$

Now if x is replaced by -5, the members of (1) are equal since each is -17, and the members of (2) are both $(-8)(-17)$. Hence -5 is a root of each equation.

The converse of theorem (4.3) is not true as seen from the fact that 3 is not a root of $3x - 2 = 5x + 8$ since $7 \neq 23$, but 3 is a root of $(x-3)(3x-2) = (x-3)(5x+8)$, since if x is replaced by 3, each member of the latter equation is equal to 0.

If $f(x) = g(x)$ is a fractional equation and $k(x)$ is the lcm of the denominators, then $k(x) \cdot f(x) = k(x) \cdot g(x)$ will contain no fractions. If the latter equation is linear, we can solve it by the methods of Sec. 4.4.

EXAMPLE 1   Solve the equation

$$\frac{x}{x+1} + \frac{5}{8} = \frac{5}{2(x+1)} + \frac{3}{4} \tag{3}$$

Solution   We shall employ theorem (4.3) as a first step in solving Eq. (3). Since the lcm of the denominators is $8(x+1)$, we shall let $k(x) = 8(x+1)$; then

$$8(x+1)\left(\frac{x}{x+1} + \frac{5}{8}\right) = 8(x+1)\left[\frac{5}{2(x+1)} + \frac{3}{4}\right] \tag{4}$$

$8x + 5(x+1) = 4(5) + 6(x+1)$ ◀ multiplying by the lcm

$8x + 5x + 5 = 20 + 6x + 6$ ◀ by the right-hand distributive axiom

$8x + 5x + 5 - 6x - 5 = 20 + 6x + 6 - 6x - 5$ ◀ adding $-6x - 5$ to each member

$7x = 21$ ◀ combining similar terms

$x = 3$ ◀ multiplying each member by $\frac{1}{7}$

By theorem (4.3) we know that each root of (3) is a root of (4), but we do not know that the converse is true. Hence, we must replace x by 3 in (3) and see if it is a root. When this is done, we see that each member is equal to $\frac{11}{8}$. Hence the root of (3) is 3, and we have

$$\left\{x \left| \frac{x}{x+1} + \frac{5}{8} = \frac{5}{2(x+1)} + \frac{3}{4} \right.\right\} = \{3\}$$

EXAMPLE 2   Find $\left\{x \left| \frac{2}{x+1} - 3 = \frac{4x+6}{x+1} \right.\right\}$.

Solution   The required set of numbers is the set of roots of

$$\frac{2}{x+1} - 3 = \frac{4x+6}{x+1} \tag{5}$$

We first employ theorem (4.3), multiply each member by $x + 1$, and get

$$2 - 3x - 3 = 4x + 6 \tag{6}$$

$2 - 3x - 3 - 4x - 2 + 3 = 4x + 6 - 4x - 2 + 3$ ◀ by (4.1) with $h(x) =$ $-4x - 2 + 3$

$\qquad\qquad -7x = 7$ ◀ combining similar terms

$\qquad\qquad\quad x = -1$ ◀ multiplying each member by $-\frac{1}{7}$

If we now replace x by -1 in (5), the left member becomes $\frac{2}{0} - 3$, which is not a number. Furthermore, the right member becomes $\frac{2}{0}$, which also is not a number. Hence, since neither member of Eq. (5) is defined when $x = -1$, we cannot accept -1 as a root. Furthermore, by theorem (4.3) each root of Eq. (5) is a root of Eq. (6), and -1 is the only root of Eq. (6). Hence we conclude that Eq. (5) has no roots. Therefore,

$$\left\{ x \left| \frac{2}{x+1} - 3 = \frac{4x+6}{x+1} \right. \right\} = \varnothing$$

where $\varnothing$ is the empty set.

Exercise 4.2　Fractional Equations

Find the solution set of the equation in each of problems 1 to 16.

1 $\dfrac{x+4}{x-2} = \dfrac{x+1}{x-3}$ 　　　　　　**2** $\dfrac{x+6}{x+4} = \dfrac{x+2}{x+1}$

3 $\dfrac{x+1}{x+7} = \dfrac{x-1}{x+1}$ 　　　　　　**4** $\dfrac{x+1}{x-1} = \dfrac{x-1}{x-2}$

5 $\dfrac{2}{x+5} - \dfrac{5}{x+7} = \dfrac{1}{(x+5)(x+7)}$

6 $\dfrac{5}{x+6} - \dfrac{3}{x+7} = \dfrac{13}{(x+7)(x+6)}$ 　　**7** $\dfrac{7}{x+2} - \dfrac{6}{x+3} = \dfrac{3}{(x+2)(x+3)}$

8 $\dfrac{2}{x+4} - \dfrac{3}{x+6} = \dfrac{5}{(x+4)(x+6)}$ 　　**9** $\dfrac{5}{x+3} - \dfrac{3}{2x-5} = \dfrac{15}{2x^2+x-15}$

10 $\dfrac{2}{x+2} - \dfrac{5}{2x+1} = \dfrac{-20}{2x^2+5x+2}$ 　　**11** $\dfrac{3}{x+1} - \dfrac{4}{3x-4} = \dfrac{2x}{3x^2-x-4}$

12 $\dfrac{2}{x-5} - \dfrac{7}{2x+5} = \dfrac{21}{2x^2-5x-25}$ 　　**13** $\dfrac{1}{x+1} + \dfrac{1}{x+5} = \dfrac{4}{2x+5}$

14 $\dfrac{1}{2x+2} + \dfrac{1}{2x-4} = \dfrac{5}{5x-7}$ 　　**15** $\dfrac{1}{2x+5} + \dfrac{2}{3x-3} = \dfrac{7}{6x-3}$

16 $\dfrac{3}{3x-7} - \dfrac{1}{2x+2} = \dfrac{1}{2x-8}$

Show that the equation in each of problems 17 to 20 has no roots.

17 $\dfrac{3}{x-2}+\dfrac{2}{x+1}=\dfrac{9}{x^2-x-2}$

18 $\dfrac{4}{x^2-4}=\dfrac{1}{x-2}+\dfrac{1}{x+2}$

19 $\dfrac{3}{x+3}-\dfrac{3}{x-1}=\dfrac{4}{x^2+2x-3}$

20 $\dfrac{5}{x^2+x-2}+\dfrac{1}{x-1}=\dfrac{1}{x+2}$

Find the solution set of the equation in each of problems 21 to 24 for the letter given after the equation.

21 $m=\dfrac{c(1-p)}{1-d},\ p$

22 $S=\dfrac{a-ar^n}{1-r},\ a$

23 $\dfrac{p}{q}=\dfrac{f}{q-f},\ f$

24 $I=\dfrac{Ne}{R+Nr},\ r$

4.6 LINEAR INEQUALITIES

Inequality   An open sentence of the form $f(x)>g(x)$ or $f(x)<g(x)$ is called an *inequality*. If $f(x)$ and $g(x)$ are polynomials of degree 1 in x or if one of them is such a polynomial and the other a constant, the inequality is *linear*.

For example, $3x>4$ and $2x+1<5x-7$ are linear inequalities.

Any element of the replacement set of the variable for which the in-
Solution   equality is a true statement is called a *solution*. The set of all solutions of an
Solution set   inequality is called the *solution set*. An inequality that is true for every element of the solution set is called an *absolute* or *unconditional inequality*. Inequalities that are not true for every element of the replacement set are
called *conditional inequalities*.

Thus, $x^2+4>0$, $x\in R$ is an unconditional inequality, and $x>2$ is a
conditional inequality.

Solving an   The procedure for finding the solution set of an inequality is called
inequality   *solving the inequality*. Two inequalities are said to be *equivalent inequalities* if
Equivalent   they have the same solution set. In solving an inequality, we replace it by a
inequalities   sequence of equivalent inequalities until we obtain one for which the solution set is obvious.

We shall now give several theorems that state conditions under which
two inequalities are equivalent.

*If $f(x)$, $g(x)$, and $h(x)$ are expressions, constants included, then for all
values of x for which they are real numbers,*

$$f(x)>g(x)$$

is equivalent to each of the following:

$f(x)+h(x)>g(x)+h(x)$	**(4.4)**
$f(x)\cdot h(x)>g(x)\cdot h(x)$ *for $x\in\{x\|h(x)>0\}$*	**(4.5)**
$f(x)\cdot h(x)<g(x)\cdot h(x)$ *for $x\in\{x\|h(x)<0\}$*	**(4.6)**

Similar statements can be made if $f(x)<g(x)$.

> ***If k is a positive constant, $f(x) > g(x)$ is equivalent to $k \cdot f(x) > k \cdot g(x)$*** **(4.5a)**
>
> ***and $f(x) < g(x)$ is equivalent to $k \cdot f(x) < k \cdot g(x)$*** **(4.5b)**
>
> ***If c is a negative constant, $f(x) > g(x)$ is equivalent to***
> $$c \cdot f(x) < c \cdot g(x)$$ **(4.6a)**
>
> ***and $f(x) < g(x)$ is equivalent to $c \cdot f(x) > c \cdot g(x)$*** **(4.6b)**

The proofs of the above theorems are similar to those of theorems (4.1) and (4.2) and are based on the additivity and multiplicative axioms. The details of the proofs are left as an exercise for the student.

EXAMPLE 1   Find the solution set of the inequality $5x - 9 > 2x + 3$.

Solution

$$5x - 9 > 2x + 3 \qquad \blacktriangleleft \text{ given}$$
$$5x - 9 - 2x + 9 > 2x + 3 - 2x + 9 \qquad \blacktriangleleft \text{ by (4.4) with } h(x) = -2x + 9$$
$$3x > 12 \qquad \blacktriangleleft \text{ combining similar terms}$$
$$x > 4 \qquad \blacktriangleleft \text{ by (4.5a) with } k = \tfrac{1}{3}$$

Hence the solution set is $\{x | x > 4\}$.

EXAMPLE 2   Find the solution set of $\frac{1}{6}x - \frac{3}{4} < \frac{3}{8}x + \frac{1}{2}$.

Solution

$$\tfrac{1}{6}x - \tfrac{3}{4} < \tfrac{3}{8}x + \tfrac{1}{2} \qquad \blacktriangleleft \text{ given}$$
$$24(\tfrac{1}{6}x - \tfrac{3}{4}) < 24(\tfrac{3}{8}x + \tfrac{1}{2}) \qquad \blacktriangleleft \text{ by (4.5b) with } k = 24$$
$$4x - 18 < 9x + 12 \qquad \blacktriangleleft \text{ by the left-hand distributive axiom}$$
$$4x - 18 - 9x + 18 < 9x + 12 - 9x + 18 \qquad \blacktriangleleft \text{ by (4.4) with } h(x) = -9x + 18$$
$$-5x < 30 \qquad \blacktriangleleft \text{ combining similar terms}$$
$$x > -6 \qquad \blacktriangleleft \text{ by (4.6b) with } c - -\tfrac{1}{5}; \text{ note that since } -\tfrac{1}{5} \text{ is negative, we reverse the inequality sign}$$

Hence the solution set is $\{x | x > -6\}$.

We call attention to the fact that aside from exceptional situations, the solution set of a conditional linear equation contains only one element, while the solution set of a linear inequality contains an infinitude of elements. For example,

$$\{x | 2x - 1 = x + 3\} = \{4\}$$
and
$$\{x | 2x - 1 > x + 3\} = \{x | x > 4\}$$

Note that any number greater than 4 is an element of $\{x | x > 4\}$.

Illustrations of exceptional situations follow:

$$\{x | x + 3 = x + 4\} = \varnothing$$
and
$$\{x | x + 3 > x + 4\} = \varnothing$$

4.7 LINEAR INEQUALITIES THAT INVOLVE ABSOLUTE VALUES

If we use the definition of absolute value, we see that an inequality of the type

$$|ax + b| < c \qquad c > 0 \tag{1}$$

requires that $ax + b$ be between c and $-c$, as indicated on the line below.†

Hence, if a value of x satisfies *both $ax + b < c$ and $ax + b > -c$*, it satisfies (1). Consequently, the solution set of (1) is the intersection of the sets

$$\{x \mid ax + b < c\} \qquad \text{and} \qquad \{x \mid ax + b > -c\}$$

Therefore, if $a > 0$, the solution set is

$$\left\{ x \middle| x < \frac{c - b}{a} \right\} \cap \left\{ x \middle| x > \frac{-c - b}{a} \right\}$$

However, if $a < 0$, the solution set is

$$\left\{ x \middle| x > \frac{c - b}{a} \right\} \cap \left\{ x \middle| x < \frac{-c - b}{a} \right\}$$

Furthermore, the inequality $|rx + t| > c$ requires that $rx + t$ represent a point that is to the right of c or one that is to the left of $-c$. Hence a value of x satisfies $|rx + t| > c$ if it satisfies *either $rx + t > c$ or $rx + t < -c$*. Thus the solution set is the union of the sets $\{x \mid rx + t > c\}$ and $\{x \mid rx + t < -c\}$. Therefore, if $r > 0$, the solution set is

$$\left\{ x \middle| x > \frac{c - t}{r} \right\} \cup \left\{ x \middle| x < \frac{-c - t}{r} \right\}$$

and if $r < 0$, the solution set is

$$\left\{ x \middle| x < \frac{c - t}{r} \right\} \cup \left\{ x \middle| x > \frac{-c - t}{r} \right\}$$

EXAMPLE 1   Solve

$$\left| \frac{x}{3} + 2 \right| < 4$$

---

† In handling conditional inequalities of this type, it may help to remember that an interpretation of absolute value is that it is the distance *without regard to direction,* from the zero of the number scale. Therefore $|ax + b|$ determines two points that are equidistant from zero. One is to the right of zero on the number scale and the other to the left of zero. Equation (1) may then be interpreted as saying that both $+(ax + b)$ and $-(ax + b)$ must lie between $-c$ and $+c$.

Solution

This inequality is satisfied if and only if both

$$\frac{x}{3} + 2 < 4 \quad \text{and} \quad \frac{x}{3} + 2 > -4$$

are satisfied. By adding -2 to each member of these inequalities, we get

$$\frac{x}{3} < 2 \quad \text{and} \quad \frac{x}{3} > -6$$

Hence, multiplying by 3 in each case, we see that the original inequality is satisfied by values of x that satisfy both $x < 6$ and $x > -18$. Therefore, the solution set is

$$\{x|-18 < x < 6\} \quad \text{or} \quad \{x|x < 6\} \cap \{x|x > -18\}$$

EXAMPLE 2

Find the solution set of the inequality

$$|-2x + 6| > 8$$

Solution

The solution set of the given inequality is the union of the solution sets of

$$-2x + 6 > 8 \quad \text{and} \quad -2x + 6 < -8$$

Hence the solution set is

$$\{x|x < -1\} \cup \{x|x > 7\}$$

EXAMPLE 3

Solve

$$\left|\frac{4x}{5} - 1\right| > 3$$

Solution

This inequality is satisfied if either

$$\frac{4x}{5} - 1 > 3 \quad \text{or} \quad \frac{4x}{5} - 1 < -3$$

is satisfied. By adding 1 to each member of these inequalities, we get

$$\frac{4x}{5} > 4 \quad \text{and} \quad \frac{4x}{5} < -2$$

Hence, by multiplying by $\frac{5}{4}$ in each case, we see that the original inequality is satisfied by values of x that are greater than 5 and by values of x that are less than $-\frac{5}{2}$, that is, by $x > 5$ and by $x < -\frac{5}{2}$.

The solution set is therefore

$$\{x|x > 5\} \cup \{x|x < -\frac{5}{2}\}$$

Exercise 4.3   Linear Inequalities

Find the solution set of the inequality in each of problems 1 to 32.

1 $3x - 1 > x + 3$ **2** $5x + 2 > 2x + 8$

3 $7x - 9 < 3x + 7$ **4** $6x + 8 < 3x + 2$

5 $2x + 5 < 5x - 1$ **6** $x + 11 < 6x + 1$

7 $4x - 8 > 9x + 7$ **8** $7 - 2x > 2x + 3$

9 $\frac{1}{4}x - \frac{2}{3} > \frac{5}{6}x + \frac{3}{4}$ **10** $\frac{2}{9}x - \frac{1}{6} < \frac{2}{3}x + \frac{1}{2}$

11 $\frac{5}{9}x + \frac{5}{6} > \frac{1}{2}x + \frac{1}{2}$ **12** $\frac{3}{4}x + \frac{5}{6} < \frac{1}{2}x - \frac{2}{3}$

13 $\frac{1}{8}x + \frac{5}{6} < \frac{7}{8}x + \frac{1}{12}$ **14** $\frac{2}{3}x + \frac{1}{4} > \frac{17}{30}x + \frac{7}{20}$

15 $\frac{5}{7}x - \frac{2}{3} < \frac{2}{3}x - \frac{5}{21}$ **16** $\frac{1}{6}x + \frac{1}{3} > \frac{1}{2}x + \frac{2}{3}$

17 $\frac{4}{7}x + \frac{3}{4} > \frac{1}{2}x + \frac{1}{4}$ **18** $\frac{3}{8}x + \frac{3}{4} > \frac{5}{16}x + \frac{1}{2}$

19 $\frac{3}{5}x + \frac{2}{3} < \frac{1}{6}x - \frac{1}{5}$ **20** $\frac{5}{7}x - \frac{1}{6} < \frac{2}{3}x + \frac{1}{2}$

21 $|x - 2| < 3$ **22** $|3x + 4| < 10$

23 $|-4x + 5| < 13$ **24** $|-2x - 7| < 9$

25 $|-3x - 8| > 4$ **26** $|-5x - 9| > 1$

27 $|2x + 3| > 1$ **28** $|4x + 11| > 3$

29 $|\frac{1}{2}x - 1| < 2$ **30** $|\frac{2}{3}x - 6| > 0$

31 $|\frac{3}{4}x + 9| > 3$ **32** $|\frac{2}{5}x + 3| < 5$

33 If $a < b$, prove that $a + b > 2a$.

34 If $a < 2b$, prove that $2a - b < a + b$.

35 If $a > b$, prove that $4a + 3b > 3a + 4b$.

36 If $a > 2b$, prove that $2a + b > a + 3b$.

4.8   NONLINEAR INEQUALITIES

There are many types of nonlinear inequalities just as there are many types of nonlinear equations. We shall deal with those nonlinear inequalities whose left members can be factored into linear factors and those that are fractions with the unknown in the denominator and probably also in the numerator. In solving such inequalities, we make use of the fact that the product or quotient of any number of positive numbers is positive, the product or quotient of an even number of negative numbers is positive, and the product or quotient of an odd number of negative numbers is negative. We shall illustrate the procedure to be followed by solving several examples.

EXAMPLE 1   Find the solution set of $(x + 1)/(x - 2) > 0$.

Solution   The fraction $(x + 1)/(x - 2)$ is positive, i.e., greater than zero, if both the numerator and denominator are positive and if both are negative. Hence, we seek the set of numbers that satisfies the two inequalities

$$x + 1 > 0 \tag{1}$$
$$x - 2 > 0 \tag{2}$$

simultaneously, and also those which satisfy

$$x + 1 < 0 \tag{3}$$
$$x - 2 < 0 \tag{4}$$

simultaneously.

If we let S_1, S_2, S_3, and S_4 stand for the solution sets of (1), (2), (3), and (4), respectively, then

$$S_1 = \{x \mid x > -1\}$$
$$S_2 = \{x \mid x > 2\}$$
$$S_3 = \{x \mid x < -1\}$$
$$S_4 = \{x \mid x < 2\}$$

The sets S_1, S_2, S_3, and S_4 are indicated on the number line in Fig. 4.1.

Now the set of numbers that satisfies *both* (1) and (2) must belong to S_1 *and* to S_2, and it is therefore the intersection, $S_1 \cap S_2$, of the two sets. From Fig. 4.1 we see that

$$S_1 \cap S_2 = \{x \mid x > 2\}$$

Likewise, the set of numbers that satisfies *both* (3) and (4) is the intersection $S_3 \cap S_4$. Again, from Fig. 4.1, we see that

$$S_3 \cap S_4 = \{x \mid x < -1\}$$

Therefore, the solution set of

$$\frac{x + 1}{x - 2} > 0$$

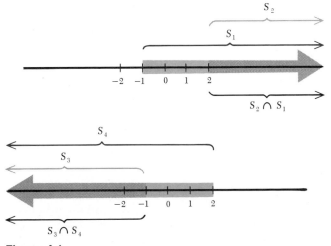

Figure 4.1

is the set of those x in $S_1 \cap S_2$ and the set of those x in $S_3 \cap S_4$. It is therefore the *union* of the two intersections. Consequently the desired solution set is

$$(S_1 \cap S_2) \cup (S_3 \cap S_4) = \{x | x > 2\} \cup \{x | x < -1\}$$

EXAMPLE 2   Find the solution set of

$$\frac{x^2 + x + 2}{x + 1} > 2 + x \tag{5}$$

Solution   We first employ theorem (4.4) to add $-x - 2$ to each member of (5) and obtain the equivalent inequalities

$$\frac{x^2 + x + 2}{x + 1} - x - 2 > 0 \tag{6}$$

$$\frac{x^2 + x + 2 - x^2 - 3x - 2}{x + 1} > 0 \qquad \blacktriangleleft \text{ adding expressions in the left member}$$

$$\frac{-2x}{x + 1} > 0 \qquad \blacktriangleleft \text{ combining similar terms}$$

Now the fraction on the left is greater than zero if

$$-2x > 0 \qquad \text{and} \qquad x + 1 > 0 \tag{7}$$

and also if

$$-2x < 0 \qquad \text{and} \qquad x + 1 < 0 \tag{8}$$

The set of numbers that satisfies the inequalities in (7) simultaneously is

$$\{x | x < 0\} \cap \{x | x > -1\} = \{x | -1 < x < 0\}$$

and the set that satisfies the inequalities in (8) is

$$\{x | x > 0\} \cap \{x | x < -1\} = \emptyset$$

since no number that is greater than zero is less than -1. Hence the solution set of (5) is

$$\{x | -1 < x < 0\} \cup \emptyset = \{x | -1 < x < 0\}$$

EXAMPLE 3   Find the solution set of $(x - 1)(x + 2)(x - 3) < 0$.

Solution   We shall make use of the fact that the product is negative if all three factors are negative and also if any one of the factors is negative and the other two are positive. Consequently, the solution set is the union of these four sets. We shall begin by finding the four sets.

If S_1 is the set for which each factor is negative, then

$$S_1 = \{x|x-1<0\} \cap \{x|x+2<0\} \cap \{x|x-3<0\}$$
$$= \{x|x<1\} \cap \{x|x<-2\} \cap \{x|x<3\}$$
$$= \{x|x<-2\}$$

since, if a number is smaller than the smallest of three numbers, it is smaller than the other two.

If S_2 is the set for which $x-1<0$ and the other two factors are positive, then

$$S_2 = \{x|x-1<0\} \cap \{x|x+2>0\} \cap \{x|x-3>0\}$$
$$= \{x|x<1\} \cap \{x|x>-2\} \cap \{x|x>3\}$$
$$= \varnothing$$

since x cannot be less than 1 and greater than 3 simultaneously.

If S_3 is the set for which the second factor is negative and the other two factors positive, then

$$S_3 = \{x|x-1>0\} \cap \{x|x+2<0\} \cap \{x|x-3>0\}$$
$$= \{x|x>1\} \cap \{x|x<-2\} \cap \{x|x>3\}$$
$$= \varnothing$$

since x cannot be less than -2 and greater than 3 simultaneously.

Finally, if S_4 is the set for which the third factor is negative and the other two positive, then

$$S_4 = \{x|x-1>0\} \cap \{x|x+2>0\} \cap \{x|x-3<0\}$$
$$= \{x|x-1>0\} \cap \{x|x-3<0\}$$

since if $x-1>0$, it follows that $x+2>0$. Therefore,

$$S_4 = \{x|x>1\} \cap \{x|x<3\}$$
$$= \{x|1<x<3\}$$

Consequently, the solution set of the given inequality is

$$S = S_1 \cup S_2 \cup S_3 \cup S_4$$
$$= \{x|x<-2\} \cup \varnothing \cup \varnothing \cup \{x|1<x<3\}$$
$$= \{x|x<-2\} \cup \{x|1<x<3\}$$

This can be put in words by saying that the solution set consists of all numbers less than -2 and all numbers between 1 and 3.

Exercise 4.4　Nonlinear Inequalities

Find the solution set of each inequality in problems 1 to 24.

1 $(x+2)(x-3)>0$ 　　　　　　　**2** $(x+4)(x-1)>0$

3 $(x - 3)(x - 5) > 0$ **4** $(x - 5)(x - 8) > 0$

5 $(x + 4)(x - 2) < 0$ **6** $(x - 1)(x - 4) < 0$

7 $(x + 2)(x + 4) < 0$ **8** $(x - 7)(x + 2) < 0$

9 $(x + 5)/(x - 2) > 0$ **10** $(x + 3)/(x - 4) > 0$

11 $(x + 2)/(x - 3) > 5$ **12** $(x - 5)/(x - 2) > 3$

13 $(11x - 1)/(3x + 2) < 4$ **14** $(9x - 1)/(4x + 3) < 2$

15 $(2x - 3)/(3x - 5) < 4$ **16** $(4x + 3)/(2x - 1) < 3$

17 $(x - 2)(2x + 3)(3x - 5) > 0$ **18** $(x + 3)(2x - 7)(3x - 5) > 0$

19 $(2x - 3)(3x + 7)(2x + 9) < 0$ **20** $(x - 4)(2x + 1)(5x - 7) < 0$

21 $\dfrac{(2x + 5)(3x - 1)}{x - 4} < 0$ **22** $\dfrac{(3x - 7)(2x + 7)}{x - 3} < 0$

23 $\dfrac{x + 3}{(2x - 1)(2x + 3)} > 0$ **24** $\dfrac{x - 5}{(2x + 3)(3x - 2)} > 0$

Exercise 4.5   Review

Find the solution set of the equation in each of problems 1 to 4.

1 $5x - 3 = 2x + 3$ **2** $3(2x + 5) - 4(2x - 17) = 77$

3 $\frac{3}{4}x - \frac{3}{5} = \frac{2}{3}x + \frac{2}{5}$ **4** $2b(x + b) + a(x - a) = ab$

5   Find $\left\{ x \left| \dfrac{2x + 3}{5} = \dfrac{x}{2} \right. \right\}$ **6**   Find $\left\{ x \left| \dfrac{x - 4}{2} - \dfrac{x + 3}{5} = \dfrac{x - 2}{10} \right. \right\}$

Find the solution set of the equation in each of problems 7 to 10.

7 $\dfrac{2}{x - 1} + \dfrac{5}{x + 2} = \dfrac{20}{(x - 1)(x + 2)}$ **8** $\dfrac{6}{x + 1} - \dfrac{7}{2x + 4} = \dfrac{42}{(x + 1)(2x + 4)}$

9 $\dfrac{2}{x + 3} + \dfrac{1}{x - 1} = \dfrac{3}{x + 1}$ **10** $\dfrac{2}{2x + 1} - \dfrac{1}{3x - 1} = \dfrac{2}{3x + 4}$

11   Show that $\dfrac{2x - 1}{x - 3} = 3 + \dfrac{5}{x - 3}$ has no roots.

12   Show that $\dfrac{3}{x - 2} + \dfrac{2}{x + 1} = \dfrac{9}{x^2 - x - 2}$ has no roots.

13   Solve $I = \dfrac{Nl}{R + Nr}$ for N.

Find the solution set of the inequality in each of problems 14 to 19.

14 $5x - 3 > 2x + 6$ **15** $2x + 5 < 4x + 9$

16 $\frac{1}{3}x + \frac{5}{6} < \frac{1}{4}x + \frac{1}{2}$ **17** $\frac{2}{5}x - 3 > \frac{1}{2}x - 4$

18 $|x + 1| < 2$ **19** $|2x + 5| > 1$

20   If $a < 2b$, prove that $3a - b > 5a - 5b$.

Find the solution set of the inequality in each of problems 21 to 28.

21 $(x + 2)(x - 1) > 0$ **22** $(x - 3)(x + 1) < 0$

23 $(x + 4)/(x + 2) < 0$ **24** $(x - 5)/(x - 2) > 0$

25 $(2x - 3)/(x - 4) < 1$ **26** $(3x + 2)/(x + 4) > 2$

27 $(x + 2)(2x + 7)(2x - 3) > 0$ **28** $(3x - 5)(x - 3)(2x + 9) < 0$

Exponents and Radicals

In Chap. 2 we stated the meaning of the notation a^n for n a positive integer, and we also defined the number a^0 for $a \neq 0$. Furthermore, we derived four rules that deal with positive integral exponents, and we used the definitions and rules to a limited extent. In areas where mathematics is used, a broader concept of an exponent is needed. Consequently, in this chapter we shall extend the definition of an exponent to include negative numbers and fractions. The extensions will be made so that the laws developed in Chap. 2 will hold. We also shall develop additional laws as needed and shall explain how to use the extended concept of an exponent and how to apply the former laws and the new ones in more complicated situations than those which occurred in Chap. 2.

5.1 NONNEGATIVE INTEGRAL EXPONENTS

For the convenience of the reader, we shall repeat the definitions and laws of exponents stated and derived in Chap. 2.

If n is a positive integer, then

$$a^n = \underbrace{a \cdot a \cdot a \cdot \cdots \cdot a}$$

$$n \text{ factors}$$

$$a^m a^n = a^{m+n} \qquad m \text{ and } n \text{ positive integers} \qquad (2.38)$$

$$(ab)^n = a^n b^n \qquad n \text{ a positive integer} \qquad (2.39)$$

$$\frac{a^m}{a^n} = a^{m-n} \qquad m \text{ and } n \text{ integers, } m > n > 0, \text{ and } a \neq 0 \qquad (2.47)$$

$$a^0 = 1 \qquad a \neq 0 \qquad (2.48)$$

We shall now derive the laws for obtaining the power of a quotient and the power of a power.

$$\left(\frac{a}{b}\right)^n = \frac{a}{b} \cdot \frac{a}{b} \cdot \frac{a}{b} \cdots \text{ to } n \text{ factors each equal to } \frac{a}{b}$$

$$= \frac{a \cdot a \cdot a \cdots \text{ to } n \text{ factors each equal to } a}{b \cdot b \cdot b \cdots \text{ to } n \text{ factors each equal to } b}$$
◄ by (2.45) the definition of the product of two or more fractions

$$= \frac{a^n}{b^n}$$
◄ by the definition of an nth power

Hence we have

Power of a quotient

$$\left(\frac{a}{b}\right)^n = \frac{a^n}{b^n} \qquad (5.1)$$

Also, $\quad (a^n)^m = a^n \cdot a^n \cdot a^n \cdots \text{ to } m \text{ factors each equal to } a^n$

$$= a^{n+n+n+\cdots \text{ to } m \text{ } n\text{'s}} \qquad ◄ \text{ by (2.38)}$$

$$= a^{mn}$$

Therefore, we have

Power of a power $$(a^n)^m = a^{mn} \qquad (5.2)$$

We shall illustrate the application of the above laws with several examples.

EXAMPLE 1

$$a^3 a^5 a^7 = a^{3+5+7} = a^{15} \qquad \blacktriangleleft \text{ by (2.38)}$$

EXAMPLE 2

$$\frac{a^9}{a^4} = a^{9-4} = a^5 \qquad \blacktriangleleft \text{ by (2.47)}$$

EXAMPLE 3

$$\frac{a^4 a^7}{a^3} = \frac{a^{11}}{a^3} = a^8 \qquad \blacktriangleleft \text{ by (2.38) and (2.47)}$$

EXAMPLE 4

$$(a^6)^4 = a^{24} \qquad \blacktriangleleft \text{ by (5.2)}$$

EXAMPLE 5

$$(a^3 b^4)^5 = a^{15} b^{20} \qquad \blacktriangleleft \text{ by (2.39) and (5.2)}$$

EXAMPLE 6

$$\left(\frac{a^7}{b^4}\right)^3 = \frac{a^{21}}{b^{12}} \qquad \blacktriangleleft \text{ by (5.1) and (5.2)}$$

The methods for obtaining the product and quotient of two similar monomials are illustrated in the following examples:

EXAMPLE 7

$$(6a^3 b^2)(8a^5 b^4) = 6 \cdot 8 \cdot a^3 a^5 b^2 b^4 \qquad \blacktriangleleft \text{ by the commutative law}$$
$$= 48a^8 b^6 \qquad \blacktriangleleft \text{ by (2.38)}$$

EXAMPLE 8

$$\frac{48x^7 y^9}{4x^2 y^5} = \frac{48}{4} \cdot \frac{x^7}{x^2} \cdot \frac{y^9}{y^5}$$
$$= 12x^5 y^4 \qquad \blacktriangleleft \text{ by (2.47)}$$

To simplify an expression involving positive integral exponents, we make all possible applications of the above laws. The following examples illustrate the procedure.

EXAMPLE 9

$$(4x^2 y^3 z^7)^4 = 4^4 (x^2)^4 (y^3)^4 (z^7)^4 \qquad \blacktriangleleft \text{ by (2.39)}$$
$$= 256x^8 y^{12} z^{28} \qquad \blacktriangleleft \text{ by (5.2)}$$

EXAMPLE 10

$$\left(\frac{5x^5y^8}{3z^4w^2}\right)^3 = \frac{5^3(x^5)^3(y^8)^3}{3^3(z^4)^3(w^2)^3} \qquad \blacktriangleleft \text{ by (5.1) and (2.39)}$$

$$= \frac{125x^{15}y^{24}}{27z^{12}w^6} \qquad \blacktriangleleft \text{ by (5.2)}$$

EXAMPLE 11

$$\left(\frac{4a^3b^4}{3c^2d^3}\right)^4\left(\frac{9c^5d}{2a^2b^5}\right)^2 = \frac{4^4(a^3)^4(b^4)^4}{3^4(c^2)^4(d^3)^4} \cdot \frac{9^2(c^5)^2d^2}{2^2(a^2)^2(b^5)^2} \qquad \blacktriangleleft \text{ by (5.1) and (2.39)}$$

$$= \frac{256a^{12}b^{16}}{81c^8d^{12}} \cdot \frac{81c^{10}d^2}{4a^4b^{10}} \qquad \blacktriangleleft \text{ by (5.2)}$$

$$= 64 \cdot \frac{a^{12}}{a^4} \cdot \frac{b^{16}}{b^{10}} \cdot \frac{c^{10}}{c^8} \cdot \frac{d^2}{d^{12}}$$

$$= \frac{64a^8b^6c^2}{d^{10}} \qquad \blacktriangleleft \text{ by (2.47)}$$

Exercise 5.1   Operations with Exponential Expressions

Perform the indicated operations and simplify.

1	2^32^4	**2**	3^53^0	**3**	3^43^1
4	5^25^3	**5**	$\dfrac{27}{2^3}$	**6**	$\dfrac{3^5}{3^0}$
7	$\dfrac{3^5}{3}$	**8**	$\dfrac{5^5}{5^2}$	**9**	$(3^2)^3$
10	$(3^3)^2$	**11**	$(4^3)^2$	**12**	$(5^4)^2$
13	$(\frac{3}{4})^2$	**14**	$(\frac{2}{5})^3$	**15**	$(3^22^1)^4$
16	$(2^45)^2$	**17**	$\left(\dfrac{a^2b^3}{c}\right)^3$	**18**	$\left(\dfrac{a^3b^2}{c^0}\right)^4$
19	$\left(\dfrac{a^2b^4c}{2d^5}\right)^2$	**20**	$\left(\dfrac{3^2a^2b^4}{6^2c^4}\right)^3$	**21**	$(2x^2y^4)(3x^3y)$

22 $(-3x^4y^2z^0)(2xyz^2)$ **23** $(-5x^0yz^2)(-2xy^2z^2)$

24 $(4x^3y^2z^4)(3xy^0z^2)$ **25** $\dfrac{6x^3y^4}{3x^2y^3}$ **26** $\dfrac{10x^5y^4}{15xy^7}$

27 $\dfrac{24x^3y^5z^7}{14xy^0z^4}$ **28** $\dfrac{35xy^3z^5}{21x^3yz^2}$

29 $(2a^2bc)^2(3a^2bc^3)^3$ **30** $(5a^3b^2c)^4(2ab^0c^2)^2$

31 $(2a^3b^3c^4)^3(3a^2bc^0)^2$ **32** $(4a^2b^4c^5)^2(2a^3b^2c)^4$

33 $\left(\dfrac{a^3b^4}{c^2}\right)^2\left(\dfrac{2c^4}{a^2b}\right)$ **34** $\left(\dfrac{3a^2}{b^3c^0}\right)^4\left(\dfrac{bc^2}{9a^4}\right)^2$

35 $\left(\dfrac{3a^2b^0}{c^3}\right)^3\left(\dfrac{c^2}{9ab^4}\right)^2$ **36** $\left(\dfrac{6a^5b^2}{c^3}\right)^2\left(\dfrac{c^4}{12ab^3}\right)^3$

37 $\left(\dfrac{6a^2b^3}{c^3d^4}\right)^3 \div \left(\dfrac{3a^3b^4}{c^4d^5}\right)^2$ **38** $\left(\dfrac{30a^5b^3}{7c^2}\right)^2 \div \left(\dfrac{15a^2b}{14c}\right)^2$

39 $\left(\dfrac{c^3}{3a^4b^2}\right)^4 \div \left(\dfrac{c^0}{6a^3b^2}\right)^3$

40 $\left(\dfrac{6b^3c^2}{4d^2a^4}\right)^2 \div \left(\dfrac{3b^4c^3}{2d^3a^2}\right)^4$

41 $\dfrac{x^{3a+1}y^{a+3}}{x^{a+2}y^{a-1}}$

42 $\dfrac{x^{2a-3}y^{b-4}}{x^{a+2}y^{b+3}}$

43 $\dfrac{x^{a+3}y^{b-4}}{x^{a+2}y^{4-b}}$

44 $\dfrac{x^{a-1}y^{b+4}}{x^{2-a}y^{b-2}}$

45 $\left(\dfrac{a^{2+b}}{a^b}\right)^c$

46 $\left(\dfrac{a^{3-b}}{a^{3-2b}}\right)^c$

47 $\dfrac{(a^{2-n}b^{n+3})^p}{(a^{2-p}b^{p+3})^n}$

48 $\dfrac{(a^{2+x}b^{2x-1})^y}{(a^{1+y}b^{2y-2})^x}$

5.2 NEGATIVE INTEGRAL EXPONENTS

In this section we shall extend the definition of a^n to include a negative integral value of n for $a \neq 0$.

If we disregard the restriction $m > n$ in law (2.47), we may write

$$\frac{a^n}{a^{n+t}} = a^{n-(n+t)} = a^{n-n-t}$$

$$= a^{-t}$$

By the fundamental principle of fractions (3.2), however,

$$\frac{a^n}{a^{n+t}} = \frac{a^n/a^n}{a^{n+t}/a^n} = \frac{a^0}{a^t} = \frac{1}{a^t}.$$

Consequently, if the law of exponents for division, $a^m/a^n = a^{m-n}$, is to hold with $m < n$, it is necessary to define a^{-t} by

Definition of
negative
exponent

$$a^{-t} = \frac{1}{a^t} \qquad \text{for } a \neq 0 \qquad\qquad (5.3)$$

We shall next prove that laws (2.38) and (2.47) hold for the above interpretation of negative exponents. To show that law (2.38) holds, we must prove that

$$a^m a^{-t} = a^{m-t} \qquad t > 0$$

and
$$a^{-r} a^{-t} = a^{-r-t} \qquad r > 0, \, t > 0$$

In order to prove the first statement, we replace a^{-t} by $1/a^t$ and get

$$a^m a^{-t} = a^m \frac{1}{a^t}$$

$$= \frac{a^m}{a^t}$$

$$= a^{m-t} \qquad \blacktriangleleft \text{ by } (2.47)$$

The above argument is not affected if m is replaced by $-r$, and so it follows that

$$a^{-r}a^{-t} = a^{-r-t}$$

To prove that (2.47) holds if the exponents are negative integers, we must prove that

$$\frac{a^{-t}}{a^{-s}} = a^{-t-(-s)}$$

We start with a^{-t}/a^{-s} and multiply the numerator and denominator by a^s and get

$$\frac{(a^{-t})(a^s)}{(a^{-s})(a^s)} = \frac{a^{-t+s}}{a^{-s+s}} \qquad \blacktriangleleft \text{ by (2.38)}$$

$$= a^{-t-(-s)} \qquad \blacktriangleleft \text{ since } a^{-s+s} = a^0 = 1$$

The above argument holds if $-t$ is replaced by m or $-s$ is replaced by n. Consequently

$$\frac{a^m}{a^{-s}} = a^{m-(-s)}$$

and

$$\frac{a^{-t}}{a^n} = a^{-t-n}$$

By use of similar arguments based on the definition of a negative exponent, the laws of positive exponents, and the properties of fractions, we can prove that laws (2.39), (5.1), and (5.2) hold for negative exponents. To illustrate the method of proof, we consider law (5.1), with $n = -t$, and prove that

$$\left(\frac{a}{b}\right)^{-t} = \frac{a^{-t}}{b^{-t}}$$

By the definition of a negative exponent,

$$\left(\frac{a}{b}\right)^{-t} = \frac{1}{\left(\dfrac{a}{b}\right)^t}$$

$$= \frac{1}{\dfrac{a^t}{b^t}} \qquad \blacktriangleleft \text{ by (5.1) since } t \text{ is positive}$$

$$= \frac{1(a^{-t})}{\left(\dfrac{a^t}{b^t}\right)(a^{-t})} \qquad \blacktriangleleft \text{ multiplying the numerator and denominator of the complex fraction by } a^{-t}$$

$$= \frac{a^{-t}}{\dfrac{a^{t-t}}{b^t}} \qquad \blacktriangleleft \text{ by (2.38)}$$

$$\left(\frac{a}{b}\right)^{-t} = \frac{a^{-t}}{\dfrac{1}{b^t}} \qquad \blacktriangleleft \text{ since } a^{t-t} = a^0 = 1$$

$$= \frac{a^{-t}}{b^{-t}} \qquad \blacktriangleleft \text{ since } 1/b^t = b^{-t}$$

Consequently, all restrictions on m and n in the statements of the laws of exponents can be discarded except that m and n be integers.

It is frequently desirable to convert a fraction whose numerator, denominator, or both involve negative exponents to an equal fraction in which all exponents are positive. We employ the fundamental principle of fractions to accomplish this purpose. For example, to convert $a^x b^{-y}/c^z d^{-w}$ to an equal fraction that involves no negative exponents, we first notice that $b^{-y} \cdot b^y = b^{-y+y} = b^0 = 1$. Hence, we multiply the given fraction by b^y/b^y and obtain

$$\frac{a^x b^{-y}}{c^z d^{-w}} \frac{b^y}{b^y} = \frac{a^x}{c^z d^{-w} b^y}$$

Similarly, if we multiply the right member by d^w/d^w, we get

$$\frac{a^x d^w}{c^z d^{-w} b^y d^w} = \frac{a^x d^w}{c^z b^y}$$

These two steps can be combined into the single operation of multiplying the given fraction by $b^y d^w/b^y d^w$ and obtaining

$$\frac{a^x b^{-y}}{c^z d^{-w}} = \frac{a^x b^{-y}}{c^z d^{-w}} \frac{b^y d^w}{b^y d^w} = \frac{a^x d^w}{c^z b^y}$$

The example suggests the following procedure for expressing a fraction in which the numerator or denominator or both are monomials that involve negative exponents as an equal fraction in which all exponents are positive.

Procedure for eliminating negative exponents

1 Form a product consisting of all factors of the numerator and denominator that have negative exponents.

2 Change the sign of each exponent in the product obtained in step 1.

3 Multiply the numerator and the denominator of the given fraction by the product obtained in step 2.

EXAMPLE 1

Convert $a^2 b^{-3} c^{-2}/x^{-1} y^3 z^{-3}$ into an equal fraction in which all exponents are positive.

Solution

The product of the factors of the numerator and denominator that have negative exponents is $b^{-3} c^{-2} x^{-1} z^{-3}$. Therefore, we multiply each member of the given fraction by $b^3 c^2 x z^3$ and get

$$\frac{a^2b^{-3}c^{-2}}{x^{-1}y^3z^{-3}}\frac{b^3c^2xz^3}{b^3c^2xz^3} = \frac{a^2b^{-3+3}c^{-2+2}xz^3}{x^{-1+1}y^3z^{-3+3}b^3c^2}$$ ◀ by the commutative law and (2.38)

$$= \frac{a^2b^0c^0xz^3}{x^0y^3z^0b^3c^2}$$

$$= \frac{a^2xz^3}{y^3b^3c^2}$$ ◀ since $b^0 = c^0 = x^0 = z^0 = 1$ by (2.48)

EXAMPLE 2   Express $2c^{-2}d^{-1}/3x^{-1}y^3$ as an equal fraction involving only positive exponents.

Solution
$$\frac{2c^{-2}d^{-1}}{3x^{-1}y^3} = \frac{2c^{-2}d^{-1}}{3x^{-1}y^3}\frac{c^2dx}{c^2dx}$$

$$= \frac{2c^{-2+2}d^{-1+1}x}{3x^{-1+1}y^3c^2d}$$ ◀ by the commutative law and (2.38)

$$= \frac{2c^0d^0x}{3x^0y^3c^2d}$$

$$= \frac{2x}{3y^3c^2d}$$ ◀ since $c^0 = d^0 = x^0 = 1$ by (2.48)

If the numerator, denominator, or both are polynomials, and if either or both the numerator and denominator involve negative exponents, we employ the principle explained above. For example, to convert $(x^{-1}+y^{-1})/(x^{-2}-y^{-2})$ to an equal fraction in which all exponents are positive, we notice that x appears with exponents -1 and -2 and y likewise appears with exponents -1 and -2. Therefore, if we multiply the fraction by x^2y^2/x^2y^2, we obtain a fraction in which the exponents of x and of y are positive. The details of the conversion process follow.

$$\frac{x^{-1}+y^{-1}}{x^{-2}-y^{-2}}\frac{x^2y^2}{x^2y^2} = \frac{x^{-1+2}y^2 + x^2y^{-1+2}}{x^{-2+2}y^2 - x^2y^{-2+2}} = \frac{xy^2 + x^2y}{y^2 - x^2} = \frac{xy(y+x)}{(y+x)(y-x)} = \frac{xy}{y-x}$$

The following examples further illustrate the method of employing the seven laws of exponents to convert an expression involving negative exponents to an equal expression with all exponents positive.

EXAMPLE 3
$$\frac{12a^{-2}b^3c^{-3}}{4a^3b^{-1}c^{-2}} = \frac{12a^{-2}b^3c^{-3}}{4a^3b^{-1}c^{-2}}\frac{a^2bc^3}{a^2bc^3}$$ ◀ multiplying by a^2bc^3/a^2bc^3

$$= \frac{12a^{-2+2}b^{3+1}c^{-3+3}}{4a^{3+2}b^{-1+1}c^{-2+3}}$$ ◀ by (2.38)

$$= \frac{12a^0b^4c^0}{4a^5b^0c}$$

$$= \frac{3b^4}{a^5c}$$ ◀ since $a^0 = b^0 = c^0 = 1$ and $\frac{12}{4} = 3$

EXAMPLE 4

$$\left(\frac{2x^{-3}y^2}{x^4z^3}\right)^{-3} = \frac{2^{-3}x^9y^{-6}}{x^{-12}z^{-9}}$$ ◄ by (5.1), (2.39), and (5.2)

$$= \frac{2^{-3}x^9y^{-6}}{x^{-12}z^{-9}}\frac{2^3y^6x^{12}z^9}{2^3y^6x^{12}z^9}$$ ◄ multiplying by $2^3y^6x^{12}z^9/2^3y^6x^{12}z^9$

$$= \frac{2^{-3+3}x^{9+12}y^{-6+6}z^9}{2^3x^{-12+12}y^6z^{-9+9}}$$ ◄ by the commutative law and (2.38)

$$= \frac{2^0x^{21}y^0z^9}{8x^0y^6z^0}$$

$$= \frac{x^{21}z^9}{8y^6}$$ ◄ by (2.48)

EXAMPLE 5

$$\left(\frac{x^{-1}-y^{-1}}{x^{-1}y^{-1}}\right)^{-2} = \left(\frac{x^{-1}-y^{-1}}{x^{-1}y^{-1}}\frac{xy}{xy}\right)^{-2}$$ ◄ multiplying the fraction inside the parentheses by xy/xy, since the negative exponent of x and y with the greatest absolute value is -1

$$= \left(\frac{x^{-1+1}y - xy^{-1+1}}{x^{-1+1}y^{-1+1}}\right)^{-2}$$ ◄ by the distributive and commutative laws and (2.38)

$$= (y-x)^{-2}$$ ◄ by (2.48)

$$= \frac{1}{(y-x)^2}$$ ◄ by (5.3)

Exercise 5.2   Elimination of Negative Exponents

Find the value of the expression in each of problems 1 to 16.

1 5^{-2}	**2** 3^{-4}	**3** 2^{-5}	**4** 7^{-3}
5 $2^{-3}2$	**6** $3^{-2}3^{-1}$	**7** $6^{-2}6^3$	**8** $7^{-4}7^2$
9 $5^{-3}/5^2$	**10** $3/3^{-2}$	**11** $4^{-3}/4^{-1}$	**12** $7^{-2}/7^{-3}$
13 $\left(\frac{2}{3}\right)^{-1}$	**14** $\left(\frac{3}{5}\right)^{-2}$	**15** $(3^{-1}2^2)^{-3}$	**16** $(3^2/5^{-2})^{-2}$

Write the expression in each of problems 17 to 24 without a denominator by making use of negative exponents if necessary.

17 $\dfrac{2a^2}{b^{-1}}$ **18** $\dfrac{3a^{-1}}{b^{-2}}$ **19** $\dfrac{5a^3}{b^2}$

20 $\dfrac{4a^{-3}}{b^4}$ **21** $\dfrac{p^3q^{-2}}{p^2q^{-3}r^2}$ **22** $\dfrac{c^{-2}d^{-1}}{c^{-5}d^2r^0}$

23 $\dfrac{2p^3q^2r}{3^{-1}p^{-1}q^4r^3}$ **24** $\dfrac{p^2q^{-2}r^3}{3p^3q^{-1}r^{-2}}$

In problems 25 to 48, make all possible combinations by use of the laws of exponents and express each result without zero or negative exponents.

25 $\dfrac{2^{-2}x^0 y^{-3} z}{3^{-1} x^{-2} y^2 z^{-3}}$

26 $\dfrac{3^{-3} x^{-4} y^{-2} z}{2^{-4} xy^{-3} z^{-2}}$

27 $\dfrac{5^{-1} x^3 y^{-3} z^0}{2^{-2} x^2 y^{-2} z^{-1}}$

28 $\dfrac{4^{-1} x^{-4} y^2 z^{-3}}{2^{-3} x^{-2} y^{-1} z^{-2}}$

29 $(a^{-3} b^2)^{-2}$

30 $(ab^{-3})^{-2}$

31 $(a^{-2} b^{-1})^4$

32 $(a^3 b^{-2})^2$

33 $\left(\dfrac{3^{-2} a^{-1} b^{-4}}{9^{-1} ab^3}\right)^{-2}$

34 $\left(\dfrac{2^{-3} a^2 b^{-2}}{4^{-2} a^{-1} b}\right)^{-3}$

35 $\left(\dfrac{a^{-1} b^2 c^{-3}}{a^4 b^{-5} c^6}\right)^2$

36 $\left(\dfrac{a^{-2} b^{-3} c^2}{a^{-3} b^{-2} c^{-1}}\right)^{-3}$

37 $a^2 + \dfrac{2}{a^{-2}}$

38 $3a^{-1} + \dfrac{2}{a}$

39 $x^{-3} - \dfrac{2}{x^3}$

40 $3a^{-1} + a^{-2}$

41 $\dfrac{x^{-3} - y^{-3}}{x^{-2} y^{-2}}$

42 $\dfrac{x^{-2} x^{-3} y^{-2}}{x^{-3} y^{-2} - y^{-2}}$

43 $\dfrac{x^{-2} y^{-1} - x^{-1} y^{-2}}{x^{-2} - y^{-2}}$

44 $\dfrac{x^{-3} y^{-2} + x^{-2} y^{-3}}{x^{-3} + y^{-3}}$

45 $(x-1)^{-2} - 2(x-1)^{-3}(x+1)$

46 $-3(x-1)^2(x-2)^{-4} + (x-1)(x-2)^{-3}$

47 $2(4x-1)^{-1}(2x+1)^{-2} + 4(4x-1)^{-2}(2x+1)^{-1}$

48 $3(2x-1)^{-2}(3x+2)^{-2} + 4(2x-1)^{-3}(3x+2)^{-1}$

5.3 RATIONAL EXPONENTS

In this section we further extend the definition of a^n to include the case in which n is the quotient of two integers, in such a way that the laws stated in Sec. 5.1 are valid for rational exponents.

If law (5.2) is valid for $n = 1/k$ and $m = k$, then $(a^{1/k})^k = a^{k/k} = a$. Hence we define $a^{1/k}$ as a number whose kth power is a. Without further restrictions, however, $a^{1/k}$ may have more than one value. For example, $16^{1/2}$ is both 4 and -4, since $4^2 = (-4)^2 = 16$. We shall remove this ambiguity by means of the following definitions.

kth root of a number, principal root

The number b is a *kth root* of a if $b^k = a$.

If a positive kth root of a constant a exists, then it is the *principal kth root of a*. If no positive kth root of a exists but there is a negative kth root, then that is the principal kth root of a.

Radical of order k, radicand, index

We use the radical $\sqrt[k]{a}$ to designate the principal kth root of a. This symbol is called a *radical of order k*. The number a is the *radicand*, and k is the *index* of the radical. If the index of the radical is not written, it is understood to be 2. For example,

$$\sqrt{16} = 4 \qquad \text{and} \qquad \sqrt[3]{-27} = -3$$

We now make use of the above definitions to define $a^{1/k}$ in terms of a root. If a is a constant then

$$a^{1/k} = \sqrt[k]{a} \qquad\qquad (5.4)$$

If a is a positive constant, then there is a positive kth root of a, and therefore $a^{1/k}$ is positive. For example, $4^{1/2} = 2$, $64^{1/3} = 4$, and $32^{1/5} = 2$. If a is a negative constant and k is an odd integer, it can be proved by the methods of Chap. 16 that only one real kth root of a exists and that this root is negative. For example, $(-8)^{1/3} = -2$, and there is no other real number whose third power is -8. Furthermore, if a is negative and k is even, no real kth root of a exists, since any even power of a real number is positive. For example, there is no real number whose square is -4, and there is no real number whose fourth power is -256, since both the square and the fourth power of a real number are positive. Since we shall deal only with real numbers in this chapter, we shall exclude the case in which the radicand is negative and the index of the radical is even. To be able to handle that case, a further extension of the number system, which will be discussed in Chaps. 11 and 16, is required.

If in (5.4) we replace a by b^j, we have $(b^j)^{1/k} = \sqrt[k]{b^j}$. Therefore, $(b^j)^{1/k}$ is a kth root of b^j. Furthermore, if we raise $(b^{1/k})^j$ to the kth power, we get

$$
\begin{aligned}
[(b^{1/k})^j]^k &= (b^{1/k})^{jk} &&\blacktriangleleft \text{ by (5.2)}\\
&= (b^{1/k})^{kj} &&\blacktriangleleft \text{ by the commutative law}\\
&= b^j &&\blacktriangleleft \text{ by the definition of } b^{1/k}
\end{aligned}
$$

Hence, $(b^j)^{1/k}$ and $(b^{1/k})^j$ are both kth roots of b^j, and it can be proved† that, except for the excluded case, the two roots have the same sign and are therefore equal. Consequently, we have

$$b^{j/k} = (b^j)^{1/k} = (b^{1/k})^j = \sqrt[k]{b^j} = (\sqrt[k]{b})^j \qquad \textbf{provided } \sqrt[k]{b} \textbf{ is real} \quad (5.5)$$

The proofs that the laws of exponents stated in Sec. 5.1 hold for rational exponents as defined in (5.4) can be established by use of (5.5) and arguments similar to the above.‡

The processes involved in working with fractional exponents are illustrated in the following examples.

---

† To prove this statement, we first assume that b is positive. Then b^j and $b^{1/k}$ are each positive. Therefore, $(b^j)^{1/k}$ and $(b^{1/k})^j$ are positive and hence are equal. If b is negative, k is odd, and j is odd, b^j and $b^{1/k}$ are negative. Therefore, $(b^j)^{1/k}$ and $(b^{1/k})^j$ are negative and therefore are equal. The argument for the case b negative, k odd, and j even can be made in a similar way.

‡ As an exercise the reader may prove that $a^{1/n}a^{1/m}$ and $a^{1/n+1/m}$ are both mnth roots of $a^m a^n$ and that, except for the excluded case, they have the same sign, thus proving that $a^{1/m}a^{1/n} = a^{1/m+1/n}$.

EXAMPLE 1 Evaluate $4^{1/2}$, $8^{2/3}$, $(-32)^{3/5}$.

Solution

$$4^{1/2} = \sqrt{4} \qquad \blacktriangleleft \text{ by (5.4)}$$
$$= 2$$
$$8^{2/3} = (\sqrt[3]{8})^2 \qquad \blacktriangleleft \text{ by (5.5)}$$
$$= 2^2 = 4$$
$$(-32)^{3/5} = (\sqrt[5]{-32})^3 \qquad \blacktriangleleft \text{ by (5.5)}$$
$$= (-2)^3 = -8$$

EXAMPLE 2 Express $3a^{1/2}b^{5/2}$ and $5a^{3/4}/b^{5/4}$ in radical form.

Solution

$$3a^{1/2}b^{5/2} = 3(ab^5)^{1/2} \qquad \blacktriangleleft \text{ by (2.39)}$$
$$= 3\sqrt{ab^5} \qquad \blacktriangleleft \text{ by (5.4)}$$
$$\frac{5a^{3/4}}{b^{5/4}} = 5\left(\frac{a^3}{b^5}\right)^{1/4} \qquad \blacktriangleleft \text{ by (5.1) with } n = \tfrac{1}{4}$$

$$= 5\sqrt[4]{\frac{a^3}{b^5}} \qquad \blacktriangleleft \text{ by (5.4)}$$

EXAMPLE 3 Find the product of $3a^{1/2}$ and $2a^{2/3}$.

Solution

$$(3a^{1/2})(2a^{2/3}) = 6a^{1/2+2/3}$$
$$= 6a^{(3+4)/6}$$
$$= 6a^{7/6}$$

EXAMPLE 4 Find the quotient of $8x^{1/2}y^{5/6}$ and $5x^{1/4}y^{1/3}$.

Solution

$$8x^{1/2}y^{5/6} \div 5x^{1/4}y^{1/3} = \tfrac{8}{5}x^{1/2-1/4}y^{5/6-1/3} = \tfrac{8}{5}x^{1/4}y^{3/6}$$
$$= \tfrac{8}{5}x^{1/4}y^{1/2}$$

EXAMPLE 5 Make all possible combinations using the laws of exponents and express the result without zero or negative exponents in $[(4x^4y^{3/4}z^2)/(9x^2y^{1/4}z)]^{1/2}$.

Solution

$$\left(\frac{4x^4y^{3/4}z^2}{9x^2y^{1/4}z}\right)^{1/2} = \left(\frac{4}{9}\frac{x^4}{x^2}\frac{y^{3/4}}{y^{1/4}}\frac{z^2}{z}\right)^{1/2}$$
$$= \left(\tfrac{4}{9}x^2y^{2/4}z\right)^{1/2}$$
$$= \tfrac{2}{3}xy^{1/4}z^{1/2}$$

EXAMPLE 6   Make all possible combinations and express the result without zero or negative exponents in $[(4x^2y^{3/4}z^{1/6})/(32x^{-1}y^0z^{-5/6})]^{-1/3}$.

Solution

$$\left(\frac{4x^2y^{3/4}z^{1/6}}{32x^{-1}y^0z^{-5/6}}\right)^{-1/3} = \left(\frac{4}{32}\frac{x^2}{x^{-1}}\frac{y^{3/4}}{y^0}\frac{z^{1/6}}{z^{-5/6}}\right)^{-1/3}$$

$$= (\tfrac{1}{8}x^3y^{3/4}z)^{-1/3}$$

$$= \frac{2}{xy^{1/4}z^{1/3}}$$

Exercise 5.3   Conversion of Radical and Exponential Expressions

Convert each number or expression in problems 1 to 24 to a form without fractional exponents or radical expressions.

1	$16^{1/4}$	**2**	$27^{1/3}$	**3**	$32^{1/5}$	**4**	$81^{1/2}$
5	$0.008^{1/3}$	**6**	$0.0001^{1/4}$	**7**	$0.0625^{1/4}$	**8**	$0.07776^{1/5}$
9	$8^{2/3}$	**10**	$16^{3/4}$	**11**	$243^{3/5}$	**12**	$32^{2/5}$
13	$64^{-2/3}$	**14**	$(-8)^{-2/3}$	**15**	$32^{-3/5}$	**16**	$625^{-3/4}$
17	$\sqrt[3]{64^2}$	**18**	$\sqrt{64^3}$	**19**	$\sqrt[5]{32^4}$	**20**	$\sqrt{0.49^3}$
21	$\sqrt{25a^2b^4}$	**22**	$\sqrt[3]{64a^3b^9}$	**23**	$\sqrt[5]{243a^{10}b^{20}}$	**24**	$\sqrt[4]{256a^{12}b^4}$

Write the expression in each of problems 25 to 32 without negative exponents and express the result in radical form.

25	$x^{1/4}y^{3/4}$	**26**	$x^{1/3}y^{2/3}$	**27**	$x^{2/5}/y^{3/5}$	**28**	$x^{2/7}/y^{4/7}$
29	$x^{3/4}/y^{-1/4}$	**30**	$x^{-2/3}/y^{1/3}$	**31**	$x^{-1/6}y^{-5/6}$	**32**	$x^{-3/7}y^{-5/7}$

Make all possible combinations in the following expressions by use of the laws of exponents and express each result without zero or negative exponents.

33 $(2x^{1/2})(5x^{1/3})$ **34** $(3x^{1/4})(2x^{2/3})$

35 $(5x^{1/5})(3x^{1/3})$ **36** $(7x^{1/5})(3x^{5/6})$

37 $\dfrac{6x^{1/2}y^{2/3}}{2x^{3/4}y^{3/5}}$ **38** $\dfrac{27x^{5/9}y^{-1/2}}{9x^{4/9}y^{-3/2}}$

39 $\dfrac{16^{3/4}x^{1/6}y^{-1/2}}{2x^{-1/6}y}$ **40** $\dfrac{32^{4/5}x^{3/2}y^{-1}}{8^{4/3}x^{2/3}y^0}$

41 $\left(\dfrac{81^{-2}p^{12/5}q^{8/3}}{16^{-3}r^{-4/5}}\right)^{1/4}$ **42** $\left(\dfrac{9^{3/2}p^{4/3}q^{1/4}}{5^{3/2}p^{-5/3}s^{-7/4}}\right)^{2/3}$

43 $\left(\dfrac{8p^9}{27q^{-3/8}}\right)^{-1/3}$ **44** $\left(\dfrac{64p^{-1}q^{2/3}}{36p^3q^0}\right)^{-1/2}$

45 $\left(\dfrac{x^{a+3b}}{x^{-a+b}}\right)^{2/(a+b)}$ **46** $\left(\dfrac{x^{a+3b}}{x^a}\right)^a \left(\dfrac{x^{a-2b}}{x^{-2b}}\right)^b$

47 $\left[(x^{1/(a+2b)})^{(a^2-4b^2)/b}\right]^{b/(a-2b)}$

48 $\left[(x^{(a+b)/(a-b)})^{(a-b)/b}\right]^{b/(a+b)}$

49 $3(2x-1)^{2/3}(x+1)^{-1/2} + 8(2x-1)^{-1/3}(x+1)^{1/2}$

50 $3(3x+2)^{2/3}(x+1)^{-1/2} + 4(3x+2)^{-1/3}(x+1)^{1/2}$

51 $3(2x+3)^{1/4}(3x-1)^{-1/2} + (3x-1)^{1/2}(2x+3)^{-3/4}$

52 $3(3x+5)^{2/3}(4x-3)^{-1/4} + 2(3x+5)^{-1/3}(4x-3)^{3/4}$

5.4 LAWS OF RADICALS

In this section we develop three useful laws of radicals, and in the remainder of the chapter we shall explain their use.

Since laws (2.39) and (5.1) are valid for $n = 1/k$, we have

$$(ab)^{1/k} = a^{1/k}b^{1/k} \qquad \text{and} \qquad \left(\frac{a}{b}\right)^{1/k} = \frac{a^{1/k}}{b^{1/k}}$$

If we convert these two equalities to radical form, we get

$$\sqrt[k]{ab} = \sqrt[k]{a}\,\sqrt[k]{b} \tag{5.6}$$

and

$$\sqrt[k]{\frac{a}{b}} = \frac{\sqrt[k]{a}}{\sqrt[k]{b}} \tag{5.7}$$

By law (5.2) with $n = 1/j$ and $m = 1/k$, we have $(a^{1/j})^{1/k} = a^{1/kj}$. Therefore,

$$\sqrt[k]{\sqrt[j]{a}} = \sqrt[kj]{a} \tag{5.8}$$

5.5 APPLICATIONS OF THE LAW $\sqrt[k]{ab} = \sqrt[k]{a}\,\sqrt[k]{b}$

We use law (5.6) to remove rational factors from the radicand, to insert factors into the radicand, and to obtain the product of two radicals of the same order.

REMOVAL OF RATIONAL FACTORS FROM THE RADICAND If the radicand of a radical of order k has factors that are kth powers, we can remove the kth roots of these powers from the radicand. For example,

$$\sqrt{125} = \sqrt{25(5)} = \sqrt{5^2}\,\sqrt{5} = 5\sqrt{5}$$
$$\sqrt[3]{256} = \sqrt[3]{64(4)} = \sqrt[3]{4^3 4} = 4\sqrt[3]{4}$$
$$\sqrt[4]{128a^6b^9} = \sqrt[4]{16a^4b^8(8a^2b)} = \sqrt[4]{(2ab^2)^4 8a^2b} = 2ab^2\sqrt[4]{8a^2b}$$

MULTIPLICATION OF RADICALS OF THE SAME ORDER To obtain the product of two radicals of the same order, we use law (5.6) read from right to left. For example, $\sqrt[3]{c}\,\sqrt[3]{d} = \sqrt[3]{cd}$. Usually, products obtained in this way should be simplified by removing all possible rational factors from the radical. The procedure is illustrated in the two following examples.

$$\sqrt{8x^3y}\,\sqrt{6x^2y^5} = \sqrt{48x^5y^6} = \sqrt{16x^4y^6(3x)} = 4x^2y^3\sqrt{3x}$$

$$\sqrt[3]{9a^5b^2}\,\sqrt[3]{81a^2b^7} = \sqrt[3]{729a^7b^9} = \sqrt[3]{3^6a^6b^9(a)} = 3^2a^2b^3\sqrt[3]{a} = 9a^2b^3\sqrt[3]{a}$$

INSERTION OF RATIONAL FACTORS INTO THE RADICAND   Frequently, it is desirable to write a radical expression that has a rational coefficient as a single radical in which the rational coefficient is absorbed in the radicand; this is done by use of the inverse of the process described in the preceding subsection. For example, in $2a\sqrt{4ab}$, we may insert $2a$ into the radicand in this way:

$$2a\sqrt{4ab} = \sqrt{4a^2}\,\sqrt{4ab} = \sqrt{16a^3b}$$

The following example further illustrates the procedure:

$$4x\sqrt[3]{2y} = \sqrt[3]{(4x)^3}\,\sqrt[3]{2y} = \sqrt[3]{64x^3(2y)} = \sqrt[3]{128x^3y}$$

5.6 APPLICATIONS OF THE LAW $\sqrt[k]{a/b} = \sqrt[k]{a}/\sqrt[k]{b}$

Law (5.7) is used for obtaining the quotient of two radicals of the same order and for rationalizing monomial denominators.

THE QUOTIENT OF TWO RADICALS OF THE SAME ORDER   To get the quotient of two radicals of the same order, we use law (5.7) read from right to left. For example, $\sqrt[4]{a}/\sqrt[4]{b} = \sqrt[4]{a/b}$. The following two examples illustrate the method to be used.

$$\frac{\sqrt{128a^3b^5}}{\sqrt{2ab^2}} = \sqrt{\frac{128a^3b^5}{2ab^2}} = \sqrt{64a^2b^3} = \sqrt{(64a^2b^2)b} = 8ab\sqrt{b}$$

$$\frac{\sqrt[3]{625x^{10}y^7z^{11}}}{\sqrt[3]{5x^2yz^4}} = \sqrt[3]{\frac{625x^{10}y^7z^{11}}{5x^2yz^4}} = \sqrt[3]{125x^8y^6z^7} = \sqrt[3]{(5x^2y^2z^2)^3x^2z}$$

$$= 5x^2y^2z^2\sqrt[3]{x^2z}$$

RATIONALIZATION OF MONOMIAL DENOMINATORS   It is frequently desirable to convert a radical with a fractional radicand to a form in which no radical appears in the denominator. This process is called *rationalizing the denominator*. If the denominator of the radicand is a monomial, we use law (5.7) for the purpose. For example, to rationalize the denominator of $\sqrt{8a/3bc^3}$, we multiply the numerator and denominator of the radicand by the expression of lowest power that will convert the denominator into a perfect square. In this case, the expression is $3bc$; hence, we have

Rationalizing the denominator

$$\sqrt{\frac{8a}{3bc^3}} = \sqrt{\frac{8a(3bc)}{3bc^3(3bc)}} = \sqrt{\frac{2^26abc}{(3bc^2)^2}} = \frac{2\sqrt{6abc}}{3bc^2}$$

As a second example, we give

$$\sqrt[3]{\frac{4xy^2}{5x^5y^7}} = \sqrt[3]{\frac{4xy^2(25xy^2)}{5x^5y^7(25xy^2)}} = \sqrt[3]{\frac{100x^2yy^3}{(5x^2y^3)^3}} = \frac{y\sqrt[3]{100x^2y}}{5x^2y^3} = \frac{\sqrt[3]{100x^2y}}{5x^2y^2}$$

5.7 CHANGING THE ORDER OF A RADICAL

If it is possible to convert a radical expression to one of lower order, it is usually advisable to do so. Such a conversion is possible if the radicand can be expressed as a power in which the exponent is a factor of the index of the radical. For example, in $\sqrt[6]{8a^6b^{12}}$, the index is $6 = 3 \cdot 2$, and the radicand is $8a^6b^{12} = (2a^2b^4)^3$. Therefore,

$$\begin{aligned}
\sqrt[6]{8a^6b^{12}} &= \sqrt[6]{(2a^2b^4)^3} \\
&= \sqrt{\sqrt[3]{(2a^2b^4)^3}} \qquad \blacktriangleleft \text{ by (5.8)} \\
&= \sqrt{2a^2b^4} \\
&= ab^2\sqrt{2}
\end{aligned}$$

This result can be obtained without (5.8) by use of fractional exponents. Thus

$$\begin{aligned}
\sqrt[6]{8a^6b^{12}} &= \sqrt[6]{(2a^2b^4)^3} \\
&= (2a^2b^4)^{3/6} \\
&= (2a^2b^4)^{1/2} \\
&= \sqrt{2a^2b^4} \\
&= ab^2\sqrt{2}
\end{aligned}$$

The same procedure can be employed to convert a radical expression in which the new index is a multiple of the given one.

EXAMPLE 1

Convert $\sqrt[3]{4xy^2}$ into an equal radical expression of order 12.

Solution

$$\begin{aligned}
\sqrt[3]{4xy^2} &= 2^{2/3}x^{1/3}y^{2/3} = 2^{8/12}x^{4/12}y^{8/12} \\
&= \sqrt[12]{2^8x^4y^8}
\end{aligned}$$

EXAMPLE 2

Convert $\sqrt[3]{2a^2b}$, $\sqrt{3ab}$, and $\sqrt[4]{4a^3b^2}$ to radical expressions of the same order.

Solution

The fractional exponents that correspond to the indices of the given radicals are $\frac{1}{3}$, $\frac{1}{2}$, and $\frac{1}{4}$; hence, any common multiple of 3, 2, and 4 can be used as the order of the converted radicals. We shall use 12 since that is the least common multiple. Therefore, we get

$$\begin{aligned}
\sqrt[3]{2a^2b} &= (2a^2b)^{1/3} = (2a^2b)^{4/12} = \sqrt[12]{2^4a^8b^4} \\
\sqrt{3ab} &= (3ab)^{1/2} = (3ab)^{6/12} = \sqrt[12]{3^6a^6b^6} \\
\sqrt[4]{4a^3b^2} &= (2^2a^3b^2)^{1/4} = (2^2a^3b^2)^{3/12} = \sqrt[12]{2^6a^9b^6}
\end{aligned}$$

In Exercise 5.4 and later work, we shall use *simplifying a radical* to mean that all possible combinations are to be made by use of the laws of exponents or radicals, all rational factors are to be removed from the radical or expression with a fractional exponent, and the result is to be expressed without zero or negative exponents or radicals in the denominator.

Exercise 5.4   Simplifying Radical Expressions

Simplify the radical expression in each of problems 1 to 16.

1 $\sqrt{512}$	**2** $\sqrt{162}$	**3** $\sqrt{150}$	**4** $\sqrt{294}$
5 $\sqrt[3]{48}$	**6** $\sqrt[3]{375}$	**7** $\sqrt[4]{224}$	**8** $\sqrt[5]{729}$
9 $\sqrt{12a^3b^5}$	**10** $\sqrt{63a^7b^3}$	**11** $\sqrt[3]{648a^5b^9}$	**12** $\sqrt[3]{686a^7b^5}$
13 $\sqrt[4]{243a^7b^9}$	**14** $\sqrt[4]{2500a^6b^7}$	**15** $\sqrt[5]{192a^6b^{12}}$	**16** $\sqrt[5]{486a^8b^5}$

Insert the number that appears before the radical into the radicand.

17 $3x\sqrt{7xy}$ **18** $2y\sqrt[3]{7x^2y}$ **19** $2a^2b\sqrt[4]{3ab^3}$ **20** $ab^2\sqrt[5]{a^2b}$

Perform the indicated operations in each of problems 21 to 36 and then simplify.

21 $\sqrt{45}\,\sqrt{20}$ **22** $\sqrt[3]{16}\,\sqrt[3]{144}$ **23** $\sqrt[3]{80}/\sqrt[3]{108}$

24 $\sqrt{180}\,\sqrt{100}$ **25** $\sqrt{3xy^3}\,\sqrt{147x^3y^5}$ **26** $\sqrt{2x^5y}\,\sqrt{200xy^3}$

27 $\sqrt[3]{625x^5y}\,\sqrt[3]{40xy^2}$ **28** $\sqrt[4]{72x^6y^7}\,\sqrt[4]{54x^2y^2}$ **29** $\dfrac{\sqrt{54x^7y^3}}{\sqrt{98x^3y}}$

30 $\dfrac{\sqrt{108x^9y^5}}{\sqrt{48x^3y}}$ **31** $\dfrac{\sqrt{162r^7y^0}}{\sqrt{98r^2y^{-3}}}$ **32** $\dfrac{\sqrt{24r^5y^3}}{\sqrt{6r^{-3}y^{-4}}}$

33 $\dfrac{\sqrt{7xy}\,\sqrt{6x^3y}}{\sqrt{21x^7y^3}}$ **34** $\dfrac{\sqrt{5x^3y^6}\,\sqrt{3xy^2}}{\sqrt{30x^5y^0}}$ **35** $\dfrac{\sqrt{10y^{-2}z}}{\sqrt{5y^0z^3}\,\sqrt{18y^3z^{-2}}}$

36 $\dfrac{\sqrt{14x^{-1}y}}{\sqrt{6x^4y^7}\,\sqrt{21x^0y^{-3}}}$

Rationalize the denominator in each of problems 37 to 44 and simplify each result.

37 $\sqrt{\dfrac{7x^3}{2y}}$ **38** $\sqrt{\dfrac{3x}{5y^5}}$ **39** $\sqrt{\dfrac{8x^5}{98y^3}}$ **40** $\sqrt{\dfrac{5x^3}{6y^5}}$

41 $\sqrt[3]{\dfrac{2x^2}{3y}}$ **42** $\sqrt[3]{\dfrac{6x}{15y^7}}$ **43** $\sqrt[4]{\dfrac{3x^3}{2y^2}}$ **44** $\sqrt[4]{\dfrac{7x^4}{8y^7}}$

Express each of problems 45 to 48 as a single radical.

45 $\sqrt{\sqrt[3]{2a}}$ **46** $\sqrt[3]{\sqrt[4]{3b}}$ **47** $\sqrt{\sqrt{11c}}$ **48** $\sqrt[3]{\sqrt[5]{ab}}$

Reduce the order of the radical in each of problems 49 to 56 and simplify.

49 $\sqrt[4]{16x^2y^6}$ **50** $\sqrt[6]{8x^6y^3}$ **51** $\sqrt[6]{64x^4y^2}$ **52** $\sqrt[10]{25x^6y^8}$

53 $\sqrt[8]{16x^4y^{12}}$ **54** $\sqrt[9]{27x^3y^6}$ **55** $\sqrt[12]{64x^9y^6}$ **56** $\sqrt[15]{32x^{10}y^5}$

Change the radicals in each of problems 57 to 60 to radicals of the same order.

57 $\sqrt{a},\ \sqrt[3]{a},\ \sqrt[6]{a}$ **58** $\sqrt[3]{a},\ \sqrt[6]{a},\ \sqrt[9]{a}$

59 $\sqrt[3]{2ab},\ \sqrt[4]{3ab^2},\ \sqrt[6]{5a^2b}$ **60** $\sqrt{ab},\ \sqrt[3]{a^2b^2},\ \sqrt[5]{a^3b^4}$

Exercise 5.5   Review

Perform the indicated operations and simplify.

1 $3^2 3^3$

2 $\dfrac{5^5}{5^3}$

3 $(2^3)^3$

4 $(\frac{3}{5})^4$

5 $(2^3 3^2)^2$

6 $\left(\dfrac{a^2 b^0 c}{d^3}\right)^2$

7 $(2x^2y^3)(3x^3y^5)$

8 $(5x^3yz^0)(2x^2y^3z)$

9 $\dfrac{12x^5y^7}{3x^2y}$

10 $(2a^3b^2c)^3(3a^2bc^0)^2$

11 $\left(\dfrac{2a^3b^3}{c^4}\right)^4\left(\dfrac{bc^8}{a^5}\right)^2$

12 $\left(\dfrac{36a^5b^3}{7c^2}\right)^2 \div \left(\dfrac{6a^4b^2}{14c}\right)^3$

13 $\dfrac{x^{2a-3}y^{b+5}}{x^{3+2a}y^{b-1}}$

14 $\left(\dfrac{a^{b+3}}{a^{b-1}}\right)^c$

Find the value of the expression in each of problems 15 to 18.

15 3^{-3}

16 $2^{-5}2^2$

17 $\dfrac{8^{-3}}{8^{-4}}$

18 $(2^{-1}3^2)^{-2}$

19 Write $\dfrac{3a^2}{b^{-2}}$ without a denominator.

20 Write $\dfrac{4a^{-3}}{b^2}$ without a denominator.

In problems 21 to 26, make all possible combinations and write the result without zero or negative exponents.

21 $\dfrac{3^{-2}x^0y^{-1}z^{-2}}{9^{-1}x^{-2}y^2z^{-3}}$ **22** $(a^{-1}b^2)^{-2}$ **23** $\left(\dfrac{2^{-4}a^{-2}b^{-1}}{4^{-3}ab^{-3}}\right)^{-2}$

24 $2a^{-1} + 3a$ **25** $\dfrac{a^{-3} + b^{-3}}{a^{-2} - b^{-2}}$

26 $-2(x + 1)(2x - 1)^{-2} + (2x - 1)^{-1}$

Convert each number or expression in problems 27 to 32 to a form without fractional exponents or radical expressions.

27 $27^{1/3}$ **28** $0.0256^{1/4}$ **29** $216^{2/3}$

30 $16^{-3/4}$ **31** $\sqrt[3]{125^2}$ **32** $\sqrt[4]{81a^0b^4c^8}$

33 Write $a^{1/4}b^{3/4}$ in radical form.

34 Write $a^{-3/5}b^{2/5}$ in radical form.

Simplify the expression in each of problems 35 to 40.

35 $3x^{1/3}2x^{1/4}$

36 $\dfrac{8x^{2/3}y^{1/2}}{4x^{1/6}y^{1/3}}$

37 $\left[\left(x^{(a-b)/(a+b)}\right)^{(a+b)/b}\right]^{2b/(a-b)}$

38 $\sqrt[3]{54}$

39 $\sqrt[4]{80a^4b^5}$

40 $\sqrt[5]{32a^5b^{15}c^4}$

41 Insert $2a^2$ under the radical sign in $\sqrt[3]{3a^{-5}}$.

Perform the indicated operations in problems 42 to 44 and simplify.

42 $\sqrt[3]{150}\ \sqrt[3]{60}$

43 $\sqrt[4]{56x^3y^5}\ \sqrt[4]{98xy^2}$

44 $\dfrac{\sqrt[3]{1,250x^5y^2}}{\sqrt[3]{432x^2y}}$

45 Rationalize the denominator in $\sqrt[3]{\dfrac{8x^4}{y}}$.

46 Express $\sqrt[4]{\sqrt{3x}}$ as a single radical.

47 Reduce the order of $\sqrt[6]{27x^6y^9}$.

48 Change $\sqrt{a}$, $\sqrt[5]{a}$, and $\sqrt[10]{a}$ to radicals of the same order.

Relations,
Functions,
Graphs

We discussed the concept of one-to-one correspondence which relates each element of one set of numbers to exactly one element of another set. We shall discuss one-to-n correspondence which leads to the concept of *relation*. Finally, we shall consider the special case of a relation that is called a *function* since this is a very important concept in mathematics.

6.1 ORDERED PAIRS OF NUMBERS

If a man earns $50/day and works 5 days, then the following tabulation shows the correspondence between the number of days worked and the number of dollars earned:

Days worked	1	2	3	4	5
Dollars earned	50	100	150	200	250

This correspondence can also be indicated by the following sets of pairs:

$$\{(1, 50), (2, 100), (3, 150), (4, 200), (5, 250)\} \tag{1}$$

Ordered pair

Each pair of numbers in (1) is an *ordered pair* and illustrates the following definition:

A pair of numbers (x, y) is an *ordered pair* if the interpretation of each number depends on its position in the pair.

An ordered pair can also be defined in this way:

If S and T are two sets of numbers, then (x, y) is an ordered pair if $x \in S$ and $y \in T$. Furthermore, (y, x) is also an ordered pair and is usually different from (x, y). Hence, the order in which the elements appear makes a difference.

Note that the first number in each pair in (1) belongs to $\{1, 2, 3, 4, 5\}$ and the second belongs to $\{50, 100, 150, 200, 250\}$.

The two numbers in an ordered pair are called the *components of the pair.*

Equality of ordered pairs

Two ordered pairs (a, b) and (c, d) are equal if and only if $a = c$ and $b = d$.

An illustration of a set of ordered pairs is the cartesian product of two sets defined below.

Cartesian products

If T and S are two sets of numbers, the cartesian product $T \times S$ is the set of all ordered pairs (x, y) such that $x \in T$ and $y \in S$.

For example, if $T = \{1, 2\}$ and $S = \{3, 4, 6\}$, then $T \times S = \{1, 2\} \times \{3, 4, 6\} = \{(1, 3), (1, 4), (1, 6), (2, 3), (2, 4), (2, 6)\}$. Similarly, $S \times T = \{3, 4, 6\} \times \{1, 2\} = \{(3, 1), (3, 2), (4, 1), (4, 2), (6, 1), (6, 2)\}$. By comparing $T \times S$ and $S \times T$, we see that each pair in the former product is a pair in the latter with the components interchanged. Hence the operation of obtaining the

125

cartesian product of two sets is not commutative, that is, in general, $T \times S \neq S \times T$.

If we let S stand for the set of all real numbers, then $S \times S$ is the set of all possible ordered pairs of real numbers. In the remainder of this chapter, we shall discuss subsets of $S \times S$ called *relations* and special types of relations called *functions*.

6.2 RELATIONS

We define a relation as follows:

Relation

If D is a set of numbers and if there exists a rule that associates each element x of D with one or more numbers y and if R is the set of all numbers y, then the set of ordered pairs (x, y) such that $x \in D$ and $y \in R$ is called a *relation* with *domain D* and *range R*. Hence, $D \times R$ is a relation.

The rule that establishes the correspondence between the elements of D and R may be an equation in two variables, a table of statistics composed of two columns, tables such as I to IV in the Appendix of this text, or any agreement that clearly pairs each number in one set with n numbers in another.

EXAMPLE 1   Write the relation established by the equation $x - y^2 = 0$, where x belongs to $D = \{0, 1, 4, 9, 16\}$.

Solution   We first solve the given equation for y^2 and get $y^2 = x$. Then $y = \pm \sqrt{x}$ since $(\pm \sqrt{x})^2 = x$. Now we assign the number 0 to x, getting $y = 0$, and we have the ordered pair $(0, 0)$. Similarly, if $x = 1$, $y = \pm 1$, and we get the ordered pairs $(1, 1)$ and $(1, -1)$. We continue this procedure by assigning each of the other numbers in D to x, calculating each corresponding value of y, and then arranging the resulting ordered pairs so that the second components appear in order of magnitude. Thus we obtain the relation

$$\{(16, -4), (9, -3), (4, -2), (1, -1), (0, 0), (1, 1), (4, 2), (9, 3), (16, 4)\}$$

This relation can be expressed in a more compact form as

$$\{(x, y) | x - y^2 = 0 \text{ and } x \in \{0, 1, 4, 9, 16\}\}$$

Note that the range of the relation is $\{-4, -3, -2, -1, 0, 1, 2, 3, 4\}$.

EXAMPLE 2   By use of Table III in the Appendix, write the relation

$$\{(x, \sqrt[3]{x}) | x \in \{-3, -2, -1, 0, 1, 2, 3\}\}$$

where the second component in each pair is correct to four digits.

Figure 6.1

Solution

By referring to Table III we see that $\sqrt[3]{3}$ correct to four digits is 1.442. Hence $\sqrt[3]{-3}$ is approximately equal to -1.442, and we therefore have the ordered pairs $(-3, -1.442)$ and $(3, 1.442)$. Using a similar procedure for the other numbers in D, we obtain

$$\{(x, \sqrt[3]{x}) \mid x \in \{-3, -2, -1, 0, 1, 2, 3\}\} = \{(-3, -1.442),$$
$$(-2, -1.260), (-1, -1), (0, 0), (1, 1), (2, 1.260), (3, 1.442)\}$$

EXAMPLE 3

If $D = \{1, 2, 3, 4\}$ and $R = \{5, 6, 7, 8, 9\}$, write the relation obtained by matching the nth number in D with the nth number in R and also with the $(n + 1)$st number in R.

Solution

The components of each ordered pair in the required relation are the numbers at the ends of the arrows in the diagram in Fig. 6.1. Hence the relation is

$$\{(1, 5), (1, 6), (2, 6), (2, 7), (3, 7), (3, 8), (4, 8), (4, 9)\}$$

6.3 FUNCTIONS

By referring to Examples 1 and 3 of Sec. 6.2, we see that couples of ordered pairs appear that have equal first components but different second components. Thus, in Example 1, we have $(16, -4)$ and $(16, 4)$, $(9, -3)$ and $(9, 3)$, $(4, -2)$ and $(4, 2)$, and $(1, -1)$ and $(1, 1)$. In Example 2, however, no two pairs have the same first component and different second components. This set is an example of a *function*. Functions are of utmost importance in mathematics, and we shall discuss their basic properties in the remainder of this section. We start with the following definition:

If D is a set of numbers, if there exists a rule such that for each element x of D exactly one number y is determined, and if R is the set of all numbers y, then the set of ordered pairs $\{(x, y) \mid x \in D \text{ and } y \in R\}$ is a *function* with the *domain* D and *range* R.

Function
Domain, Range

As in the case of relations, the rule that establishes the correspondence may be an equation in two variables or any other agreement or device that associates each element of D with exactly one element of R. In this chapter we shall confine our discussion to functions in which the rule is an algebraic equation that is usually in the form $y = f(x)$.

127

EXAMPLE   Find the set of ordered pairs $\{(x, y)\}$ if $y = x^2 - 2x - 3$ and $D = \{x | x$ is an integer and $1 \le x \le 4\}$.

Solution   We first note that $D = \{1, 2, 3, 4\}$; furthermore, the corresponding number pairs (x, y) are

$$x = 1, y = 1^2 - (2 \times 1) - 3 = 1 - 2 - 3 = -4$$
$$x = 2, y = 2^2 - (2 \times 2) - 3 = 4 - 4 - 3 = -3$$
$$x = 3, y = 3^2 - (2 \times 3) - 3 = 9 - 6 - 3 = 0$$
$$x = 4, y = 4^2 - (2 \times 4) - 3 = 16 - 8 - 3 = 5$$

Hence $\{(x, y)\} = \{(1, -4), (2, -3), (3, 0), (4, 5)\}$.

Independent
variable
Dependent
variable

In the function $\{(x, y) | y = f(x)\}$ defined by $y = f(x)$, x is called the *independent* variable and y is the *dependent* variable. In other words, the independent variable in a function is the variable whose replacement set is the set of first numbers in the ordered pairs in the function, and the dependent variable is the variable whose replacement set is the set of second numbers in these ordered pairs. If the domain of the function is not specified, it is understood to be the set of real numbers.

6.4 FUNCTIONAL NOTATION

It has been customary for some time to designate the function defined by $y = f(x)$ by the letter f, and either of the following notations may be used for this purpose:

$$f = \{(x, y) | y = f(x)\}$$
$$f = \{[x, f(x)]\}$$

The second expression for f illustrates the fact that it is not necessary to introduce the letter y for the dependent variable, since $f(x)$ stands for the second number in the ordered pair with the first number x. Thus in the function $f = \{[x, f(x)]\}$, f designates the function and $f(x)$ is the value of f for a specified replacement for x. For example, if $y = 3x^2 - 2x + 4$, we may designate the function f in any of the following ways:

$$f = \{(x, y) | y = 3x^2 - 2x + 4\}$$
$$f = \{[x, f(x)] | f(x) = 3x^2 - 2x + 4\}$$
$$f = \{(x, 3x^2 - 2x + 4)\}$$

Furthermore, the value of f for $x = 5$ is $f(5) = (3 \times 5^2) - (2 \times 5) + 4 = 75 - 10 + 4 = 69$.

If more than one function is involved in a particular discussion, it is customary to designate one of them by f and the others by letters other than f. Thus the functions defined by $y = h(x)$, $y = g(x)$, and $y = k(x)$ are $h = \{[x, h(x)]\}$, $g = \{[x, g(x)]\}$, and $k = \{[x, k(x)]\}$, respectively.

EXAMPLE 1

If $f(x) = \dfrac{x-2}{x+1}$, find the function values $f(2)$, $f(\frac{1}{2})$, and $f(-\frac{3}{4})$.

Solution

$$f(2) = \frac{2-2}{2+1} = \frac{0}{3} = 0$$

$$f(\tfrac{1}{2}) = \frac{\tfrac{1}{2}-2}{\tfrac{1}{2}+1}$$

$$= \frac{2(\tfrac{1}{2}-2)}{2(\tfrac{1}{2}+1)} \qquad \blacktriangleleft \text{ by Eq. (4.4)}$$

$$= \frac{1-4}{1+2} = \frac{-3}{3} = -1$$

$$f(-\tfrac{3}{4}) = \frac{-\tfrac{3}{4}-2}{-\tfrac{3}{4}+1}$$

$$= \frac{4(-\tfrac{3}{4}-2)}{4(-\tfrac{3}{4}+1)} \qquad \blacktriangleleft \text{ by Eq. (4.4)}$$

$$= \frac{-3-8}{-3+4} = -11$$

EXAMPLE 2

If $f(x) = x^2 - x - 3$, $g(x) = (x^2-1)/(x+2)$, and $F(x) = f(x) + g(x)$, find $F(2)$.

Solution

$$F(2) = f(2) + g(2)$$

$$= 2^2 - 2 - 3 + \frac{2^2-1}{2+2}$$

$$= 4 - 2 - 3 + \tfrac{3}{4}$$

$$= -1 + \tfrac{3}{4}$$

$$= -\tfrac{1}{4}$$

EXAMPLE 3

If $D = \{x \mid x \text{ is an integer and } -2 \le x \le 1\}$, find the function $\{[x, f(x)] \mid f(x) = x^3 - 3 \text{ and } x \in D\}$.

Solution

$$D = \{-2, -1, 0, 1\}$$
$$f(-2) = (-2)^3 - 3 = -8 - 3 = -11$$
$$f(-1) = (-1)^3 - 3 = -1 - 3 = -4$$
$$f(0) = 0^3 - 3 = -3$$
$$f(1) = 1^3 - 3 = -2$$

Consequently,

$$f = \{[x, f(x)] \mid f(x) = x^3 - 1, x \text{ is an integer, and } -2 \le x \le 1\}$$
$$= \{(-2, -11), (-1, -4), (0, -3), (1, -2)\}$$

EXAMPLE 4 If $f(x) = 3x + 4$ and $D = \{x|-1 \leq x \leq 3\}$, find the range R of $f(x)$.

Solution We shall first prove that the function value of $3x + 4$ increases as x increases. If $X > x$, then

$$3X > 3x \qquad \blacktriangleleft \text{ by Eq. (2.25)}$$

and it follows that

$$3X + 4 > 3x + 4 \qquad \blacktriangleleft \text{ by Eq. (2.24)}$$

Consequently, if x belongs to D, the function value $f(x) = 3x + 4$ is least when $x = -1$ and greatest when $x = 3$. Hence, since $f(-1) = -3 + 4 = 1$ and $f(3) = 9 + 4 = 13$, $R = \{y|1 \leq y \leq 13\}$.

Occasionally, the independent variable in an equation in two variables is expressed in terms of a third variable. For example, if $y = f(x) = x^2 + 1$ and $x = g(t) = t - 1$, then we can express y in terms of t by replacing x in $f(x)$ by $g(t) = t - 1$. Thus, we obtain

$$y = f(x) = f[g(t)] = F(t) = (t - 1)^2 + 1 = t^2 - 2t + 2$$

Now if $D = \{0, 1, 2, 3, 4\}$ and $t \in D$, we shall consider the three functions

$$\begin{aligned}
g &= \{(t, x)|x = g(t) = t - 1, t \in D\} \\
&= \{(0, -1), (1, 0), (2, 1), (3, 2), (4, 3)\} \\
f &= \{(x, y)|y = f(x) = x^2 + 1, x \in \{-1, 0, 1, 2, 3\} \\
&= \{(-1, 2), (0, 1), (1, 2), (2, 5), (3, 10)\} \\
F &= \{(t, y)|y = f(x) = f[g(t)] = F(t) = t^2 - 2t + 2, t \in D\} \\
&= \{(0, 2), (1, 1), (2, 2), (3, 5), (4, 10)\}
\end{aligned}$$

Note that the domain of g and F is D, the domain of f is the range of g, and the range of F is the range of f. This example illustrates the following definition:

Composite function If $f = \{(x, y)|y = f(x)\}$ and $g = \{(t, x)|x = g(t)\}$, then $F = \{(t, y)|y = F(t) = f[g(t)]\}$ is a composite function whose components are f and g and is often represented by $f \circ g$. In order to be able to find F under the above conditions, we must have x in the domain of g and $g(x)$ in the domain of f. Consequently, the domain of $f \circ g$ consists of those elements x in the domain of g such that $g(x)$ belongs to the domain of f.

EXAMPLE 5 If $y = f(x) = \dfrac{x^2 - 2}{x^2 + 4}$ and $x = g(t) = t + 1$, obtain the equation $y = F(t) = f[g(t)]$.

Solution

$$y = F(t) = f[g(t)]$$
$$= \frac{(t+1)^2 - 2}{(t+1)^2 + 4}$$
$$= \frac{t^2 + 2t + 1 - 2}{t^2 + 2t + 1 + 4}$$
$$= \frac{t^2 + 2t - 1}{t^2 + 2t + 5}$$

Exercise 6.1 Relations, Functions, Functional Notation

Find the cartesian product indicated in each of problems 1 to 4.

1 $\{3, 5, 7\} \times \{4, 6, 13\}$
2 $\{c, a, t\} \times \{g, o\}$
3 $\{x|x \in D\} \times \{x|x + 2 \in D\}$ where $D = \{-1, 0, 1\}$
4 $\{x|x^3 \leq 8, x \text{ a positive integer}\} \times \{x^2 \leq 9, x \text{ a positive integer}\}$
5 Find the set of ordered pairs obtained by pairing the nth element of $S = \{1, 3, 5, 7\}$ with the $(n + 1)$st element of $T = \{10, 8, 6, 4, 2\}$.
6 Find the set of ordered pairs obtained by pairing the nth element of $S = \{m, a, t, h\}$ with $(4 - n)$th element of $T = \{w, h, i, z\}$ for $n = 1, 2, 3$.
7 Write the set of ordered pairs obtained by pairing each of 1, 2, 3, 4, 5, and 6 on the face of a clock with the diametrically opposite number.
8 Write the set of ordered pairs obtained by pairing each of 4, 6, and 9 with each integer that differs from it by 5.
9 Is the set of ordered pairs $\{(x, \sqrt{x})|x \in \{1, 4, 9\}\}$ a function or merely a relation? Why?
10 Is the set of ordered pairs $\{(x, x^2)|x \in \{1, 4, 9\}\}$ a function or merely a relation? Why?
11 Is the set of ordered pairs $\{(x, y)|y^2 = x, x \in \{0, 1, 4, 9\}\}$ a function or merely a relation? Why?
12 Is the set of ordered pairs $\{(x, y)|y^2 - 2x + 1, x \in 0, 4, 7.5, 12\}$ a function or merely a relation? Why?

Write out each set of ordered pairs described in problems 13 to 20 and state whether or not it is a function.

13 $\{(x, 2x - 1)|x \text{ is an integer and } -1 \leq x \leq 3\}$

14 $\{(x, \frac{x}{3} + 2)|x \text{ is divisible by 3 and } 2 < x < 15\}$

15 $\{(x, y)|y = \frac{3}{5}x + 3, x \text{ is a multiple of 5 and } -6 < x < 17\}$
16 $\{(x, y)|y = 5x - 4, x \text{ is an integer and } -3 \leq x \leq 2\}$
17 $\{(x, y)|y = x^2 - 3x - 2, x \text{ is an integer and } -2 < x < 1\}$

18   $\{(x, y)|y = 2x^2 - 3x - 5, x$ is an integer and $-4 \le x \le 0\}$
19   $\{(x, y)|y^2 = 16 - 3x^2, x$ is an integer and $-2 \le x \le 2\}$
20   $\{(x, y)|y^2 = 3x^2 + 1, x$ is an odd integer and $0 < x \le 4\}$
21   If $f(x) = 4x - 3$, find $f(-1)$, $f(0)$, and $f(2)$.
22   If $g(x) = 7 - 2x$, find $g(-4)$, $g(3)$, and $g(7)$.
23   If $h(x) = x^2 + 3x + 4$, find $h(-2)$ and $h(3)$.
24   If $q(x) = 2 - 5x - x^2$, find $q(-1)$ and $q(2)$.
25   If $f(x) = 4x + 5$, find $f(x + 1) - f(x)$ and $f(1)$.
26   If $g(x) = x^2 - x + 1$, find $g(x + 2) - g(x)$ and $g(2)$.

27   If $f(x) = x^2 + 2x - 3$, find $\dfrac{f(x + h) - f(x)}{h}$.

28   If $f(x) = 2x^2 - x + 1$, find $\dfrac{f(x + h) - f(x)}{h}$.

29   If $f(x) = \dfrac{x + 2}{2x + 1}$ and $x = g(t) = 2t - 1$, find $F(t) = f[g(t)]$.

30   If $f(x) = 3x + 9$ and $x = g(t) = 2t - 3$, find $F(t) = f[g(t)]$.

31   If $f(x) = \dfrac{x^2 - 1}{x^2 + 1}$, $x = g(t) = \sqrt{t}$, and $F(t) = f[g(t)]$, find $F(1)$.

32   If $f(x) = \dfrac{\sqrt{3x^2 + 4}}{x - 1}$, $x = g(t) = 2t + 1$, and $F(t) = f[g(t)]$, find $F(\tfrac{1}{2})$.

Find the range of the function or relation for the specified domain in each of problems 33 to 36.

33   $f = \{(x, 3x + 4)\}$, $D = \{2, 3, 4, 5\}$
34   $f = \{(x, 2x - 3)\}$, $D = \{1, 2, 4, 8\}$
35   $f = \{(x, \sqrt{x^2 + 3})\}$, $D = \{-3, -1, 0, 1\}$
36   $f = \{(x, \sqrt{31 - x^2})\}$, $D = \{-\sqrt{30}, \sqrt{6}, \sqrt{15}, \sqrt{22}\}$
37   If $D = \{1, 2, 3, 4\}$, $f = \{(x, y)|y = \dfrac{x^2 + 2}{x + 1}$, and $x \in D\}$, find the number pairs that make up f.
38   Find the number pairs that make up f if $f = \{(x, y)|y = \dfrac{1}{x^2 - x}$ and $x \in D\}$, where $D = \{2, 3, 4, 5\}$.
39   Find the number pairs that make up F if $F = \{[t, F(t)]|F(t) = f[g(t)]\}$ where $x = g(t) = 2t - 3$, $f(x) = 3x + 1$, $x \in D$, and $D = \{1, 3, 5, 7\}$.
40   If $D = \{-2, 0, 2, 4\}$, $f(x) = 5x + 2$, $x = g(t) = t - 13$, and $F = \{[t, F(t)]|F(t) = f[g(t)]\}$, find the number pairs that constitute F provided $x \in D$.

In problems 41 to 44, find $f \cap g$.

41   $f = \{(x, x + 1)\}$, $g = \{(x, 2x - 1)\}$, $D = \{1, 2, 3, 4\}$
42   $f = \{(x, 2x + 3)\}$, $g = \{(x, 2x^2 + 7x)\}$, $D = \{-3, 0, 0.5, 2\}$
43   $f = \{(x, 3x^2 - 2x + 4)\}$, $g = \{(x, 3x + 2)\}$, $D = \{-1, 0, \tfrac{2}{3}, 1\}$
44   $f = \{(x, 3x^2 + 2x - 17)\}$, $g = \{(x, x^2 + 3x - 2)\}$, $D = \{-\tfrac{5}{2}, 0, 1, 3\}$

6.5 THE RECTANGULAR COORDINATE SYSTEM

In this section we shall introduce a device for associating an ordered pair of numbers with a point in a plane. Invented by the French mathematician and philosopher René Descartes (1596–1650), it is called the rectangular or cartesian coordinate system.

    In order to set up this system, we construct two perpendicular number lines in the plane and choose a suitable scale on each. For convenience these lines are horizontal and vertical, and the unit length on each is the same (see Fig. 6.2), although neither restriction is necessary. The two lines are called the *coordinate axes,* the horizontal line being the *X* axis and the vertical line the *Y* axis. The intersection of the two lines is the *origin,* designated by the letter *O.* The coordinate axes divide the plane into four sections called *quadrants.* These quadrants are numbered I, II, III, and IV counterclockwise, as indicated in Fig. 6.2*a.*

    Next it is agreed that horizontal distances measured to the right from the *Y* axis are positive, and horizontal distances measured to the left are negative. Similarly, vertical distances measured upward from the *X* axis are positive, and vertical distances measured downward are negative. These distances, because of their signs, are called *directed distances.* Finally, we agree that the first number in an ordered pair of numbers represents the directed distance from the *Y* axis to a point and the second number in the pair represents the directed distance from the *X* axis to the point. It follows then that an ordered pair of numbers uniquely determines the position of a point in the plane. For example, (4, 1) determines the point that is 4 units to the right of the *Y* axis and 1 unit above the *X* axis. This point is designated by *Q* in Fig. 6.2*b.* Similarly, the ordered pair (−5, −1) determines the point *S* in Fig. 6.2*b* that is 5 units to the left of the *Y* axis and 1 unit below the *X* axis. Conversely, each point in the plane determines a unique ordered pair of numbers. For example, the point *P* in Fig. 6.2*b* is 3 units to the right of the *Y* axis and 2 units below the *X* axis, and so *P* determines the ordered pair (3, −2). A plane in which the coordinate axes have been constructed is called a *cartesian plane.*

Coordinate axes

X axis   Y axis

Origin

Directed distance

Cartesian plane

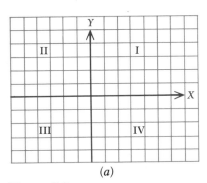

(a)

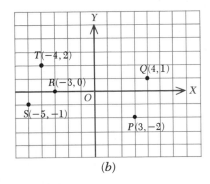

(b)

Figure 6.2

Coordinates
Abscissa
Ordinate

Plotting a point

The two numbers in an ordered pair that is associated with a point in the cartesian plane are called the *coordinates* of the point. The first number is called the *abscissa* of the point, and it is the directed distance from the *Y* axis to the point. The second number in the pair is the *ordinate* of the point, and it represents the directed distance from the *X* axis to the point. The procedure for locating a point in the plane by means of its coordinates is called *plotting* the point. The notation $P(a, b)$ means that P is the point whose coordinates are (a, b). In order to plot the point $T(-4, 2)$, we count 4 units to the left of the origin on the *X* axis and then upward 2 units and thus arrive at the point. Similarly, the point $R(-3, 0)$ is 3 units to the left of the origin and on the *X* axis. The general point and its coordinates are written $P(x, y)$.

6.6 THE GRAPH OF A FUNCTION

By use of the rectangular coordinate system, we can obtain a geometric representation, or a geometric "picture" of a function. For this purpose, we require that each ordered pair of numbers (x, y) of a function be the coordinates of a point in the cartesian plane with x as the abscissa and y as the ordinate. We then define the graph of a function as follows:

Graph
of a function

The *graph of a function* is the set of all points (x, y) in the plane such that the ordered pair (x, y) is an element of the function.

The graphs of most functions we shall discuss in this chapter are smooth† continuous curves. When we say that the graph of a function is a curve, we mean that the point determined by each ordered pair of numbers in the function is on the curve and that the coordinates of each point on the curve are an ordered pair of numbers in the function.

We shall illustrate the procedure for obtaining the graph of a function by explaining the steps in the construction of the graph of the function defined by $y = x^2 - 3x - 1$ for $-2 \leq x \leq 5$. Note that this function is

$$\{(x, y) \mid y = x^2 - 3x - 1 \text{ and } x \text{ belongs to } D = \{x \mid -2 \leq x \leq 5\}\}$$

The first step is to assign each of the integers in D to x and then calculate each corresponding value of y, using the defining equation $y = x^2 - 3x - 1$. Before doing this, however, it is advisable to make a table like the one below in which to record the corresponding values.

---

† At present we are not in a position to give a rigorous definition of a "smooth continuous curve." For our purposes, however, the following *description* will suffice. A smooth continuous curve contains no breaks or gaps, and there are no sudden or abrupt changes in its direction.

Now we shall calculate the corresponding value of y when x is assigned the numbers $-2, -1, 0, 1, 2, 3, 4$, and 5.

$$y = \begin{cases} (-2)^2 - 3(-2) - 1 = 9 & \text{for } x = -2 \\ (-1)^2 - 3(-1) - 1 = 3 & \text{for } x = -1 \\ 0^2 - 3(0) - 1 = -1 & \text{for } x = 0 \end{cases}$$

Continuing this process, we obtain the additional ordered pairs $(1, -3)$, $(2, -3)$, $(3, -1)$, $(4, 3)$, and $(5, 9)$ and enter the results in the table:

x	-2	-1	0	1	2	3	4	5
y	9	3	-1	-3	-3	-1	3	9

Now we plot the points (x, y) thus determined, as shown in Fig. 6.3. Because there is an ordered pair in the function for every intermediate real value of x, we connect the plotted points with a smooth curve. This is the portion of the graph defined by $y = x^2 - 3x - 1$ over the specified domain.

We now investigate the nature of the graph as x increases through values greater than 5 and decreases through values less than -2. For this purpose we first replace x in $y = x^2 - 3x - 1$ by $5 + h$ and get $y = h^2 + 7h + 9$. From the latter equation we see that as h increases from zero, y increases from 9. Hence the graph extends upward and to the right from the point $(5, 9)$. Similarly, if we replace x by $-2 - k$, we get $y = k^2 + 7k + 9$, and we therefore conclude that as k increases from zero, y increases from 9. Hence the graph extends upward and to the left from the point $(-2, 9)$.

Zero of a function   A *zero* of a function defined by $y = f(x)$ is the value of the independent variable x for which $y = 0$. Hence the zeros of a function are the abscissas

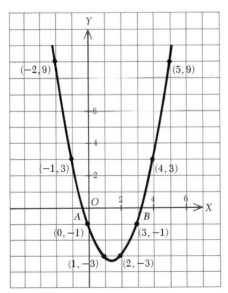

Figure 6.3

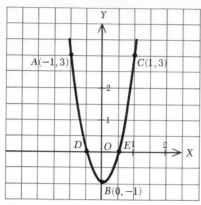

Figure 6.4

of the points where the graph crosses the X axis. The zeros of many classes of functions can be obtained by algebraic methods, but we must depend upon graphical methods for others. The zeros of the function $\{(x, y)\,|\,y = x^2 - 3x - 1\}$ are the abscissas of the points A and B in Fig. 6.3 and are approximately -0.3 and 3.3.

If the domain of a function is not specified, it is assumed to be the real number system. In such cases, to get a set of ordered pairs for constructing the graph, it is usually advisable to start by assigning consecutive small integers to x and continue the process until a sufficient number of points are obtained to determine the nature of the graph. At times the points obtained by assigning consecutive integers to x are too far apart to enable one to sketch the curve. For example, in the function defined by $y = 4x^2 - 1$, if we assign -1, 0, and 1 to x, we get the pairs $(-1, 3)$, $(0, -1)$ and $(1, 3)$. These pairs determine the points $A, B,$ and C in Fig. 6.4, and it is evident that these points alone do not show the nature of the graph. We can get two additional points by assigning $-\frac{1}{2}$ and $\frac{1}{2}$ to x and obtaining the pairs $(-\frac{1}{2}, 0)$ and $(\frac{1}{2}, 0)$. When these points are plotted, the curve in Fig. 6.4 can be sketched.

6.7 LINEAR FUNCTIONS

Linear function

It is proved in analytic geometry that if the domain is the set of real numbers, the graph of the function $\{(x, y)\,|\,y = ax + b\}$ is a straight line. Such a function is called a *linear function,* and its graph is completely determined by two points. If the graph does not pass through the origin and is not parallel to either axis, the points where the graph crosses the axes are readily determined by assigning 0 to x in the equation $y = ax + b$ and obtaining the point $(0, b)$ and then by assigning 0 to y and solving for x and obtaining $(-b/a, 0)$. These points determine the graph, but it is advisable to calculate a third point as a check. The abscissa of the point where the

136

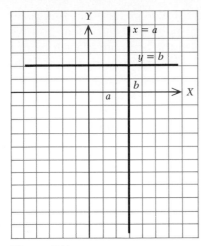

Figure 6.5

Figure 6.6

X intercept   line crosses the *X* axis is called the *X intercept,* and the ordinate of the point
Y intercept   where it crosses the *Y* axis is the *Y intercept.*

The graph of $\{(x, y)\,|\,y = ax\}$ passes through the origin, since $(0, 0)$ satisfies the equation $y = ax$. Hence we must assign a number other than zero to *x* to get a second point to determine the graph and a third number as a check.

The graph of $\{(x, y)\,|\,x = a\}$ is the line parallel to the *Y* axis at the directed distance of *a* units from it. The graph of $\{(x, y)\,|\,y = b\}$ is the line parallel to the *X* axis and at the directed distance of *b* units from it. These graphs are shown in Fig. 6.5.

EXAMPLE   Construct the graph of the function defined by $y = 3x - 9$.

Solution   We find the intercepts by assigning 0 to *x* and solving for *y* and by assigning 0 to *y* and solving for *x*. We shall find a third point by assigning 4 to *x* and solving for *y*. Thus we get the following table of corresponding numbers:

x	0	3	4
y	−9	0	3

We now plot the points determined by these pairs of numbers, draw a straight line through them, and obtain the graph in Fig. 6.6.

Exercise 6.2   Graphs

1 Plot the points determined by the following ordered pairs of numbers:
   $(2, 5)$, $(3, 6)$, $(1, -2)$, $(-4, 3)$, $(0, 2)$, $(-5, 0)$, $(-3, -7)$, $(-6, 1)$, $(4, 0)$, $(0, -5)$, $(0, 0)$, $(7, -2)$, $(-6, 5)$.

137

2 Describe the line on which each of the following sets of points is located: (*a*) the two coordinates of each point are equal and positive; (*b*) the abscissa of each point is zero; (*c*) the ordinate of each point is zero; (*d*) the two coordinates of each point are numerically equal but of opposite sign.

3 Describe the lines determined by the following sets of conditions: (*a*) the ordinate of each point is 6; (*b*) the abscissa of each point is −5; (*c*) the abscissa of each point is 9; (*d*) the ordinate of each point is −4.

4 If *n* is a negative number, state the quadrant in which each of the following points is located: $(n, 1)$, $(3, n)$, $(2, -n)$, $(-3, n)$, $(n, -2)$, $(-5, -n)$, $(-n, -6)$, $(-n, 4)$.

Construct the graph of the function defined by the equation in each of problems 5 to 24 and estimate all zeros in 5 to 20.

5 $y = x + 3$ 6 $y = x - 2$
7 $y = 2x - 1$ 8 $y = 3x + 4$
9 $y = -2x + 7$ 10 $y = -3x + 4$
11 $y = -x - 5$ 12 $y = -2x - 3$
13 $y = x^2$ 14 $y = -x^2$
15 $y = -x^2 + x$ 16 $y = x^2 + 5$
17 $y = x^2 - 3x + 3$ 18 $y = -x^2 + 5x - 7$
19 $y = -x^2 + x + 4$ 20 $y = x^2 + 8x + 13$
21 $y = \sqrt{25 - x^2}$, $D = \{x | 0 \leq x \leq 5\}$
22 $y = -\sqrt{16 - x^2}$, $D = \{x | 0 \leq x \leq 4\}$
23 $y = -\sqrt{36 - x^2}$, $D = \{x | -6 \leq x \leq 0\}$
24 $y = \sqrt{49 - x^2}$, $D = \{x | -7 \leq x \leq 0\}$

In each of problems 25 to 28, find the intercepts of the indicated function and then construct the graph in the given domain.

25 $\{(x, y) | y = 2x - 6\}$, $D = \{x | -3 \leq x \leq 5\}$
26 $\{(x, y) | y = -3x + 7\}$, $D = \{x | -4 \leq x \leq 6\}$
27 $\{(x, y) | y = -4x - 3\}$, $D = \{x | -5 \leq x \leq 3\}$
28 $\{(x, y) | y = 3x + 4\}$, $D = \{x | -6 \leq x < 3\}$

6.8 SOME SPECIAL FUNCTIONS

In this section we shall discuss examples of functions whose graphs are not continuous curves. As a first example, we shall consider the function

$$f = \{(x, y) | y = x + 2, x \leq 3\} \cup \{(x, y) | y = -x + 4, x > 3\}$$

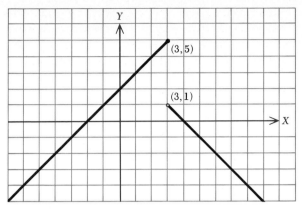

Figure 6.7

This graph is defined by the equations

$$y = x + 2 \qquad \text{for} \qquad x \leq 3 \qquad\qquad (1)$$
$$y = -x + 4 \qquad \text{for} \qquad x > 3 \qquad\qquad (2)$$

The graph of $y = x + 2$ is a straight line, and its intercepts are -2 and 2. Furthermore, if $x = 3$, the largest value in the domain, then $y = 5$; hence, $(3, 5)$ is on the graph. Since the domain is $D = \{x | x \leq 3\}$, the graph does not extend to the right of $(3, 5)$ and is a ray that extends downward and to the left from $(3, 5)$ and including $(3, 5)$.

We can find similarly that the graph of $y = -x + 4$ consists of a ray that extends down and to the right from $(3, 1)$ but does not include $(3, 1)$. The graphs are shown in Fig. 6.7.

Bracket function   An important function in the theory of numbers is the *bracket function* $\{(x, y) | y = [x]\}$, where the notation $[x]$ means the greatest integer that is less than or equal to x. Hence, in the equation $y = [x]$, if $x = \frac{1}{2}$, then $y = 0$, since 0 is the greatest integer less than $\frac{1}{2}$. Similarly, if $x = 2\frac{1}{2}$, then $y = [2\frac{1}{2}] = 2$. Furthermore, if n is an integer, $y = [n] = n$. Hence, by the above and similar arguments, we have the following corresponding values of x and y for the bracket function:

$$y = \begin{cases} 0 & 0 \leq x < 1 \\ 1 & 1 \leq x < 2 \\ 2 & 2 \leq x < 3 \\ \cdots & \cdots\cdots\cdots \\ n & n \leq x < n+1 \qquad \text{for } n \text{ an integer} \end{cases}$$

Consequently the graph of the bracket function is a set of horizontal line segments, four of which are shown in Fig. 6.8.

139

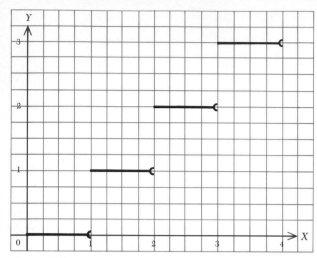

Figure 6.8

6.9 INVERSE FUNCTIONS

We studied relations in Sec. 6.2 and functions in Sec. 6.3 and shall now consider the two functions

$$f = \{(1, 2), (3, 4), (5, 6)\} \quad \text{and} \quad g = \{(1, 2), (3, 4), (5, 4)\}$$

Both have $\{1, 3, 5\}$ as the domain, whereas the range of f is $\{2, 4, 6\}$ and of g is $\{2, 4\}$.

If we interchange the first and second elements in f, we obtain

$$F = \{(2, 1), (4, 3), (6, 5)\}$$

and they constitute a function with domain $\{2, 4, 6\}$ and range $\{1, 3, 5\}$. If, however, we interchange the first and second elements in g, we get

$$\{(2, 1), (4, 3), (4, 5)\}$$

and this is a relation but not a function since two pairs have the same first element 4 and different second elements 3 and 5.

Inverse relation
Inverse function
  The set of ordered pairs obtained by interchanging the elements in each ordered pair of a function is called the *inverse relation*. Furthermore, if the inverse relation is a function, it is called the *inverse function*. If the function is designated by f, then we use f^{-1} to indicate the inverse. Consequently, if $f = \{(2, 3), (4, 5), (6, 7)\}$, then $f^{-1} = \{(3, 2), (5, 4), (7, 6)\}$.

It follows from the definition of a function that *if a function f is such that no two of its ordered pairs with different first elements have the same second elements, then corresponding to the function f, there exists the inverse function f^{-1}.* The inverse function is obtained by interchanging first and second elements of each ordered pair in f.

140

If we designate a function by

$$f = \{(x, y) \,|\, y = f(x)\}$$

then $y = f(x)$ defines the function, and $x = f(y)$ defines its inverse. If the latter equation is solvable for y, the solution is ordinarily put in the form $y = f^{-1}(x)$, and we designate the inverse relation by

$$f^{-1} = \{(x, y) \,|\, y = f^{-1}(x)\}$$

and it is the inverse function provided $y = f^{-1}(x)$ defines a function.

EXAMPLE 1   Find the inverse of

$$f = \{(x, y) \,|\, y = f(x) = 2x + 4\}$$

and sketch the graphs of the function and its inverse.

Solution   In order to find the defining equation of the inverse of the function defined by $y = 2x + 4$, we must solve the equation for x and then interchange x and y. Thus, solving for x, we get $x = \frac{1}{2}y - 2$, and interchanging x and y, we get $y = \frac{1}{2}x - 2$. Consequently,

$$f^{-1} = \{(x, y) \,|\, y = f^{-1}(x) = \tfrac{1}{2}x - 2\}$$

is the inverse of f, and it is a function. Both graphs are shown in Fig. 6.9.

EXAMPLE 2   If $F = \{(x, y) \,|\, y = F(x) = \sqrt{x^2 + 16}\}$, find F^{-1}.

Solution   The defining equation for F is

$$y = \sqrt{x^2 + 16} \tag{1}$$

and we solve it for x in terms of y as follows:

$$y^2 = x^2 + 16 \qquad \blacktriangleleft \text{ equating the squares of the members of (1)}$$

$$x = \sqrt{y^2 - 16} \qquad \blacktriangleleft \text{ solving for } x \tag{2}$$

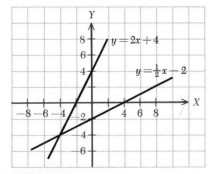

Figure 6.9

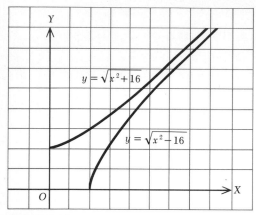

Figure 6.10

Note that since $x \geq 0$, we are concerned only with the principal square root of $y^2 - 16$. Furthermore, since x is real, $y \geq 4$ in (2). Now we interchange x and y in (2) and obtain the defining equation for F^{-1}. It is

$$y = \sqrt{x^2 - 16} \qquad \text{for } x \geq 4 \tag{3}$$

Hence, $\qquad F^{-1} = \{(x, y) \mid y = F^{-1}(x) = \sqrt{x^2 - 16} \text{ and } x \geq 4\}$

Note that the domain and range of F are $\{x \mid x \geq 0\}$ and $\{y \mid y \geq 4\}$, respectively. Furthermore the domain and range of F^{-1} are $\{x \mid x \geq 4\}$ and $\{y \mid y \geq 0\}$.

Now if x is an element of the intersection $\{x \mid x \geq 0\} \cap \{x \mid x \geq 4\}$ of the domains of F and F^{-1}, then $F(x)$ is an element of the domain of F^{-1}, and $F^{-1}(x)$ is an element in the domain of F. Furthermore, for each x in this intersection,

$$F[F^{-1}(x)] = \sqrt{(\sqrt{x^2 - 16})^2 + 16}$$
$$= \sqrt{x^2 - 16 + 16}$$
$$= \sqrt{x^2} = x$$

and
$$F^{-1}[F(x)] = \sqrt{(\sqrt{x^2 + 16})^2 - 16}$$
$$= \sqrt{x^2 + 16 - 16}$$
$$= \sqrt{x^2} = x$$

Hence, in the specified domain, $F[F^{-1}(x)] = F^{-1}[F(x)]$.

The graphs of the functions defined by Eqs. (1) and (3) are shown in Fig. 6.10. In this figure, it can be seen that the graph of the inverse function is situated in the same position relative to the X axis as the graph of the function is situated with respect to the Y axis.

142

Exercise 6.3 Special Functions and Inverses

1 Plot the graph of $T \times S$ if $T = \{1, 2, 4\}$ and $S = \{2, 4, 5\}$.
2 Plot the graph of $\{(x, y) | y = x\} \cap (T \times S)$ where $S = T = \{0, 1, 2\}$.
3 Plot the graph of $\{(x, y) | y \geq x\} \cap (T \times S)$ where $T = \{-2, -1, 0\}$ and $S = \{0, 1, 2\}$.
4 Plot the graph of $\{(x, y) | y \leq x\} \cap (S \times T)$ where $S = \{-2, -1, 3\}$ and $T = \{-2, -1, 0\}$.

Construct the graph of the function in each of problems 5 to 16.

5 $\begin{cases} y = x \text{ for } x \geq 0 \\ y = x + 1 \text{ for } x < 0 \end{cases}$

6 $\begin{cases} y = x + 3 \text{ for } x \geq 1 \\ y = x + 1 \text{ for } x < 1 \end{cases}$

7 $\begin{cases} f(x) = \sqrt{16 - x^2} \text{ for } 0 \leq x \leq 4 \\ f(x) = -x \qquad \text{ for } x < 0 \end{cases}$

8 $\begin{cases} f(x) = x^2 + 1 \text{ for } x \geq -1 \\ f(x) = x + 1 \text{ for } x < -1 \end{cases}$

9 $\{(x, y) | y = |x|\}$
10 $\{(x, y) | y = |x| + 1\}$
11 $\{(x, y) | y = |x| + x\}$
12 $\{(x, y) | y = |x| - x\}$
13 $\{(x, y) | y = [x] + 3\}, D = \{1 \leq x \leq 7\}$
14 $\{(x, y) | y = [x] - 2\}, D = \{0 \leq x \leq 6\}$
15 $\{(x, y) | y = [x - 2]\}, D = \{-4 \leq x \leq 1\}$
16 $\{(x, y) | y = 2[x]\}, D = \{-1 \leq x \leq 3\}$

Find the equation which defines the inverse of the relation defined by the equation in each of problems 17 to 20. State whether the inverse is a function or a relation and give the domain.

17 $y^2 = 9 - x^2, 0 \leq x \leq 3$
18 $y^2 = 8 - 2x^2, 0 \leq x \leq 2$
19 $y^2 = 3x - 2, x \geq \frac{2}{3}$
20 $(y - 1)^2 = 4(x - 2), x \geq 2$

Find the inverse of the function in each of problems 21 to 28 and find the domain of the inverse.

21 $f = \{(x, x + 1) | x \geq 1\}$
22 $f = \{(x, 2x - 1) | x \geq -2\}$
23 $f = \{(x, y) | y = 3x - 2, x \geq -2\}$
24 $f = \{(x, y) | y = -x + 3, x \geq 1\}$

25 $f = \{(x, y) \,|\, y = \dfrac{x + 4}{x},\ x > 0\}$

26 $f = \{(x, y) \,|\, y = \dfrac{x - 3}{x},\ x > 3\}$

27 $f = \{(x, y) \,|\, y = \dfrac{x}{x - 1},\ x > 1\}$

28 $f = \{(x, y) \,|\, y = \dfrac{x - 1}{x + 2},\ x > -1\}$

Exercise 6.4   Review

1   Find $S \times T$ if $S = \{a, e, i\}$ and $T = \{t, n\}$.

2   Find the set of ordered pairs obtained by pairing the nth element of $\{s, a, m, e\}$ with the $(5 - n)$th element of $\{b, r, a, t, s\}$ for $n = 1, 2, 3, 4$.

3   Is the set of ordered pairs $\{(x, x^3) \,|\, x \in \{0, 1, 2, 3\}$ a function?

4   Is the set of ordered pairs $\{(x, y) \,|\, y^2 = x - 1,\ x = 1, 5, 10\}$ a function? Why?

5   Write out the set of ordered pairs $\{(x, 3x + 2) \,|\, x$ is an integer and $-2 \le x < 2\}$.   Do these pairs form a function?

6   Does the set of ordered pairs determined by $\{(x, y) \,|\, y^2 = x - 1$ with x an integer and $1 < x \le 5\}$ form a function?   Why?

7   If $f(x) = 3x - 4$, find $[f(x + h) - f(x)]/h$

8   If $f(x) = \dfrac{x + 1}{2x - 3}$ and $x = g(t) = t + 2$, find $F(t) = f[g(t)]$.

9   Find the range of $f = \{(x, y) \,|\, y = 2x + 5\}$ if its domain is $D = \{-1, 0, 2, 3\}$.

10 If the common domain of $f = \{x, x - 2\}$ and $g = \{x, 2x^2 - 3x - 2\}$ is $D = \{-1, 0, 2, 3\}$, find $f \cap g$.

11 Describe the line determined by each of the following sets of conditions: (*a*) the ordinate of each point on it is 3; (*b*) the abscissa of each point is negative and numerically equal to the ordinate of the point which is positive.

12 Construct the graph of the function defined by $y = -3x + 1$.

13 Construct the graph of the function defined by $y = x^2 - 2x - 3$.

14 Construct the graph of the function defined by $y = -\sqrt{9 - x^2}$, $D = \{x \,|\, 0 \le x \le 3\}$.

15 Find the intercept and construct the graph of $\{(x, y) \,|\, y = 2x + 7\}$ for $D = \{-5 \le x \le 2\}$.

16 Plot the graph of $S \times T$ for $S = \{-2, 1, 3\}$ and $T = \{-1, 4\}$.

17 Plot the graph of $y = \begin{cases} x, & \text{for } x \ge 3 \\ -x + 1, & \text{for } x < 3. \end{cases}$

18 Sketch the graph of $\{(x, y) \,|\, y = |x| + 2\}$ for $-2 \le x \le 3$.

19 Sketch the graph of $\{(x, y) \,|\, y = [x] + 2\}$ for $-2 \le x \le 3$.

20 Find the inverse of $\{(x, y) | y = x + 2, x \geq 3\}$ and the domain of this inverse.

21 Find the inverse of $\{(x, y) | y = \dfrac{x + 3}{2x - 1}, x \geq 1\}$ and the domain of the inverse.

22 Find the inverse of the function defined by $y = \sqrt{16 - x^2}$ with $0 \leq x \leq 4$. Is this inverse a function or a relation?

Angular Measure and the Trigonometric Functions

Several units are used in connection with linear measurements. For example, the foot, inch, yard, and mile are used in the English system, whereas the meter, along with fractions and multiples thereof, is used in the metric system; furthermore, the Spanish vara is used in many of the old records of the Southwest. Similarly, we use several units for measuring angles. The oldest and most commonly used units are the degree and the radian; however, a right angle, a revolution, and a mil are often used. The radian is used almost exclusively in pure mathematics, and the degree is employed in navigation and surveying. We shall discuss both of these units in this chapter.

We shall also define and discuss six ratios that are associated with an angle since they are the basis for trigonometry. They are called the *trigonometric ratios*.

7.1 THE DISTANCE FORMULA

We shall make use of the pythagorean theorem and of the distance between two points that are equidistant from a coordinate axis in developing a formula for the distance between any two points $P_1(x_1, y_1)$ and $P_2(x_2, y_2)$. If two points are the same distance from the X axis as shown in Fig. 7.1, then the distance between them is $|x_2 - x_1|$ since x_2 is the distance and direction from the Y axis to P_2, and x_1 is the distance and direction from the Y axis to P_1. Similarly, the distance between two points $P_1(x_1, y_1)$ and $P_2(x_1, y_2)$ that are equidistant from the X axis is $|y_2 - y_1|$. For example, the distance between $(2, 6)$ and $(7, 6)$ is $|7 - 2| = 5$ and the distance between $(5, 2)$ and $(-3, 2)$ is $|-3 - 5| - 8$.

We shall now consider any two points $P_1(x_1, y_1)$ and $P_2(x_2, y_2)$ as shown in Fig. 7.2 and find the distance between them. The point $P_3(x_2, y_1)$ is also shown in the figure and is determined by the intersection of a line parallel to the X axis through P_1 and a line parallel to the Y axis through P_2; hence, its coordinates are as shown. Furthermore, the lengths of P_1P_3 and P_3P_2 are as indicated in the figure since they are parallel to the X and Y axes, respectively. Consequently, since the square of $|x_2 - x_1|$ is equal to

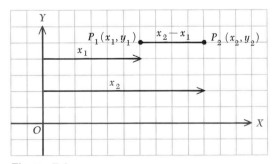

Figure 7.1

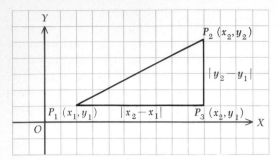

Figure 7.2

$(x_2 - x_1)^2$, we find, by use of the pythagorean theorem, that

$$(P_1P_2)^2 = (P_1P_3)^2 + (P_3P_2)^2$$
$$= (x_2 - x_1)^2 + (y_2 - y_1)^2$$

Therefore, *the distance **d** between* $P_1(x_1, y_1)$ *and* $P_2(x_2, y_2)$ *is*

Distance formula
$$\boldsymbol{d = \sqrt{(x_2 - x_1)^2 + (y_2 - y_1)^2}} \qquad \textbf{(7.1)}$$

7.2 ANGLES

Ray
Angle

Initial side
Terminal side
Vertex
Positive angle
Negative angle

A part of a line that originates at a fixed point and extends indefinitely far in one direction is called a *ray*. If two rays are drawn from the same point, an *angle* is formed. We shall often think of an angle as being formed by rotating one ray about the end point of a stationary ray as shown in Fig. 7.3. The stationary ray is called the *initial side,* the final position of the rotating ray is the *terminal side* of the angle, and the fixed point is known as the *vertex* of the angle. The angle is positive or negative depending on whether the rotation is in a counterclockwise or clockwise direction. In the figure, AB is the initial side, AC is the terminal side, and θ is a positive angle since the rotation is in a counterclockwise direction.

The magnitude or size of an angle is determined by the amount of rotation that is necessary to move a ray from the initial to the terminal side.

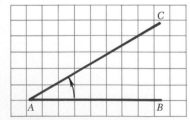

Figure 7.3

148

7.3 UNITS OF ANGULAR MEASURE

In measuring the length of a line segment, we determine the number of times a chosen unit can be laid off along the segment. Similarly, in measuring an angle, we determine the number of times a chosen unit is contained in the angle. There are several units that are used in measuring angles. They include the mil, right angle, revolution, point, degree, and radian.

Mil   The *mil* is used by artillery gunners. It is the central angle subtended by an arc that is $\frac{1}{6,400}$ of the circumference of the circle. It is a convenient unit because an angle of 1 mil subtends an arc of approximately 1 yard on a circle of radius 1,000 yards.

Point   The right angle was divided into eight equal parts by compass navigators, and each part was called a *point*. This unit was adequate for steering vessels. Astronomers through the ages have needed a smaller unit, and they were responsible for the development of the *sexagesimal system*, in which the degree is the unit. The degree is defined as follows: A *degree* is

Degree an angle which, if placed with its vertex at the center of a circle, subtends an arc equal to $\frac{1}{360}$ of the circumference. The degree is divided into 60 minutes, and a minute is divided into 60 seconds. The degree and subdivisions of it are used in most practical work as well as in surveying and navigation.

Radian The *circular system* used in advanced mathematics and in most scientific work employs the *radian* as a unit. A *radian* is an angle which, if placed with its vertex at the center of a circle, subtends an arc equal in length to the radius. A radian is shown in Fig. 7.4.

Exercise 7.1 Angles and Distances

Construct the angles given in each of problems 1 to 8, label each initial and terminal side, and indicate the direction of rotation.

1 120°, −60°, 45°   **2** 150°, 300°, −135°

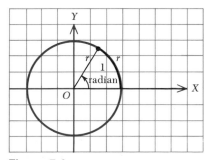

Figure 7.4

3 $-90°$, $330°$, $225°$ **4** $-240°$, $270°$, $-45°$
5 $480°$, $180°$, $-315°$ **6** $540°$, $-120°$, $585°$
7 $360°$, $-570°$, $-405°$ **8** $750°$, $-420°$, $-675°$

Plot the points whose coordinates are given in each of problems 9 to 12.

9 $(2, 3)$, $(4, -2)$, $(0, 5)$ **10** $(4, -1)$, $(-3, -2)$, $(4, 0)$
11 $(-5, -1)$, $(3, -7)$, $(-4, 0)$ **12** $(-6, -2)$, $(-8, 3)$, $(0, -3)$

Use $(-2, 5)$ as the starting point, draw the directed segments described in each of problems 13 to 20, and find the coordinates of each terminal point.

13 2 units to the right and 3 units upward
14 5 units to the left and 4 units upward
15 3 units to the right and 2 units downward
16 4 units to the left and 5 units downward
17 6 units to the left, 2 units downward, 3 units to the right
18 3 units to the right, 4 units upward, 1 unit to the left
19 5 units upward, 2 units to the right, 4 units downward
20 2 units downward, 3 units to the left, 6 units upward

Draw the line segment between the pair of points given in each of problems 21 to 24 and find the length of each segment.

21 $(2, 4)$, $(5, 8)$ **22** $(-3, 5)$, $(1, 2)$
23 $(2, 9)$, $(7, -3)$ **24** $(-4, -6)$, $(4, 9)$

If the three points whose coordinates are given in each of problems 25 to 28 are the vertices of a triangle, find the lengths of the sides.

25 $(2, 3)$, $(5, 7)$, $(-1, 4)$ **26** $(3, -4)$, $(7, -1)$, $(8, 5)$
27 $(-3, 2)$, $(9, 7)$, $(6, 3)$ **28** $(-4, 0)$, $(11, -8)$, $(-1, -3)$

7.4   DEGREES AND RADIANS IN TERMS OF ONE ANOTHER

We shall obtain a relation between degrees and radians in order to be able to change from one unit of angular measure to the other. For this purpose, we use the fact that there are $360°$ in a circle and the fact that the radius can be laid off 2π times on the circumference of a circle. Consequently, a circumference subtends a central angle of 2π radians. Therefore, 2π radians is equal to $360°$; hence,

$$\pi \text{ radians} = 180° \tag{7.2}$$

By use of this fundamental relation between degrees and radians, we can get the number of degrees in a radian and the number of radians in a degree. Thus,

Relations between degrees and radians

$$1 \text{ radian} = \frac{180}{\pi} \text{ degrees} \qquad \blacktriangleleft \text{ dividing (7.2) by } \pi \qquad (7.3)$$

and

$$1 \text{ degree} = \frac{\pi}{180} \text{ radians} \qquad \blacktriangleleft \text{ dividing (7.2) by 180} \qquad (7.4)$$

It is not necessary to memorize relations (7.3) and (7.4), since they are readily obtained from (7.2). Furthermore, an angle can be changed from degrees to radians by expressing it as a fractional part of 180° and then replacing 180° by π radians. Thus,

$$30° = \tfrac{30}{180}(180°) = \tfrac{1}{6}(180°) = \tfrac{1}{6}\pi \text{ radians}$$
$$70° = \tfrac{70}{180}(180°) = \tfrac{7}{18}(180°) = \tfrac{7}{18}\pi \text{ radians}$$

If an angle is expressed in degrees, minutes, and seconds, it can be written in terms of radians by first expressing it in terms of degrees only and then proceeding as above. If this is done, we see that

$$16°32'36'' = 16°1,956''$$
$$= 16\frac{1,956}{3,600} \text{ degrees} \qquad \blacktriangleleft 1° = 3,600''$$
$$= \left(16\frac{1,956}{3,600}\right)\frac{\pi}{180} \text{ radians} \qquad \blacktriangleleft 1° = \frac{\pi}{180} \text{ radians}$$
$$= \frac{16.543}{180}(180°)$$
$$\doteq 0.09191 \,(3.142 \text{ radians}) \qquad \blacktriangleleft \pi \doteq 3.142$$
$$= 0.2888 \text{ radian}$$

We can change from radians to degrees without making use of (7.3) by expressing the angle as a fractional part of π radians and then replacing π radians by 180°. For example,

$$\frac{\pi}{4} \text{ radians} = \tfrac{1}{4}\pi \text{ radians} = \tfrac{1}{4}(180°) = 45°$$

$$1.2 \text{ radians} = \frac{1.2}{\pi}\pi \text{ radians} = \frac{1.2}{\pi}(180°) = \frac{216°}{\pi}$$

If at any time an approximation to the nearest second is acceptable, we may use 1 radian = 57°17'45'' since

$$1 \text{ radian} = \frac{180°}{\pi}$$
$$\doteq \frac{180°}{3.14159} \qquad \blacktriangleleft \text{ using } \pi = 3.14159$$
$$\doteq 57.2958°$$
$$= 57°17'45''$$

If the angle is given as a fractional part of a radian, it is more convenient to use 57.2958° than 57°17′45″ for 1 radian. Thus,

$$
\begin{aligned}
0.346 \text{ radian} &\doteq 0.346(57.2958)° \\
&\doteq 19.8243° \qquad \blacktriangleleft \text{ to six figures} \\
&= 19° + (0.8243)60′ \\
&= 19°49.458′ \\
&= 19°49′ + 0.458(60″) \\
&= 19°49′27.48″ \\
&\doteq 19°49′27″ \qquad \blacktriangleleft \text{ to the nearest second}
\end{aligned}
$$

From this point we shall omit the word "radian" if the angle is expressed as a number of radians. Thus, we shall write 3 for the angle to indicate that it is 3 radians.

Exercise 7.2   Angular Measure

Express each angle in problems 1 to 24 as a constant times π radians.

1	36°	**2**	30°	**3**	9°
4	12°	**5**	315°	**6**	27°
7	60°	**8**	75°	**9**	12°15′
10	20°40′	**11**	11°15′	**12**	6°15′
13	3°20′	**14**	18°45′	**15**	32°24′
16	8°24′	**17**	2°5′10″	**18**	16°8′24″
19	1°5′42″	**20**	10°7′30″	**21**	3.1 right angles
22	4.7 right angles	**23**	1.7 revolutions	**24**	3.4 revolutions

Express the angle given in each of problems 25 to 40 in terms of degrees, minutes, and seconds.

25	$\pi/4$	**26**	$\pi/12$	**27**	$\pi/15$
28	$\pi/36$	**29**	$2\pi/5$	**30**	$3\pi/4$
31	$7\pi/12$	**32**	$5\pi/9$	**33**	$5\pi/32$
34	$4\pi/27$	**35**	$7\pi/36$	**36**	$9\pi/64$
37	2.6 radians	**38**	3.7 radians	**39**	1.3 radians
40	5.9 radians				

7.5   ARC LENGTH IN TERMS OF THE CENTRAL ANGLE

In order to find the arc length intercepted by a specified central angle, we need only make use of the definition of a radian. Thus, if on a circle of radius r a central angle of θ radians intercepts an arc of length s, then

$$\frac{s}{r} = \theta$$

since there is 1 radian in the central angle for each time the radius can be laid off along the arc. Consequently, on multiplying each member by r, we see that the arc length intercepted on a circle of radius r by a central angle of θ radians is

Arc length

$$s = r\theta \qquad\qquad (7.5)$$

EXAMPLE

What length of arc is subtended by a central angle of 75° on a circle 13.7 inches in radius?

Solution

In order for (7.5) to be applicable, we must express the angle in radians. Thus,

$$75° = \tfrac{75}{180}(180°) = \tfrac{5}{12}\pi \text{ radians}$$

and by use of (7.5) we have

$$s = 13.7\,\frac{5\pi}{12}\text{ inches} = 17.9\text{ inches}$$

7.6 LINEAR AND ANGULAR SPEEDS

Angular speed
Linear speed

If a ray revolves at a uniform rate in a plane about its origin, the angle through which it revolves in a unit of time is called its *angular speed* and the distance a point on the ray moves in a unit of time is called its *linear speed*. The angular speed of the second hand of a watch is $360° = 2\pi$ radians $= 1$ revolution per minute. If the hand is 0.5 inch in length, the linear speed of its tip is $2\pi(0.5) = \pi$ inches per minute by use of (7.5).

If we designate the angular speed per second of a ray that moves through an angle of θ radians in t seconds by ω, we have

$$\omega = \frac{\theta}{t} \qquad\qquad (7.6)$$

Furthermore, if P is a point on the ray that is revolving about O with $OP = r$, we can use $s = r\theta$ and (7.6) to obtain an expression for the linear speed v in terms of r and ω. Dividing each member of $s = r\theta$ by t gives $s/t = r\theta/t$. Therefore,

$$v = r\omega \qquad\qquad (7.7)$$

Linear speed in terms of angular speed

since $s/t = v$ and $\theta/t = \omega$, where s is distance, t is time, and v is speed. Therefore, *the linear speed of a point on a revolving ray is equal to the angular speed of the ray times the distance from the center of rotation to the point.*

EXAMPLE   The rear wheel of a wagon has a radius of 2.5 feet and is revolving at 16 revolutions per minute. What is the linear speed of a point on the rim, in miles per hour?

Solution   We must express the angular speed in radians per unit of time to use (7.7). Thus, 16 revolutions per minute = $16(2\pi)$ radians per minute = 32π radians per minute. Therefore, $v = r\omega$ becomes

$$v = 2.5(32\pi) \text{ feet per minute}$$
$$= 80\pi \text{ feet per minute}$$
$$= 80\pi(60) \text{ feet per hour} \qquad \blacktriangleleft \; 1 \text{ hour} = 60 \text{ minutes}$$
$$= \frac{80\pi(60)}{5{,}280} \text{ miles per hour} \qquad \blacktriangleleft \; 1 \text{ mile} = 5{,}280 \text{ feet}$$
$$= \frac{10\pi}{11} = 2.86 \text{ miles per hour}$$

Exercise 7.3   Arc Length and Speeds

The first number in each of problems 1 to 8 is a measure of a central angle, and the second is the radius of the circle. Find the intercepted arc to three digits in problems 1 to 4 and to four digits in problems 5 to 8.

1   0.234 radian, 6.37 centimeters
2   2.07 radians, 4.31 centimeters
3   3.71 radians, 8.63 centimeters
4   0.763 radian, 87.3 centimeters
5   106°25′, 3.251 feet
6   82°33′, 26.44 feet
7   115°32′, 98.43 feet
8   57°42′, 11.77 feet

The arc length is followed by the radius in each of problems 9 to 16. Find the central angle in radians to three digits in problems 9 to 12 and to four digits in problems 13 to 16.

9   3.75 meters, 5.04 meters
10   28.3 meters, 78.7 meters
11   0.926 feet, 0.489 feet
12   0.858 feet, 1.73 feet
13   98.3 yards, 172 yards
14   803 yards, 476 yards
15   37.1 centimeters, 104 centimeters
16   7.43 centimeters, 19.6 centimeters

17 Through how many radians does the second hand of a watch move in 31 seconds?

18 Through how many radians does the hour hand of a grandmother clock move in 3 hours and 36 minutes?

19 The minute hand of a town clock is 4 feet long. How fast does its tip move?

20 How far does the tip of a 4-inch clock hand move in 2 hours and 54 minutes?

21 A bucket is drawn from a well by pulling a rope over a pulley. Find the radius of the pulley if the bucket is raised 63.4 inches while the pulley is turned through 3.47 revolutions.

22 A curve on a highway subtends an angle of 34° on a circle of radius 1,700 feet. How long will it take a car at 45 miles per hour to round the curve?

23 An engine has drive wheels 5 feet in diameter and is traveling 60 miles per hour. Find the angular velocity of the drive wheels in radians per second.

24 Find the latitude and longitude of a point that is 3,000 miles south of the equator and 3,229 miles east of the meridian through Greenwich, England. The longitude of Greenwich is 0°, and we shall use the radius of the earth as 4,000 miles.

7.7 STANDARD POSITION OF AN ANGLE

Standard position

In order to define the ratios of trigonometry that are associated with an angle, it is desirable to place the angle in a specified position. We say that an angle is in *standard position* relative to the coordinate axes if its vertex is at the origin and its initial side is along the positive X axis. The two angles θ and α are in standard position in Fig. 7.5.

Coterminal angles

If two angles in standard position have the same terminal side, they are called *coterminal angles*. Thus θ and α in Fig. 7.5 are coterminal angles. We say that an angle is in the quadrant in which its terminal side lies; both θ and α are in the second quadrant.

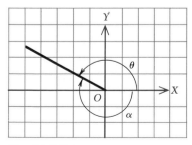

Figure 7.5

The undirected distance between the origin and a point on the ter-
Radius vector   minal side of an angle is called the *radius vector* of the point.

7.8 THE DEFINITION OF THE TRIGONOMETRIC RATIOS

The coordinates and radius vector of a point on the terminal side of an angle are used in defining the six ratios on which trigonometry is based. Before proceeding further, we shall see that the ratio of any two of the abscissa, ordinate, and radius vector of a point on the terminal side of an angle does not depend on the particular point that is chosen.

Let θ be an angle in standard position, as shown in Fig. 7.6, and let P and Q be two points on the terminal side. Now drop the perpendiculars PR and QS to the X axis and thus obtain the similar triangles ORP and OSQ. The ratios of the corresponding sides of these triangles are then equal, and we have

$$\frac{OR}{OP} = \frac{OS}{OQ} \qquad \frac{RP}{OP} = \frac{SQ}{OQ} \qquad \frac{RP}{OR} = \frac{SQ}{OS}$$

$$\frac{OP}{OR} = \frac{OQ}{OS} \qquad \frac{OP}{RP} = \frac{OQ}{SQ} \qquad \frac{OR}{RP} = \frac{OS}{SQ}$$

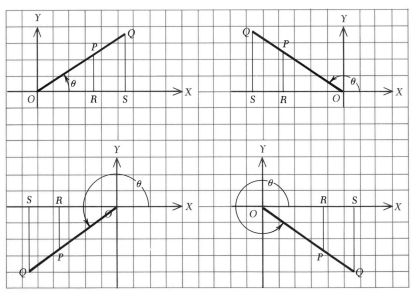

Figure 7.6

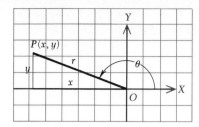

Figure 7.7

Furthermore, *OR*, *RP*, and *OP* are the abscissa, the ordinate, and the radius vector of *P*; and *OS*, *SQ*, and *OQ* are the coordinates and radius vector of *Q*. Consequently, the values of the six ratios that can be formed from the abscissa, ordinate, and radius vector of a point on the terminal side of an angle in standard position are independent of the particular point chosen; hence, they depend solely on the angle. They are called the *trigonometric ratios* associated with the angle, are given names, and are defined and abbreviated as follows. If θ is any angle in standard position, as in Fig. 7.7, and if *x*, *y*, and *r* are the abscissa, ordinate, and radius vector, respectively, of any point on its terminal side, then

The ratios of an angle

$$\text{sine } \theta = \frac{y}{r} = \sin \theta \qquad \qquad \text{cotangent } \theta = \frac{x}{y} = \cot \theta, \ y \neq 0$$

$$\text{cosine } \theta = \frac{x}{r} = \cos \theta \qquad \qquad \text{secant } \theta = \frac{r}{x} = \sec \theta, \ x \neq 0$$

$$\text{tangent } \theta = \frac{y}{x} = \tan \theta, \ x \neq 0 \qquad \text{cosecant } \theta = \frac{r}{y} = \csc \theta, \ y \neq 0$$

Signs of the ratios

The sign of each trigonometric ratio of an angle is determined by the usual convention of signs in division. Thus, the tangent of a second-quadrant angle is negative since, for a point in the second quadrant, the abscissa is negative and the ordinate is positive. Similarly, we find that all trigonometric ratios of first-quadrant angles, the sine and cosecant of second-quadrant angles, the tangent and cotangent of third-quadrant angles, and the cosine and secant of fourth-quadrant angles are positive.

The angle θ is not necessarily less than 2π radians, since the ray *OP* can make any real number of revolutions about *O* before coming to rest. Hence θ can represent any number of degrees or radians.

The definition of the trigonometric ratios was given earlier in this section with angles measured in degrees and radians, as the domain. We shall now extend the domain of the trigonometric ratios to all real numbers by accepting the definition that *a trigonometric ratio of any real number is equal to the same ratio of that number of radians.*

By use of this definition, we conclude that $\sin \pi/6 = \sin (\pi/6 \text{ radians})$ and $\tan 2 = \tan (2 \text{ radians})$.

157

7.9 RELATION BETWEEN THE ABSCISSA, ORDINATE, AND RADIUS VECTOR

Radius vector

If $P(x, y)$ is any point in the cartesian plane, then the undirected distance between the origin and P is called the *radius vector* of P, as pointed out in Sec. 7.7, and it will be represented by r. If in Fig. 7.8 the line PQ is drawn perpendicular to the X axis, then $OQ = x$ and $QP = y$ either are the sides of a right triangle with $OP = r$ as its hypotenuse or are the negatives of the sides of such a triangle. Now, since the square of a number and the square of its negative are equal, we see that

$$x^2 + y^2 = r^2 \tag{7.8}$$

Relation of abscissa, ordinate, and radius vector

for *any* point $P(x, y)$. This may be stated in words as follows: *The sum of the squares of the abscissa and ordinate of a point is equal to the square of the radius vector of the point.* Thus if the point is $(8, 15)$, then

$$r^2 = 8^2 + 15^2 = 64 + 225 = 289$$

and hence the radius vector is $\sqrt{289} = 17$. We used only 17 instead of ± 17 because the radius vector was defined as a positive number.

EXAMPLE

If points A and B in Fig. 7.9 are on the same line through the origin, find the abscissa and radius vector of B provided its ordinate is 10 and the coordinates of A are as indicated by $A(12, 5)$.

Solution

Since triangles OCA and ODB are similar, we have

$$\frac{OD}{OC} = \frac{DB}{CA}$$

Consequently, putting in the lengths of these segments, we get

$$\frac{x}{12} = \frac{10}{5}$$

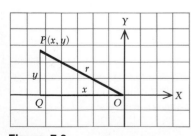

Figure 7.8

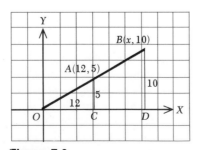

Figure 7.9

Therefore, $5x = 120$ and $x = 24$. We now obtain the length of OB by use of (7.8). Thus

$$r^2 = 24^2 + 10^2 = 676$$
$$r = 26$$

Exercise 7.4   Trigonometric Ratios

Find the one of x, y, and r that is missing in each of problems 1 to 16 by use of $x^2 + y^2 = r^2$.

1	$x = 3$, $y = 4$	**2**	$x = 5$, $y = -12$
3	$x = -15$, $y = 8$	**4**	$x = -7$, $y = 24$
5	$x = -5$, $r = 13$	**6**	$x = 15$, $r = 17$
7	$x = 7$, $r = 25$	**8**	$x = -3$, $r = 5$
9	$y = -8$, $r = 17$	**10**	$y = -24$, $r = 25$
11	$y = 4$, $r = 5$	**12**	$y = 12$, $r = 13$

13 $x = 8$, $r = 17$, $P(x, y)$ in the first quadrant
14 $x = -7$, $r = 25$, $P(x, y)$ in the third quadrant
15 $y = 4$, $r = 7$, $P(x, y)$ in the second quadrant
16 $y = -7$, $r = 11$, $P(x, y)$ in the fourth quadrant

In each of problems 17 to 24, A and B are on the same line through the origin. Find the unknown coordinate and radius vector of B in each problem.

17	$A(5, 12)$, $B(10, y)$	**18**	$A(8, -15)$, $B(-4, y)$
19	$A(-7, -24)$, $B(21, y)$	**20**	$A(-3, 4)$, $B(-6, y)$
21	$A(12, -5)$, $B(x, 15)$	**22**	$A(15, 8)$, $B(5, y)$
23	$A(-4, 3)$, $B(x, 9)$	**24**	$A(-5, -12)$, $B(x, 24)$

Draw an angle in standard position that satisfies the conditions in each of problems 25 to 32 and find the values of the trigonometric ratios of the angle in each case.

25 $P(3, -4)$ is on the terminal side
26 $P(-8, 15)$ is on the terminal side
27 $P(6, 8)$ is on the terminal side
28 $P(-24, -7)$ is on the terminal side
29 $P(3, y)$, $y < 0$, is 5 units out on the terminal side
30 $P(x, -15)$, $x > 0$, is 17 units out on the terminal side
31 $P(7, y)$, $y > 0$, is 25 units out on the terminal side
32 $P(x, -15)$, $x < 0$, is out 17 units on the terminal side

7.10 THE TRIGONOMETRIC FUNCTIONS, THEIR DOMAIN AND RANGE

In this section we shall consider six ordered pairs of numbers that can be associated with a point P on the circumference of a circle. We begin by drawing a unit circle as in Fig. 7.10, choosing a point $P(x, y)$ on the circumference, drawing a ray through the center O and P, and constructing perpendicular lines AP and BC to the X axis. If we designate the length of the arc BP by a, then the radius OP is 1. Thus a is the measure of the length of the arc in linear units and of the central angle in radians. Furthermore, $OA = x$, $AP = y$, $\sin a = y/1$, and $\cos a = x/1$. Therefore, $x = \cos a$ and $y = \sin a$ and, since the triangles OAP and OBC are similar, it follows that

$$\frac{BC}{OB} = \frac{y}{x} = \tan a$$

and

$$\frac{OC}{OB} = \frac{OP}{OA} = \frac{1}{x} = \frac{1}{\cos a} = \sec a \qquad \blacktriangleleft \text{ since } OP = 1$$

Finally, since $OB = 1$, we have

$$BC = \tan a \qquad \text{and} \qquad OC = \sec a$$

We can now form the following pairs of numbers that are associated with the point $P(x, y)$ on the circumference of the unit circle about O:

$$(a, y) = (a, \sin a) \tag{1}$$
$$(a, x) = (a, \cos a) \tag{2}$$
$$(a, \overline{BC}) = (a, y/x) = (a, \tan a) \tag{3}$$
$$(a, \overline{OC}) = (a, 1/x) = (a, \sec a) \tag{4}$$

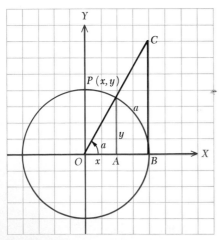

Figure 7.10

(Notice that the line *BC* of Fig. 7.10 which has length tan *a*, is actually a tangent line to the circle.)

We can obtain two additional pairs by use of Fig. 7.11. Since triangles *OED* and *PAO* are similar, we have

$$\frac{ED}{OE} = \frac{OA}{AP} = \frac{x}{y} = \cot a$$

and
$$\frac{OD}{OE} = \frac{OP}{AP} = \frac{1}{y} = \frac{1}{\sin a} = \csc a \qquad \blacktriangleleft \text{ since } OP = 1 \text{ and } y = \sin a$$

Since *OE* = 1, it follows that *ED* = cot *a* and *OD* = csc *a*. Thus, we have two more pairs of numbers associated with *P*. They are

$$(a, \overline{ED}) = (a, \cot a) \qquad\qquad\qquad (5)$$

and
$$(a, \overline{OD}) = (a, \csc a) \qquad\qquad\qquad (6)$$

We showed in Sec. 7.8 that the value of a trigonometric ratio depends on the coordinates of a point *P* on the terminal side of the angle θ but is independent of the choice of *P*, and that θ can represent any real number of radians. Therefore, the argument in this section is not affected if *P* makes any integral number of revolutions about *O* before coming to rest at the point indicated in the figure. Consequently, the arc length traced by *P* and the number of radians in the angle generated by *OP* may be any real number. This number is positive if the rotation is counterclockwise and negative if it is clockwise.

We shall consider the ordered pair $(a, y) = (a, \sin a)$ as *P* moves around the unit circle. As this happens, *a* varies over the set of real numbers and $y = \sin a$ varies over the set of numbers from -1 to 1, inclusive, since $|y| \le 1$. Furthermore, there is exactly one value of *y* corresponding to each value of *a*. Therefore, we have a set of ordered pairs of real numbers $\{(a, \sin a)\}$ that satisfies the definition of a function, and it is

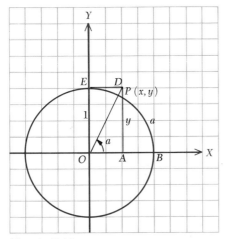

Figure 7.11

called the *sine function*. The *domain* is the set of all real numbers, and the *range* is $\{y|-1 \leq y \leq 1\}$.

We can show similarly that as P moves around the circle, each of the other five pairs of numbers associated with P generates a function. They are called *trigonometric functions*. We shall tabulate these functions and show the domain and range of each. The tabulated values are justified below the table. In the table R stands for the set of all real numbers.

Domain and range of the trigonometric functions

Function name	Function	Domain	Range
sine	$\{(a, \sin a)\}$	R	$\{y|-1 \leq y \leq 1\}$
cosine	$\{(a, \cos a)\}$	R	$\{y|-1 \leq y \leq 1\}$
tangent	$\{(a, \tan a)\}$	$R - \{n\pi/2 | n \text{ is an odd integer}\}$	R
cotangent	$\{(a, \cot a)\}$	$R - \{n\pi | n \text{ is an integer}\}$	R
secant	$\{(a, \sec a)\}$	$R - \{n\pi/2 | n \text{ is an odd integer}\}$	$R - \{y|-1 < y < 1\}$
cosecant	$\{(a, \csc a)\}$	$R - \{n\pi | n \text{ is an integer}\}$	$R - \{y|-1 < y < 1\}$

Values of $x = \cos a$ and $y = \sin a$ exist for all values of a; hence, the domain of the sine function and of the cosine function is R. Furthermore, $-1 \leq x \leq 1$ and $-1 \leq y \leq 1$; hence, the range is as specified in the table.

There is no value of $\tan a$ or $\sec a$ for $x = 0$ since $\tan a = y/x$ and $\sec a = 1/x$. Consequently, we must delete $n\pi/2$ for n is an odd integer from R to obtain the domain of the tangent and secant functions. If, in $\tan a = y/x$, $y = 0$, then $\tan a = 0$. As $|y|$ increases from zero toward 1, $|x|$ decreases from 1 toward zero. Consequently, $|y|/|x|$ increases continuously from zero and approaches infinity. Thus, if x and y are of the same sign, then $\tan a$ varies continuously over the set of real positive numbers, and if they have opposite signs, then $\tan a$ varies continuously over the sets of negative integers. Therefore, the domain and range of $\tan a$ are as given in the table.

It can be shown similarly that the domain and range of $\cot a$ are as tabulated above.

We shall now consider the secant function and determine its domain and range. Since y takes on all values from -1 to 1, inclusive, it follows that $1/|y|$ is never less than 1; furthermore, as $|y|$ decreases from 1 toward zero, $1/y$ increases continuously over the set of all real numbers that are greater than 1 if y is positive and decreases continuously over the set of real numbers that are less than -1 if y is negative. Hence, the range of $\sec a$ is as in the table. We must delete all odd multiples of $\pi/2$ from R to obtain the domain of $\sec a$ since $y = 0$ and no value exists for $1/y$ if a is an odd multiple of $\pi/2$.

We can show similarly that the domain and range of $\csc a$ are as tabulated.

Trigonometric
function
value

We shall refer to the second number in each trigonometric function as the *trigonometric function value* since the second number in each ordered pair in a function is called the function value. Hereafter, we may refer to sin *a* as a trigonometric function value or as a trigonometric ratio. Similar statements are true for cos *a*, tan *a*, cot *a*, sec *a*, and csc *a*.

7.11 GIVEN ONE FUNCTION VALUE, TO FIND THE OTHERS

If we know the value of one of the trigonometric ratios of an angle and the quadrant in which the angle lies, we can find the value of each of the other five by use of Eq. (7.8) and the definition of the functions.

EXAMPLE

If $\tan \theta = \frac{5}{12}$ and θ is in the third quadrant, find the value of each of the other functions.

Solution

We know that the terminal side of θ is in the third quadrant and that there is a point on the terminal side whose ordinate is -5 and whose abscissa is -12, as shown in Fig. 7.12. Therefore, by making use of the relation $r^2 = x^2 + y^2$, we have

$$r^2 = (-12)^2 + (-5)^2 = 169$$

Hence,
$$r = 13$$

Consequently, the function values are

$$\sin \theta = -\frac{5}{13} \qquad \cos \theta = -\frac{12}{13}$$

$$\tan \theta = \frac{5}{12}, \text{ as given} \qquad \cot \theta = \frac{-12}{-5} = \frac{12}{5}$$

$$\sec \theta = \frac{13}{-12} = -\frac{13}{12} \qquad \csc \theta = \frac{13}{-5} = -\frac{13}{5}$$

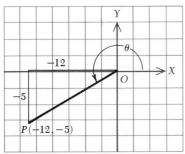

Figure 7.12

If the value of a ratio is given but the quadrant in which the angle lies is not specified, there are two angles to be considered. For example, if $\tan \theta = \frac{5}{12}$, as in the above example, there is an angle θ in the first quadrant and an angle θ in the third quadrant. Furthermore, there is a point $Q(12, 5)$ on the terminal side of the first-quadrant angle and a point $P(-12, -5)$ on the terminal side of the third-quadrant angle. Moreover, $OQ = OP = r = 13$. Therefore, the values of the functions of the angle can be found by use of the definitions. There are two values for each of four of the other functions but only one for the function whose definition uses the same quantities as the given function.

7.12 FUNCTION VALUES OF 30°, 45°, AND THEIR MULTIPLES

The computation of the function values of angles in general is beyond the scope of an elementary book on trigonometry. We can, however, find the function values of 30°, 45°, and their multiples by using the definition of the trigonometric ratios of an angle and some theorems from plane geometry.

We first find the function values of 45°. If we chose any point P on the terminal side of a 45° angle that is in standard position and drop a perpendicular to the X axis striking it at Q, we know that $OQ = QP$ because they are sides opposite equal angles; hence, the abscissa and ordinate of P are equal. If we choose them to be 1, then the radius vector r is seen to be $\sqrt{2}$ by use of relation (7.8). Now, by applying the definition of the trigonometric ratios to the angle in Fig. 7.13, we have

Function values of 45°

$$\sin 45° = \frac{1}{\sqrt{2}} = \frac{\sqrt{2}}{2} \qquad \cos 45° = \frac{1}{\sqrt{2}} = \frac{\sqrt{2}}{2}$$

$$\tan 45° = \frac{1}{1} = 1 \qquad \cot 45° = \frac{1}{1} = 1$$

$$\sec 45° = \frac{\sqrt{2}}{1} = \sqrt{2} \qquad \csc 45° = \frac{\sqrt{2}}{1} = \sqrt{2}$$

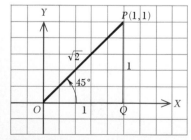

Figure 7.13

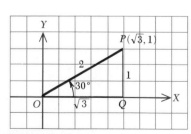

Figure 7.14

We now find the function values of 30°. If we draw a 30° angle in standard position, as in Fig. 7.14, and lay off $OP = 2$ on the terminal side, then the perpendicular PQ to the X axis is of length 1 because the side opposite the 30° angle in a 30° right triangle is one-half as long as the hypotenuse. Now, by using relation (7.8), we find that $OQ = \sqrt{3}$; hence, the coordinates of P are as given in Fig. 7.14. Therefore, the use of the definition of the trigonometric ratios leads to

Function values of 30°

$$\sin 30° = \frac{1}{2} \qquad\qquad \cos 30° = \frac{\sqrt{3}}{2}$$

$$\tan 30° = \frac{1}{\sqrt{3}} = \frac{\sqrt{3}}{3} \qquad \cot 30° = \frac{\sqrt{3}}{1} = \sqrt{3}$$

$$\sec 30° = \frac{2}{\sqrt{3}} = \frac{2\sqrt{3}}{3} \qquad \csc 30° = \frac{2}{1} = 2$$

In order to find the values of the ratios for 60°, we shall draw a 60° angle in standard position, choose a point P that is 2 units out on the radius vector r, and drop a perpendicular PQ to the X axis, as in Fig. 7.15. Then $OQ = 1$, since OPQ is a 30° angle. Now, by using relation (7.8), we find that $QP = \sqrt{3}$; hence, the coordinates of P are as given in Fig. 7.15. Finally, by use of the definition of the trigonometric ratios we have

Function values of 60°

$$\sin 60° = \frac{\sqrt{3}}{2} \qquad\qquad \cos 60° = \frac{1}{2}$$

$$\tan 60° = \frac{\sqrt{3}}{1} = \sqrt{3} \qquad \cot 60° = \frac{1}{\sqrt{3}} = \frac{\sqrt{3}}{3}$$

$$\sec 60° = \frac{2}{1} = 2 \qquad \csc 60° = \frac{2}{\sqrt{3}} = \frac{2\sqrt{3}}{3}$$

The values of the trigonometric ratios of multiples of 30° and 45° that are not also multiples of 90° can be found by methods similar to those used above. For example, in order to find the values of the ratios for 135°, we begin by drawing the angle in standard position, as in Fig. 7.16, choosing a

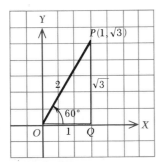

Figure 7.15

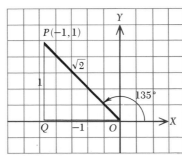

Figure 7.16

point P on it, dropping a perpendicular PQ to the X axis, and noticing that angle QOP is 45°. Therefore, OQ and QP are the same length but OQ is in the negative direction. Consequently, if we let $QP = 1$, then $OP = -1$ and $OP = \sqrt{2}$ by relation (7.8). Now,

Function values of 135°

$$\sin 135° = \frac{1}{\sqrt{2}} = \frac{\sqrt{2}}{2} \qquad \cos 135° = \frac{-1}{\sqrt{2}} = -\frac{\sqrt{2}}{2}$$

$$\tan 135° = \frac{1}{-1} = -1 \qquad \cot 135° = \frac{-1}{1} = -1$$

$$\sec 135° = \frac{\sqrt{2}}{-2} = -\sqrt{2} \qquad \csc 135° = \frac{\sqrt{2}}{1} = \sqrt{2}$$

7.13 FUNCTION VALUES OF QUADRANTAL ANGLES

Quadrantal angle

If the terminal side of an angle in standard position coincides with a co-ordinate axis, we say that the angle is a *quadrantal angle*. Consequently, 0°, 90°, 180°, 270°, and angles that are coterminal with any one of these angles are quadrantal angles. Either the abscissa or the ordinate of a point on the terminal side of a quadrantal angle is zero. Therefore, in substituting in the definitions of the ratios, we have zero in some numerators and in some denominators. If we recall that zero divided by a nonzero number is zero and that we cannot divide by zero, we shall have no difficulty in dealing with quadrantal angles.

Figure 7.17 shows an angle of 180° in standard position. The point P on the terminal side is chosen to be 1 unit out from the origin; hence, its coordinates are shown in the figure. Now, applying the definitions and recalling the necessary facts about division if zero is involved, we have

$$\sin 180° = \frac{0}{1} = 0 \qquad \cos 180° = \frac{-1}{1} = -1$$

$$\tan 180° = \frac{0}{-1} = 0 \qquad \cot 180° = \frac{-1}{0}, \text{ not defined}$$

$$\sec 180° = \frac{1}{-1} = -1 \qquad \csc 180° = \frac{1}{0}, \text{ not defined}$$

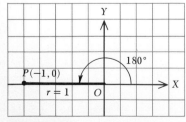

Figure 7.17

Exercise 7.5   Function Values

Construct an angle in standard position that satisfies the conditions in each of problems 1 to 8 and find the other function values of the angle.

1 $\sin \theta = \frac{4}{5}$, θ in the first quadrant
2 $\cos \theta = \frac{5}{13}$, θ in the fourth quadrant
3 $\tan \theta = \frac{8}{15}$, θ in the third quadrant
4 $\csc \theta = \frac{25}{7}$, θ in the second quadrant
5 $\cot \theta = -\frac{5}{12}$, θ in the fourth quadrant
6 $\sec \theta = -\frac{13}{12}$, θ in the second quadrant
7 $\cos \theta = -\frac{3}{5}$, θ in the third quadrant
8 $\sin \theta = -\frac{8}{17}$, θ in the fourth quadrant

Construct two positive angles less than $360°$ in standard position that satisfy the condition in each of problems 9 to 12 and find all missing function values.

9 $\tan \theta = \frac{24}{7}$ **10** $\sec \theta = \frac{17}{15}$
11 $\sin \theta = -\frac{12}{13}$ **12** $\cot \theta = -\frac{4}{3}$

Construct the angles given in problems 13 to 16, select a point on the terminal side of each, and find the function values.

13 $60°$, $135°$, $210°$ **14** $315°$, $330°$, $240°$
15 $180°$ $-225°$, $120°$ **16** $270°$, $-60°$, $-150°$

Make use of the function values of $30°$, $45°$, and their multiples to show that the statement in each of problems 17 to 28 is true. Note that the square of $\sin x$ is written as $\sin^2 x$ and that a similar notation is used for the other function values.

17 $\sin^2 240° + \cos^2 120° = 1$ **18** $1 + \tan^2 240° = \sec^2 300°$
19 $1 + \cot^2 (-120°) = \csc^2 300°$ **20** $\sec^2 150° - \tan^2 330° = 1$
21 $\sin 240° = 2 \sin 60° \sin 330°$ **22** $\csc 150° = \csc 120° \tan 240°$
23 $\csc 120° - \cot 240° = \cot 60°$
24 $\cot 225° = \cot 90° - \csc 270°$
25 $\cos 210° = \cos 90° \cos 240° + \sin 270° \sin 120°$
26 $\cos 330° + \cos 240° \tan 210° = \cot 240°$
27 $\tan 300° = \dfrac{\tan 60° + \cot 210°}{1 - \tan 300° \cot 330°}$
28 $\dfrac{\cos 235°}{\sin 135°} + \dfrac{\sin 315°}{\cos 135°} = 3 \sin 180°$

Decide which of the following statements are true and which are false.

29 $\sin 30° + \sin 90° = \sin 120°$

30 $2 \cos 45° = \cos 90°$

31 $\sin 90° = \cos^2 330° + \cos 300° \sin 150°$

32 $\cos 210° = \cos 90° \cos 240° + \sin 270° \sin 120°$

33 $4 \sin^3 30° = 3 \sin 150° + \cos 180°$

34 $\cos 240° = \cos 210° \cos 30° - \sin^2 150°$

35 $\cos \theta = 1.003$

36 $\sin \theta = -3.07$

37 $\cos 420° = \sqrt{\dfrac{1 + \sin 210°}{2}}$

38 $\cos 210° = \sqrt{\dfrac{1 - \cos 120°}{2}}$

39 $\dfrac{\sin 225°}{\sin 135°} - \dfrac{\cos 315°}{\cos 135°} = 2 \sin 540°$

40 $\cot 240° - \cot 330° = \dfrac{\sec 300°}{\sin 90° \sin 120°}$

Prove the statement in each of problems 41 to 44.

41 If $0 < \theta < \pi/2$, then $\sin \theta < \tan \theta$

42 If $0 < \theta < \pi/4$, then $\sin \theta < \cos \theta$

43 If $0 < \theta < \pi/2$, then $\csc \theta < \cot \theta$

44 If θ is acute, then $\sec \theta < \tan \theta$

Exercise 7.6 Review

Use $(3, -1)$ as the starting point, draw the directed segments described in problems 1 and 2, then find the coordinates of each terminal point.

1 3 units to the left, 4 units downward

2 5 units to the right, 3 units upward, 2 units to the left

3 Find the length of the segment between $(3, 2)$ and $(6, -2)$.

4 Find the length of the segment between $(-2, 1)$ and $(2, 3)$.

5 Find the length of the sides of the triangle with vertices at $(3, 5)$, $(0, 9)$, and $(-2, -7)$.

Express each angle in problems 6 to 8 as a constant times π radians.

6 $54°$ **7** $10°20'$ **8** 1.7 revolutions

Express each angle in problems 9 to 11 in degrees, minutes, and seconds.

9 $\pi/15$ **10** $7\pi/64$ **11** 2.3 radians

12 If the central angle is 0.573 radian in a circle of radius 4.79 centimeters, find the length of the intercepted arc.

13 If a central angle intercepts an arc of 7.92 centimeters on a circle of radius 5.83 centimeters, find the angle.

14 The minute hand of a grandfather clock is 13.2 centimeters long. How fast does its tip move?

15 Find the latitude and longitude of a point that is 1,720 miles north of the equator and 740 miles west of the meridian through Greenwich. Use 4,000 miles as the radius of the earth.

16 Find y if $P(x, y)$ is in the second quadrant and $x = -15$ and $r = 17$.

In problems 17 and 18, A and B are on the same line through the origin. Find the unknown coordinate and radius vector of B.

17 $A(12, 5)$, $B(-6, y)$ **18** $A(8, -15)$, $B(-16, y)$

Find the six trigonometric ratios of an angle θ in standard position that has the point given in each of problems 19, 20, and 21 on its terminal side.

19 $P(-4, 3)$ **20** $P(15, -8)$ **21** $P(-5, -12)$

22 Find the other function values if $\sin \theta = \frac{5}{13}$ and θ is in the second quadrant.

23 Find the other function values if $\tan \theta = -\frac{3}{4}$ and θ is in the fourth quadrant.

24 Find the other function values if $\csc \theta = \frac{17}{8}$.

25 Draw 60°, 150°, and 315° in standard position and find the function values of each.

26 Prove that $1 + \tan^2 315° = \sec^2 135°$.

27 Prove that $\cos 210° = -\cos 300° \cos 210° + \sin 120° \sin 330°$.

28 Prove that $\sin \theta > \cos \theta$ for $\pi < \theta < 5\pi/4$.

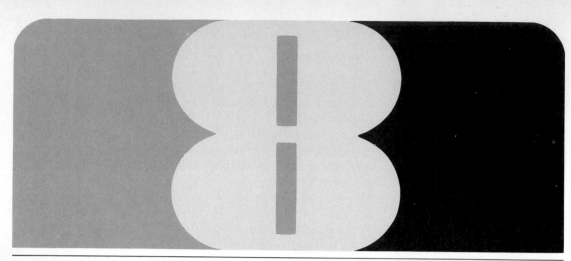

Fundamental Identities

Since there are six trigonometric ratios among the three numbers that are the measures of the abscissa, ordinate, and radius vector of a point, it should not be surprising that there are fundamental relations between the ratios. This chapter shall be devoted to the development and use of eight such relations between the function values. These relations are called *fundamental identities,* and they fall naturally into three sets because of the way in which they are derived. They can be used to change the form of a trigonometric expression. This is very important in trigonometry, calculus, and other branches of pure and applied mathematics.

8.1 EIGHT FUNDAMENTAL IDENTITIES

The trigonometric function values can be grouped in three pairs so that the numbers in each pair are reciprocal. Thus, by use of the definition of the ratios involved, we have

$$\sin \theta \csc \theta = \frac{y}{r} \frac{r}{y} = 1$$

Consequently, we know that

Reciprocal relations

$$\sin \theta \csc \theta = 1 \tag{8.1}$$

We can show similarly that

$$\cos \theta \sec \theta = 1 \tag{8.2}$$
and
$$\tan \theta \cot \theta = 1 \tag{8.3}$$

The reader should be able to solve each of these three equations for either of the function values in it since the forms so obtained are often quite useful.

By using the definitions again, we get

$$\frac{\sin \theta}{\cos \theta} = \frac{y/r}{x/r} = \frac{y}{x} = \tan \theta$$
and
$$\frac{\cos \theta}{\sin \theta} = \frac{x/r}{y/r} = \frac{x}{y} = \cot \theta$$

Consequently, we know that

$$\tan \theta = \frac{\sin \theta}{\cos \theta} \tag{8.4}$$
and
$$\cot \theta = \frac{\cos \theta}{\sin \theta} \tag{8.5}$$

In order to obtain the third set of fundamental identities, we shall use the relation

$$x^2 + y^2 = r^2 \tag{1}$$

If we divide each member of (1) by r^2, we have

$$\left(\frac{x}{r}\right)^2 + \left(\frac{y}{r}\right)^2 = \left(\frac{r}{r}\right)^2 = 1$$

Now, by use of the definitions of $\cos \theta$ and $\sin \theta$, we obtain

Pythagorean
relations

$$\mathbf{\cos^2 \theta + \sin^2 \theta = 1} \qquad\qquad (8.6)$$

$$1 + \mathbf{\tan^2 \theta = \sec^2 \theta} \qquad\qquad (8.7)$$

and $$1 + \mathbf{\cot^2 \theta = \csc^2 \theta} \qquad\qquad (8.8)$$

can be obtained by first dividing each member of (1) by x^2 and then by y^2 and applying the definitions each time.

8.2 CHANGING THE FORM OF A TRIGONOMETRIC EXPRESSION

The form of a trigonometric expression can be materially changed by use of the fundamental identities. The desirable form depends on the use to which the expression is to be put, and there is no one procedure for changing to the desired form. A considerable amount of practice is required if one is to become adept at problems of this type. Frequently, the required process involves performing the indicated algebraic operations before using one or more of the fundamental identities.

EXAMPLE 1   Change $4 + (\tan \theta - \cot \theta)^2$ to $\sec^2 \theta + \csc^2 \theta$.

Solution   If we square the binomial in the first expression, we have

$$
\begin{aligned}
4 + (\tan \theta - \cot \theta)^2 &= 4 + \tan^2 \theta - 2 \tan \theta \cot \theta + \cot^2 \theta \\
&= 4 + \tan^2 \theta - 2 + \cot^2 \theta \qquad \blacktriangleleft \text{ using (8.3)} \\
&= 2 + \tan^2 \theta + \cot^2 \theta \\
&= 1 + \tan^2 \theta + 1 + \cot^2 \theta \qquad \blacktriangleleft 2 = 1 + 1 \\
&= \sec^2 \theta + \csc^2 \theta \qquad \blacktriangleleft \text{ using (8.7) and (8.8)}
\end{aligned}
$$

EXAMPLE 2   Change $\tan \theta \ (\sin \theta + \cot \theta \cos \theta)$ to $\sec \theta$.

Solution $\tan \theta \ (\sin \theta + \cot \theta \cos \theta) = \tan \theta \sin \theta + \tan \theta \cot \theta \cos \theta$

$\qquad\qquad\qquad\qquad\qquad\qquad \blacktriangleleft$ performing the indicated operation

$$
\begin{aligned}
&= \tan \theta \sin \theta + \cos \theta \qquad \blacktriangleleft \text{ using (8.3)} \\
&= \frac{\sin \theta}{\cos \theta} \sin \theta + \cos \theta \qquad \blacktriangleleft \text{ using (8.4)} \\
&= \frac{\sin^2 \theta + \cos^2 \theta}{\cos \theta} \qquad \blacktriangleleft \text{ getting a common denominator} \\
&= \frac{1}{\cos \theta} \qquad \blacktriangleleft \text{ using (8.6)} \\
&= \sec \theta \qquad \blacktriangleleft \text{ using (8.2)}
\end{aligned}
$$

Exercise 8.1 Changing the Form of a Trigonometric Expression

Change the first expression into the second in each of problems 1 to 24.

1 $\sin \theta \cot \theta$, $\cos \theta$

2 $\sec \theta \cot \theta$, $\csc \theta$

3 $\tan \theta \cos \theta$, $\sin \theta$

4 $\sin \theta \sec \theta$, $\tan \theta$

5 $\dfrac{\sin \theta}{\tan \theta}$, $\cos \theta$

6 $\dfrac{\csc \theta}{\sec \theta}$, $\cot \theta$

7 $\dfrac{\cos \theta}{\cot \theta}$, $\sin \theta$

8 $\dfrac{\tan \theta}{\sin \theta}$, $\sec \theta$

9 $\tan \theta + \cot \theta$, $\sec \theta \csc \theta$

10 $\sin \theta + \cos \theta \cot \theta$, $\csc \theta$

11 $\sec \theta - \cos \theta$, $\sin \theta \tan \theta$

12 $\tan \theta + \sec \theta$, $(1 + \sin \theta) \sec \theta$

13 $\sin \theta \, (\csc \theta + \sin \theta \sec^2 \theta)$, $\sec^2 \theta$

14 $(1 + \cos \theta)(1 - \cos \theta)$, $\sin^2 \theta$

15 $(\sec \theta + 1)(\sec \theta - 1)$, $\tan^2 \theta$

16 $\cos \theta (\sec \theta + \cos \theta \csc^2 \theta)$, $\csc^2 \theta$

17 $\dfrac{\sec^2 \theta - 1}{\sin \theta}$, $\sec \theta \tan \theta$

18 $\dfrac{1 + \tan^2 \theta}{\csc^2 \theta}$, $\tan^2 \theta$

19 $\dfrac{1 - \cos^2 \theta}{\tan \theta}$, $\cos \theta \sin \theta$

20 $\dfrac{\csc^2 \theta}{1 + \tan^2 \theta}$, $\cot^2 \theta$

21 $\dfrac{\sec \theta}{\sin \theta} - \dfrac{\sin \theta}{\cos \theta}$, $\cot \theta$

22 $\dfrac{\csc \theta}{\cos \theta} - \dfrac{\cos \theta}{\sin \theta}$, $\tan \theta$

23 $\dfrac{\tan \theta}{\csc \theta} + \dfrac{\sin \theta}{\tan \theta}$, $\sec \theta$

24 $\dfrac{\cos \theta}{\cot \theta} + \dfrac{\cot \theta}{\sec \theta}$, $\csc \theta$

8.3 IDENTITIES

Identity

In Chap. 4, we defined an equation and stated that the solution set may be only certain values of the variable or every value for which both numbers are defined. The first type is called a conditional equation, and the second type is called an *identity*. The conditional equation was discussed in Chap. 4 and will be further discussed in Chaps. 11, 12, 17, and 24. The identity will be considered in this chapter and again in Chap. 15. The eight relations that were derived in Sec. 8.1 are identities.

A desirable procedure for proving that an equation is an identity is to reduce one member to the same form as the other member. There is no one method for performing this reduction but the following suggestions should prove helpful.

Suggestions for proving identities

1 In most problems, work with the more complicated member.

2 If a member contains one or more indicated operations, work with that member.

3 If one member contains only one function value and the other contains two or more, express the function values in the second member in terms of the one in the first member by use of identities.

4   If the denominator of a member contains only one term and the numerator is the sum or difference of several terms, express the number as the sum or difference of several terms which have the several numerators as numerator and the common denominator as denominator; e.g., use $(a + b)/c = a/c + b/c$.

5   Factor any member that is factorable.

6   Multiply the numerator and denominator of one or more terms by the same factor in order to get a common denominator.

7   If none of the above procedures seem to be applicable, express the function values in the more complicated member in terms of sines and cosines or in terms of the function, if only one, that appears in the other member.

EXAMPLE 1   Prove that

$$\frac{\cos A}{\csc A - 1} + \frac{\cos A}{\csc A + 1} = 2 \tan A$$

is an identity.

Solution   The left member is the more complicated; hence, we shall work with it and begin by performing the indicated addition.   The common denominator is $\csc^2 A - 1$, and by using it we have

$$\frac{\cos A}{\csc A - 1} + \frac{\cos A}{\csc A + 1} = \frac{\cos A(\csc A + 1) + \cos A(\csc A - 1)}{\csc^2 A - 1}$$

$$= \frac{\cos A \csc A + \cos A + \cos A \csc A - \cos A}{\csc^2 A - 1}$$

$$= \frac{2 \cos A \csc A}{\cot^2 A} \qquad \blacktriangleleft \text{ removing parentheses}$$

$$\qquad\qquad\qquad\qquad \blacktriangleleft \text{ collecting and using (8.8)}$$

$$= \frac{(2 \cos A)/(\sin A)}{\cot^2 A} \qquad \blacktriangleleft \text{ using (8.1)}$$

$$= \frac{2 \cot A}{\cot^2 A} \qquad \blacktriangleleft \text{ using (8.5)}$$

$$= \frac{2}{\cot A} \qquad \blacktriangleleft \text{ factoring out cot } A$$

$$= 2 \tan A \qquad \blacktriangleleft \text{ using (8.3)}$$

EXAMPLE 2   Prove that $\cos^4 B - \sin^4 B = \cos^2 B - \sin^2 B$ is an identity.

Solution   If we factor the left member, we have

$$\cos^4 B - \sin^4 B = (\cos^2 B + \sin^2 B)(\cos^2 B - \sin^2 B)$$
$$= \cos^2 B - \sin^2 B \qquad \blacktriangleleft \text{ using (8.6)}$$

EXAMPLE 3

Prove that

$$\frac{1}{\sec A - \tan A} = \sec A + \tan A$$

is an identity.

Solution

If we multiply numerator and denominator of the left member of the given equation by $\sec A + \tan A$, we have

$$\frac{1}{\sec A - \tan A} = \frac{\sec A + \tan A}{(\sec A - \tan A)(\sec A + \tan A)}$$

$$= \frac{\sec A + \tan A}{\sec^2 A - \tan^2 A} \quad \blacktriangleleft \text{performing the indicated operation}$$

$$= \sec A + \tan A \quad \blacktriangleleft \text{using (8.7)}$$

EXAMPLE 4

Prove that

$$\frac{\cos^3 x - \cos x + \sin x}{\cos x} = \tan x - \sin^2 x$$

is an identity.

Solution

If we put each term of the numerator separately over the denominator, we have

$$\frac{\cos^3 x - \cos x + \sin x}{\cos x} = \frac{\cos^3 x}{\cos x} - \frac{\cos x}{\cos x} + \frac{\sin x}{\cos x}$$

$$= \cos^2 x - 1 + \tan x \quad \blacktriangleleft \text{using (8.4)}$$

$$= -\sin^2 x + \tan x \quad \blacktriangleleft \text{using (8.6)}$$

Exercise 8.2   Trigonometric Identities

Prove that each of the following equations is an identity.

1 $\sin x(\csc x - \sin x) = \cos^2 x$ 2 $\csc x(\csc x - \sin x) = \cot^2 x$
3 $\cos x(\sec x - \cos x) = \sin^2 x$ 4 $\tan x(\tan x + \cot x) = \sec^2 x$
5 $(\sec x + \tan x)(\sec x - \tan x) = 1$
6 $(\csc x + \cot x)(\csc x - \cot x) = 1$
7 $(1 + \sec x)(\sec x - 1) = \tan^2 x$ 8 $(\cos x + 1)(-\cos x + 1) = \sin^2 x$
9 $\dfrac{\cos x - \sin x}{\sin x} = \cot x - 1$ 10 $\dfrac{\sec^2 x - 1}{\tan x} = \tan x$

11 $\dfrac{1 + \sin x - \sin^2 x}{\cos x} = \cos x + \tan x$

12 $\dfrac{1 + \cot x}{\cot x} = \tan x + \csc^2 x - \cot^2 x$

13   $\sec^4 x - \tan^4 x = \sec^2 x + \tan^2 x$

14   $\cos^4 x - \cos^2 x + \sin^2 x = \sin^4 x$

15   $2 \sin^2 x - \cos^2 x + 1 = 3 \sin^2 x$

16   $\sec^2 x \tan^2 x - \tan^2 x = \tan^4 x$

17   $\dfrac{\cos x}{\sin x} + \dfrac{\sin x}{\cos x} = \sec x \csc x$ 18 $\dfrac{\sin x}{\csc x} + \dfrac{\cos x}{\sec x} = \sin x \csc x$

19   $\dfrac{1 + \sec x}{\tan x} - \dfrac{\tan x}{\sec x} = \dfrac{1 + \sec x}{\sec x \tan x}$ 20 $\dfrac{1 + \cot x}{\cot x \csc x} - \dfrac{1 - \cot x}{\csc x} = \sec x$

21   $\dfrac{1 - \cos x}{\sin x} + \dfrac{\sin x}{1 - \cos x} = 2 \csc x$

22   $\dfrac{\sec x + \tan x}{\sin x} - \dfrac{\sin x}{\sec x - \tan x} = \cos x + \cot x$

23   $\dfrac{1 - \cos x}{\csc x} = \dfrac{\sin^3 x}{1 + \cos x}$

24   $\dfrac{\sec x - \tan x}{\cos x \cot x} = \dfrac{\sec x}{\csc x + 1}$

25   $\dfrac{\cot y (\cos x - \sin x) + \cos y (1 + \sin x \csc y)}{\cos x + \sin y} = \cot y$

26   $\dfrac{(\sin x + \cos y)^2 + (\cos x + \sin y)(\cos x - \sin y)}{\sin x + \cos y} = 2 \cos y$

27   $\dfrac{\tan x + \tan y}{\cot x + \cot y} = \tan x \tan y$

28   $\dfrac{\tan x (\sin y - \cos x) + \sin x (1 + \sec x \cos y)}{\sin y + \cos y} = \tan x$

29   $\dfrac{1 + \cos x}{1 - \cos x} - \dfrac{\csc x - \cot x}{\csc x + \cot x} = 4 \csc x \cot x$

30   $\dfrac{1 + \sin x}{1 - \sin x} - \dfrac{\sec x - \tan x}{\sec x + \tan x} = \dfrac{4 \csc x}{\csc^2 x - 1}$

31   $\dfrac{\sin^3 x - \cos^3 x}{\sin x - \cos x} = 1 + \sin x \cos x$

32   $\dfrac{\cos x}{\csc x - \sin x} - \dfrac{\sin x}{\sec x + \cos x} = \dfrac{\tan x}{1 + \cos^2 x}$

In problems 33 to 36 the sign to be used depends on the location of the angle. Note the symbol used for quadrant.

33   $\pm \sqrt{\dfrac{1 + \cos x}{1 - \cos x}} = \dfrac{\sin x}{1 - \cos x}$ for x in Q_1 or Q_2 use $+$
                                                        for x in Q_3 or Q_4 use $-$

34   $\pm \sqrt{\dfrac{\sec x - \tan x}{\sec x + \tan x}} = \sec x - \tan x$ for x in Q_1 or Q_4 use $+$
                                                        for x in Q_2 or Q_3 use $-$

35   $\pm \sqrt{\dfrac{\csc x + 1}{\csc x - 1}} = \dfrac{1 + \sin x}{\cos x}$ for x in Q_1 or Q_4 use $+$
                                                        for x in Q_2 or Q_3 use $-$

36   $\pm \sqrt{\dfrac{1 - \sin x}{1 + \sin x}} = \sec x - \tan x$ for x in Q_1 or Q_4 use $+$
                                                        for x in Q_2 or Q_3 use $-$

Exercise 8.3   Review

Reduce the first expression to the second in each of problems 1 to 8.

1 $\cos A \csc A$, $\cot A$

2 $\csc A \tan A$, $\sec A$

3 $\csc A - \sin A$, $\cos A \cot A$

4 $\cot A(\cos A + \tan A \sin A)$, $\csc A$

5 $\dfrac{\sec^2 A - 1}{\sin A}$, $\sec A \tan A$

6 $\dfrac{\sec A}{\sin A} - \dfrac{\sin A}{\cos A}$, $\cot A$

7 $\dfrac{\cot A}{1 - \sin^2 A}$, $\sec A \csc A$

8 $\dfrac{\tan A}{\sin A} + \dfrac{\csc A}{\tan A}$, $\sec A \csc^2 A$

Prove that each of the following equations is an identity.

9 $\dfrac{1}{1 - \cos A} + \dfrac{1}{1 + \cos A} = 2 \csc^2 A$

10 $\dfrac{1}{\sec A - \tan A} + \dfrac{1}{\sec A + \tan A} = 2 \sec A$

11 $\dfrac{\cot A \cos A}{1 - \sin A} = 1 + \csc A$

12 $\dfrac{\sin^2 A}{\sec A - 1} - \cos A = \cos^2 A$

13 $\dfrac{1 + \sin A}{1 - \sin A} - \dfrac{\sec A - \tan A}{\sec A + \tan A} = \dfrac{4 \csc A}{\csc^2 A - 1}$

14 $\dfrac{\sec A \tan^2 A + \sec^2 A - \sec^3 A}{(\sec A - 1)^2} = \dfrac{1}{1 - \cos A}$

15 $\dfrac{\cot A + \cot B}{\tan A + \tan B} = \cot A \cot B$

16 $\dfrac{\cot A(\cos B - \sin B) + \cos A(1 + \sin B \csc A)}{\sin A + \cos B} = \cot A$

Reductions, Use of Tables

It is often desirable to be able to express a function value of a given angle in terms of a function value of a positive acute angle and then to find the latter value by use of a table and, if necessary, by interpolation. We shall now see how to do this.

9.1 THE REFERENCE ANGLE

The concept of the reference angle is used in expressing a function value of an angle in terms of the function value of a positive acute angle.

Reference angle

If a given angle other than an integral multiple of $\pi/2 = 90°$ is in standard position, then the positive acute angle between the X axis and the terminal side of the given angle is called the *reference* or *related* angle; furthermore, if the given angle is an integral multiple of $\pi/2$, then the reference angle is 0 or $\pi/2$ according to whether the terminal side coincides with the X or Y axis.

In Fig. 9.1, A is the given angle in each case, and R is its reference angle. If A is in the first or third quadrant, then the X axis is the initial side, and if A is in the second or fourth quadrant, then the X axis is the terminal side. The reference angle is always measured in a positive direction. Thus, 34° is the reference angle of 146°, and 73° is the related angle of 253° as shown in Fig. 9.2.

In order to find the relation between the function values of an angle

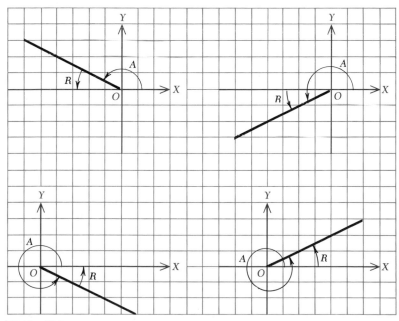

Figure 9.1

179

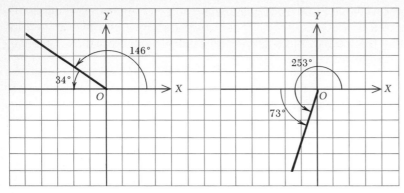

Figure 9.2

and those of the reference angle, we make use of the fact that any positive angle θ less than $360°$ can be expressed as R, $180° - R$, $180° + R$, or $360° - R$, where R is a positive acute angle. Furthermore, since the function values of two coterminal angles are equal, we need only find the desired relation for angles between $0°$ and $360°$.

In Fig. 9.3, we have an angle in each quadrant along with its reference angle R. Since $\theta = R$ is a first-quadrant angle, the coordinates of any point $P(x, y)$ on its terminal side are positive. We choose the points $P(x, y)$, $P_2(x_2, y_2)$, $P_3(x_3, y_3)$, and $P_4(x_4, y_4)$ on the terminal sides of the angles and at a distance r from the origin and drop perpendicular lines to the X axis. These lines meet the X axis at Q, Q_2, Q_3, and Q_4. The right triangles OQP, OQ_2P_2, OQ_3P_3, and OQ_4P_4 are all congruent; hence, corresponding sides are

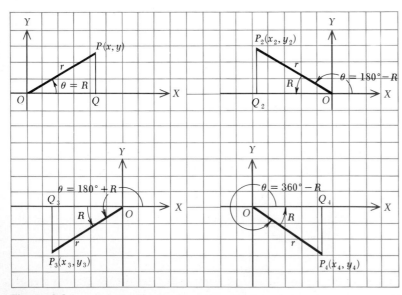

Figure 9.3

180

equal. Now taking the directional aspects of coordinates into consideration, we see that $x_2 = x_3 = -x$, $x_4 = x$ and $y_2 = y$, $y_3 = y_4 = -y$. Therefore, by
using the definition of the function values, we can construct the following
table.

	sin	*cos*	*tan*	*cot*	*sec*	*csc*
$\theta = R$	$\dfrac{y}{r}$	$\dfrac{x}{r}$	$\dfrac{y}{x}$	$\dfrac{x}{y}$	$\dfrac{r}{x}$	$\dfrac{r}{y}$
$\theta = 180° - R$	$\dfrac{y}{r}$	$\dfrac{-x}{r}$	$\dfrac{y}{-x}$	$\dfrac{-x}{y}$	$\dfrac{r}{-x}$	$\dfrac{r}{y}$
$\theta = 180° + R$	$\dfrac{-y}{r}$	$\dfrac{-x}{r}$	$\dfrac{-y}{-x}$	$\dfrac{-x}{-y}$	$\dfrac{r}{-x}$	$\dfrac{r}{-y}$
$\theta = 360° - R$	$\dfrac{-y}{r}$	$\dfrac{x}{r}$	$\dfrac{-y}{x}$	$\dfrac{x}{-y}$	$\dfrac{r}{x}$	$\dfrac{r}{-y}$

If we consider the sine of any angle in the table, we find that, except
possibly for algebraic sign, it has the same value as the sine of the related
angle R. A similar statement is true for each of the other function values;
hence, we have the following theorem which enables us to find a function
value of any angle by use of a table of function values of angles from $0°$ to
$90°$ and which is called the *reference-angle theorem.*

**Reference-angle
theorem**
*Any trigonometric function value of any given angle is numerically equal to
the same function value of the reference angle. The algebraic sign of each function
value is determined by the quadrant in which the given angle lies.*

The reference angle of $138°$ is $42°$, since that is the positive acute angle
between the X axis and the terminal side of the given angle; furthermore,
since $138°$ is a second-quadrant angle, $\sin 138°$ is positive and $\cos 138°$ is
negative. Consequently,

$$\sin 138° = +\sin 42°$$
$$\cos 138° = -\cos 42°$$

As another illustration, we see that

$$\sin 237° = -\sin 57°$$

since $57°$ is the reference angle and the sine of any third-quadrant angle is
negative. As a further use of the reference angle, we point out that

$$\tan 324° = -\tan 36°$$

since $36°$ is the reference angle and the tangent of any fourth-quadrant
angle is negative. Finally,

$$\sec (-145°) = -\sec 35°$$

since $35°$ is the reference angle and the secant of any third-quadrant angle
is negative.

9.2 FUNCTION VALUES OF NEGATIVE ANGLES

We can find a function value of a negative angle by use of the reference-angle theorem, but it is sometimes desirable to express it in terms of the given function value of a numerically equal positive angle. In order to derive such a relation, we shall consider a positive angle θ in standard position along with the negative angle $-\theta$, as shown in Fig. 9.4. We shall choose points $P(x, y)$ and $P_1(x_1, y_1)$ at a distance r from the origin and on the terminal sides of θ and $-\theta$, respectively. Therefore, $x_1 = x$ and $y_1 = -y$. Consequently, using the definition of the function values,

$$\sin(-\theta) = \frac{y_1}{r} = \frac{-y}{r} = -\sin\theta$$

$$\cos(-\theta) = \frac{x_1}{r} = \frac{x}{r} = +\cos\theta$$

$$\tan(-\theta) = \frac{y_1}{x_1} = \frac{-y}{x} = -\tan\theta$$

We now have the following theorem, since the other three function values are the reciprocals of these three.

Functions of negative angles

The cosine and secant of a negative angle are equal to the cosine and secant, respectively, of the numerically equal positive angle. Each of the other four function values of a negative angle is the negative of the same function value of the numerically equal positive angle.

By use of this theorem, we see that

$$\sin(-76°) = -\sin 76°$$
$$\cos(-81°) = +\cos 81°$$
$$\tan(-6°) = -\tan 6°$$

Exercise 9.1 Reference-Angle Theorem

In each of problems 1 to 8, construct the angle and its reference angle.

1 162°	**2** 347°	**3** 529°	**4** 223°
5 −81°	**6** −134°	**7** −244°	**8** −676°

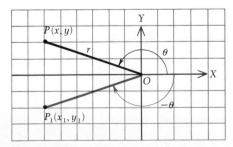

Figure 9.4

Express the function value of the angle in each of problems 9 to 20 in terms of the given function value of the reference angle.

9	sin 106°	**10**	cos 314°	**11**	tan 222°
12	cot 419°	**13**	sec 238°	**14**	csc 162°
15	sin 641°	**16**	cos 308°	**17**	tan (−83°)
18	cot (−118°)	**19**	sec (−207°)	**20**	csc (−555°)

Express each of the following in terms of the given function of a positive angle that is numerically equal to the given angle and then in terms of the same function of the reference angle.

21	cos (−100°)	**22**	tan (−193°)	**23**	cot (−286°)
24	sec (−385°)	**25**	csc (−227°)	**26**	sin (−341°)
27	cos (−403°)	**28**	tan (−84°)	**29**	cot (−298°)
30	sec (−486°)	**31**	csc (−196°)	**32**	sin (−808°)

9.3 FUNCTION VALUES OF GIVEN ANGLES

By means of the material of the last two sections, we are able to express a trigonometric function value of any angle in terms of the same function value of a positive acute angle. We shall now see how to use a table in order to find a function value of a positive acute angle. Table II in the back of this book gives the approximate values of the natural sines, cosines, tangents, and cotangents of certain angles correct to four decimal places, and it also gives the logarithms of these function values. The angles are listed at intervals of 10′ from 0° to 90°, with angles from 0° to 45° on the left of the page and those from 45° to 90° on the right. The function names that are to be used with the angles on the left are given at the top of the page, and those to be used with angles on the right are printed at the bottom. This arrangement is possible for reasons that are developed in Sec. 19.2.

EXAMPLE 1

Find the value of sin 34°30′.

Solution

If we look in Table II across from 34°30′ and under Sines, we find .5664; hence, sin 34°30′ = .5664.

EXAMPLE 2

Find the value of cos 24°20′.

Solution

If we look in Table II across from 24°20′ and under Cosines, we find .9112; hence, cos 24°20′ = .9112.

EXAMPLE 3   Find the value of sin 52°50′.

Solution   The value of sin 52°50′ is .7969, since this is the value in Table II across from 52°50′ on the right of the page and above Sines.

EXAMPLE 4   Evaluate tan 58°30′.

Solution   We know that tan 58°30′ = 1.6319, since 1.6319 is the number in Table II across from 58°30′ and above Tangents.

EXAMPLE 5   Evaluate sin (−141°).

Solution
$$\sin (-141°) = -\sin 141° \qquad \blacktriangleleft \text{ sine of a minus angle}$$
$$= -\sin 39° \qquad \blacktriangleleft \text{ reference-angle theorem}$$
$$= -.6293$$

9.4 GIVEN A FUNCTION VALUE OF AN ANGLE, TO FIND THE ANGLE

If we know a function value of an angle and want to find the angle, we look in the column of Table II that has the function name at the top or bottom of the page until we find the value. Then the angle is on the left or right according as the function name is at the top or bottom of the page.

EXAMPLE 1   Find θ if tan θ = .6371.

Solution   If tan θ = .6371, then θ = 32°30′, since .6371 is an entry in the column below Tangents and 32°30′ is the angle across from it on the left of the page in Table II.

EXAMPLE 2   What is the value of θ if cos θ = .4014?

Solution   If cos θ = .4014, then θ = 66°20′, since .4014 is in the column above Cosines and 66°20′ is the angle across from it on the right of the page.

9.5 INTERPOLATION

Quite often we need a function value for an angle that is not listed in the table or want the angle that corresponds to a ratio that is not in the table.

Under such circumstances, we resort to a procedure known as linear interpolation, which we now illustrate.

EXAMPLE 1

Find the value of tan 38°46′ by use of interpolation.

Solution

Since 38°46′ is between 38°40′ and 38°50′, we assume that tan 38°46′ is between tan 38°40′ and tan 38°50′. In fact, since 38°46′ is six-tenths of the way from 38°40′ toward 38°50′, we assume that tan 38°46′ is six-tenths of the way from tan 38°40′ = .8002 toward tan 38°50′ = .8050. These two values differ by .0048, and .6(.0048) = .00288 = .0029. Consequently, beginning at tan 38°40′ = .8002 and going .0029 in the direction of tan 38°50′ = .8050, we see that tan 38°46′ is .8002 + .0029 = .8031.

If the diagrammatic form given below is used, the procedures described in Example 1 are readily followed.

$$10'\left[6'\begin{bmatrix}\text{angle} & \text{tangent} \\ 38°40' & .8002 \\ 38°46' & \\ 38°50' & .8050\end{bmatrix}c\right].0048$$

$$c = \tfrac{6}{10}(.0048) = .0029$$

$$\tan 38°46' = .8002 + .0029 = .8031$$

The correction c was added because tan θ increases from $\theta = 38°40'$ to $\theta = 38°50'$.

EXAMPLE 2

Find cos 73°13′.

Solution

We shall find cos 73°13′ and shall use only the diagrammatic form. The two entries nearest cos 73°13′ are cos 73°10′ and cos 73°20′, and they will be used in the interpolation.

$$10'\left[3'\begin{bmatrix}\text{angle} & \text{cosine} \\ 73°10' & .2896 \\ 73°13' & \\ 73°20' & .2868\end{bmatrix}c\right].0028$$

$$c = \tfrac{3}{10}(.0028)$$
$$= .0008$$
$$\cos 73°13' = .2896 - .0008$$
$$= .2888$$

The correction was subtracted because cos θ decreases from $\theta = 73°10'$ to $\theta = 73°20'$.

EXAMPLE 3   Find θ if $\sin \theta = .6212$.

Solution   Making the usual diagram, we have

$$.0023 \left[.0010 \left[\begin{array}{cc} sine & angle \\ .6202 & 38°20' \\ .6212 & \theta \\ .6225 & 38°30' \end{array} \right]c \right]10'$$

$$c = \frac{.0010}{.0023}\,(10') = \frac{10}{23}\,(10')$$
$$= 4'$$
$$\theta = 38°20' + 4'$$
$$= 38°24'$$

Exercise 9.2   The Use of Tables

Find the indicated function value in each of problems 1 to 12.

1	$\tan 36°30'$	**2**	$\cot 70°10'$	**3**	$\sin 5°40'$
4	$\cos 53°50'$	**5**	$\tan 45°0'$	**6**	$\cot 18°30'$
7	$\sin 74°10'$	**8**	$\cos 84°40'$	**9**	$\tan 136°20'$
10	$\cot 234°50'$	**11**	$\sin 197°20'$	**12**	$\cos 309°50'$

Find the acute angle A in each of problems 13 to 24.

13	$\cot A = 11.06$	**14**	$\sin A = .2334$	**15**	$\cos A = .2840$
16	$\tan A = .4040$	**17**	$\cot A = .4950$	**18**	$\sin A = .8339$
19	$\cos A = .8557$	**20**	$\tan A = 6.561$	**21**	$\cot A = 1.303$
22	$\sin A = .7071$	**23**	$\cos A = .7698$	**24**	$\tan A = .6168$

Find the indicated function value of the angle in each of problems 25 to 32.

25	$\sin 71°12'$	**26**	$\cos 32°17'$	**27**	$\tan 41°43'$
28	$\cot 68°34'$	**29**	$\sin 126°46'$	**30**	$\cos 249°58'$
31	$\tan 333°33'$	**32**	$\cot 444°22'$		

Find the acute angle A in each of problems 33 to 40.

33	$\cos A = .2308$	**34**	$\tan A = .4773$	**35**	$\cot A = 2.654$
36	$\sin A = .6161$	**37**	$\cos A = .3259$	**38**	$\tan A = 7.703$
39	$\cot A = .8273$	**40**	$\sin A = .9104$		

Exercise 9.3   Review

1   Express cos 207° in terms of the cosine of the reference angle.
2   Express tan 168° in terms of the tangent of the reference angle.
3   Express sec (−346°) in terms of the secant of the reference angle.
4   Express sin (−174°) in terms of the sine of the reference angle.
5   Express sin 249° in terms of a function of a positive angle less than 45°.
6   Express cos 301° in terms of a function of an angle between zero and 45°.
7   Express cot (−258°) in terms of a function of an angle between zero and 45°.
8   Express csc (−83°) in terms of a function of a positive angle less than 45°.

Find the indicated function value in each of problems 9 to 14.

9 cos 35°20′	10 tan 117°30′	11 sin 228°40′
12 tan 171°23′	13 sin 274°37′	14 cos 472°31′

Find the acute angle in each of problems 15 to 23.

15 sin A = .9974	16 cos A = .2840	17 tan A = 1.446
18 cos A = .9261	19 tan A = .8243	20 sin A − .7030
21 tan A = 3.472	22 sin A = .2105	23 cos A = .1753

Functions
of a
Composite
Angle

In our previous work, we have dealt with function values of a single angle. There are circumstances under which we need to be able to express function values of the sum or difference of two angles, and function values of twice an angle or half an angle in terms of function values of the given angle. This chapter shall be devoted to the development of such formulas.

10.1 THE COSINE OF THE DIFFERENCE OF TWO ANGLES

We shall use Fig. 10.1 in deriving the formula for the cosine of the difference of two angles. We begin the figure by drawing the angles A and B in standard position with their terminal sides intersecting the unit circle at P_2 and P_3; hence, angle P_2OP_3 is equal to $B - A$. We now designate the intersection of the X axis and the unit circle by P_1 and select P_4 on the unit circle so that angle P_4OP_1 is equal to $B - A$. Finally, draw chords P_2P_3 and P_4P_1. They are of equal length since they are intercepted by equal angles. The coordinates of P_2, P_3, and P_4 are as shown in the figure. If we apply the distance formula, we find that

$$
\begin{aligned}
(P_2P_3)^2 &= (\cos A - \cos B)^2 + (\sin A - \sin B)^2 \\
&= \cos^2 A - 2\cos A \cos B + \cos^2 B + \sin^2 A - 2\sin A \sin B + \sin^2 B \\
&= 2 - 2(\cos A \cos B + \sin A \sin B) \qquad \blacktriangleleft \text{ by (8.6)} \qquad (1)
\end{aligned}
$$

and

$$
\begin{aligned}
(P_4P_1)^2 &= [\cos (A - B) - 1]^2 + [\sin (A - B) - 0]^2 \\
&= \cos^2 (A - B) - 2[\cos (A - B)] + 1 + \sin^2 (A - B) \qquad (2) \\
&= 2 - 2\cos (A - B) \qquad \blacktriangleleft \text{ since } \cos^2 (A - B) + \\
&\qquad\qquad\qquad\qquad\qquad\qquad\qquad \sin^2 (A - B) = 1
\end{aligned}
$$

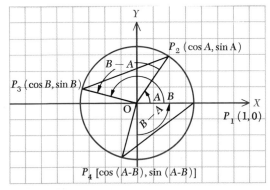

Figure 10.1

We can equate the right members of (1) and (2) since the left members are equal. Thus, we find that

$$\cos{(A - B)} = \cos A \cos B + \sin A \sin B \qquad (10.1)$$

Cosine of a difference   This equation is known as the identity for the cosine of the difference of two angles and can be put in words by saying that *the cosine of the differ-ence of two angles is the product of their sines added to the product of their cosines.* The angle may be expressed in degrees or radians, and a function value of a given number of radians is equal to that function value of the given num-ber; hence, it follows that (10.1) is true for any two real numbers A and B.

Equation (10.1) may be used for changing the form of a trigonometric equation or for numerical calculation.

EXAMPLE   Find the exact value of $\cos 15°$.

Solution   We shall think of $15°$ as $45° - 30°$ so as to be able to use (10.1). If this is done, we have

$$\begin{aligned}
\cos 15° &= \cos{(45° - 30°)} \\
&= \cos 45° \cos 30° + \sin 45° \sin 30° \\
&= \frac{\sqrt{2}}{2}\frac{\sqrt{3}}{2} + \frac{\sqrt{2}}{2}\frac{1}{2} \\
&= \frac{\sqrt{2}\,(\sqrt{3} + 1)}{4}
\end{aligned}$$

This is the exact value, and an approximation can be obtained by using $\sqrt{2}$ and $\sqrt{3}$ in decimal form.

10.2   FUNCTION VALUES OF $\frac{\pi}{2} - B$

We shall replace A by $\pi/2$ in (10.1) in order to find a formula for $\cos{(\pi/2 - B)}$. Thus,

$$\cos\left(\frac{\pi}{2} - B\right) = \cos\frac{\pi}{2}\cos B + \sin\frac{\pi}{2}\sin B$$

$$= \sin B \qquad \blacktriangleleft \text{ since } \cos\frac{\pi}{2} = 0 \text{ and } \sin\frac{\pi}{2} = 1$$

$\cos\left(\frac{\pi}{2} - B\right)$ Therefore, $\cos\left(\frac{\pi}{2} - B\right) = \sin B$ (10.2)

for any angle or real number B.

If, in (10.2), B is replaced by $\pi/2 - B$, we get

$$\cos\left[\frac{\pi}{2} - \left(\frac{\pi}{2} - B\right)\right] = \sin\left(\frac{\pi}{2} - B\right)$$

Since $\pi/2 - (\pi/2 - B) = B$, we have

$\sin\left(\dfrac{\pi}{2} - B\right)$ $$\sin\left(\frac{\pi}{2} - B\right) = \cos B \qquad\qquad\qquad (10.3)$$

for any angle or real number B.

Function values of $\dfrac{\pi}{2} - B$

The identities for the other function values of $\pi/2 - B$ can be obtained by use of (10.2), (10.3), and the ratio and reciprocal relations of Sec. 8.1. If this is done, we find that *any trigonometric function value of an angle is equal to the corresponding cofunction of the complementary angle.*

The function values of the negative of an angle were found in Sec. 9.2 and can be found by use of (10.2), (10.3), and the ratio and reciprocal relations. For example, if we replace A by zero in (10.1), we get

$$\cos(0 - B) = \cos 0 \cos B + \sin 0 \sin B$$
$$= \cos B \qquad \blacktriangleleft \text{ since } \cos 0 = 1 \text{ and } \sin 0 = 0$$

We shall now illustrate the use of (10.2) and (10.3) with two examples.

EXAMPLE 1 $$\cos 62° = \cos(90° - 28°) = \sin 28°$$

EXAMPLE 2 $$\tan 37° = \tan(90° - 53°) = \frac{\sin(90° - 53°)}{\cos(90° - 53°)}$$
$$= \frac{\cos 53°}{\sin 53°} = \cot 53°$$

10.3   THE COSINE OF A SUM

Since (10.1) is true for all angles, it is true if we replace B by $-B$. If that is done, we have

$$\cos[A - (-B)] = \cos(A + B)$$
$$= \cos A \cos(-B) + \sin A \sin(-B)$$
$$= \cos A \cos B - \sin A \sin B \qquad \blacktriangleleft \text{ since } \cos(-B) = \cos B$$
$$\text{and } \sin(-B) = -\sin B$$

Consequently, we know that

Cosine of a sum $$\cos(A + B) = \cos A \cos B - \sin A \sin B \qquad\qquad (10.4)$$

This can be put in words by saying that *the cosine of the sum of two angles is the product of their cosines minus the product of their sines.*

EXAMPLE   Find the exact value of cos 75°.

Solution   If we think of 75° as 45° + 30° and apply (10.4), we get

$$\cos 75° = \cos (45° + 30°)$$
$$= \cos 45° \cos 30° - \sin 45° \sin 30°$$
$$= \frac{\sqrt{2}}{2} \frac{\sqrt{3}}{2} - \frac{\sqrt{2}}{2} \frac{1}{2}$$
$$= \frac{\sqrt{2}(\sqrt{3} - 1)}{4}$$

10.4 THE COSINE OF TWICE AN ANGLE

If we replace B by A in (10.4), it becomes

$$\cos (A + A) = \cos A \cos A - \sin A \sin A$$

Consequently, $\cos 2A = \cos^2 A - \sin^2 A$

Now, replacing $\sin^2 A$ by $1 - \cos^2 A$, as justified by (8.6), we have

$$\cos 2A = 2 \cos^2 A - 1$$
$$= 1 - 2 \sin^2 A$$

by a second use of (8.6).   We now have three forms for the cosine of twice an angle.   They are

Cosine of $$\mathbf{\cos 2A = \cos^2 A - \sin^2 A} \qquad\qquad (\mathbf{10.5}a)$$
twice an angle $$\mathbf{= 2 \cos^2 A - 1} \qquad\qquad (\mathbf{10.5}b)$$
$$\mathbf{= 1 - 2 \sin^2 A} \qquad\qquad (\mathbf{10.5}c)$$

EXAMPLE   Find the exact value of cos 120°.

Solution   If we use 120° as 2(60°) and apply (10.5a), we get

$$\cos 120° = \cos 2(60°)$$
$$= \cos^2 60° - \sin^2 60°$$
$$= \left(\frac{1}{2}\right)^2 - \left(\frac{\sqrt{3}}{2}\right)^2$$
$$= \tfrac{1}{4} - \tfrac{3}{4} = -\tfrac{1}{2}$$

10.5 THE COSINE OF HALF AN ANGLE

If we solve $(10.5b)$ for $\cos A$, we have

$$\cos A = \pm\sqrt{\frac{1 + \cos 2A}{2}}$$

Now replacing A by $\theta/2$, we find that

Cosine of
half an angle

$$\cos \frac{1}{2}\theta = \pm\sqrt{\frac{1 + \cos \theta}{2}} \qquad (10.6)$$

This equation is the identity for the cosine of half an angle in terms of the cosine of the angle. The algebraic sign to be used in a particular case is determined by the quadrant in which $\frac{1}{2}\theta$ lies.

EXAMPLE Find $\cos 105°$.

Solution Since $105°$ is a second-quadrant angle, we must use the minus sign with (10.6). Therefore

$$\cos 105° = -\sqrt{\frac{1 + \cos 210°}{2}}$$

$$= -\sqrt{\frac{1 - \sqrt{3}/2}{2}}$$

$$= -\frac{\sqrt{2 - \sqrt{3}}}{2}$$

10.6 IDENTITIES

The most important use of the fundamental identities developed in Chap. 8 and in Secs. 10.1 to 10.5 and to be developed later in this chapter is in changing the form of a trigonometric expression or proving that an equation is an identity. The procedures are essentially those outlined in Sec. 8.3.

EXAMPLE Prove that

$$\frac{\cos 2\theta}{\cos \theta} = \frac{1 - \tan^2 \theta}{\sec \theta}$$

is an identity.

Solution

We shall convert the left member to the form of the right member. We will begin by using the double-angle formula. Thus,

$$\frac{\cos 2\theta}{\cos \theta} = \frac{\cos^2 \theta - \sin^2 \theta}{\cos \theta}$$

$$= \frac{1 - \dfrac{\sin^2 \theta}{\cos^2 \theta}}{\dfrac{\cos \theta}{\cos^2 \theta}} \qquad \blacktriangleleft \text{ dividing numerator and denominator by } \cos^2 \theta$$

$$= \frac{1 - \tan^2 \theta}{\dfrac{1}{\cos \theta}} \qquad \blacktriangleleft \quad \dfrac{\sin \theta}{\cos \theta} = \tan \theta$$

$$= \frac{1 - \tan^2 \theta}{\sec \theta} \qquad \blacktriangleleft \quad \dfrac{1}{\cos \theta} = \sec \theta$$

Exercise 10.1   Cosine of a Composite Angle

Find the function value called for in each of problems 1 to 4 by use of identities (10.1) to (10.6) and the trigonometric ratios for 30°, 45°, and their multiples.

1   cos 15°, with 15° = 60° − 45°
Show this is equal to the value obtained by use of the half-angle formula.

2   cos 105°, with 105° = 60° + 45°
Show this is equal to the value obtained by use of the half-angle formula.

3   cos 60°, with 60° = 2(30°)

4   cos 22°30′, with 22°30′ = ½(45°)

Find cos $(A + B)$ and cos $(A − B)$ if A, B, and their function values are as given in problems 5 to 8.

5   $\cos A = \frac{3}{5}, 0 < A < \pi/2$; $\cos B = -\frac{4}{5}, \pi/2 < B < \pi$

6   $\cos A = -\frac{5}{13}, \pi < A < 3\pi/2$; $\sin B = \frac{3}{5}, \pi/2 < B < \pi$

7   $\sin A = -\frac{8}{17}, 3\pi/2 < A < 2\pi$; $\cos B = \frac{12}{13}, 0 < B < \pi/2$

8   $\sin A = \frac{4}{5}, \pi/2 < A < \pi$; $\sin B = -\frac{7}{25}, 3\pi/2 < B < 2\pi$

If A and one of its function values are as given in problems 9 to 12, find $\cos 2A$ and $\cos \frac{1}{2}A$.

9   $\sin A = \frac{12}{13}, \ \pi/2 < A < \pi$

10 $\sin A = -\frac{4}{5}, \ 3\pi/2 < A < 2\pi$

11 $\cos A = -\frac{8}{17}, \ \pi < A < 3\pi/2$

12 $\cos A = -\frac{24}{25}, \ \pi/2 < A < \pi$

Use the identities for the cosine of the sum or difference of two angles to prove the statement in each of problems 13 to 16.

13 $\cos (\pi + \theta) = -\cos \theta$ 14 $\cos (\pi - \theta) = -\cos \theta$

15 $\cos \left(\dfrac{3\pi}{2} + \theta\right) = \sin \theta$ 16 $\cos \left(\dfrac{\pi}{2} + \theta\right) = -\sin \theta$

Change the first expression in each of problems 17 to 28 to the second.

17 $\cos 4\theta \cos \theta + \sin 4\theta \sin \theta, \ \cos 3\theta$

18 $\cos 7\theta \cos 5\theta + \sin 7\theta \sin 5\theta, \ \cos 2\theta$

19 $\cos \theta \cos 2\theta - \sin \theta \sin 2\theta, \ \cos 3\theta$

20 $\cos 6\theta \cos 3\theta - \sin 6\theta \sin 3\theta, \ \cos 9\theta$

21 $\cos 3\theta \cos \theta - \sin 3\theta \sin \theta, \ 2 \cos^2 2\theta - 1$

22 $\cos 5\theta \cos 3\theta + \sin 5\theta \sin 3\theta, \ 1 - 2 \sin^2 \theta$

23 $\cos^2 4\theta + 4 \sin^2 2\theta \cos^2 2\theta, \ 1$

24 $\cos 2\theta + 2 \sin^2 \theta, \ 1$

25 $\cos 4\theta + \cos 2\theta, \ 2 \cos^2 \theta \ (4 \cos^2 \theta - 3)$

26 $(1 + 2 \sin 2\theta)(1 - \sin 2\theta), \ \cos 4\theta + \sin 2\theta$

27 $(2 \cos \theta - 1)(\cos \theta + 1), \ \cos 2\theta + \cos \theta$

28 $\cos 2\theta + \cos \theta, \ (2 \cos \theta + 1)(\cos \theta - 1)$

Prove that the equation in each of problems 29 to 36 is an identity.

29 $2 \sin^2 \theta + \cos 2\theta = 1$

30 $\cos^2 3\theta - \sin^2 3\theta = \cos 6\theta$

31 $\cos 2\theta + 4 \cos \theta + 3 = 8 \cos^4 \dfrac{\theta}{2}$

32 $\cos 5\theta \cos \theta + \sin 5\theta \sin \theta = 1 - 4 \sin^2 \theta \cos^2 \theta - \sin^2 2\theta$

33 $\dfrac{1 - \cos 2\theta}{\sin \theta \cos \theta} = 2 \tan \theta$

34 $\dfrac{\cos 2\theta}{\sin \theta \cos \theta} = \dfrac{1 - \tan^2 \theta}{\tan \theta}$

35 $\dfrac{\cos 2\theta}{\cos \theta} = \dfrac{2 - \sec^2 \theta}{\sec \theta}$

36 $\dfrac{\sin \theta \cos \theta}{\cos 2\theta} = \dfrac{\cot \theta}{\csc^2 \theta - 2}$

10.7 THE SINE OF THE SUM OF TWO ANGLES

If we use (10.2) with B replaced by $A + B$, we have

$$\sin (A + B) = \cos \left[\frac{\pi}{2} - (A + B) \right]$$

$$= \cos \left[\left(\frac{\pi}{2} - A \right) - B \right]$$

$$= \cos \left(\frac{\pi}{2} - A \right) \cos B + \sin \left(\frac{\pi}{2} - A \right) \sin B \qquad \blacktriangleleft \text{ by (10.1)}$$

$$= \sin A \cos B + \cos A \sin B \qquad\qquad\qquad \blacktriangleleft \text{ by (10.2)}$$
$$\text{and (10.3)}$$

Consequently, we know that

$\sin (A + B)$ $$\mathbf{\sin (A + B) = \sin A \cos B + \cos A \sin B} \qquad (10.7)$$

10.8 THE SINE OF TWICE AN ANGLE

If in (10.7) we let $B = A$, we have

$$\sin (A + A) = \sin A \cos A + \cos A \sin A$$

$\sin 2A$ Hence, $$\mathbf{\sin 2A = 2 \sin A \cos A} \qquad (10.8)$$

10.9 THE SINE OF HALF AN ANGLE

If we solve (10.5c) for $\sin A$, we get

$$\sin A = \pm \sqrt{\frac{1 - \cos 2A}{2}}$$

Now, by replacing A by $\frac{1}{2}\theta$, we have

$\sin \frac{1}{2}\theta$ $$\mathbf{\sin \tfrac{1}{2}\theta = \pm \sqrt{\frac{1 - \cos \theta}{2}}} \qquad (10.9)$$

This is the formula for the sine of half an angle. The algebraic sign to be used is determined by the quadrant in which $\frac{1}{2}\theta$ lies.

10.10 THE SINE OF THE DIFFERENCE OF TWO ANGLES

If we replace B by $-C$ in (10.7), the equation becomes

$$\sin (A - C) = \sin [A + (-C)]$$
$$= \sin A \cos (-C) + \cos A \sin (-C)$$

Consequently, we see that

$\sin (A - C)$ $\qquad$ $$\sin (A - C) = \sin A \cos C - \cos A \sin C \qquad \textbf{(10.10)}$$

since $\cos (-C) = \cos C$ and $\sin (-C) = -\sin C$.

EXAMPLE

Express $\sin 3A$ in terms of $\sin A$.

Solution

$$\sin 3A = \sin (2A + A)$$
$$= \sin 2A \cos A + \cos 2A \sin A \qquad \blacktriangleleft \text{ by (10.7)}$$
$$= 2 \sin A \cos A \cos A + (2 \cos^2 A - 1) \sin A \qquad \blacktriangleleft \text{ by (10.8) and (10.5}b)$$
$$= (4 \cos^2 A - 1) \sin A \qquad \blacktriangleleft \text{ collecting and factoring}$$
$$= [4(1 - \sin^2 A) - 1] \sin A \qquad \blacktriangleleft \text{ since } \cos^2 A = 1 - \sin^2 A$$
$$= 3 \sin A - 4 \sin^3 A$$

Exercise 10.2   Sine of a Composite Angle

Find the function value called for in each of problems 1 to 4 by use of the identities developed in this chapter and the trigonometric ratios for $30°$, $45°$, and their multiples.

1   $\sin 15°$, with $15° = 45° - 30°$
2   $\sin 75°$, with $75° = 45° + 30°$
3   $\sin 120°$, with $120° = 2(60°)$
4   $\sin 22°30'$, with $22°30' = \frac{1}{2}(45°)$

Find $\sin (A + B)$ and $\sin (A - B)$ if A, B, and their function values are as given in problems 5 to 8.

5   $\sin A = \frac{4}{5}$, $0 < A < \pi/2$; $\sin B = -\frac{3}{5}$, $\pi < B < 3\pi/2$
6   $\sin A = \frac{8}{17}$, $\pi/2 < A < \pi$; $\cos B = -\frac{5}{13}$, $\pi/2 < B < \pi$
7   $\cos A = -\frac{7}{25}$, $\pi < A < 3\pi/2$; $\sin B = -\frac{4}{5}$, $3\pi/2 < B < 2\pi$
8   $\cos A = \frac{12}{13}$, $3\pi/2 < A < 2\pi$; $\cos B = \frac{15}{17}$, $0 < B < \pi/2$

If A and one of its function values are as given in problems 9 to 12, find $\sin 2A$ and $\sin \frac{1}{2}A$.

9 $\cos A = \frac{7}{25}, 0 < A < \pi/2$
10 $\cos A = -\frac{8}{17}, \pi/2 < A < \pi$
11 $\sin A = -\frac{3}{5}, \pi < A < 3\pi/2$
12 $\sin A = -\frac{12}{13}, 3\pi/2 < A < 2\pi$

Prove each of the following by use of the identities for the sine of the sum and the difference of two angles.

13 $\sin(\pi - \theta) = \sin\theta$ **14** $\sin(2\pi - \theta) = -\sin\theta$
15 $\sin(\pi + \theta) = -\sin\theta$ **16** $\sin(\pi/2 + \theta) = \cos\theta$

Change the first expression in each of problems 17 to 28 to the second.

17 $\sin 3\theta \cos 2\theta + \cos 3\theta \sin 2\theta, \sin 5\theta$
18 $\sin 4\theta \cos 3\theta + \cos 4\theta \sin 3\theta, \sin 7\theta$
19 $\sin 5\theta \cos 2\theta - \cos 5\theta \sin 2\theta, \sin 3\theta$
20 $\sin 4\theta \cos\theta - \cos 4\theta \sin\theta, \sin 3\theta$
21 $\sin 3\theta \cos\theta + \cos 3\theta \sin\theta, 2 \sin 2\theta \ (2\cos^2\theta - 1)$
22 $\sin 6\theta \cos 2\theta - \cos 6\theta \sin 2\theta, 4\sin\theta \cos\theta(1 - 2\sin^2\theta)$
23 $\dfrac{\sin 2\theta}{\sin\theta} - \dfrac{\cos 2\theta}{\cos\theta}, \sec\theta$
24 $\dfrac{\sin 6\theta}{\cos 3\theta} + \dfrac{\cos 6\theta}{\sin 3\theta}, \csc 3\theta$
25 $\cot\theta + \tan\theta, 2\csc 2\theta$
26 $\cot 2\theta - \cot 4\theta, \csc 4\theta$
27 $\sin 4\theta, 4\sin\theta \cos\theta \ (1 - 2\sin^2\theta)$
28 $\sin 4\theta + \cos 2\theta, (4\sin\theta \cos\theta + 1)(2\cos^2\theta - 1)$

Prove that the statement in each of problems 29 to 36 is an identity.

29 $\dfrac{\sin^2 2\theta}{1 + \cos 2\theta} = 2\sin^2\theta$
30 $\cot\theta - \tan\theta = 2\cot 2\theta$
31 $\sin 2\theta - \tan\theta \cos 2\theta = \dfrac{\sin 2\theta}{2\cos^2\theta}$
32 $\dfrac{2\sin 2\theta}{(1 + \cos 2\theta)^2} = \sec^2\theta \tan\theta$
33 $\sin 5\theta + \sin 3\theta = 8\sin\theta \cos^2\theta \cos 2\theta$
      SUGGESTION: Use $5\theta = 4\theta + \theta$ and $3\theta = 4\theta - \theta$.
34 $\sin 6\theta + \sin 2\theta = 8\sin\theta \cos\theta \cos^2 2\theta$
35 $\sin 3\theta - \sin\theta = 2(2\cos^2\theta - 1) \sin\theta$
36 $\sin 10\theta - \sin 2\theta = 8\cos 6\theta \sin\theta \cos\theta \ (\cos^2\theta - \sin^2\theta)$

10.11 THE TANGENT OF THE SUM OF TWO ANGLES

In order to derive an identity for the tangent of the sum of two angles, we shall make use of the formulas for the sine and cosine of the sum of two angles and the fact that the tangent of an angle is the quotient of the sine and cosine of the angle. Thus,

$$\tan (A + B) = \frac{\sin (A + B)}{\cos (A + B)}$$

$$= \frac{\sin A \cos B + \cos A \sin B}{\cos A \cos B - \sin A \sin B} \qquad \blacktriangleleft \text{ by (10.7) and (10.4)}$$

$$= \frac{\dfrac{\sin A \cos B}{\cos A \cos B} + \dfrac{\cos A \sin B}{\cos A \cos B}}{\dfrac{\cos A \cos B}{\cos A \cos B} - \dfrac{\sin A \sin B}{\cos A \cos B}} \qquad \blacktriangleleft \text{ dividing each term by } \cos A \cos B$$

Now by performing the indicated divisions and replacing $(\sin A)/(\cos A)$ by $\tan A$ as well as $(\sin B)/(\cos B)$ by $\tan B$, we see that

$\tan (A + B)$
$$\tan (A + B) = \frac{\tan A + \tan B}{1 - \tan A \tan B} \qquad \qquad \textbf{(10.11)}$$

10.12 THE TANGENT OF TWICE AN ANGLE

To obtain a formula for the tangent of twice an angle, we replace B by A in (10.11). Thus,

$$\tan (A + A) = \frac{\tan A + \tan A}{1 - \tan A \tan A}$$

Consequently,

$\tan 2A$
$$\tan 2A = \frac{2 \tan A}{1 - \tan^2 A} \qquad \qquad \textbf{(10.12)}$$

10.13 THE TANGENT OF HALF AN ANGLE

In order to derive an expression for the tangent of half an angle in terms of the angle, we shall begin with the identity that gives the tangent of an angle as the sine of the angle divided by its cosine. Thus,

$$\tan \tfrac{1}{2}\theta = \frac{\sin \tfrac{1}{2}\theta}{\cos \tfrac{1}{2}\theta} \tag{1}$$

$$= \frac{2 \sin^2 \tfrac{1}{2}\theta}{2 \sin \tfrac{1}{2}\theta \cos \tfrac{1}{2}\theta} \qquad \blacktriangleleft \text{ multiplying by } (2 \sin \tfrac{1}{2}\theta)/(2 \sin \tfrac{1}{2}\theta)$$

If we now use (10.9) to replace the numerator by $1 - \cos \theta$ and (10.8) to replace the denominator by $\sin \theta$, we see that

$\tan \tfrac{1}{2}\theta$

$$\boldsymbol{\tan \tfrac{1}{2}\theta = \frac{1 - \cos \theta}{\sin \theta}} \tag{10.13a}$$

If we multiply the numerator and denominator of (10.13a) by $1 + \cos \theta$, replace $1 - \cos^2 \theta$ by $\sin^2 \theta$, and take the common factor $\sin \theta$ from the numerator and denominator, we get

$\tan \tfrac{1}{2}\theta$

$$\boldsymbol{\tan \tfrac{1}{2}\theta = \frac{\sin \theta}{1 + \cos \theta}} \tag{10.13b}$$

Thus, there are two convenient formulas for the tangent of one-half an angle.

EXAMPLE

If we replace θ by $60°$ in (10.13b), we have

$$\tan \tfrac{1}{2}(60°) = \frac{\sin 60°}{1 + \cos 60°}$$

$$= \frac{\sqrt{3}/2}{1 + \tfrac{1}{2}}$$

$$= \frac{\sqrt{3}/2}{\tfrac{3}{2}}$$

$$= \frac{\sqrt{3}}{3}$$

10.14 THE TANGENT OF THE DIFFERENCE OF TWO ANGLES

If in (10.11) we replace B by $-C$, we get

$$\tan (A - C) = \frac{\tan A + \tan (-C)}{1 - \tan A \tan (-C)}$$

Now, by replacing $\tan (-C)$ by $-\tan C$, we find that

$\tan (A - C)$

$$\boldsymbol{\tan (A - C) = \frac{\tan A - \tan C}{1 + \tan A \tan C}} \tag{10.14}$$

EXAMPLE

Prove that

$$\tan\left(\theta + \frac{\pi}{3}\right) + \tan\left(\theta - \frac{\pi}{3}\right) = \frac{8 \tan \theta}{1 - 3 \tan^2 \theta}$$

is an identity.

Solution

If we use the identities for the tangent of the sum and the tangent of the difference of two angles, we have

$$\tan\left(\theta + \frac{\pi}{3}\right) + \tan\left(\theta - \frac{\pi}{3}\right)$$

$$= \frac{\tan \theta + \tan (\pi/3)}{1 - \tan \theta \tan (\pi/3)} + \frac{\tan \theta - \tan (\pi/3)}{1 + \tan \theta \tan (\pi/3)}$$

$$= \frac{\tan \theta + \sqrt{3}}{1 - \sqrt{3} \tan \theta} + \frac{\tan \theta - \sqrt{3}}{1 + \sqrt{3} \tan \theta} \qquad \blacktriangleleft \text{ since } \tan (\pi/3) = \sqrt{3}$$

$$= \frac{(\tan \theta + \sqrt{3})(1 + \sqrt{3} \tan \theta) + (\tan \theta - \sqrt{3})(1 - \sqrt{3} \tan \theta)}{(1 - \sqrt{3} \tan \theta)(1 + \sqrt{3} \tan \theta)}$$

$$\qquad\qquad\qquad\qquad \blacktriangleleft \text{ reducing to a common denominator}$$

$$= \frac{8 \tan \theta}{1 - 3 \tan^2 \theta} \qquad \blacktriangleleft \text{ performing the indicated multiplications and combining terms}$$

10.15 AN APPLICATION FROM ANALYTIC GEOMETRY

We saw, in Sec. 7.11, how to evaluate the other trigonometric ratios if one of them and the quadrant in which the angle lies are known. We shall now find how to evaluate the sine and cosine of an acute angle if the cotangent of twice the angle is known. This problem arises in analytic geometry and can be solved by use of several sets of identities. We shall use

$$\cot 2\theta = \frac{1 - \tan^2 \theta}{2 \tan \theta} \qquad (1)$$

and

$$\tan \theta = \frac{\sin \theta}{\cos \theta} \qquad (2)$$

The first of these enables us to find $\tan \theta$ if $\cot 2\theta$ is known and the second helps us to find $\sin \theta$ and $\cos \theta$ if $\tan \theta$ is known. In order to do the latter, we must realize that if $\tan \theta = b/a$, it does not always follow that $\sin \theta = b$ and $\cos \theta = a$ but rather that, for some value of k, $\sin \theta = kb$ and $\cos \theta = ka$ since $b/a = kb/ka$. Squaring each member of $\sin \theta = kb$ and $\cos \theta = ka$ and adding gives us $1 = k^2(a^2 + b^2)$, and it follows that $k = 1/\sqrt{a^2 + b^2}$. Thus, if we know that $\tan \theta = b/a$, we may write

$$\sin \theta = \frac{b}{\sqrt{a^2 + b^2}} \quad \text{and} \quad \cos \theta = \frac{a}{\sqrt{a^2 + b^2}} \qquad (3)$$

EXAMPLE   Find $\sin \theta$ and $\cos \theta$ if $\cot 2\theta = \frac{5}{12}$ with θ acute.

Solution   If we substitute the given value for $\cot 2\theta$ in (1), we get

$$\frac{5}{12} = \frac{1 - \tan^2 \theta}{2 \tan \theta}$$

Hence, $5 \tan \theta = 6 - 6 \tan^2 \theta$

$$6 \tan^2 \theta + 5 \tan \theta - 6 = 0$$

$$\tan \theta = \tfrac{2}{3}, -\tfrac{3}{2}$$

We must use $\frac{2}{3}$ because θ is acute. Since $\sqrt{2^2 + 3^2} = \sqrt{13}$, the use of (3) gives $\sin \theta = 2/\sqrt{13}$ and $\cos \theta = 3/\sqrt{13}$.

Exercise 10.3   Tangent of a Composite Angle

Evaluate the function value called for in each of problems 1 to 4 by use of the fundamental identities and the function values of 30°, 45°, and their multiples.

1   $\tan 15°$, with $15° = 60° - 45°$
2   $\tan 120°$, with $120° = 2(60°)$
3   $\tan 30°$, with $30° = \frac{1}{2}(60°)$
4   $\tan 75°$, with $75° = 30° + 45°$

Find $\tan (A + B)$ and $\tan (A - B)$ if A, B, and their function values are as given in problems 5 to 8.

5   $\tan A = \frac{3}{4}, 0 < A < \pi/2; \tan B = -\frac{5}{12}, \pi/2 < B < \pi$
6   $\tan A = \frac{8}{15}, \pi < A < 3\pi/2; \sin B = -\frac{3}{5}, \pi < B < 3\pi/2$
7   $\cos A = \frac{15}{17}, 3\pi/2 < A < 2\pi; \sin B = \frac{8}{17}, \pi/2 < B < \pi$
8   $\sin A = -\frac{7}{25}; \pi < A < 3\pi/2; \cot B = -\frac{12}{5}, \pi/2 < B < \pi$

If A and one of its function values are as given in problems 9 to 12, find $\tan 2A$ and $\tan \frac{1}{2}A$.

9    $\tan A = \frac{15}{8}, 0 < A < \pi/2$
10   $\cot A = -\frac{7}{24}, \pi/2 < A < \pi$
11   $\sec A = \frac{5}{3}, 3\pi/2 < A < 2\pi$
12   $\csc A = -\frac{13}{5}, \pi < A < 3\pi/2$

Prove that the statements in problems 13 to 16 are identities.

13 $\tan (\pi + \theta) = \tan \theta$ **14** $\tan (\pi - \theta) = -\tan \theta$

15 $\tan (2\pi - \theta) = -\tan \theta$ **16** $\tan \left(\dfrac{3\pi}{2} + \theta \right) = -\cot \theta$

Change the first expression to the second in each of problems 17 to 24.

17 $\tan (\theta + 45°) + \tan (\theta - 45°)$, $2 \tan 2\theta$
18 $\tan (\theta + 45°) + \tan (\theta - 45°)$, $2 \sec 2\theta$
19 $\cot 4\theta + \tan 2\theta$, $\csc 4\theta$
20 $\tan 4\theta - \tan 2\theta$, $\tan 2\theta \sec 4\theta$
21 $\tan \theta \tan \frac{1}{2}\theta$, $\sec \theta - 1$
22 $\cot \theta - \tan \theta$, $2 \cot 2\theta$
23 $\cot 2\theta + \tan \theta$, $\csc 2\theta$
24 $\cot \theta - \cot 3\theta$, $2 \cos \theta \csc 3\theta$

Prove that the statement in each of problems 25 to 32 is an identity.

25 $\dfrac{\tan 4\theta - \tan 2\theta}{1 + \tan 4\theta \tan 2\theta} = \dfrac{2 \tan \theta}{1 - \tan^2 \theta}$

26 $\dfrac{\tan 3\theta + \tan \theta}{1 - \tan 3\theta \tan \theta} = \dfrac{2 \tan 2\theta}{1 - \tan^2 2\theta}$

27 $\tan 3\theta = \dfrac{3 \tan \theta - \tan^3 \theta}{1 - 3 \tan^2 \theta}$

28 $\tan 4\theta = \dfrac{4 \tan \theta - 4 \tan^3 \theta}{1 - 6 \tan^2 \theta + \tan^4 \theta}$

29 $\tan 2\theta = \dfrac{2}{\cot \theta - \tan \theta}$

30 $\tan \left(\dfrac{\theta}{2} + \dfrac{\pi}{4} \right) = \dfrac{1 + \sin \theta}{\cos \theta}$

31 $\tan \left(\dfrac{\pi}{4} - \theta \right) = \dfrac{1 - \tan \theta}{1 + \tan \theta}$

32 $\tan \left(\dfrac{\pi}{4} + \theta \right) = \dfrac{1 + \tan \theta}{1 - \tan \theta}$

Find $\sin A$ and $\cos A$ if A is a positive acute angle and if $\cot 2A$ has the values given in problems 33 to 40.

33 $\cot 2A = \frac{4}{3}$ **34** $\cot 2A = \frac{5}{12}$
35 $\cot 2A = \frac{15}{8}$ **36** $\cot 2A = \frac{24}{7}$
37 $\cot 2A = -\frac{8}{15}$ **38** $\cot 2A = -\frac{4}{3}$
39 $\cot 2A = -\frac{12}{5}$ **40** $\cot 2A = -\frac{24}{7}$

10.16 SUMMARY OF FORMULAS

We here collect the fundamental identities derived in this chapter so that they will be readily available for reference.

$$\cos(A - C) = \cos A \cos C + \sin A \sin C \tag{10.1}$$

$$\cos\left(\frac{\pi}{2} - B\right) = \sin B \tag{10.2}$$

$$\sin\left(\frac{\pi}{2} - B\right) = \cos B \tag{10.3}$$

$$\cos(A + B) = \cos A \cos B - \sin A \sin B \tag{10.4}$$

$$\cos 2A = \cos^2 A - \sin^2 A \tag{10.5a}$$

$$= 1 - 2\sin^2 A \tag{10.5c}$$

$$= 2\cos^2 A - 1 \tag{10.5b}$$

$$\cos \tfrac{1}{2}\theta = \pm\sqrt{\frac{1 + \cos\theta}{2}} \tag{10.6}$$

$$\sin(A + B) = \sin A \cos B + \cos A \sin B \tag{10.7}$$

$$\sin 2A = 2\sin A \cos A \tag{10.8}$$

$$\sin \tfrac{1}{2}\theta = \pm\sqrt{\frac{1 - \cos\theta}{2}} \tag{10.9}$$

$$\sin(A - C) = \sin A \cos C - \cos A \sin C \tag{10.10}$$

$$\tan(A + B) = \frac{\tan A + \tan B}{1 - \tan A \tan B} \tag{10.11}$$

$$\tan 2A = \frac{2\tan A}{1 - \tan^2 A} \tag{10.12}$$

$$\tan \tfrac{1}{2}\theta = \frac{1 - \cos\theta}{\sin\theta} \tag{10.13a}$$

$$= \frac{\sin\theta}{1 + \cos\theta} \tag{10.13b}$$

$$\tan(A - C) = \frac{\tan A - \tan C}{1 + \tan A \tan C} \tag{10.14}$$

Exercise 10.4   Identities

Prove that each of the following equations is an identity.

1   $\dfrac{\cos 3x}{\sin x} + \dfrac{\sin 3x}{\cos x} = 2\cos 2x \csc 2x$

2 $\dfrac{\cos 4x}{\sin x} - \dfrac{\sin 4x}{\cos x} = 2 \cos 5x \csc 2x$

3 $\dfrac{\sin 5x}{\sin x} - \dfrac{\cos 5x}{\cos x} = 4 \cos 2x$

4 $\dfrac{\cos 3x}{\cos 2x} + \dfrac{\sin 3x}{\sin 2x} = \dfrac{2 \sin 5x}{\sin 4x}$

5 $\cos x + \sin x \tan y = \sec y \cos (x - y)$

6 $\sin x - \cos x \cot y = -\csc y \cos (x + y)$

7 $\sin x + \cos x \tan y = \sec y \sin (x + y)$

8 $\cos x - \sin x \cot y = \csc y \sin (y - x)$

9 $\cot y - \cot x = \dfrac{\sin (x - y)}{\sin x \sin y}$

10 $\cos x + \sin x \cot y = \sin (x + y) \csc y$

11 $\sin y (\cot y - \tan x) = \cos (x + y) \sec x$

12 $\cos y - \sin y \tan x = \sec x \cos (x + y)$

13 $\csc y - 2 \sin y = 2 \cos y \cot 2y$

14 $1 + \csc^2 x \cos 2x = \cot^2 x$

15 $\csc 2x + \cot 2x = \cot x$

16 $\csc 2x - \cot 2x = \tan x$

17 $\dfrac{1 + \sin 2x}{\cos 2x} = \dfrac{1 + \tan x}{1 - \tan x}$

18 $1 + \dfrac{\cos 3x}{\cos x} = 2 \cos 2x$

19 $\cot 2x = \dfrac{\csc x - 2 \sin x}{2 \cos x}$

20 $\dfrac{(1 + \cot x)^2}{1 + \cot^2 x} = 1 + \sin 2x$

21 $\sin 3x = 3 \sin x - 4 \sin^3 x$

22 $\sin 4x = 4 \sin x \cos x (1 - 2 \sin^2 x)$

23 $\cos 3x = 4 \cos^3 x - 3 \cos x$

24 $\cos 4x = 8 \cos^4 x - 7 \cos^2 x + \sin^2 x$

25 $2 \sin \frac{1}{2} (x + y) \cos \frac{1}{2} (x - y) = \sin x + \sin y$

26 $2 \cos \frac{1}{2} (x + y) \cos \frac{1}{2} (x - y) = \cos x + \cos y$

27 $2 \cos \frac{1}{2} (x + y) \sin \frac{1}{2} (x - y) = \sin x - \sin y$

28 $2 \sin \frac{1}{2} (x + y) \sin \frac{1}{2} (x - y) = -\cos x + \cos y$

29 $\dfrac{1 + 2 \sin 2x + \cos 2x}{2 + \sin 2x - 2 \cos 2x} = \cot x$

30 $\dfrac{\sin x - \sin 3x + \sin 2x}{2(1 - \cos x)} = \sin x + \sin 2x$

31 $\tan (x - y) + \tan y = \sec (x - y) \sin x \sec y$

32 $\cot x - \cot (x - y) = \csc (y - x) \sin y \csc x$

Exercise 10.5   Review

1   Find the value of sin 15°, cos 15°, and tan 15° by use of the half-angle formulas.

2   Find the value of sin 120°, cos 120°, and tan 120° by use of double-angle formulas.

3   Find the value of sin 105°, cos 75°, and tan 105° by use of formulas for functions of the sum of two angles.

4   If $\sin A = \frac{3}{5}$, $0 < A < \pi/2$; $\cos B = \frac{8}{17}$, $3\pi/2 < B < 2\pi$, find $\cos(A-B)$, $\cos(A+B)$, and $\cos 2A$.

5   If $\sin A = \frac{15}{17}$, $\pi/2 < A < \pi$; $\cos B = -\frac{3}{5}$, $\pi < B < 3\pi/2$, find $\sin(A-B)$, $\sin(A+B)$, and $\sin 2A$.

6   If $\sin A = -\frac{12}{13}$, $\pi < A < 3\pi/2$; $\cos B = -\frac{8}{17}$, $\pi/2 < B < \pi$, find $\tan(A-B)$, $\tan(A+B)$, and $\tan 2A$.

Change the first expression in each of problems 7 to 11 to the second.

7 $\cos 7\theta \cos 5\theta + \sin 7\theta \sin 5\theta$, $2\cos^2\theta - 1$

8 $\sin 5\theta \cos\theta - \cos 5\theta \sin\theta$, $4\sin\theta\cos\theta\,(1 - 2\sin^2\theta)$

9 $(1 + 2\sin 3\theta)(1 - \sin 3\theta)$, $\cos 6\theta + \sin 3\theta$

10 $(-1 + 4\sin\theta\cos\theta)(\cos^2\theta - \sin^2\theta)$, $\sin 4\theta - \cos 2\theta$

11 $\cot\theta - \tan\theta$, $2\cot 2\theta$

Prove that each of the following equations is an identity.

12 $\dfrac{\sin 2x}{\cos 2x} - \dfrac{\sin x}{\cos x} = \tan x \sec 2x$

13 $\cos x + \sin x \cot y = \sin(x+y)\csc y$

14 $2\cos x - \sec x = \dfrac{\sec x}{\sec 2x}$

15 $\dfrac{\cos 2x}{1 + \sin 2x} = \dfrac{1 - \tan x}{1 + \tan x}$

16 $1 + \dfrac{\sin 3x}{\sin x} = 4\cos^2 x$

17 $\dfrac{\tan 4x - \tan 2x}{1 + \tan 4x \tan 2x} = \dfrac{2\sin x}{2\cos x - \sec x}$

18 $\dfrac{2\sin 2x}{(1 + \cos 2x)^2} = \tan x + \tan^3 x$

19 $8\cos^4 x = 3 + 4\cos 2x + \cos 4x$

20 $\dfrac{\sin 2x + \sin\dfrac{\pi}{3}}{\cos 2x + \cos\dfrac{\pi}{3}} = \tan\left(x + \dfrac{\pi}{6}\right)$

21 $\quad 1 + \cos x \cos 3x - \sin x \sin 3x = 2 \cos^2 2x$

22 $\quad \dfrac{\tan x}{\tan 4x} = \dfrac{2 \cos^2 2x - 1}{2 \cos 2x \; (\cos 2x + 1)}$

23 $\quad \csc 4x - \cot 4x = \dfrac{2 \tan x}{1 - \tan^2 x}$

24 $\quad \csc 2x + \cot 2x = \cot x$

25 $\quad \tan A + \tan B + \tan C = \tan A \tan B \tan C, \qquad$ if $A + B + C = \pi$

Quadratic Equations

In previous chapters we have dealt with problems that can be solved by means of equations involving only the first power of the unknown. Practical problems and theoretical considerations, however, frequently lead to equations in which the unknown appears to a degree greater than 1. In this chapter we shall discuss equations in which the unknown appears to the second degree. Such an equation is called a quadratic equation.

11.1 DEFINITIONS

We shall define a quadratic equation as follows:

Quadratic equation

The equation $f(x) = g(x)$ is a *quadratic equation* in x if (1) one of $f(x)$ and $g(x)$ is a polynomial of second degree in x and the other is a polynomial of first degree in x or is a constant, or (2) both $f(x)$ and $g(x)$ are polynomials of second degree in x with the coefficients of x^2 not equal. Such an equation can be put in the form $ax^2 + bx + c = 0$, where a, b, and c are constants.

Examples of quadratic equations are

$$5x^2 - 2x + 3 = 2x^2 + x - 6$$
$$3x^2 - 5x + 8 = 6x + 3$$
$$7x^2 - 4x = 5$$

Before discussing a method for solving the general quadratic equation, we note that we obtain the solution set of an equation of the type

$$ax^2 - b = 0 \qquad (1)$$

by first solving for x^2 and obtaining

$$x^2 = \frac{b}{a}$$

Then

$$x = \pm\sqrt{\frac{b}{a}} \qquad \blacktriangleleft \text{ since } (\pm\sqrt{b/a})^2 = b/a$$

$$= \pm\sqrt{\frac{ab}{a^2}}$$

$$= \pm\frac{\sqrt{ab}}{a}$$

Hence the solution set of Eq. (1) is $\{\sqrt{ab}/a, -\sqrt{ab}/a\}$.

We shall prove in Sec. 11.5 that any given quadratic equation is equivalent to an equation of the type

$$(x + d)^2 = k \qquad (2)$$

Consequently, by the definition of a square root, an element in the solution set of (2) must satisfy the equation $x + d = \pm\sqrt{k}$ since $(\pm\sqrt{k})^2 = k$.

209

Therefore,

$$x = -d \pm \sqrt{k}$$

and the solution set of Eq. (2) is $\{-d+\sqrt{k}, -d-\sqrt{k}\}$. Hence we have the following conclusion:

> *If the equation $(x + d)^2 = k$ is given, then $x + d = \pm\sqrt{k}$, and the solution set of the equation is $\{-d + \sqrt{k}, -d - \sqrt{k}\}$*   **(11.1)**

Thus there are two elements in the solution set of a quadratic equation, or in other words, a quadratic equation has two roots. It may happen, however, that the two elements in the solution set are equal. For example, if in Eq. (2), $k = 0$, the solution set is $\{-d, -d\}$.

Before discussing the general procedure for solving a quadratic equation, we shall explain a method that is applicable to a wide variety of equations.

11.2 SOLUTION BY FACTORING

If a given quadratic equation is equivalent to one of the type

$$(ax + b)(cx + d) = 0 \tag{1}$$

we obtain the solution set by the following method.

If the product of two or more factors is equal to zero, then at least one of the factors is zero by (2.35). Therefore, if S is the solution set of (1), then each element in S is a replacement for x such that either $ax + b = 0$ or $cx + d = 0$. Hence if

$$S_1 = \{x|ax + b = 0\} \qquad \text{and} \qquad S_2 = \{x|cx + d = 0\}$$

then

$$S = S_1 \cup S_2$$

To obtain S_1 we set $ax + b = 0$, solve for x, and get $x = -b/a$. Consequently,

$$S_1 = \left\{-\frac{b}{a}\right\}$$

Similarly, setting $cx + d = 0$ and solving for x, we get $x = -d/c$. Therefore

$$S_2 = \left\{-\frac{d}{c}\right\}$$

and it follows that

$$S = S_1 \cup S_2 = \left\{-\frac{b}{a}\right\} \cup \left\{-\frac{d}{c}\right\} = \left\{-\frac{b}{a}, -\frac{d}{c}\right\}$$

As the above discussion illustrates, the procedure for solving a quadratic equation by the factoring method consists of the following steps:

1 From the given equation, obtain an equivalent equation in which the right member is zero.

Solution
by factoring

2 Factor the left member of the equation obtained in step 1.

3 Set each factor obtained in step 2 equal to zero and solve the resulting equation for the letter that stands for the unknown.

4 The elements of the solution set are the two roots obtained in step 3.

EXAMPLE 1 Solve the equation $2x^2 = x + 6$ by the factoring method.

Solution

$$
\begin{aligned}
2x^2 &= x + 6 & &\blacktriangleleft \text{ given equation} \\
2x^2 - x - 6 &= 0 & &\blacktriangleleft \text{ adding } -x - 6 \text{ to each member} \\
(2x + 3)(x - 2) &= 0 & &\blacktriangleleft \text{ factoring left member} \\
2x + 3 &= 0 & &\blacktriangleleft \text{ setting the first factor equal to 0} \\
x &= -\tfrac{3}{2} & &\blacktriangleleft \text{ solving for } x \\
x - 2 &= 0 & &\blacktriangleleft \text{ setting second factor equal to 0} \\
x &= 2 & &\blacktriangleleft \text{ solving for } x
\end{aligned}
$$

Consequently the solution set is $\{-\tfrac{3}{2}, 2\}$.

The solution set can be verified by replacing x in the given equation by each element in the set.

EXAMPLE 2 Solve the equation $5y^2 = 6y$ by the factoring method.

Solution

$$
\begin{aligned}
5y^2 &= 6y & &\blacktriangleleft \text{ given equation} \\
5y^2 - 6y &= 0 & &\blacktriangleleft \text{ adding } -6y \text{ to each member} \\
y(5y - 6) &= 0 & &\blacktriangleleft \text{ factoring left member} \\
y &= 0 & &\blacktriangleleft \text{ setting first factor equal to 0} \\
5y - 6 &= 0 & &\blacktriangleleft \text{ setting second factor equal to 0} \\
y &= \tfrac{6}{5} & &\blacktriangleleft \text{ solving for } y
\end{aligned}
$$

Therefore the solution set is $\{0, \tfrac{6}{5}\}$.

Check If we replace y by 0 in the given equation, each member becomes zero. If we replace y by $\tfrac{6}{5}$, we obtain $5(\tfrac{36}{25}) = \tfrac{36}{5}$.

Exercise 11.1 Solution by Factoring

Find the solution set of the quadratic in each of problems 1 to 8.

1 $x^2 - 4 = 0$ **2** $x^2 - 36 = 0$

$$3 \quad 9x^2 - 1 = 0 \qquad\qquad 4 \quad 16x^2 - 1 = 0$$
$$5 \quad 4x^2 - 81 = 0 \qquad\qquad 6 \quad 36x^2 - 121 = 0$$
$$7 \quad 9x^2 - 169 = 0 \qquad\qquad 8 \quad 49x^2 - 144 = 0$$

Find the solution sets of the following quadratics by use of factoring.

$9 \quad x^2 - 3x + 2 = 0$	$10 \quad x^2 - 7x + 12 = 0$
$11 \quad x^2 - x - 6 = 0$	$12 \quad x^2 + 2x - 8 = 0$
$13 \quad x^2 = 5x$	$14 \quad x^2 + x = 2$
$15 \quad x^2 = -x + 12$	$16 \quad x^2 = 2x + 8$
$17 \quad 2x^2 + 5x = 3$	$18 \quad 3x^2 = x + 2$
$19 \quad 2x^2 + 6 = -7x$	$20 \quad 4x^2 - 11x = 3$
$21 \quad 7x = 12 - 12x^2$	$22 \quad 6x^2 - 15 = x$
$23 \quad 14x^2 + 19x = 3$	$24 \quad 12x^2 = 17x + 5$
$25 \quad 6x^2 - 35 = 11x$	$26 \quad 10x^2 + 6 = 19x$
$27 \quad 34x + 8 = -21x^2$	$28 \quad 18x^2 = 15 - 17x$
$29 \quad 13x = 35x^2 - 12$	$30 \quad 30x^2 - 2 = 7x$
$31 \quad 42x^2 - 10 = 23x$	$32 \quad 35x^2 = x + 12$
$33 \quad 24x^2 = -38x - 15$	$34 \quad 30x^2 + 103x = 7$
$35 \quad 42x^2 + 55x = 75$	$36 \quad 36x^2 = 97x - 36$
$37 \quad x^2 + cx - 6c^2 = 0$	$38 \quad x^2 - dx - 12d^2 = 0$
$39 \quad c^2x^2 + cx - 2 = 0$	$40 \quad 2x^2 + 5cx - 3c^2 = 0$

11.3 COMPLEX NUMBERS

In Sec. 11.1 we stated that every quadratic equation is equivalent to an equation of the type $(x + d)^2 = k$, and in (11.1) that the roots of such an equation are $-d \pm \sqrt{k}$. If k is negative, $\sqrt{k}$ is not a real number, and we must ignore such numbers as $\sqrt{-5}$ or define an additional set or system of numbers. We choose the latter course. If $k = -n$, $n > 0$, we define $\sqrt{-n}$ as $\sqrt{-1}\sqrt{n}$. We let i stand for $\sqrt{-1}$ and write $\sqrt{-n} = \sqrt{-1}\sqrt{n} = i\sqrt{n}$. Similarly $-\sqrt{-n} = -i\sqrt{n}$. Therefore,

$$\pm\sqrt{-n} = \pm\sqrt{n}\,i \qquad n > 0 \qquad\qquad \textbf{(11.2)}$$

EXAMPLE $\quad \sqrt{-25} = 5i, \sqrt{-7} = \sqrt{7}i$, and $-\sqrt{-36} = -6i$.

Imaginary number The early mathematicians did not understand such numbers and called them *imaginary*. In the eighteenth century, however, Gauss and Argand devised a geometrical interpretation of these numbers, which since then have become very useful in mathematics, physics, electrical engineering, and electronics.

Complex number   ***A number of the type a + bi, where a and b are real numbers, is called***
a complex number   **(11.3)**

Pure imaginary

If neither a nor b is 0, $a + bi$ is also called an imaginary number. If $a = 0$ and $b \neq 0$, $a + bi$ is a *pure imaginary* number. If, however, $a \neq 0$ and $b = 0$, $a + bi = a$ is a real number. Therefore {real numbers} $\subset$ {complex numbers}.

We shall discuss complex numbers more fully in Chap. 16 but now call attention to a cyclic property of the powers of i. Since $i = \sqrt{-1}$, $i^2 = -1$, $i^3 = i(i^2) = i(-1) = -i$, $i^4 = i^2(i^2) = -1(-1) = 1$, $i^5 = i(i^4) = i$, and so on.

11.4   SOLUTION BY COMPLETING THE SQUARE

As we pointed out in Sec. 11.1, the solution set of an equation of the type

$$(x + d)^2 = k$$

is readily obtained. Hence, if we can obtain an equation of this type that is equivalent to the given equation, we can obtain the solution set of the given equation without difficulty. We now discuss a method for accomplishing this. Since we shall employ theorems (4.1) to (4.3) repeatedly in the application of this method, the reader is advised to review these theorems now.

We first examine the conditions under which a quadratic trinomial of the type $x^2 + bx + c$ is a perfect square. By (2.51)

$$(x + d)^2 = x^2 + 2dx + d^2$$

From this identity, we see that the constant term in the quadratic trinomial at the right of the equality sign is the *square of one-half the coefficient of x*. Consequently:

Condition for a perfect square

A quadratic trinomial in x with the coefficient of x^2 equal to 1 is a perfect square if the *constant term is the square of one-half the coefficient of x*.

EXAMPLE 1 $x^2 - 6x + 9$ is a perfect square since $9 = (-\frac{6}{2})^2$. Furthermore $x^2 - 6x + 9 = (x - 3)^2$.

EXAMPLE 2 $x^2 - \frac{1}{3}x + \frac{1}{36}$ is a perfect square since $\frac{1}{36} = [\frac{1}{2}(\frac{1}{3})]^2$ and $x^2 - \frac{1}{3}x + \frac{1}{36} = (x - \frac{1}{6})^2$.

The next example illustrates the use of this information in solving a quadratic equation.

EXAMPLE 3   Solve the equation

$$x = 3 - 2x^2 \qquad (1)$$

by completing the square.

Solution   We first obtain an equation of the type $x^2 + bx = c$ that is equivalent to (1). The first step in this procedure is to employ (4.1), add $2x^2$ to each member of (1), and obtain

$$2x^2 + x = 3 \qquad (2)$$

Now we employ (4.3), multiply each member of (2) by $\frac{1}{2}$, and get

$$x^2 + \tfrac{1}{2}x = \tfrac{3}{2} \qquad (3)$$

The next step is to *complete the square* by adding the square of one-half the coefficient of x, or $[\frac{1}{2}(\frac{1}{2})]^2 = \frac{1}{16}$ to each member of (3), and get

$$x^2 + \tfrac{1}{2}x + \tfrac{1}{16} = \tfrac{3}{2} + \tfrac{1}{16}$$

or

$$x^2 + \tfrac{1}{2}x + \tfrac{1}{16} = \tfrac{25}{16}$$

Now express this equation in the form

$$(x + \tfrac{1}{4})^2 = (\tfrac{5}{4})^2 \qquad \blacktriangleleft \text{ since } x^2 + \tfrac{1}{2}x + \tfrac{1}{16} = (x + \tfrac{1}{4})^2 \text{ and } \tfrac{25}{16} = (\tfrac{5}{4})^2$$

We now employ (11.1) to equate the square roots of the members of the last equation and prefix the square root of $(\frac{5}{4})^2$ by the plus and minus signs and thus obtain

$$x + \tfrac{1}{4} = \pm \tfrac{5}{4}$$

We now solve this pair of linear equations for x and get

$$x = \pm \tfrac{5}{4} - \tfrac{1}{4}$$
$$= \tfrac{5}{4} - \tfrac{1}{4} \text{ and } -\tfrac{5}{4} - \tfrac{1}{4}$$
$$x = 1 \text{ and } -\tfrac{3}{2}$$

Hence the solution set is $\{1, -\frac{3}{2}\}$.

   The solution set can be checked by replacing x in (1) by 1 and then by $-\frac{3}{2}$.

   Example 3 illustrates the following steps in the process of solving a quadratic equation by completing the square:

Completing the   **1** By adding the appropriate polynomial, monomial, or constant to
square   each member of the given equation and then rearranging terms,
                    obtain an equivalent equation of the type $ax^2 + bx = c$.

2 Multiply each member of the equation in step 1 by $1/a$ to obtain an equation in which the coefficient of x^2 is 1.

3 Add the square of one-half the coefficient of x to each member of the equation obtained in step 2 and simplify the right member.

4 Equate the square roots of the members of the equation in step 3, prefixing the square root of the right member with plus and minus signs. This procedure yields two linear equations.

5 Solve the two linear equations obtained in step 4 for x. The two roots thus obtained are the elements of the solution set.

6 Check the solution set by substituting each element for x in the given equation.

EXAMPLE 4

Solve the equation $4x^2 = 4x + 11$ by completing the square.

Solution

$$4x^2 = 4x + 11 \qquad \blacktriangleleft \text{ given equation}$$
$$4x^2 - 4x = 11 \qquad \blacktriangleleft \text{ adding } -4x \text{ to each member}$$
$$x^2 - x = \tfrac{11}{4} \qquad \blacktriangleleft \text{ multiplying each member by } \tfrac{1}{4}$$
$$x^2 - x + (-\tfrac{1}{2})^2 = \tfrac{11}{4} + (-\tfrac{1}{2})^2 \qquad \blacktriangleleft \text{ adding } [\tfrac{1}{2}(-1)]^2 \text{ to each member}$$

$$x^2 - x + \tfrac{1}{4} = \tfrac{12}{4}$$
$$(x - \tfrac{1}{2})^2 = 3$$

$$x - \tfrac{1}{2} = \pm\sqrt{3} \qquad \blacktriangleleft \text{ equating the square roots of the members}$$

$$x = \tfrac{1}{2} \pm \sqrt{3} \qquad \blacktriangleleft \text{ solving for } x$$

Since $\sqrt{3}$ is an irrational number, the above roots can be simplified no further, and the solution set is $\{\tfrac{1}{2} + \sqrt{3}, \tfrac{1}{2} - \sqrt{3}\}$.

Check

We replace x in the given equation by $\tfrac{1}{2} \pm \sqrt{3}$ and get for the left member

$$4(\tfrac{1}{2} \pm \sqrt{3})^2 = 4[\tfrac{1}{4} \pm 2(\tfrac{1}{2}\sqrt{3}) + (\sqrt{3})^2] = 4(\tfrac{1}{4} \pm \sqrt{3} + 3) = 13 \pm 4\sqrt{3}$$

Similarly, for the right member,

$$4(\tfrac{1}{2} \pm \sqrt{3}) + 11 = 2 \pm 4\sqrt{3} + 11 = 13 \pm 4\sqrt{3}$$

Hence, the proposed solution checks since the two members are equal.

An approximate value of the above roots accurate to three decimal places is $\tfrac{1}{2} \pm \sqrt{3} = 0.5 \pm 1.732 = 2.232$ and -1.232.

EXAMPLE 5

Solve the equation $x^2 + 8 = 4x$ by completing the square.

Solution

$$x^2 + 8 = 4x \qquad \blacktriangleleft \text{ given equation}$$
$$x^2 - 4x = -8 \qquad \blacktriangleleft \text{ adding } -4x - 8 \text{ to each member}$$
$$x^2 - 4x + (-2)^2 = -8 + (-2)^2 \qquad \blacktriangleleft \text{ adding } [\tfrac{1}{2}(-4)]^2 \text{ to each member}$$
$$x^2 - 4x + 4 = -4$$
$$(x - 2)^2 = -4$$
$$x - 2 = \pm\sqrt{-4}$$
$$x - 2 = \pm 2i \qquad \blacktriangleleft \text{ by (11.2)}$$
$$x = 2 \pm 2i$$

Therefore the solution set is $\{2 + 2i, \ 2 - 2i\}$.

Check   If we replace x by $2 \pm 2i$ in the given equation, we obtain

$$(2 \pm 2i)^2 + 8 = 4 \pm 8i + 4i^2 + 8 = 4 \pm 8i - 4 + 8 = \pm 8i + 8$$

for the left member, and

$$4(2 \pm 2i) = 8 \pm 8i$$

for the right.   Hence, the proposed solution checks.

Exercise 11.2   Solution by Completing the Square

Find the solution set of the equation in each of problems 1 to 40 by completing the square.

1	$x^2 + 2x - 3 = 0$	**2**	$x^2 - 2x - 15 = 0$
3	$x^2 + 4x - 21 = 0$	**4**	$x^2 - 4x - 12 = 0$
5	$x^2 - 5x + 6 = 0$	**6**	$x^2 - 7x + 12 = 0$
7	$x^2 - 8x + 15 = 0$	**8**	$x^2 - 5x - 14 = 0$
9	$x^2 - 4x + 5 = 0$	**10**	$x^2 + 2x + 2 = 0$
11	$x^2 + 6x + 13 = 0$	**12**	$x^2 - 8x + 25 = 0$
13	$2x^2 - 7x + 3 = 0$	**14**	$3x^2 + 10x - 8 = 0$
15	$5x^2 + 13x - 6 = 0$	**16**	$4x^2 + 5x - 6 = 0$
17	$6x^2 + 7x - 3 = 0$	**18**	$6x^2 - 5x = 6$
19	$12x^2 - x = 6$	**20**	$10x^2 + 21x = -9$
21	$x^2 - 2x - 6 = 0$	**22**	$x^2 - 4x + 1 = 0$
23	$x^2 + 4x + 2 = 0$	**24**	$x^2 - 6x + 4 = 0$
25	$4x^2 + 4x - 1 = 0$	**26**	$9x^2 - 12x + 1 = 0$
27	$9x^2 - 18x + 4 = 0$	**28**	$16x^2 + 16x - 1 = 0$
29	$x^2 + (b - 2c)x - 2bc = 0$	**30**	$x^2 - (b + 3c)x + 3bc = 0$
31	$2x^2 + (a - 4b)x - 2ab = 0$	**32**	$3x^2 + (6b - a)x - 2ab = 0$
33	$x^2 + 2ax + a^2 + b^2 = 0$	**34**	$x^2 + 4ax + 4a^2 - 9b^2 = 0$
35	$4x^2 - 4ax + a^2 + b^2 = 0$	**36**	$9x^2 + 12bx + 4b^2 + a^2 = 0$
37	$a^2x^2 - abx - 2b^2 = 0$	**38**	$b^2x^2 + abx - 6a^2 = 0$
39	$abx^2 + (b^2 - a^2)x = ab$	**40**	$3abx^2 + (6b^2 - a^2)x = 2ab$

11.5   THE QUADRATIC FORMULA

In this section we derive a formula that will give us the roots of a quadratic equation.   The quadratic equation

$$ax^2 + bx + c = 0 \tag{11.4}$$

is said to be in *standard form.* By use of theorems (4.1) to (4.3) and the commutative axiom, we can obtain an equation in standard form that is equivalent to any given quadratic equation.

EXAMPLE 1 Obtain the quadratic equation in standard form that is equivalent to $4x - 3 = 5x^2$, and identify a, b, and c in (11.4).

Solution

$$4x - 3 = 5x^2 \quad \blacktriangleleft \text{ given equation}$$
$$4x - 3 - 5x^2 = 5x^2 - 5x^2 \quad \blacktriangleleft \text{ adding } -5x^2 \text{ to each member}$$
$$-5x^2 + 4x - 3 = 0 \quad \blacktriangleleft \text{ by the commutative axiom}$$

This is the required equation with $a = -5$, $b = 4$, and $c = -3$.

We can now derive a formula by solving Eq. (11.4) by completing the square;

$$ax^2 + bx + c = 0 \quad \blacktriangleleft \text{ Eq. (11.4)}$$
$$ax^2 + bx = -c \quad \blacktriangleleft \text{ adding } -c \text{ to both members}$$
$$x^2 + \frac{bx}{a} = -\frac{c}{a} \quad \blacktriangleleft \text{ dividing each member by } a$$
$$x^2 + \frac{bx}{a} + \left(\frac{b}{2a}\right)^2 = -\frac{c}{a} + \frac{b^2}{4a^2} \quad \blacktriangleleft \text{ adding } [\tfrac{1}{2}(b/a)]^2 \text{ to each member}$$
$$\left(x + \frac{b}{2a}\right)^2 = \frac{b^2 - 4ac}{4a^2}$$
$$x + \frac{b}{2a} = \pm\frac{\sqrt{b^2 - 4ac}}{2a}$$
$$x = -\frac{b}{2a} \pm \frac{\sqrt{b^2 - 4ac}}{2a}$$

Since the two denominators in the right member are the same, we can write

Quadratic formula

$$x = \frac{-b \pm \sqrt{b^2 - 4ac}}{2a} \qquad (11.5)$$

Consequently the solution set of Eq. (11.4) is

$$\left\{ \frac{-b + \sqrt{b^2 - 4ac}}{2a}, \frac{-b - \sqrt{b^2 - 4ac}}{2a} \right\}$$

Equation (11.5) is known as the *quadratic formula.* It expresses the roots of Eq. (11.4) in terms of the constant term and the coefficients of x^2 and x. By properly matching the coefficients and the constant term in any quadratic equation with those in (11.4), we can determine the values of a,

b, and c, and then we can use the formula to get the roots of the equation. We shall illustrate the use of (11.5) with two examples.

EXAMPLE 2   Solve the equation $3x^2 - 5x + 2 = 0$ by means of the quadratic formula.

Solution   To solve $3x^2 - 5x + 2 = 0$ by the quadratic formula, we compare the equation with (11.4) and see that $a = 3$, $b = -5$, and $c = 2$. Hence we substitute these values in (11.5) and get

$$x = \frac{-(-5) \pm \sqrt{(-5)^2 - 4(3)(2)}}{2(3)}$$

$$= \frac{5 \pm \sqrt{25 - 24}}{6} = \frac{5 \pm 1}{6}$$

$$= \tfrac{6}{6} \text{ and } \tfrac{4}{6}$$

$$x = 1 \text{ and } \tfrac{2}{3}$$

Hence the solution set is $\{1, \tfrac{2}{3}\}$.

The reader should verify that the elements of $\{1, \tfrac{2}{3}\}$ are the roots of the given equation.

EXAMPLE 3   Solve the equation $4x^2 = 8x - 7$ by means of the quadratic formula.

Solution   The first step in solving the given equation is to convert it to an equivalent equation in which the terms are in the same order as in (11.4). For this purpose, we add $-8x + 7$ to each member and obtain $4x^2 - 8x + 7 = 0$. By comparing this equation with (11.4), we see that $a = 4$, $b = -8$, and $c = 7$. If we substitute these values in (11.5), we get

$$x = \frac{-(-8) \pm \sqrt{(-8)^2 - 4(4)(7)}}{2(4)}$$

$$= \frac{8 \pm \sqrt{64 - 112}}{8}$$

$$= \frac{8 \pm \sqrt{-48}}{8}$$

$$= \frac{8 \pm 4i\sqrt{3}}{8}$$

$$x = \tfrac{1}{2}(2 \pm i\sqrt{3})$$

Hence the solution set is $\{\tfrac{1}{2}(2 + i\sqrt{3}), \tfrac{1}{2}(2 - i\sqrt{3})\}$.

The reader should verify that these two complex numbers are the roots of the given equation.

11.6 EQUATIONS IN QUADRATIC FORM

The unknown in a quadratic equation may be any quantity. This unknown quantity must, however, enter to the second power and may enter to the first power. Thus, if $f(x)$ is the unknown, then

$$a[f(x)]^2 + b[f(x)] + c = 0$$

Equation in quadratic form

where a, b, and c are constants is a quadratic equation. Such an equation is ordinarily said to be in *quadratic form*. If the unknown $f(x)$ in an equation in quadratic form is linear or quadratic, we can find the values of x which satisfy the given equation after solving it for $f(x)$.

EXAMPLE 1   Solve the equation $(x^2 - 3x)^2 - 2(x^2 - 3x) - 8 = 0$ for x.

Solution   The given equation is a quadratic provided we think of $x^2 - 3x$ as the unknown. We shall do that and solve for $x^2 - 3x$ by use of the quadratic formula. Thus,

$$x^2 - 3x = \frac{-(-2) \pm \sqrt{(-2)^2 - 4(1)(-8)}}{2(1)}$$

$$= \frac{2 \pm 6}{2} = 4, -2$$

We now find the desired values of x by solving $x^2 - 3x = 4$ and $x^2 - 3x = -2$. From the former, we get $x = 4$ and -1, and from the latter, we find that $x = 2$ and 1. Therefore, $\{4, -1, 2, 1\}$ is the solution set of the given equation.

EXAMPLE 2   Solve the equation $x^4 - 5x^2 - 36 = 0$ for x.

Solution   The given equation is a quadratic provided we consider x^2 as the unknown since the equation is then

$$(x^2)^2 - 5(x^2) - 36 = 0$$

Consequently, $(x^2 - 9)(x^2 + 4) = 0$ ◄ factoring

and $x^2 = 9, -4$

Therefore, $x = \pm 3, \pm 2i$

and the solution set of the given equation is $\{3, -3, 2i, -2i\}$.

Exercise 11.3   Quadratic Formula; Equations in Quadratic Form

Find the solution set of the equation in each of problems 1 to 24.

1 $x^2 - 8x - 5 = 0$ **2** $x^2 + 4x - 6 = 0$

3 $2x^2 - 6x - 7 = 0$ **4** $6 = 3x^2 - 2x$

5 $11x^2 - 8x = 2$ **6** $2x^2 - 6x - 13 = 0$

7 $x^2 + 4x - 15 = 0$ **8** $x^2 - 14 = -8x$

9 $2x^2 + 10x + 17 = 0$ **10** $4x^2 + 8x = 5$

11 $2x^2 - 2x + 1 = 0$ **12** $x^2 + 4x + 13 = 0$

13 $35x^2 + 18x = -8$ **14** $30x^2 - 17x - 35 = 0$

15 $x^2 - 5x = 7$ **16** $x^2 - 7x + 11 = 0$

17 $b^2x^2 - abx - 2a^2 = 0$ **18** $3cdx^2 + (6d^2 - c^2)x = 2cd$

19 $6x^2 + 3x + a = -2ax$ **20** $2a^2x^2 - 2ax - 1 = 0$

21 $c^2d^2x^2 + 6cdx + 15 = 0$ **22** $2p^2x^2 - 6px + 5 = 0$

23 $10b^2x^2 + 6bx + 1 = 0$ **24** $5d^2x^2 - 3bdx + b^2 = 0$

Change the following equations to quadratic form and find the solution set of each.

25 $x^4 - 5x^2 + 4 = 0$ **26** $x^4 - 10x^2 + 9 = 0$

27 $x^4 - 2x^2 + 1 = 0$ **28** $x^4 - 40x^2 + 144 = 0$

29 $x^4 + 8x^2 - 9 = 0$ **30** $x^4 + 5x^2 + 4 = 0$

31 $x^4 - 5x^2 - 36 = 0$ **32** $x^4 + 13x^2 + 36 = 0$

33 $6x^{-2} - 5x^{-1} + 1 = 0$ **34** $20x^{-2} - 13x^{-1} + 2 = 0$

35 $x^{-2} + 3x^{-1} - 10 = 0$ **36** $5x^{-2} + 14x^{-1} - 3 = 0$

37 $x^8 - 17x^4 + 16 = 0$ **38** $x^8 - 82x^4 + 81 = 0$

39 $16x^8 - 17x^4 + 1 = 0$ **40** $81x^8 - 97x^4 + 16 = 0$

41 $(x^2 - 3)^2 - 6(x^2 - 3) + 5 = 0$

42 $(x^2 + 1)^2 - 12(x^2 + 1) + 20 = 0$

43 $(x^2 - 3x)^2 - 2(x^2 - 3x) - 8 = 0$

44 $(2x^2 + 5x)^2 + (2x^2 + 5x) - 6 = 0$

45 $\left(\dfrac{2x - 1}{x + 3}\right)^2 - 4\left(\dfrac{2x - 1}{x + 3}\right) + 3 = 0$

46 $\left(\dfrac{3x + 2}{2x + 1}\right)^2 - 3\left(\dfrac{3x + 2}{2x + 1}\right) + 2 = 0$

47 $3\left(\dfrac{x + 5}{2x - 3}\right)^2 - 4\left(\dfrac{x + 5}{2x - 3}\right) + 1 = 0$

48 $2\left(\dfrac{3x + 4}{2x - 3}\right)^2 + 7\left(\dfrac{3x + 4}{2x - 3}\right) + 6 = 0$

49 $\dfrac{x + 2}{x - 1} - 1 - 2\left(\dfrac{x - 1}{x + 2}\right) = 0$

50 $\dfrac{2x - 3}{2x + 1} - 1 - 2\left(\dfrac{2x + 1}{2x - 3}\right) = 0$

51 $\dfrac{3x-9}{2x+3} - 4 + 3\left(\dfrac{2x+3}{3x-9}\right) = 0$ **52** $2\left(\dfrac{x+4}{3x-1}\right) - 7 + 3\left(\dfrac{3x-1}{x+4}\right) = 0$

11.7 RADICAL EQUATIONS

A radical equation is one in which either or both members contain a radical that has an unknown in the radicand. If the radicals are of the second order, we can solve the equation by the method explained in this section. The method depends upon the following theorem:

Theorem for solving radical equations

Proof ▶

Any root of a given equation is also a root of the equation obtained by equating the squares of the members of the given equation.

If r is a root of the equation $f(x) = g(x)$, then $f(r) = g(r)$, and it follows that

$$f(r) - g(r) = 0$$

If we multiply each member of this equation by $f(r) + g(r)$ and equate the products, we have

$$[f(r) - g(r)]\,[f(r) + g(r)] = 0[f(r) + g(r)]$$

Now performing the indicated multiplication and noting that the product on the right is 0, we have

$$[f(r)]^2 - [g(r)]^2 = 0$$

Finally, we add $[g(r)]^2$ to each member and get

$$[f(r)]^2 = [g(r)]^2$$

The converse of the above theorem is not true. For example, if we equate the squares of the members of

$$\sqrt{x^2 - 4x} = 3x$$

we get

$$x^2 - 4x = 9x^2$$

The solution set of $x^2 - 4x = 9x^2$ is $\{0, -\frac{1}{2}\}$. If $x = 0$ is substituted in $\sqrt{x^2-4x} = 3x$, each member becomes zero, and $x = 0$ is therefore a root. If, however, $x = -\frac{1}{2}$ is substituted in $\sqrt{x^2 - 4x} = 3x$, the right member is equal to $-\frac{3}{2}$, and the left member is zero or positive since it is the principal square root of $x^2 - 4x$; hence it cannot be equal to $-\frac{3}{2}$. If we keep in mind that a radical of the second order denotes the principal root and must be zero or positive, we may be able to tell in advance that a radical equation has no roots. For example, in each of the equations $\sqrt{x^2 - 5x} = -3$ and $\sqrt{x+3} + \sqrt{x-2} = -5$ the left member is positive or zero and the right member is negative. Therefore, neither equation has a root.

If an equation contains three or fewer radicals and all radicals are of the second order, the procedure for solving it consists of the following:

Steps in solving
radical equations

1 Obtain an equation that is equivalent to the given equation and that has one radical and no other term in one member. This process is called *isolating a radical*.

2 Equate the squares of the members of the equation obtained in step 1.

3 If the equation obtained in step 2 contains one or more radicals, repeat the process until an equation free of radicals is obtained. The equation obtained is called the *rationalized equation*.

4 Solve the rationalized equation.

5 Substitute the numbers obtained in step 4 in the original equation in order to determine which of them satisfy the original equation.

We shall illustrate the process with two examples.

EXAMPLE 1 Solve the equation $\sqrt{5x-11} - \sqrt{x-3} = 4$.

Solution

$$\sqrt{5x-11} - \sqrt{x-3} = 4 \qquad \blacktriangleleft \text{ given equation}$$

$$\sqrt{5x-11} = \sqrt{x-3} + 4 \qquad \blacktriangleleft \text{ isolating } \sqrt{5x-11}$$

$$5x - 11 = x - 3 + 8\sqrt{x-3} + 16$$
$$\blacktriangleleft \text{ equating squares of the members}$$

$$5x - 11 - x + 3 - 16 = 8\sqrt{x-3} \qquad \blacktriangleleft \text{ isolating } 8\sqrt{x-3}$$

$$4x - 24 = 8\sqrt{x-3} \qquad \blacktriangleleft \text{ combining similar terms}$$

$$x - 6 = 2\sqrt{x-3} \qquad \blacktriangleleft \text{ dividing each member by 4}$$
$$x^2 - 12x + 36 = 4(x-3) \qquad \blacktriangleleft \text{ squaring each member}$$
$$x^2 - 12x + 36 = 4x - 12 \qquad \blacktriangleleft \text{ by the distributive axiom}$$
$$x^2 - 16x + 48 = 0 \qquad \blacktriangleleft \text{ adding } -4x + 12 \text{ to each member}$$
$$(x - 12)(x - 4) = 0 \qquad \blacktriangleleft \text{ factoring left member}$$
$$x = 12 \qquad \blacktriangleleft \text{ setting } x - 12 = 0 \text{ and solving}$$
$$x = 4 \qquad \blacktriangleleft \text{ setting } x - 4 = 0 \text{ and solving}$$

Check

$$\sqrt{60-11} - \sqrt{12-3} = \sqrt{49} - \sqrt{9} \qquad \blacktriangleleft \text{ substituting 12 for } x \text{ in the left member of the given equation}$$

$$= 7 - 3 = 4$$

Hence 12 is a root.

$$\sqrt{20-11} - \sqrt{4-3} = \sqrt{9} - \sqrt{1} \qquad \blacktriangleleft \text{ substituting 4 for } x \text{ in the left member of the given equation}$$

$$= 3 - 1$$
$$= 2$$

Consequently, since the right member of the given equation is not 2, 4 is not a root, and the solution set of the given equation is $\{12\}$.

EXAMPLE 2 Solve the equation $\sqrt{x+1} + \sqrt{2x+3} - \sqrt{8x+1} = 0$.

Solution

$$\sqrt{x+1} + \sqrt{2x+3} - \sqrt{8x+1} = 0 \qquad \blacktriangleleft \text{ the given equation}$$

$$\sqrt{x+1} + \sqrt{2x+3} = \sqrt{8x+1} \qquad \blacktriangleleft \text{ isolating } \sqrt{8x+1}$$

$$x+1 + 2\sqrt{(x+1)(2x+3)} + 2x+3 = 8x+1 \qquad \blacktriangleleft \text{ equating the squares of the members}$$

$$2\sqrt{(x+1)(2x+3)} = 8x+1 - x - 1 - 2x - 3 \qquad \blacktriangleleft \text{ isolating } 2\sqrt{(x+1)(2x+3)}$$

$$2\sqrt{2x^2 + 5x + 3} = 5x - 3 \qquad \blacktriangleleft \text{ simplifying the radicand and combining similar terms}$$

$$4(2x^2 + 5x + 3) = 25x^2 - 30x + 9 \qquad \blacktriangleleft \text{ squaring each member}$$

$$8x^2 + 20x + 12 = 25x^2 - 30x + 9 \qquad \blacktriangleleft \text{ by the distributive axiom}$$

$$17x^2 - 50x - 3 = 0 \qquad \blacktriangleleft \text{ adding } -25x^2 + 30x - 9 \text{ to each member and dividing by } -1$$

$$x = \frac{50 \pm \sqrt{2500 + 204}}{34} \qquad \blacktriangleleft \text{ by the quadratic formula}$$

$$= \frac{50 \pm \sqrt{2704}}{34}$$

$$= \frac{50 \pm 52}{34}$$

$$= \tfrac{102}{34} \text{ and } -\tfrac{2}{34}$$

$$x = 3$$

$$x = -\tfrac{1}{17}$$

Check

$$\sqrt{3+1} + \sqrt{6+3} - \sqrt{24+1} = \sqrt{4} + \sqrt{9} - \sqrt{25} \qquad \blacktriangleleft \text{ substituting 3 for } x \text{ in the left member of the given equation}$$

$$= 2 + 3 - 5$$

$$= 0$$

Therefore, since the right member of the given equation is also zero, 3 is a root.

$$\sqrt{-\frac{1}{17}+1} + \sqrt{-\frac{2}{17}+3} - \sqrt{-\frac{8}{17}+1} = \sqrt{\frac{16}{17}} + \sqrt{\frac{49}{17}} - \sqrt{\frac{9}{17}}$$ ◄ substituting $-\frac{1}{17}$ for x in the given equation

$$= \frac{4}{\sqrt{17}} + \frac{7}{\sqrt{17}} - \frac{3}{\sqrt{17}}$$

$$= \frac{8}{\sqrt{17}}$$

Consequently, since the right member of the given equation is not $8/\sqrt{17}$, $-\frac{1}{17}$ is not a root, and therefore the solution set of the given equation is $\{3\}$.

Exercise 11.4   Radical Equations

Find the solution set of the equation in each of problems 1 to 36.

1 $\sqrt{2x+3} = \sqrt{3x}$

2 $\sqrt{4x+5} = \sqrt{6x-5}$

3 $\sqrt{7x+11} = \sqrt{1-3x}$

4 $\sqrt{1-4x} = \sqrt{11+x}$

5 $x-2 = \sqrt{3x-2}$

6 $2x-5 = \sqrt{11x+4}$

7 $\sqrt{5x+1} = x+1$

8 $\sqrt{3x+10} = x+2$

9 $\sqrt{x^2+3x+5} = \sqrt{5-x}$

10 $\sqrt{x^2+4x+7} = \sqrt{1-x}$

11 $\sqrt{5x+6} = \sqrt{2x^2+3x+2}$

12 $\sqrt{13x+3} = \sqrt{3x^2+10x+3}$

13 $\sqrt{5x^2-6x+9} = 2x+3$

14 $\sqrt{3x^2-5x+3} = 9+2x$

15 $3x-4 = \sqrt{2x^2+4x-5}$

16 $4x-7 = \sqrt{5x^2+x-3}$

17 $\sqrt{2x+3} = \sqrt{4x-1}+1$

18 $\sqrt{3x+8} = \sqrt{6x-1}+2$

19 $\sqrt{6x-5} + \sqrt{2x-2} = 3$

20 $\sqrt{9x-2} + \sqrt{3x+2} = 4$

21 $\sqrt{x^2+3x} - \sqrt{x^2+2x-2} = 1$

22 $\sqrt{2x^2+3x-5} + \sqrt{2x^2-x-2} = 5$

23 $\sqrt{3x^2+x+2} = \sqrt{3x^2+x-1}+1$

24 $\sqrt{4x^2+x+3} = \sqrt{4x^2+2x-5}+1$

25 $\sqrt{2x+3} + \sqrt{x-2} = \sqrt{5x+1}$

26 $\sqrt{5x+1} - \sqrt{2x} = \sqrt{3x+1}$

27 $\sqrt{7x+2} - \sqrt{3x-2} = \sqrt{6x-8}$

28 $\sqrt{9x+7} - \sqrt{2x-1} = \sqrt{7x+2}$

29 $\dfrac{\sqrt{2x+3}+3}{\sqrt{3x+4}+1} = 2$

30 $\dfrac{\sqrt{7-x}+1}{\sqrt{x+3}+1} = 2$

31 $\dfrac{\sqrt{3x+4}+2}{\sqrt{2x+1}-1}=3$ **32** $\dfrac{\sqrt{7-3x}-1}{\sqrt{x+7}-1}=3$

33 $\sqrt{ax+3a^2}-\sqrt{2a^2-ax}=a$

34 $\sqrt{5bx-b^2}+b=\sqrt{7bx+2b^2}$

35 $\sqrt{ax-a^2}-\sqrt{bx-2ab}=a$

36 $\sqrt{ax-2ab}+\sqrt{bx-2b^2+a^2}=a$

11.8 NATURE OF THE ROOTS OF A QUADRATIC EQUATION

In this section and the next we shall discuss methods that enable us to obtain information about the roots of a quadratic equation without solving the equation. We use the standard form

$$ax^2+bx+c=0 \tag{11.4}$$

and the quadratic formula

$$x=\frac{-b\pm\sqrt{b^2-4ac}}{2a} \tag{11.5}$$

for this purpose. We let the letter D represent the radicand in (11.5); thus

$$D=b^2-4ac \tag{11.6}$$

Discriminant
As we shall see, the value of D enables us to determine the nature of the roots of a quadratic equation. For this reason $D=b^2-4ac$ is called the *discriminant* of the equation.

If we let r represent the root $(-b+\sqrt{b^2-4ac})/2a$ and s represent the root $(-b-\sqrt{b^2-4ac})/2a$, then

$$r=\frac{-b+\sqrt{D}}{2a} \quad \text{and} \quad s=\frac{-b-\sqrt{D}}{2a} \tag{11.7}$$

If a, b, and c in Eq. (11.4) are rational numbers, we can have D equal zero, be negative, or be positive.

1 If $D=0$, then $r=s=-b/2a$ and the roots of (11.4) are rational and equal.

2 If $D<0$, then $\sqrt{D}$ is a pure imaginary number and r and s are imaginary.

3 If $D>0$, two situations may exist: (1) if D is a perfect square, then $\sqrt{D}$ is a rational number and it follows that r and s are rational; (2) if D is not a perfect square, then $\sqrt{D}$ is an irrational number and

r and s are therefore irrational. In either case, r and s are not equal since $(-b + \sqrt{D})/2a \neq (-b - \sqrt{D})/2a$.

We summarize these conclusions below.

Discriminant, $D = b^2 - 4ac$	Roots
$D = 0$	Rational and equal
$D > 0$ and a perfect square	Rational and unequal
$D > 0$ and not a perfect square	Irrational and unequal
$D < 0$	Imaginary

EXAMPLE 1   Compute the value of the discriminant and then determine the nature of the roots of each of the following four equations: $4x^2 - 12x + 9 = 0$, $3x^2 - 7x - 6 = 0$, $5x^2 + 2x - 9 = 0$, and $x^2 + 3x + 5 = 0$.

Solution $$4x^2 - 12x + 9 = 0 \qquad D = (-12)^2 - 4(4)(9) = 144 - 144 = 0$$

Hence the roots are rational and equal.

$$3x^2 - 7x - 6 = 0 \qquad D = (-7)^2 - 4(3)(-6) = 49 + 72 = 121 = 11^2$$

Hence the roots are rational and unequal.

$$5x^2 + 2x - 9 = 0 \qquad D = 2^2 - 4(5)(-9) = 4 + 180 = 184$$

Hence the roots are irrational and unequal.

$$x^2 + 3x + 5 = 0 \qquad D = 3^2 - 4(1)(5) = 9 - 20 = -11$$

Hence the roots are imaginary.

If in (11.4) a, b, and c are real but not necessarily rational, the information we derive from D is less specific:

1   If $D = 0$, then the roots are real and equal. (In this case $r = s = -b/2a$, but we know only that b and a are real.)

2   If $D > 0$, then the roots are real and unequal.

3   If $D < 0$, then the roots are imaginary.

EXAMPLE 2   Compute the value of the discriminant and determine the nature of the roots in each of the following three equations: $4x^2 - 4\sqrt{5}x + 5 = 0$, $\sqrt{3}x^2 - 6x + \sqrt{12} = 0$, and $\sqrt{2}x^2 + 3x + \sqrt{5} = 0$.

Solution $$4x^2 - 4\sqrt{5}x + 5 = 0 \qquad D = 80 - 80 = 0$$

Hence the roots are real and equal.

$$\sqrt{3}x^2 - 6x + \sqrt{12} = 0 \qquad D = 36 - 24 = 12$$

Hence the roots are real and unequal.

$$\sqrt{2}x^2 + 3x + \sqrt{5} = 0 \qquad D = 9 - 4\sqrt{10} < 0$$

Hence the roots are imaginary.

11.9 THE SUM AND PRODUCT OF THE ROOTS

By use of (11.7) we can show that the sum and the product of the roots of a quadratic equation are simple combinations of the coefficients in the equation. For example, the sum of the two roots is

$$r + s = \frac{-b + \sqrt{D}}{2a} + \frac{-b - \sqrt{D}}{2a} = \frac{-2b}{2a} = -\frac{b}{a}$$

Consequently,

Sum of roots
$$r + s = -\frac{b}{a} \tag{11.8}$$

Furthermore,

$$\begin{aligned}
rs &= \frac{-b + \sqrt{D}}{2a} \frac{-b - \sqrt{D}}{2a} \\
&= \frac{b^2 - D}{4a^2} - \frac{b^2 - b^2 + 4ac}{4a^2} \qquad \blacktriangleleft \text{ since } D - b^2 - 4ac \\
&= \frac{4ac}{4a^2} = \frac{c}{a}
\end{aligned}$$

Therefore,

Product of roots
$$rs = \frac{c}{a} \tag{11.9}$$

Sum and product of roots

Consequently, since r and s are the two roots of the equation $ax^2 + bx + c = 0$, it follows that *the sum of the two roots of a quadratic equation is equal to the negative of the quotient of the coefficients of x and x^2, and the product of the two roots is the quotient of the constant term and the coefficient of x^2.*

EXAMPLE 1

Find the sum and the product of the roots in each of the following three equations: $x^2 - 3x + 2 = 0$, $2x^2 + 8x - 5 = 0$, and $\sqrt{2}x^2 + 5x - \sqrt{8} = 0$.

Solution

In the following tabulation $-b/a$ is the sum of the roots, and c/a is the product of the roots.

Equation	Sum of roots	Product of roots
$x^2 - 3x + 2 = 0$	$-\dfrac{b}{a} = \dfrac{-(-3)}{1} = 3$	$\dfrac{c}{a} = \dfrac{2}{1} = 2$
$2x^2 + 8x - 5 = 0$	$-\dfrac{b}{a} = \dfrac{-8}{2} = -4$	$\dfrac{c}{a} = \dfrac{-5}{2}$
$\sqrt{2}x^2 + 5x - \sqrt{8} = 0$	$-\dfrac{b}{a} = \dfrac{-5}{\sqrt{2}}$	$\dfrac{c}{a} = \dfrac{-\sqrt{8}}{\sqrt{2}} = -\sqrt{4} = -2$

The theorem provides a rapid method for checking the roots of an equation. For example, suppose the equation is $6x^2 - x - 12 = 0$ and the possible roots $\frac{3}{2}$ and $-\frac{4}{3}$ are obtained. By the above theorem the sum of the roots is $\frac{1}{6}$ and the product is $-\frac{12}{6} = -2$. Therefore, since

$$\frac{3}{2} - \frac{4}{3} = \frac{9 - 8}{6} = \frac{1}{6} \quad \text{and} \quad \frac{3}{2} \frac{-4}{3} = -2$$

the roots are correct.

As we shall next demonstrate, we can use the above theorem to form a quadratic equation that has two specified numbers as roots. We first divide each member of the equation $ax^2 + bx + c = 0$ by a and obtain $x^2 + bx/a + c/a = 0$. If we now replace b/a by $-(r + s)$ and c/a by rs, the equation becomes

$$x^2 - (r + s)x + rs = 0 \tag{1}$$

EXAMPLE 2

Obtain quadratic equations that have the following solution sets: $\{5, -7\}$; $\{-\frac{1}{2}, -\frac{2}{3}\}$; and $\{2 + \sqrt{3}, 2 - \sqrt{3}\}$.

Solution

In each of the following examples the equation is formed so that it has the roots indicated by use of (1).

Roots are $5, -7$:

$$x^2 - (5 - 7)x + (5)(-7) = 0$$
$$x^2 + 2x - 35 = 0$$

Roots are $-\frac{1}{2}, -\frac{2}{3}$:

$$x^2 - (-\tfrac{1}{2} - \tfrac{2}{3})x + (-\tfrac{1}{2})(-\tfrac{2}{3}) = 0$$
$$x^2 + \frac{7x}{6} + \frac{2}{6} = 0 \qquad \blacktriangleleft \text{ simplifying}$$
$$6x^2 + 7x + 2 = 0 \qquad \blacktriangleleft \text{ multiplying each member by 6}$$

Roots are $2 + \sqrt{3}, 2 - \sqrt{3}$:

$$x^2 - (2 + \sqrt{3} + 2 - \sqrt{3})x + (2 + \sqrt{3})(2 - \sqrt{3}) = 0$$
$$x^2 - 4x + (4 - 3) = 0$$
$$x^2 - 4x + 1 = 0$$

We shall now show that $x - r$ and $x - s$ are factors of $ax^2 + bx + c$ where r and s are its roots.

$$ax^2 + bx + c = a\left(x^2 + \frac{b}{a}x + \frac{c}{a}\right)$$
$$= a[x^2 + (-r - s)x + rs] \qquad \blacktriangleleft \text{ replacing } b/a \text{ by}$$
$$\qquad\qquad\qquad\qquad\qquad\qquad -r - s \text{ and } c/a \text{ by } rs$$
$$= a(x - r)(x - s)$$

Hence,

$$ax^2 + bx + c = a(x - r)(x - s) \qquad (11.10)$$

where $\{r, s\}$ is the solution set of the equation $ax^2 + bx + c = 0$

Factorability test

If a, b, and c are rational and if $b^2 - 4ac$ is a perfect square, then the trinomial $ax^2 + bx + c$ is factorable into the product of two linear binomials with rational coefficients and they can be obtained by theorem (11.10).

EXAMPLE 3

Factor $80x^2 - 67x - 300$.

Solution

This trinomial is factorable into rational factors since the coefficients are rational numbers and $b^2 - 4ac = (-67)^2 - 4(80)(-300) = 317^2$ is a perfect square. We shall obtain the solution set of $80x^2 - 67x - 300 = 0$ by use of the quadratic formula.

$$x = \frac{67 \pm \sqrt{4{,}489 + 96{,}000}}{160} = \frac{67 \pm \sqrt{100{,}489}}{160} = \frac{67 \pm 317}{160}$$

$$= \tfrac{12}{5} \text{ and } -\tfrac{25}{16}$$

Then, since $a = 80$, we have, by (11.10),

$$80x^2 - 67x - 300 = 80(x - \tfrac{12}{5})(x + \tfrac{25}{16}) = (5x - 12)(16x + 25)$$

EXAMPLE 4

Factor $2x^2 - 2x + 5$.

Solution

We solve the equation $2x^2 - 2x + 5 = 0$ by use of the quadratic formula and get

$$x = \frac{2 \pm \sqrt{4 - 40}}{4} = \frac{2 \pm \sqrt{-36}}{4} = \frac{2 \pm 6i}{4}$$

$$= \frac{1 + 3i}{2} \text{ and } \frac{1 - 3i}{2}$$

Therefore, since $a = 2$, we have, by (11.10),

$$2x^2 - 2x + 5 = 2\left(x - \frac{1 + 3i}{2}\right)\left(x - \frac{1 - 3i}{2}\right)$$

Exercise 11.5   Use of the Discriminant

Calculate the discriminant, determine the nature of the roots, and find their sum and product in each of problems 1 to 20.

1	$x^2 - 5x + 6 = 0$	**2**	$x^2 + 3x + 2 = 0$
3	$12x^2 - 5x - 2 = 0$	**4**	$6x^2 + 7x - 3 = 0$
5	$x^2 - 6x + 9 = 0$	**6**	$x^2 + 8x + 16 = 0$
7	$9x^2 + 12x + 4 = 0$	**8**	$4x^2 + 20x + 25 = 0$
9	$x^2 - 6x - 9 = 0$	**10**	$x^2 + 8x - 16 = 0$
11	$4x^2 + 2x - 5 = 0$	**12**	$7x^2 - 3x - 2 = 0$
13	$4x^2 + 2x + 5 = 0$	**14**	$7x^2 - 3x + 2 = 0$
15	$x^2 - 2x + 9 = 0$	**16**	$x^2 + 8x + 17 = 0$
17	$x^2 + \sqrt{6}x + 1 = 0$	**18**	$\sqrt{2}x^2 - \sqrt{5}x - \sqrt{2} = 0$
19	$x^2 - 2\sqrt{2}x - 2 = 0$	**20**	$x^2 - 2\sqrt{2}x + 2 = 0$

Use (11.8) and (11.9) to find which of the following statements are true and which are false. If false is given as the answer, state why.

21   The roots of $x^2 - x - 6 = 0$ are 3 and -2
22   The roots of $x^2 + x - 20 = 0$ are -5 and 4
23   The roots of $x^2 + 4x + 3 = 0$ are -3 and -1
24   The roots of $x^2 - 7x + 12 = 0$ are 4 and 3
25   The roots of $x^2 - 4x + 5 = 0$ are 5 and -1
26   The roots of $x^2 + x + 6 = 0$ are -3 and 2
27   The roots of $x^2 + 7x - 10 = 0$ are -2 and -5
28   The roots of $x^2 - 5x - 4 = 0$ are 1 and 4
29   The roots of $6x^2 + x - 1 = 0$ are $\frac{1}{2}$ and $-\frac{1}{3}$
30   The roots of $9x^2 + 2x + 2 = 0$ are $(1 + i)/3$ and $(1 - i)/3$
31   The roots of $4x^2 + 4x + 1 = 0$ are $\frac{1}{2}(2 + \sqrt{3})$ and $\frac{1}{2}(2 - \sqrt{3})$
32   The roots of $4x^2 + 6x + 5 = 0$ are $\frac{1}{2}(3 + \sqrt{2})$ and $\frac{1}{2}(3 - \sqrt{2})$

Find, by use of (11.8) and (11.9), the quadratic that has the solution set given in each of problems 33 to 40.

33	$\{2, 1\}$	**34**	$\{-3, 5\}$
35	$\{-\frac{2}{3}, \frac{4}{5}\}$	**36**	$\{-\frac{1}{3}, -\frac{2}{7}\}$
37	$\{2 + \sqrt{5}, 2 - \sqrt{5}\}$	**38**	$\{3 + \sqrt{7}, 3 - \sqrt{7}\}$
39	$\{2 + 6i, 2 - 6i\}$	**40**	$\{4 + i, 4 - i\}$

Test the equation given in each of problems 41 to 48 for factorability into factors with rational coefficients and give the factors or the reason for there not being any.

41	$12x^2 - 17x + 6 = 0$	**42**	$10x^2 + x - 2 = 0$

43 $5x^2 + 3x + 2 = 0$ **44** $2x^2 + 3x + 10 = 0$

45 $x^2 - 4x - 1 = 0$ **46** $x^2 - 6x + 5 = 0$

47 $4x^2 - 4x + 5 = 0$ **48** $4x^2 - 12x + 7 = 0$

11.10 GRAPH OF THE QUADRATIC FUNCTION

It is proved in analytic geometry that the graph of the function

$$\{(x, y) \mid y = ax^2 + bc + c, \, a \neq 0\} \tag{1}$$

is a parabola that opens upward if $a > 0$ and downward if $a < 0$. The vertex of the parabola is the lowest point on the graph if it opens upward and the highest point on the graph if it opens downward. In constructing the graph, the first step is to ascertain the coordinates of the vertex. We now show how this is done.

The equation that defines the function (1) is

$$y = ax^2 + bx + c \tag{2}$$

and can be converted to the form

$$y = a\left(x + \frac{b}{2a}\right)^2 + \left(c - \frac{b^2}{4a}\right) \tag{3}$$

Least and greatest value

Since $(x + b/2a)^2 \geq 0$, the value of y is least if $x + b/2a = 0$ for $a > 0$, and then $x = -(b/2a)$ and $y = c - b^2/4a$. Similarly, if $a < 0$, then $a(x + b/2a)^2 \leq 0$, and it follows that the value of y is greatest for $x = -b/2a$. Therefore, in either case, the coordinates of the vertex are $(-b/2a, \, c - b^2/4a)$.

After the coordinates of the vertex are determined, it is then necessary to obtain the coordinates of a few points on each side of the vertex in order to sketch the graph.

EXAMPLE 1

Construct the graph of the function $\{(x, y) \mid y = x^2 - 2x - 3\}$.

Solution

The equation that defines this function is

$$y = x^2 - 2x - 3$$

and we have $a = 1$, $b = -2$, and $c = -3$. Since $a > 0$, the graph is a parabola opening upward. Furthermore, $-b/2a = \frac{2}{2} = 1$ and $c - b^2/4a = -3 - \frac{4}{4} = -4$. Hence the coordinates of the vertex are $(1, -4)$. We now obtain four points to the left of the vertex and four points to the right of it by assigning the integers from -3 to 5 to x and computing each corresponding value of y. Thus we get the following table:

x	-3	-2	-1	0	1	2	3	4	5
y	12	5	0	-3	-4	-3	0	5	12

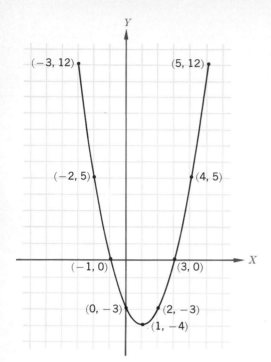

Figure 11.1

We next plot the points determined by the above table, draw a smooth curve through them, and obtain the graph shown in Fig. 11.1.

EXAMPLE 2   Construct the graph of the function

$$\{(x, y)\,|\,y = -\tfrac{1}{4}x^2 - x + 8\}$$

Solution   Here $a = -\tfrac{1}{4}$, $b = -1$, and $c = 8$, and since $a < 0$, the graph is a parabola opening downward. Also, $-(b/2a) = 1/(-\tfrac{1}{2}) = -2$ and $c - b^2/4a = 8 - \tfrac{1}{-1} = 9$. Hence the coordinates of the vertex are $(-2, 9)$. We obtain four points to the left of the vertex and four points to the right of it by assigning the consecutive even integers from -10 to 6 to x. After each corresponding value of y is computed, we have the following table:

x	-10	-8	-6	-4	-2	0	2	4	6
y	-7	0	5	8	9	8	5	0	-7

When the points whose coordinates are the pairs of corresponding numbers in the above table are plotted and a smooth curve is drawn through them, we obtain the graph shown in Fig. 11.2.

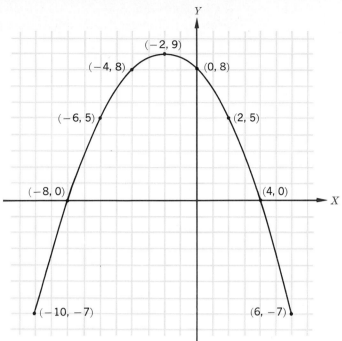

Figure 11.2

11.11 QUADRATIC INEQUALITIES

In this section we shall consider a procedure for obtaining the solution set of an inequality of the type

$$ax^2 + bx + c > 0 \tag{1}$$

and we shall use the following equation for the purpose.

$$y = ax^2 + bx + c \tag{2}$$

The graph of (2) is the graph of the quadratic function discussed in Sec. 11.10. Since $y > 0$ at all points on the graph above the X axis, the set of abscissas of the former points is the solution set of (1).

The graph crosses the X axis at the points whose ordinates are zero and only at them. Consequently, the abscissas of these points are the roots of the equation

$$ax^2 + bx + c = 0 \tag{3}$$

We shall denote the roots of (3) by r and s and assume that $r > s$.

Now, if $a > 0$, the graph of (2) is a parabola that opens upward and crosses the X axis at $(r, 0)$ and $(s, 0)$. Consequently, the graph is above the X axis and $y > 0$ if $x < s$ or $x > r$. Therefore, the solution set of (1) is $\{x|x < s\} \cup \{x|x > r\}$.

If, however, $a < 0$, the graph is a parabola that opens downward and is above the X axis if $s < x < r$. Therefore, in this case, the solution set of (1) is $\{x|s < x < r\}$.

It can be shown in a manner similar to the above discussion that the solution set of $ax^2 + bx + c < 0$ is $\{x|s < x < r\}$ if $a > 0$ and $\{x|x < s\} \cup \{x|x > r\}$ if $a < 0$.

EXAMPLE 1

Find the solution set of $2x^2 + 5x - 3 > 0$.

Solution

We first solve the quadratic equation

$$2x^2 + 5x - 3 = 0$$

by the quadratic formula and get

$$x = \frac{-5 \pm \sqrt{25 + 24}}{4}$$
$$= \frac{-5 \pm 7}{4}$$
$$= \tfrac{1}{2} \text{ or } -3$$

Since the coefficient of x^2 is positive, the graph of $y = 2x^2 + 5x - 3$ opens upward and is above the X axis if $x < -3$ or $x > \tfrac{1}{2}$. Hence the solution set is $\{x|x < -3\} \cup \{x|x > \tfrac{1}{2}\}$.

EXAMPLE 2

Find the solution set of $-3x^2 > 8x + 4$.

Solution

By adding $-8x - 4$ to each member of the inequality, we obtain the equivalent inequality

$$-3x^2 - 8x - 4 > 0$$

Next we solve the equation

$$-3x^2 - 8x - 4 = 0$$

and get

$$x = \frac{8 \pm \sqrt{64 - 48}}{-6}$$
$$= \frac{8 \pm 4}{-6}$$
$$= -2 \text{ or } -\tfrac{2}{3}$$

Since $a = -3$ is negative, the graph of $y = -3x^2 - 8x - 4$ opens downward and is above the X axis if $-2 < x < -\tfrac{2}{3}$. Hence the required solution set is

$$\{x|-2 < x < -\tfrac{2}{3}\}$$

EXAMPLE 3   Solve the inequality $2x^2 < x + 10$.

Solution   By adding $-x - 10$ to each member, we obtain the equivalent inequality

$$2x^2 - x - 10 < 0$$

The roots of $2x^2 - x - 10 = 0$ are

$$x = \frac{1 \pm \sqrt{1 + 80}}{4}$$
$$= \tfrac{5}{2} \text{ or } -2$$

Since $a = 2$ is positive, the graph of $y = 2x^2 - x - 10$ opens upward and is below the X axis if $-2 < x < \tfrac{5}{2}$. Hence the solution set is $\{x | -2 < x < \tfrac{5}{2}\}$.

If the roots of (3) are imaginary, the graph of (2) does not cross the X axis and is entirely above it if $a > 0$ and entirely below it if $a < 0$. Consequently, if $a > 0$, the solution set of (1) is the set of all real numbers and the solution set of (2) is $\varnothing$. Similarly, if $a < 0$, the solution sets of (1) and (2) are, respectively, $\varnothing$ and the set of all real numbers.

EXAMPLE 4   Solve the inequality $x^2 - 4x + 13 < 0$.

Solution   The roots of equation

$$x^2 - 4x + 13 = 0$$

are
$$x = \frac{4 \pm \sqrt{16 - 52}}{2}$$
$$= \frac{4 \pm \sqrt{-36}}{2}$$
$$= \frac{4 \pm 6i}{2}$$
$$= 2 + 3i \text{ or } 2 - 3i$$

Since these roots are imaginary and $a = 1$ is positive, the graph of $y = x^2 - 4x + 13$ is entirely above the X axis. Hence the solution set of the given inequality is $\varnothing$.

Exercise 11.6   Quadratic Functions and Inequalities

Find the coordinates of the vertex of the parabola in each of problems 1 to 12, find the direction it opens, and sketch the graph.

1 $\{(x, y) | y = x^2 - 3x - 1\}$ **2** $\{(x, y) | y = x^2 + 2x + 2\}$

3 $\{(x, y) | y = x^2 + 3x - 3\}$ **4** $\{(x, y) | y = x^2 - 2x - 2\}$
5 $\{(x, y) | y = -2x^2 + x + 2\}$ **6** $\{(x, y) | y = -2x^2 + 3x + 1\}$
7 $\{(x, y) | y = -3x^2 - 2x + 3\}$ **8** $\{(x, y) | y = -3x^2 + 4x - 2\}$
9 $\{(x, y) | y = 3x^2 - 6x - 1\}$ **10** $\{(x, y) | y = 5x^2 + 10x + 3\}$
11 $\{(x, y) | y = -3x^2 + 12x - 5\}$ **12** $\{(x, y) | y = -2x^2 + 8x - 7\}$

Find the solution set of the inequality in each of problems 13 to 32.

13 $x^2 - 5x - 6 > 0$ **14** $x^2 - 6x + 5 > 0$
15 $x^2 + x - 6 > 0$ **16** $x^2 + 7x + 12 > 0$
17 $x^2 - 3x + 2 < 0$ **18** $x^2 - 8x + 15 < 0$
19 $x^2 - 3x - 10 < 0$ **20** $x^2 - 4x - 21 < 0$
21 $-x^2 + 4x - 3 < 0$ **22** $-x^2 + 7x - 10 < 0$
23 $-x^2 - x + 6 > 0$ **24** $-x^2 - 4x + 5 > 0$
25 $x^2 - 6x + 9 < 0$ **26** $x^2 - 8x + 16 < 0$
27 $x^2 - 4x + 5 > 0$ **28** $x^2 - 10x + 26 < 0$
29 $3x^2 > 2x - 5$ **30** $-11x > 2x^2 + 12$
31 $6x^2 - 6 < 5x$ **32** $12 < 6x^2 + x$

Exercise 11.7   Review

Find the solution set of the equation in each of problems 1 to 6 by factoring and by completing the square.

1 $x^2 + 2x - 3 = 0$ **2** $6x^2 - 7x - 3 = 0$
3 $12x^2 + 11x + 2 = 0$ **4** $6x^2 - 19x + 15 = 0$
5 $2x^2 - cx - 3c^2 = 0$ **6** $3x^2 + 5ax + 2a^2 = 0$

Find the solution set of the equation in each of problems 7 to 12 by completing the square and by formula.

7 $9x^2 + 12x + 1 = 0$ **8** $4x^2 + 18x + 9 = 0$
9 $2x^2 - bx - 3b^2 = 0$ **10** $2x^2 - 7cx + 6c^2 = 0$
11 $4x^2 + 12x + 13 = 0$ **12** $9x^2 + 24x + 25 = 0$

Find the solution set of the equation in each of problems 13 to 20.

13 $x^4 - 10x^2 + 9 = 0$ **14** $x^4 + 8x^2 - 9 = 0$
15 $2\left(\dfrac{3x+1}{x+2}\right)^2 + 5\left(\dfrac{3x+1}{x+2}\right) - 3 = 0$ **16** $\dfrac{x-2}{2x+3} - 5 + 6\left(\dfrac{2x+3}{x-2}\right) = 0$
17 $\sqrt{5x+1} - \sqrt{2x+10} = 0$ **18** $\sqrt{x^2 - 2x - 6} = 2x - 7$
19 $\sqrt{x+2} + \sqrt{x-3} = \sqrt{3x+4}$
                          20 $\dfrac{\sqrt{3x-2} + 2}{\sqrt{5x-1} + 1} = 1$

Find the discriminant and the nature, sum, and product of the roots in each of problems 21 to 24.

21 $x^2 + 7x + 12 = 0$ **22** $x^2 - 8x + 16 = 0$

23 $2x^2 + 3x + 4 = 0$ **24** $x^2 + 3\sqrt{2}x + 4 = 0$

25 Find the equation whose roots are 4 and $-\frac{3}{4}$.

26 Find the equation whose roots are $3 + 5i$ and $3 - 5i$.

27 Test (a) $x^2 + 15x + 44 = 0$ and (b) $x^2 + 15x + 45 = 0$ for factorability into factors with rational coefficients.

28 Find the coordinate of the vertex and sketch $\{(x, y) \mid y = x^2 - 3x + 4\}$.

29 Find the coordinate of the vertex and sketch $\{(x, y) \mid y = -2x^2 + 5x + 3\}$.

30 Show that the roots of $cx^2 + bx + a = 0$ are the reciprocals of those of $ax^2 + bx + c = 0$.

31 Show that the two roots of $ax^2 + bx + a = 0$ are reciprocals.

32 Find the sum of the reciprocals of the roots of a quadratic if the sum and product are s and p, respectively.

33 Show that the root of $4x + 5 = 0$ is the average of the roots of $2x^2 + 5x - 3 = 0$.

Systems of Equations and of Inequalities

In Chap. 4 we solved stated problems by first expressing the conditions of the problem as an equation in one unknown. Frequently a problem is of such a nature that it is necessary to introduce more than one unknown and to employ more than one equation in order to obtain the solution. The major portion of this chapter will be devoted to the discussion of systems of equations in two unknowns and to the method for obtaining a set of ordered pairs of numbers that will satisfy both equations.

12.1 EQUATIONS IN TWO VARIABLES

In the equation

$$2x + 3y = 12 \tag{1}$$

the two variables are x and y. Here we may replace x by an arbitrary real number and then find the corresponding replacement for y, and since this operation requires only subtraction and division, the replacement for y is a real number. Therefore, the replacement set for x and for y is the set of real numbers. Thus x and y are variables, and we call (1) an equation in two variables.

We now consider the two equations

$$x + y = 6$$
$$x - y - 2$$

and seek values of x and y that satisfy the two equations simultaneously. By use of methods discussed later, we can say that $x = 4$, $y = 2$ is the only pair of values that satisfies these equations simultaneously. Hence there is only one replacement for x and one corresponding replacement for y. Therefore, we consider x and y as unknowns whose values are determined when the equations are solved. Consequently, we speak of two equations with two letters (usually x and y) to be replaced by numbers as two equations in two unknowns.

Solution
Solution set

If a single equation in two variables x and y is satisfied when x is replaced by the first of an ordered pair of numbers and y is replaced by the second, then the ordered pair is called a *solution* of the equation. The set of solutions is the *solution set* of the equation. Usually the solution set of a single equation in two variables contains an unlimited number of solution pairs. To illustrate the ideas involved, we shall solve Eq. (1) for y in terms of x and get

$$y = 4 - \frac{2x}{3} \tag{2}$$

If we replace x in (2) successively by 4, 5, 6, and 7, we get the following corresponding replacements for y: $\frac{4}{3}, \frac{2}{3}, 0$, and $-\frac{2}{3}$. Consequently, each of $(4, \frac{4}{3})$, $(5, \frac{2}{3})$, $(6, 0)$, and $(7, -\frac{2}{3})$ is a solution pair of (1). Furthermore, since we can replace x in (4) by any arbitrary number and calculate the corresponding replacement for y, the number of solution pairs of Eq. (1) is unlimited.

12.2 GRAPH OF AN EQUATION IN TWO VARIABLES

In Chap. 6 we explained the method for associating an ordered pair of numbers with a point in the cartesian plane and discussed the method for obtaining the graph of a function. In most of the functions discussed in Chap. 6, the rule that established the correspondence between the elements of each ordered pair was an equation in two variables, and we obtained the set of ordered pairs in the function by finding the set of solution pairs of the equation. The methods employed in this section will be similar to those used in Chap. 6.

Graph of an equation in two variables

The graph of an equation in two variables is the totality of points in the cartesian plane such that the coordinates of each is a solution pair of the equation.

In this chapter we shall consider only linear and quadratic equations in two variables, using information from analytic geometry that enables us to determine the general nature of the graph of a particular equation by considering only the degree of the equation and the coefficients in it.
LINEAR EQUATIONS IN TWO VARIABLES An equation of the type

Linear equation

$$ax + by = c \tag{1}$$

is a linear equation in two variables. It is proved in analytic geometry that the graph of Eq. (1) is a straight line. Since a straight line is fully determined by the position of two points on it, we need only find two solution pairs of Eq. (1) in order to construct the graph. If $c \neq 0$, these two solution pairs are easily obtained by assigning zero to x and solving for y and then by assigning zero to y and solving for x. As a check, it is advisable to get a third solution pair in which neither x nor y is zero. If $c = 0$, then a solution pair is $(0, 0)$, and the graph passes through the origin. In this case it is advisable to assign a positive and a negative number to x and compute each corresponding value of y, and then we have the coordinates of three points on the line. In the illustrative examples in the following sections we shall discuss the graphs of several linear equations and explain the methods for constructing them.
QUADRATIC EQUATIONS IN TWO VARIABLES The general form of a quadratic equation in two variables is

$$Ax^2 + Bxy + Cy^2 + Dx + Ey = F \tag{2}$$

It is proved in analytic geometry that the graph of Eq. (2) is either a circle, an ellipse, a hyperbola, or a parabola. In certain degenerate cases, the graph of (2), if it exists, is a straight line, two straight lines, or a single point. For example, the graph of $x^2 + y^2 = 0$ is the single point $(0, 0)$, since no other pair of numbers satisfies the equation. Furthermore the graph of $x^2 + y^2 = -1$ does not exist since x^2 and y^2 are nonnegative numbers. In this chapter we shall deal with the special cases of Eq. (2) listed below:

Equation		Graph
$ax^2 + by^2 = c$	where $\begin{cases} a > 0 \\ b > 0 \\ c > 0 \end{cases}$	An ellipse if $a \neq b$ and a circle if $a = b$
$ax^2 - by^2 = c$	where $\begin{cases} a > 0 \\ b > 0 \\ c \neq 0 \end{cases}$	A hyperbola
$axy = b$	where $b \neq 0$	A hyperbola
$y = ax^2 + bx + c$	where $a \neq 0$	A parabola
$x = ay^2 + by + c$	where $a \neq 0$	A parabola

The graphs of these equations are shown in Fig. 12.1.

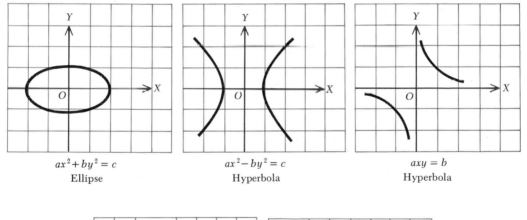

$ax^2 + by^2 = c$ $ax^2 - by^2 = c$ $axy = b$
Ellipse   Hyperbola   Hyperbola

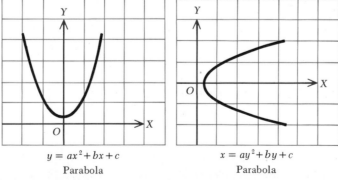

$y = ax^2 + bx + c$ $x = ay^2 + by + c$
Parabola   Parabola

Figure 12.1

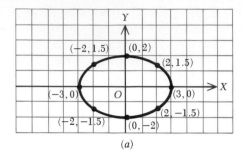

(a)

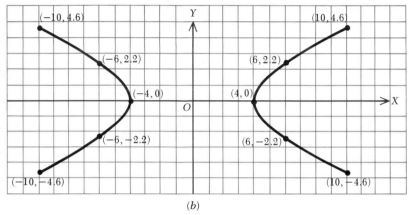

(b)

Figure 12.2

If the nature of the graph is known, it is necessary to obtain a relatively few solution pairs of the equation in order to construct the graph. The method for constructing the graph of a parabola was explained in Example 1 of Sec. 11.10. In the following examples we illustrate the procedure for obtaining the graph of an ellipse and a hyperbola.

EXAMPLE 1   Construct the graph of $4x^2 + 9y^2 = 36$.

Solution   Since the equation is of the type $ax^2 + by^2 = c$, the graph is an ellipse. We solve the equation for y and get

$$y = \pm\tfrac{2}{3}\sqrt{9 - x^2}$$

From this equation we calculate the following table of corresponding values of x and y:

x	-3	-2	-1	0	1	2	3
y	0	±1.5	±1.9	±2	±1.9	±1.5	0

When the points determined by the pairs of numbers in this table are

242

plotted and a smooth curve is drawn through them, we obtain the ellipse in Fig. 12.2*a*.

EXAMPLE 2

Construct the graph of $x^2 - 4y^2 = 16$.

Solution

The equation is of type $ax^2 - by^2 = c$, and so the graph is a hyperbola. We construct the graph as follows. Solve the equation for y and get

$$y = \pm\tfrac{1}{2}\sqrt{x^2 - 16}$$

Assign the set of numbers $\{-10, -8, -6, -5, -4, 4, 5, 6, 8, 10\}$ to x, calculate each corresponding value of y, and tabulate these values. Thus

x	-10	-8	-6	-5	-4	4	5	6	8	10
y	±4.6	±3.5	±2.2	±1.5	0	0	±1.5	±2.2	±3.5	±4.6

Plot the points, draw a smooth curve through them and get the hyperbola in Fig. 12.2*b*.

Exercise 12.1   Graphs of Equations in Two Variables

Construct the graph of the function defined by the equation in each of problems 1 to 40.

1	$2x - y = 7$	**2**	$4x - y = 10$
3	$3x + y = -9$	**4**	$x + y = 5$
5	$x - y = -4$	**6**	$x - 3y = 12$
7	$x + 2y = 6$	**8**	$x + 4y = -8$
9	$3x + 2y = -12$	**10**	$2x + 3y = 18$
11	$3x - 4y = 12$	**12**	$4x - 5y = 10$
13	$x - y = 0$	**14**	$x + y = 0$
15	$2x + y = 0$	**16**	$x - 3y = 0$
17	$x^2 - y = 5$	**18**	$6x^2 + y = 11$
19	$2x + y^2 = 4$	**20**	$3x - y^2 = -3$
21	$y = x^2 - 4x + 3$	**22**	$y = -x^2 + 2x + 3$
23	$x = y^2 - 2y - 1$	**24**	$x = -y^2 - 3y - 2$
25	$x^2 + y^2 = 16$	**26**	$x^2 + y^2 = 9$
27	$x^2 - y^2 = 4$	**28**	$y^2 - x^2 = 25$
29	$x^2 + 4y^2 = 4$	**30**	$9x^2 + y^2 = 9$
31	$4x^2 + 9y^2 = 36$	**32**	$25x^2 + 4y^2 = 100$
33	$x^2 - 4y^2 = 4$	**34**	$9x^2 - y^2 = 9$
35	$-4x^2 + 9y^2 = 36$	**36**	$25x^2 - 4y^2 = -100$
37	$xy = 6$	**38**	$xy = -8$
39	$4x^2 + 3xy = 9$	**40**	$2x + xy = 10$

12.3 SIMULTANEOUS SOLUTION BY GRAPHICAL METHODS

Simultaneous
solution set

The simultaneous solution set of two equations in x and y is the set of ordered pairs of numbers such that each equation is a true statement if x is replaced by the first number in each pair of the set and y is replaced by the second.

Since the pair of coordinates of each point on the graph of an equation in two variables is a solution pair of the equation, the pair of coordinates of each *point of intersection* of the graphs of two equations is a solution pair of each equation. Consequently, if S is the simultaneous solution set of two equations in two variables, if S_1 is the solution set of the first equation, and if S_2 is the solution set of the second, then

$$S = S_1 \cap S_2$$

Of course if the graphs do not intersect, the simultaneous solution set is the empty set $\varnothing$.

We can therefore obtain the simultaneous solution set of two equations in two unknowns by constructing the graphs of the equations and then estimating the coordinates of their points of intersection. We illustrate the method with three examples.

EXAMPLE 1

Find graphically the simultaneous solution set of the equations

$$3x - 4y = 8 \tag{1}$$
$$2x + 5y = 10 \tag{2}$$

Solution

Since each of the given equations is linear, their graphs are straight lines. We obtain the coordinates of three points on each graph by assigning 0 and 4 to x and 0 to y in each equation, solving for the other variable, thus getting the following tables of corresponding values:

x	0	$\frac{8}{3}$	4
y	-2	0	1

for Eq. (1) and

x	0	5	4
y	2	0	$\frac{2}{5}$

for Eq. (2). We next plot the points whose coordinates are the pairs of corresponding numbers in each table, draw a straight line through each set, and get the lines in Fig. 12.3. By inspection, we see that these lines intersect at a point whose coordinates to the nearest tenth are (3.5, 0.6).

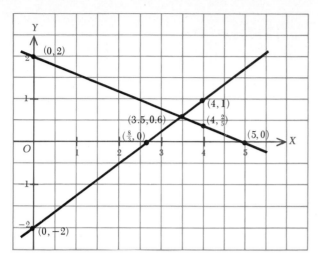

Figure 12.3

Therefore we say that the simultaneous solution set of Eqs. (1) and (2) appears to be $\{(3.5, 0.6)\}$. If we replace x and y in the given equations by 3.5 and 0.6, respectively, we get

$$3(3.5) - 4(0.6) = 10.5 - 2.4 = 8.1 \qquad \blacktriangleleft \text{ from (1)}$$
$$2(3.5) + 5(0.6) = 7 + 3 = 10 \qquad \blacktriangleleft \text{ from (2)}$$

Since the right members of (1) and (2) are 8 and 10, respectively, we are fairly safe in assuming that the simultaneous solution set $(3.5, 0.6)$ is correct to one decimal place.

EXAMPLE 2

Find graphically the simultaneous solution set of

$$5x^2 + 8y^2 = 220 \tag{3}$$
$$3x + 4y = 14 \tag{4}$$

Solution

Equation (3) is of the type $ax^2 + by^2 = c$, and its graph is therefore an ellipse. We obtain the intercepts and additional points by assigning -6, -2, 0, 2, and 6 to x and zero to y. The corresponding pairs of numbers thus obtained are:

x	-6.6	-6	-2	0	2	6	6.6
y	0	± 2.2	± 5	± 5.2	± 5	± 2.2	0

We plot the points whose coordinates are the pairs of corresponding numbers in this table, draw a smooth curve through them, and thus obtain the ellipse in Fig. 12.4.

The graph of Eq. (4) is the straight line in Fig. 12.4 that is determined

245

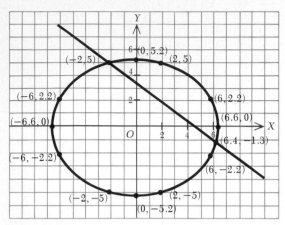

Figure 12.4

by the points whose coordinates are the corresponding numbers in the table below.

x	0	2	$4\frac{2}{3}$
y	$3\frac{1}{2}$	2	0

From the figure we see that the coordinates of the upper intersection of the two graphs appear to be $(-2, 5)$, and we estimate that the coordinates of the lower intersection to one decimal place are $(6.4, -1.3)$. Therefore, insofar as we are able to estimate from the graph, the simultaneous solution set of Eqs. (3) and (4) to one decimal place is $\{(-2, 5), (6.4, -1.3)\}$.

We test the accuracy of this solution set by replacing x and y in each given equation by the appropriate number in each solution pair. Using $(-2, 5)$, we have

$$5(-2)^2 + 8(5)^2 = 20 + 200 = 220 \qquad \blacktriangleleft \text{ from (3)}$$
$$3(-2) + 4(5) = -6 + 20 = 14 \qquad \blacktriangleleft \text{ from (4)}$$

Consequently, the solution pair $(-2, 5)$ is exact. Similarly, using $(6.4, -1.3)$ and calculating each result to one decimal place, we get

$$5(6.4)^2 + 8(-1.3)^2 = 204.8 + 13.5 = 218.3 \qquad \blacktriangleleft \text{ from (3)}$$
$$3(6.4) + 4(-1.3) = 19.2 - 5.2 = 14 \qquad \blacktriangleleft \text{ from (4)}$$

Hence, since the right members of (3) and (4) are 220 and 14, we have reason to believe that our estimate $(6.4, -1.3)$ is correct to one decimal place. By algebraic methods that will be explained later, we can obtain the solution set $\{(-2, 5), (\frac{122}{19}, -\frac{25}{19})\}$, and to one decimal place the latter pair is equal to $(6.4, -1.3)$.

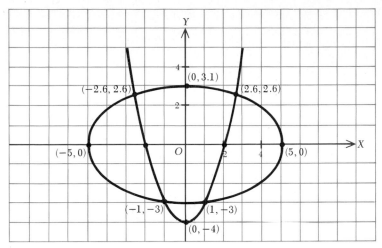

Figure 12.5

If the graphs of the equations in a given system do not intersect, the graphical method will not reveal the solution set since only real number pairs are obtained graphically. By algebraic methods to be discussed later, we can show that either the simultaneous solution set is the empty set ∅ or the elements of the set are imaginary numbers. Furthermore, except in certain cases,† if a straight line intersects one of the curves in Fig. 12.1 at all, it will intersect it at two points. Thus, in general, a system consisting of a linear and a quadratic equation will have two solution pairs in the solution set.

EXAMPLE 3   Obtain the simultaneous solution set of the equations

$$y = x^2 - 4 \tag{5}$$
$$3x^2 + 8y^2 = 75 \tag{6}$$

by the graphical method.

Solution   Equation (5) is of the type $y = ax^2 + bx + c$, with $a = 1 > 0$, $b = 0$, and $c = -4$. Hence the graph is a parabola opening upward (see Fig. 12.1). We construct the parabola by means of the following table of corresponding values and show the graph in Fig. 12.5.

x	-3	-2	-1	0	1	2	3
y	5	0	-3	-4	-3	0	5

---

† The exceptional cases occur when the line is tangent to the curve or when it is parallel to the axis of symmetry of a parabola. The axis of symmetry of a parabola is the line L that bisects every chord of the parabola that is perpendicular to L.

247

Equation (6) is of the type $ax^2 + by^2 = c$, with $a = 3$, $b = 8$, and $c = 75$. The graph is therefore an ellipse (see Fig. 12.1). We construct the graph by means of the corresponding numbers tabulated below, and thus obtain the ellipse in Fig. 12.5.

x	-5	-4	-2	-1	0	1	2	4	5
y	0	±1.8	±2.8	±3	±3.1	±3	±2.8	±1.8	0

The graphs intersect at points whose coordinates appear to be $(1, -3)$, $(-1, -3)$, $(-2.6, 2.6)$, and $(2.6, 2.6)$. Consequently the solution set of the two equations is $\{(1, 3), (-1, -3), (-2.6, 2.6), (2.6, 2.6)\}$.

12.4   INDEPENDENT, INCONSISTENT, AND DEPENDENT EQUATIONS

The graphs of two linear equations in two unknowns may be two intersecting straight lines, two parallel lines, or two coincident lines. If the two lines intersect, the simultaneous solution set contains only one ordered pair of numbers and the equations are said to be *independent*. If the two lines are parallel, there is no point of intersection; therefore no solution pair of one equation is a solution pair of the other, and so the simultaneous solution set is the empty set $\varnothing$. In this case the equations are called *inconsistent*. If the two lines coincide, then every solution pair of one equation is a solution pair of the other and the equations are said to be *dependent*. Figure 12.6a to c illustrates the three possibilities.

We now derive a simple criterion that enables us to decide whether two linear equations in two unknowns are independent, inconsistent, or dependent. We consider the equations

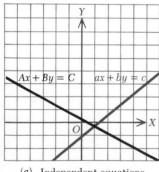

(a) Independent equations
$$\frac{A}{a} \neq \frac{B}{b}$$

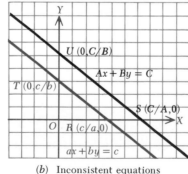

(b) Inconsistent equations
$$\frac{A}{a} = \frac{B}{b} \neq \frac{C}{c}$$

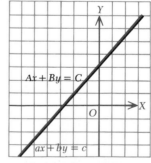

(c) Dependent equations
$$\frac{A}{a} = \frac{B}{b} = \frac{C}{c}$$

Figure 12.6

$$ax + by = c \tag{1}$$
$$Ax + By = C \tag{2}$$

The graphs of (1) and (2) intersect the X axis at $R(c/a, 0)$ and $S(C/A, 0)$, respectively, and the Y axis at the points $T(0, c/b)$ and $U(0, C/B)$, respectively, as illustrated in Fig. 12.6b. We can see that the graphs of Eqs. (1) and (2) are parallel if the segments RT and SU are parallel, and these two segments are parallel if and only if the triangles ORT and OSU are similar. The triangles are similar if and only if $OR/OS = OT/OU$.

Now $OR/OS = c/a \div C/A = Ac/aC$, and $OT/OU = c/b \div C/B = Bc/bC$. Consequently, $OR/OS = OT/OU$ if and only if $Ac/aC = Bc/bC$. If we multiply each member of the last equation by C/c, we obtain $A/a = B/b$. Therefore, the graphs of (1) and (2) are parallel if and only if $A/a = B/b$. If in addition to $A/a = B/b$ we have $A/a = B/b = C/c = k$, then $A = ak$ and $C = ck$. Hence, $C/A = ck/ak = c/a$, and the points R and S coincide. Therefore, since the graphs are parallel and have one point in common, they coincide. Hence we have the following theorem:

Independent, inconsistent, and dependent systems

Two linear equations $ax + by = c$ and $Ax + By = C$ are independent if and only if $A/a \neq B/b$, inconsistent if and only if $A/a = B/b \neq C/c$, and dependent if and only if $A/a = B/b = C/c$.

The following three examples illustrate the application of the above theorem.

The equations

$$2x - 3y = 4$$
$$5x + 2y = 8$$

are independent, since $\frac{2}{5} \neq -\frac{3}{2}$.

The equations

$$3x - 9y = 1$$
$$2x - 6y = 2$$

are inconsistent, since $\frac{3}{2} = -9/(-6) \neq \frac{1}{2}$.

The equations

$$2x - 4y = 12$$
$$3x - 6y = 18$$

are dependent, since $\frac{2}{3} = -4/(-6) = \frac{12}{18}$.

Exercise 12.2 Graphical Solution of Simultaneous Equations

Obtain the solution set of each pair of independent equations in problems 1 to 32 by use of the graphical method. Determine each number to one decimal place. If a pair is not independent, state whether it is inconsistent or dependent.

1	$2x + 3y = 12$	**2**	$3x + y = -3$
	$x - 3y = -3$		$x - y = -5$
3	$x + 2y = 6$	**4**	$x + 2y = 4$
	$3x - 2y = 6$		$3x + 4y = 9$
5	$2x + 4y = -7$	**6**	$3x + 4y = 2$
	$x - 3y = 9$		$2x - y = 5$
7	$4x + 5y = 6$	**8**	$5x + 4y = -8$
	$2x - 5y = -6$		$6x - 9y = -7$
9	$x - 2y = 3$	**10**	$2x + y = 7$
	$3x - 6y = 9$		$4x + 2y = 14$
11	$4x + 5y = 6$	**12**	$3x + 4y = 8$
	$8x + 10y = 12$		$6x + 8y = 16$
13	$2x + y = 2$	**14**	$3x - 2y = 4$
	$6x + 3y = 5$		$6x - 4y = 5$
15	$5x - 3y = 1$	**16**	$9x + 12y = 21$
	$10x - 6y = -2$		$6x + 8y = 15$
17	$3x - y = 1$	**18**	$x^2 = -2y + 11$
	$y = 2x^2 - 5$		$y = x + 5$
19	$x^2 + 9y^2 = 16$	**20**	$x^2 + 4y^2 = 25$
	$x - 3y = -2$		$x - 2y = -1$
21	$2y = x^2 - 12$	**22**	$y = x^2 - 3$
	$x = y^2$		$y^2 = 3x$
23	$y = x^2 - 2$	**24**	$y = x^2 - x - 1$
	$y^2 = 2x$		$(y - 2)^2 = 2x$
25	$x^2 + y^2 = 25$	**26**	$x^2 + y^2 = 169$
	$y^2 = 5x + 1$		$y^2 = 5x - 35$
27	$4x^2 - 5y^2 = -16$	**28**	$3x^2 - 13y^2 = -108$
	$y^2 = 4x$		$y^2 = 3x$
29	$2x^2 + y^2 = 33$	**30**	$4x^2 + y^2 = 100$
	$x^2 + y^2 = 29$		$12x^2 - 5y^2 = 12$
31	$2x^2 + y^2 = 11$	**32**	$3x^2 + y^2 = 12$
	$9x^2 - 4y^2 = -27$		$5x^2 - 2y^2 = -13$

12.5 ALGEBRAIC METHODS

The first step in the algebraic procedure for solving two equations in two unknowns is to combine the two equations in such a way as to obtain one equation in one unknown whose roots are each one of the numbers in an ordered pair of the simultaneous solution set. This process is called

Eliminating an unknown

eliminating an unknown. After one number in each solution pair is determined, the other can be obtained by substitution. In subsequent sections we shall discuss several methods of elimination.

12.6 ELIMINATION BY ADDITION OR SUBTRACTION

In this section we employ axioms (2.5) and (2.6), which are restated below for convenience.

$$\text{If } a = b, \text{ then } a + c = b + c \qquad (2.5)$$
$$\text{If } a = b, \text{ then } ac = bc \qquad (2.6)$$

We shall also use the theorem

$$\textbf{If } \boldsymbol{a = b} \textbf{ and } \boldsymbol{c = d} \textbf{, then } \boldsymbol{a \pm c = b \pm d} \qquad \textbf{(12.1)}$$

as is readily proved by two applications of (2.5).

TWO EQUATIONS OF THE TYPE $ax + by = c$ The procedure for obtaining the simultaneous solution set of two equations of the type $ax + by = c$ is explained in the following example.

EXAMPLE 1

Obtain the simultaneous solution set of the equations

$$3x + 4y = -6 \qquad (1)$$
$$5x + 6y = -8 \qquad (2)$$

Solution

We eliminate one of the unknowns to obtain an equation in one unknown whose root is one of the numbers in a simultaneous solution pair. We arbitrarily choose to eliminate y and first notice that the lcm of the coefficients of y in (1) and (2) is 12. Now if a simultaneous solution pair (x, y) exists, then by (2.5), (x, y) is also a simultaneous solution pair of the two equations

$$
\begin{array}{ll}
9x + 12y = -18 & \blacktriangleleft \text{ multiplying each member of (1) by 3} \\
\underline{10x + 12y = -16} & \blacktriangleleft \text{ multiplying each member of (2) by 2} \\
-x \quad\quad = -2 & \blacktriangleleft \text{ by (12.1), equating the differences of} \\
& \quad\text{ the corresponding members of (3) and (4)} \\
x = 2 & \blacktriangleleft \text{ multiplying each member by } -1
\end{array}
$$

Consequently the first number in (x, y) is 2. We obtain the second number by replacing x by 2 in either (1) or (2) and solving for y. We shall choose (1) and get

$$3(2) + 4y = -6$$
$$y = -3$$

Hence $(x, y) = (2, -3)$.

Check

Replacing x by 2 and y by -3 in (1) and (2), we get

$$6 - 12 = -6 \quad \blacktriangleleft \text{ from (1)}$$
$$10 - 18 = -8 \quad \blacktriangleleft \text{ from (2)}$$

Therefore the simultaneous solution set is $\{(2, -3)\}$.

Two Equations of the Type $Ax^2 + Cy^2 = F$ The procedure for obtaining the simultaneous solution set of two equations of the type $Ax^2 + Cy^2 = F$ is the same as that used in Example 1 except that, in general, we should obtain four solution pairs instead of just one. We shall illustrate the process in Example 2.

EXAMPLE 2

Obtain the simultaneous solution set of the equations

$$2x^2 + 3y^2 = 21 \tag{5}$$
$$3x^2 - 4y^2 = 23 \tag{6}$$

Solution

We arbitrarily select y as the unknown to be eliminated and proceed:

$$8x^2 + 12y^2 = 84 \quad \blacktriangleleft \text{ by (2.6), multiplying (5) by 4} \tag{7}$$
$$\underline{9x^2 - 12y^2 = 69} \quad \blacktriangleleft \text{ multiplying (6) by 3} \tag{8}$$
$$17x^2 \qquad = 153 \quad \blacktriangleleft \text{ equating the sums of the corresponding}$$
$$\qquad\qquad\qquad\qquad \text{members of (7) and (8)}$$
$$x^2 \qquad = 9 \quad \blacktriangleleft \text{ solving for } x^2$$
$$x = \pm 3 \quad \blacktriangleleft \text{ equating the square roots of } x^2 \text{ and 9}$$

We now replace x by 3 and -3 in Eq. (5) and solve for y. In either case we get

$$18 + 3y^2 = 21$$
$$y = \pm 1$$

Consequently, the solution set is $\{(3, 1), (3, -1), (-3, 1), (-3, 1)\}$.

Check

If we replace x by 3 and y by 1 in each of the given equations, we get a check of the first pair:

$$18 + 3 = 21 \quad \blacktriangleleft \text{ from (5)}$$
$$27 - 4 = 23 \quad \blacktriangleleft \text{ from (6)}$$

The other solution pairs can be checked in a similar manner.

Two Equations of the Type $Ax^2 + Cy^2 + Dx = F$ or $Ax^2 + Bxy + Dx = F$ The first step in the process of solving two equations of the type $Ax^2 + Cy^2 + Dx = F$ is to eliminate y^2 by addition or subtraction. Next, we solve the resulting equation for x. Finally, we replace x in one of the given equations by the roots obtained in the second step, solve the resulting equation for y, and then form the simultaneous solution pairs by writing a value of x as the first number in each pair and the corresponding value of y as the second. The procedure is illustrated in Example 3.

EXAMPLE 3

Find the simultaneous solution set of the equations

$$3x^2 - 2y^2 - 6x = -23 \qquad (9)$$
$$x^2 + y^2 - 4x = 13 \qquad (10)$$

Solution

Since each of the given equations contains one term in y^2 and no other term involving y, we eliminate y^2 and then complete the process of solving as follows:

$$\begin{aligned}
3x^2 - 2y^2 - 6x &= -23 \qquad &\blacktriangleleft \text{ Eq. (9) recopied} \qquad (9)\\
\underline{2x^2 + 2y^2 - 8x} &= \underline{26} \qquad &\blacktriangleleft \text{ multiplying Eq. (10) by 2} \qquad (11)\\
5x^2 \qquad - 14x &= 3 \qquad &\blacktriangleleft \text{ equating the sums of the left}\\
& &\text{and right members of (9) and (11)}
\end{aligned}$$

$$\begin{aligned}
5x^2 - 14x - 3 &= 0 \qquad &\blacktriangleleft \text{ adding } -3 \text{ to each member}\\
x &= 3 \text{ and } -\tfrac{1}{5} \qquad &\blacktriangleleft \text{ by the quadratic formula}\\
9 + y^2 - 12 &= 13 \qquad &\blacktriangleleft \text{ replacing } x \text{ by 3 in (10)}\\
y^2 &= 16 \qquad &\blacktriangleleft \text{ adding 3 to each member}\\
y &= \pm 4
\end{aligned}$$

Hence, if $x = 3$, then $y = \pm 4$. We continue by replacing x by $-\tfrac{1}{5}$.

$$\begin{aligned}
(-\tfrac{1}{5})^2 + y^2 - 4(-\tfrac{1}{5}) &= 13 \qquad &\blacktriangleleft \text{ replacing } x \text{ by } -\tfrac{1}{5} \text{ in (10)}\\
\tfrac{1}{25} + y^2 + \tfrac{4}{5} &= 13 \qquad &\blacktriangleleft \text{ performing the indicated}\\
& &\text{operations}\\
1 + 25y^2 + 20 &= 325 \qquad &\blacktriangleleft \text{ multiplying each member by 25}\\
25y^2 &= 304 \qquad &\blacktriangleleft \text{ adding } -21 \text{ to each member}
\end{aligned}$$

$$y = \pm\sqrt{\frac{304}{25}}$$

$$= \pm\frac{4\sqrt{19}}{5}$$

Consequently, if $x = -\tfrac{1}{5}$, $y = \pm 4\sqrt{19}/5$. Therefore the simultaneous solution set is

$$\{(3, 4), (3, -4), (-\tfrac{1}{5}, 4\sqrt{19}/5), (-\tfrac{1}{5}, -4\sqrt{19}/5)\}$$

Each solution pair can be checked by replacing x and y in the given equations by the appropriate number from the solution pair.

We solve two equations of the type $Ax^2 + Bxy + Dx = F$ by first eliminating xy and then solving the resulting equation for x. The corresponding values of y then can be found by substitution. The method is illustrated in Example 4.

EXAMPLE 4 Obtain the solution set of

$$x^2 + 4xy - 7x = 12 \qquad\qquad (12)$$
$$3x^2 - 4xy + 4x = 15 \qquad\qquad (13)$$

Solution Since the sum of the xy terms in the left members of Eqs. (12) and (13) is zero, we proceed as follows:

$$4x^2 - 3x = 27 \qquad \blacktriangleleft \text{ equating the sums of the corresponding}$$
$$\text{members of (12) and (13)}$$

$$4x^2 - 3x - 27 = 0$$
$$x = 3 \text{ and } -\tfrac{9}{4} \qquad \blacktriangleleft \text{ by use of the quadratic formula}$$

We find the second numbers in the solution pairs as follows:

$$9 + 12y - 21 = 12 \qquad \blacktriangleleft \text{ replacing } x \text{ by 3 in (12)}$$
$$12y = 24 \qquad \blacktriangleleft \text{ adding 12 to each member}$$
$$y = 2$$

Hence one simultaneous solution pair is $(3, 2)$. Then

$$\tfrac{81}{16} - 9y + \tfrac{63}{4} = 12 \qquad \blacktriangleleft \text{ replacing } x \text{ by } -\tfrac{9}{4} \text{ in (12)}$$
$$81 - 144y + 252 = 192 \qquad \blacktriangleleft \text{ multiplying each member by 16}$$
$$-144y = -141 \qquad \blacktriangleleft \text{ adding } -333 \text{ to each member}$$
$$y = \tfrac{141}{144} = \tfrac{47}{48}$$

Therefore a second simultaneous solution pair is $(-\tfrac{9}{4}, \tfrac{47}{48})$, and the simultaneous solution set is

$$\{(3, 2), (-\tfrac{9}{4}, \tfrac{47}{48})\}$$

Check Using $(3, 2)$, we have

$$9 + 24 - 21 = 33 - 21 = 12 \qquad \blacktriangleleft \text{ from (12)}$$
$$27 - 24 + 12 = 3 + 12 = 15 \qquad \blacktriangleleft \text{ from (13)}$$

Similarly $(-\tfrac{9}{4}, \tfrac{47}{48})$ checks.

The procedure for solving two equations of the type $Ax^2 + Cy^2 + Ey = F$ is the same as that used in Example 3 except that we eliminate x^2 as the first step and solve the resulting equation for y. To solve two equations of the type $Bxy + Cy^2 + Ey = F$, we proceed as in Example 4. In this case, however, the equation obtained after eliminating xy involves the variable y instead of x.

Exercise 12.3   Elimination by Addition and Subtraction

Find the solution set of the pair of equations in each of the following problems by use of the addition and subtraction method.   In problems 13 to 16 begin by solving for $1/x$ and $1/y$.

1 $3x - y = 1$
   $x + 2y = 5$

2 $x + 3y = 11$
   $2x + 5y = 19$

3 $2x + 3y = 3$
   $3x + y = 8$

4 $5x + 2y = 1$
   $2x + y = 0$

5 $4x + 3y = -1$
   $8x + 6y = -2$

6 $6x + 5y = -6$
   $2x - 3y = -2$

7 $9x - 2y = 12$
   $5x + 6y = 28$

8 $2x - 7y = 12$
   $5x + 21y = -47$

9 $4x + 3y = 19$
   $3x - 2y = -7$

10 $5x - 7y = -3$
   $2x + 3y = -7$

11 $3x - 5y = 1$
   $6x - 10y = 3$

12 $4x - 3y = 6$
   $3x + 2y = -4$

13 $\dfrac{1}{x} + \dfrac{2}{y} = 1$

   $\dfrac{3}{x} + \dfrac{10}{y} = 4$

14 $\dfrac{3}{x} - \dfrac{6}{y} = 2$

   $\dfrac{4}{x} - \dfrac{10}{y} = 3$

15 $\dfrac{2}{x} + \dfrac{1}{y} = 7$

   $\dfrac{1}{x} - \dfrac{2}{y} = -4$

16 $\dfrac{3}{x} + \dfrac{1}{y} = -1$

   $\dfrac{2}{x} + \dfrac{3}{y} = 11$

17 $2x^2 + 4y^2 = 19$
   $3x^2 - 8y^2 = 25$

18 $x^2 + 2y^2 = 3$
   $3x^2 - y^2 = 2$

19 $4x^2 + 3y^2 = 4$
   $8x^2 + 5y^2 = 7$

20 $9x^2 + y^2 = 2$
   $18x^2 + 3y^2 = 5$

21 $2x^2 + 3y^2 = 7$
   $3x^2 + 2y^2 = 8$

22 $2x^2 + 4y^2 = 9$
   $3x^2 + 8y^2 = 14$

23 $3x^2 - 5y^2 = -8$
   $2x^2 - y^2 = 4$

24 $3x^2 + 2y^2 = 14$
   $2x^2 + 3y^2 = 11$

25 $x^2 - y^2 + 3x = 7$
   $2x^2 + 3y^2 - 6x = 5$

26 $6x^2 + 5y^2 + 2x = 3$
   $5x^2 + 3y^2 - 4x = -2$

27 $x^2 + 4y^2 - 2x = 1$
   $3x^2 + 8y^2 - 6x = 2$

28 $2x^2 + y^2 - 3x = 3$
   $3x^2 + 2y^2 + x = 16$

29 $9x^2 - 41y^2 + 60y = 100$
   $x^2 - 5y^2 + 8y = 12$

30 $2x^2 + 2y^2 + 3y = 1$
   $3x^2 - 2y^2 + 3y = 1$

31 $4x^2 + 3y^2 - 3y = 2$
   $3x^2 - 2y^2 + 5y = -1$

32 $2x^2 + 3y^2 - 7y = 2$
   $8x^2 + 5y^2 - 9y = 2$

33 $x^2 + 2xy - x = 4$
   $2x^2 - xy + 3x = 3$

34 $x^2 - 4xy + 2x = -2$
   $3x^2 - 4xy - 2x = -6$

35 $x^2 + xy - 3x = -2$
   $2x^2 - 3xy - x = -4$

36 $4x^2 + 5xy + 2x = 0$
   $2x^2 + xy + 3x = 3$

37 $3y^2 - 5xy + y = 4$ 38 $y^2 - 2xy = -2$
    $2y^2 - 4xy + y = 2$ $3y^2 - 5xy - y = -6$
39 $y^2 + xy + 2y = 6$ 40 $5y^2 - 5xy + 4y = -18$
    $2y^2 - xy + 2y = 1$ $48y^2 + 25xy - 3y = 48$

12.7 ELIMINATION BY SUBSTITUTION

If in a system of two equations in two unknowns, one equation can be solved for one unknown in terms of the other, this unknown can be eliminated by substitution. If the unknowns are x and y and one equation can be solved for y in terms of x, then such a solution is in the form $y = f(x)$. We then replace y in the other equation by $f(x)$ and obtain an equation involving only x. Designate this equation by $F(x) = 0$. We solve $F(x) = 0$ for x; each root obtained will be the first number in a pair in the simultaneous solution set. We complete the process by substituting each root of $F(x) = 0$ in $y = f(x)$ and thus obtain each corresponding value of y. Finally, we arrange the corresponding values of x and y in pairs, with the first number in each pair as the value of x, and thus obtain the simultaneous solution set.

We illustrate the method with four examples. In the first example, the method is applied to two linear equations, and in the second it is applied to a system containing a linear and a quadratic equation. The third and fourth examples illustrate the method for solving a system of two quadratics in two unknowns in which one equation is easily solvable for one unknown in terms of the other.

EXAMPLE 1   Solve the equations

$$5x + 3y = 5 \qquad (1)$$
$$4x + \ y = 11 \qquad (2)$$

simultaneously by substitution.

Solution   We first notice that Eq. (2) is readily solvable for y in terms of x, and the solution is

$$y = 11 - 4x \qquad (3)$$

We now replace y by $11 - 4x$ in Eq. (1) and complete the process of solving as follows:

$5x + 3(11 - 4x) = 5$ ◄ replacing y by $11 - 4x$ in (1)
$5x + 33 - 12x = 5$ ◄ by the distributive axiom
$-7x = -28$ ◄ adding -33 to each member and combining terms
$x = 4$ ◄ dividing by -7

We now replace x by 4 in (3) and get

$$y = 11 - 16$$
$$y = -5$$

Therefore the simultaneous solution set is $\{(4, -5)\}$.

Check   Replacing x and y by 4 and -5, respectively, in each of the given equations, we get

$$20 - 15 = 5 \qquad \blacktriangleleft \text{ from (1)}$$
$$16 - 5 = 11 \qquad \blacktriangleleft \text{ from (2)}$$

EXAMPLE 2   Obtain the simultaneous solution set of

$$x^2 + 2y^2 = 54 \qquad\qquad (4)$$
$$2x - y = -9 \qquad\qquad (5)$$

Solution   Equation (5) is readily solvable for y in terms of x, and so we proceed as follows:

$$y = 2x + 9 \qquad\qquad \blacktriangleleft \text{ solving (5) for } y \text{ in terms of } x \quad (6)$$
$$x^2 + 2(2x + 9)^2 = 54 \qquad\qquad \blacktriangleleft \text{ replacing } y \text{ by } 2x + 9 \text{ in (4)}$$
$$x^2 + 2(4x^2 + 36x + 81) = 54 \qquad\qquad \blacktriangleleft \text{ squaring } 2x + 9$$
$$x^2 + 8x^2 + 72x + 162 = 54 \qquad\qquad \blacktriangleleft \text{ by the distributive axiom}$$
$$9x^2 + 72x + 108 = 0 \qquad\qquad \blacktriangleleft \text{ adding } -54 \text{ to each member and}$$
$$\text{combining terms}$$
$$x^2 + 8x + 12 = 0 \qquad\qquad \blacktriangleleft \text{ dividing by 9}$$
$$(x + 6)(x + 2) = 0 \qquad\qquad \blacktriangleleft \text{ factoring}$$
$$x = -6 \qquad\qquad \blacktriangleleft \text{ setting } x + 6 = 0 \text{ and solving}$$
$$x = -2 \qquad\qquad \blacktriangleleft \text{ setting } x + 2 = 0 \text{ and solving}$$

To find y, we proceed as follows:

$$y = 2(-6) + 9 = -3 \qquad\qquad \blacktriangleleft \text{ replacing } x \text{ by } -6 \text{ in (6)}$$
$$y = 2(-2) + 9 = 5 \qquad\qquad \blacktriangleleft \text{ replacing } x \text{ by } -2 \text{ in (6)}$$

Therefore the simultaneous solution set is $\{(-6, -3), (-2, 5)\}$.

Check   Using $(-6, -3)$, we obtain

$$36 + 18 = 54 \qquad\qquad \blacktriangleleft \text{ from (4)}$$
$$-12 + 3 = -9 \qquad\qquad \blacktriangleleft \text{ from (5)}$$

Similarly, using $(-2, 5)$, we get

$$4 + 50 = 54 \qquad\qquad \blacktriangleleft \text{ from (4)}$$
$$-4 - 5 = -9 \qquad\qquad \blacktriangleleft \text{ from (5)}$$

EXAMPLE 3   Obtain the simultaneous solution set of

$$4x^2 - 2xy - y^2 = -5 \tag{7}$$
$$y + 1 = -x^2 - x \tag{8}$$

Solution

$$y = -x^2 - x - 1 \quad \blacktriangleleft \text{ solving (8) for } y \tag{9}$$
$$4x^2 - 2x(-x^2 - x - 1) - (-x^2 - x - 1)^2 = -5$$
$$\qquad\qquad \blacktriangleleft \text{ replacing } y \text{ by } -x^2 - x - 1 \text{ in (7)} \tag{10}$$
$$4x^2 + 2x^3 + 2x^2 + 2x - x^4 - x^2 - 1 - 2x^3 - 2x^2 - 2x = -5$$
$$\qquad\qquad \blacktriangleleft \text{ performing the indicated operations in (10)} \quad (11)$$
$$-x^4 + 3x^2 + 4 = 0 \quad \blacktriangleleft \text{ adding 5 to each member of (11) and} \tag{12}$$
$$\qquad\qquad \text{combining similar terms}$$

Equation (12) is in quadratic form, and we solve it as follows:

$$x^4 - 3x^2 - 4 = 0 \quad \blacktriangleleft \text{ dividing each member of (12) by } -1$$
$$(x^2 - 4)(x^2 + 1) = 0 \quad \blacktriangleleft \text{ factoring}$$
$$x^2 = 4 \quad \blacktriangleleft \text{ setting } x^2 - 4 = 0 \text{ and solving for } x^2$$
$$x = \pm 2 \quad \blacktriangleleft \text{ equating the square roots}$$
$$x^2 = -1 \quad \blacktriangleleft \text{ setting } x^2 + 1 = 0 \text{ and solving for } x^2$$
$$x = \pm\sqrt{-1} \quad \blacktriangleleft \text{ equating the square roots}$$
$$x = \pm i \quad \blacktriangleleft \text{ since } \sqrt{-1} = i$$

We obtain the value of y corresponding to each of the four values of x by use of Eq. (9) as follows:

$$y = -2^2 - 2 - 1 = -7 \quad \blacktriangleleft \text{ replacing } x \text{ by } 2$$
$$y = -(-2)^2 - (-2) - 1 = -3 \quad \blacktriangleleft \text{ replacing } x \text{ by } -2$$
$$y = -i^2 - i - 1 \quad \blacktriangleleft \text{ replacing } x \text{ by } i$$
$$= 1 - i - 1 = -i \quad \blacktriangleleft \text{ since } i^2 = -1$$
$$y = -(-i)^2 - (-i) - 1 \quad \blacktriangleleft \text{ replacing } x \text{ by } -i$$
$$= 1 + i - 1 = i$$

Consequently the solution set is

$$\{(2, -7), (-2, -3), (i, -i), (-i, i)\}$$

The solution set can be checked in the usual manner and will be left as an exercise for the student.

EXAMPLE 4   Obtain the solution set of the following system of equations by substitution:

$$xy = 2 \tag{13}$$
$$15x^2 + 4y^2 = 64 \tag{14}$$

Solution   Equation (13) can be solved easily for y in terms of x, and the solution is

$$y = \frac{2}{x} \tag{15}$$

We now replace y by $2/x$ in Eq. (14) and complete the process of solving as follows:

$$15x^2 + 4\left(\frac{2}{x}\right)^2 = 64$$

$$15x^2 + 4\,\frac{4}{x^2} = 64 \qquad \blacktriangleleft \text{ squaring } 2/x$$

$$15x^4 + 16 = 64x^2 \qquad \blacktriangleleft \text{ multiplying each member by } x^2$$

$$15x^4 - 64x^2 + 16 = 0 \qquad \blacktriangleleft \text{ adding } -64x^2 \text{ to each member and arranging terms}$$

$$(x^2 - 4)(15x^2 - 4) = 0 \qquad \blacktriangleleft \text{ factoring}$$

$$x^2 = 4 \qquad \blacktriangleleft \text{ setting } x^2 - 4 = 0 \text{ and solving for } x^2$$

$$x = \pm 2$$

$$x^2 = \tfrac{4}{15} \qquad \blacktriangleleft \text{ setting } 15x^2 - 4 = 0 \text{ and solving for } x^2$$

$$x = \pm \frac{2}{\sqrt{15}}$$

$$= \pm \frac{2\sqrt{15}}{15}$$

To obtain y, we replace x by all four of its values:

$$y = \frac{2}{\pm 2} \qquad \blacktriangleleft \text{ replacing } x \text{ by } \pm 2 \text{ in (15)}$$

$$= \pm 1$$

$$y = \frac{2}{\pm 2\sqrt{15}/15} \qquad \blacktriangleleft \text{ replacing } x \text{ by } \pm 2\sqrt{15}/15 \text{ in (15)}$$

$$= \pm \sqrt{15}$$

Therefore the simultaneous solution set is

$$\{(2, 1),\ (-2, -1),\ (2\sqrt{15}/15,\ \sqrt{15}),\ (-2\sqrt{15}/15,\ -\sqrt{15})\}$$

The solution set can be checked by the usual method and will be left as an exercise for the student.

Exercise 12.4 Elimination by Substitution

Find the solution set of the pair of equations in each of the following equations by use of the substitution method.

1 $\quad 2x + y = 7$
$\quad\quad 5x - 3y = 1$

2 $\quad 2x + 3y = 9$
$\quad\quad 5x - y = 14$

3 $\quad 3x - y = 7$
$\quad\quad 2x + 3y = 1$

4 $\quad 5x + y = -4$
$\quad\quad 3x + 7y = 4$

5   $x + 2y = 7$
     $3x + 4y = 17$

6   $3x + 5y = -1$
     $x - 3y = -5$

7   $3x - 4y = 2$
     $x - 5y = -3$

8   $x + 3y = 15$
     $3x - 2y = 1$

9   $3x + 4y = 5$
     $5x - 2y = 4$

10   $5x - 3y = 9$
     $4x + 9y = 11$

11   $2x + 3y = 2$
     $4x - 9y = -1$

12   $2x + 5y = -2$
     $4x + 4y = -1$

13   $3x - y = 8$
     $3x^2 - y^2 = 26$

14   $3x - 2y = 5$
     $3x^2 - 2y^2 = 19$

15   $2x - y = -1$
     $16x^2 - 3y^2 = -11$

16   $4x - y = 11$
     $8x^2 + 5y^2 = 77$

17   $ax^2 - by^2 = ab^2 - a^2b$
     $ax - by = 0$

18   $x^2 + a^2y^2 = 10b^2$
     $x + ay = 2b$

19   $m^2x^2 + y^2 = b^2$
     $y = mx + b$

20   $2x^2 - 3y^2 = -a^2$
     $2x + y = 3a$

21   $y^2 - 6xy = -20$
     $y + 4 = 2x^2 + 3x$

22   $3x^2 - 12xy = 0$
     $x + 1 = 3y^2 + 2y$

23   $y^2 - 6xy = -32$
     $y + 2 = x^2 + 3x$

24   $y^2 - 2xy = 40$
     $y + 11 = 2x^2 + x$

25   $x^2 - 2xy + y^2 = 144$
     $y = 3x^2 + x$

26   $x^2 - 2xy + y^2 = 1$
     $x + 2 = y^2 + y$

27   $x^2 - 2xy + y^2 = 9$
     $y^2 + y = 4 + x$

28   $2x^2 - 12xy + y^2 = -72$
     $y = 2x^2 + 6x$

29   $x^2 - 2xy - y + x = 68$
     $x = 2y^2 + y$

30   $x^2 - 4xy - 6y + 3x = 3$
     $x = 4y^2 + 2y$

31   $x^2 - 8xy - 4y + x = -14$
     $x = y^2 + 4y$

32   $x^2 + 2xy + y + x = 3$
     $x = 2y^2 - y$

33   $x^2 + 2y^2 = 9$
     $xy = 2$

34   $2x^2 - 3y^2 = 2$
     $xy = -20$

35   $2x^2 + 3y^2 = 21$
     $xy = 3$

36   $3x^2 - 4y^2 = 32$
     $xy = -8$

37   $x^2 + xy - y^2 = -1$
     $xy = 2$

38   $4x^2 - 6xy + 2y^2 = 1$
     $4xy = 3$

39   $x^2 + 2xy - y^2 = 1$
     $xy = 2$

40   $4x^2 - 6xy + y^2 = 1$
     $2xy = 3$

12.8   ELIMINATION BY A COMBINATION OF METHODS

Frequently, the computation involved in solving the system of equations

$$F(x, y) = K \tag{1}$$
$$f(x, y) = k \tag{2}$$

by substitution is very tedious. In such cases, it may be more efficient to use a method that depends upon a theorem in analytic geometry. It states that the graph of

$$mF(x, y) + nf(x, y) = mK + nk \qquad (3)$$

passes through the intersections of the graphs of Eqs. (1) and (2). Hence the solution set of the system (1) and (2) is the same as the solution set of the system composed of either (1) or (2) and (3) for all nonzero replacements of m and n.

Now if we can so determine m and n in (3) that the resulting equation is easily solvable for either unknown in terms of the other, we can obtain the solution set of (1) and (2) by solving either (1) or (2) with (3) by substitution. We shall discuss two classes of systems of quadratic equations in two unknowns that can be solved by this method.

TWO EQUATIONS OF THE TYPE $Ax^2 + Bxy + Cy^2 = D$ The first step in solving two equations of the type $Ax^2 + Bxy + Cy^2 = D$ is to eliminate the constant terms by addition or subtraction. That is, we combine the two equations in such a way as to obtain an equation in which the constant term is zero. We then solve this equation for one unknown in terms of the other and complete the process by substitution.

For example, if we solve the equation for y in terms of x, we usually obtain two equations of the type $y = rx$ and $y = tx$, where r and t are constants. We then replace y in one of the given equations successively by rx and tx and solve each resulting equation for x. We then find the value of y that corresponds to each root thus obtained by replacing x in $y = rx$ and $y = tx$ by the root. We shall illustrate the process with an example.

EXAMPLE 1   Find the simultaneous solution set of the equations

$$3x^2 - 4xy + 2y^2 = 3 \qquad (4)$$
$$2x^2 - 6xy + \ y^2 = -6 \qquad (5)$$

Solution   We eliminate the constant terms:

$$
\begin{array}{ll}
6x^2 - \ 8xy + 4y^2 = \ \ 6 & \blacktriangleleft \text{ multiplying (4) by 2} \qquad (6)\\
\underline{2x^2 - \ 6xy + \ y^2 = -6} & \blacktriangleleft \text{ Eq. (5) recopied}\\
8x^2 - 14xy + 5y^2 = \ \ 0 & \blacktriangleleft \text{ equating the corresponding sums} \quad (7)
\end{array}
$$

We now solve Eq. (7) for y in terms of x by the quadratic formula with $a = 5$, $b = -14x$, and $c = 8x^2$, and obtain

$$y = \frac{14x \pm \sqrt{196x^2 - 160x^2}}{10}$$

$$= \frac{14x \pm 6x}{10} = \frac{20x}{10} \text{ and } \frac{8x}{10}$$

Consequently

$$y = 2x \tag{8}$$

$$y = \frac{4x}{5} \tag{9}$$

We continue the process by replacing y in either (4) or (5) separately by $2x$ and $4x/5$ and solving the resulting equation for x. Substituting $y = 2x$ in (4), we get

$$3x^2 - 4x(2x) + 2(2x)^2 = 3$$
$$3x^2 - 8x^2 + 8x^2 = 3 \quad \blacktriangleleft \text{ performing the indicated operations}$$
$$3x^2 = 3 \quad \blacktriangleleft \text{ combining terms}$$
$$x^2 = 1$$
$$x = \pm 1$$

We now replace x in (8) by ± 1 and get $y = 2(\pm 1) = \pm 2$. Hence two pairs in the solution set are $(1, 2)$ and $(-1, -2)$.

Next we replace y by $4x/5$ in (4) and solve the resulting equation for x. Thus we obtain

$$3x^2 - 4x\,\frac{4x}{5} + 2\left(\frac{4x}{5}\right)^2 = 3$$

$$3x^2 - \frac{16x^2}{5} + \frac{32x^2}{25} = 3$$
$$75x^2 - 80x^2 + 32x^2 = 75 \quad \blacktriangleleft \text{ multiplying each member by 25}$$
$$27x^2 = 75$$
$$x^2 = \tfrac{75}{27} = \tfrac{25}{9}$$
$$x = \pm \tfrac{5}{3}$$

Finally, we replace x by $\pm \tfrac{5}{3}$ in (9) and get $y = \tfrac{4}{5}(\pm \tfrac{5}{3}) = \pm \tfrac{4}{3}$. Therefore, two additional solution pairs are $(\tfrac{5}{3}, \tfrac{4}{3})$ and $(-\tfrac{5}{3}, -\tfrac{4}{3})$.

Consequently the complete simultaneous solution set is $\{(1, 2), (-1, -2), (\tfrac{5}{3}, \tfrac{4}{3}), (-\tfrac{5}{3}, -\tfrac{4}{3})\}$, which can be checked by the usual methods.

If the constant term in either of the given equations is zero, as in the system

$$2x^2 - 3xy - 2y^2 = 0$$
$$x^2 + 2xy + 5y^2 = 17$$

it is not necessary to perform the first step in Example 1. We immediately solve the first equation for y in terms of x, or x in terms of y, and then complete the solution by the method of substitution.

TWO EQUATIONS OF THE TYPE $Ax^2 + Ay^2 + Dx + Ey = F$ If each of the given equations is of the type $Ax^2 + Ay^2 + Dx + Ey = F$, we can eliminate the second-degree terms by addition or subtraction and obtain a linear equation in x and y. We can then solve this equation with one of the given equations by the method illustrated in Example 2 of Sec. 12.8.

EXAMPLE 2   Obtain the simultaneous solution set of

$$3x^2 + 3y^2 + x - 2y = 20 \qquad (10)$$
$$2x^2 + 2y^2 + 5x + 3y = 9 \qquad (11)$$

Solution

$6x^2 + 6y^2 + 2x - 4y = 40$	◀ multiplying (10) by 2 (12)
$6x^2 + 6y^2 + 15x + 9y = 27$	◀ multiplying (11) by 3 (13)
$ -13x - 13y = 13$	◀ equating the differences of (14) the corresponding members of (12) and (13)

Now we solve Eq. (14) simultaneously with Eq. (11) and complete the process of solving as indicated below.

$y = -x - 1$	◀ solving (14) for y (15)
$2x^2 + 2(-x-1)^2 + 5x + 3(-x-1) = 9$	◀ replacing y by $-x - 1$ in (11)
$2x^2 + 2x^2 + 4x + 2 + 5x - 3x - 3 - 9 = 0$	◀ performing the indicated operations and adding -9 to each member
$4x^2 + 6x - 10 = 0$	◀ combining similar terms
$2x^2 + 3x - 5 = 0$	◀ dividing by 2
$x = 1, -\frac{5}{2}$	◀ by the quadratic formula

We find the corresponding values of y as follows:

$y = -1 - 1 = -2$	◀ replacing x by 1 in (15)
$y = \frac{5}{2} - 1 = \frac{3}{2}$	◀ replacing x by $-\frac{5}{2}$ in (15)

Consequently the solution set is $\{(1, -2), (-\frac{5}{2}, \frac{3}{2})\}$ and can be checked in the usual manner.

12.9   SYMMETRIC EQUATIONS

Symmetric   An equation in two variables is *symmetric* if the equation is not altered when
equation   the variables are interchanged. For example,

$$Ax^2 + Bxy + Ay^2 + Dx + Dy = F \qquad (1)$$

is symmetric since interchanging x and y does not change the equation.
        If we transform Eq. (1) by means of the equations

$$x = u + v \qquad (2)$$
$$y = u - v$$

we get

$$(2A + B)u^2 + (2A - B)v^2 + 2Du = F \qquad (3)$$

Therefore, the transformation (2) will convert two symmetric equations to two equations of the type (3), which can be solved by the method of Example 3, Sec. 12.6. We illustrate the procedure with the following example.

EXAMPLE   Obtain the simultaneous solution set of the equations

$$7x^2 + 2xy + 7y^2 - 8x - 8y = 108 \tag{4}$$
$$3x^2 - 2xy + 3y^2 + 4x + 4y = 68 \tag{5}$$

Solution   We first replace x by $u + v$ and y by $u - v$ in each of (1) and (2), divide by 4, and get

$$4u^2 + 3v^2 - 4u = 27 \qquad \blacktriangleleft \text{from (4)} \tag{6}$$
$$u^2 + 2v^2 + 2u = 17 \qquad \blacktriangleleft \text{from (5)} \tag{7}$$

Since each of (6) and (7) contains one term involving v^2 and no other term involving v, we can eliminate v^2 by addition or subtraction and complete the procedure of solving as follows:

$$8u^2 + 6v^2 - 8u = 54 \qquad \blacktriangleleft \text{multiplying (6) by 2} \tag{8}$$
$$\underline{3u^2 + 6v^2 + 6u = 51} \qquad \blacktriangleleft \text{multiplying (7) by 3} \tag{9}$$
$$5u^2 \qquad\quad - 14u = 3 \qquad \blacktriangleleft \text{equating the differences} \tag{10}$$
$$\qquad\qquad\qquad\qquad\qquad \text{of the corresponding}$$
$$\qquad\qquad\qquad\qquad\qquad \text{members of (8) and (9)}$$

$$5u^2 - 14u - 3 = 0 \qquad \blacktriangleleft \text{adding } -3 \text{ to each}$$
$$\qquad\qquad\qquad\qquad\qquad \text{member of (10)}$$

$$u = \frac{14 \pm \sqrt{196 + 60}}{10} \qquad \blacktriangleleft \text{by the quadratic formula}$$

$$u = \frac{14 \pm 16}{10} = 3 \text{ and } -\tfrac{1}{5}$$

We now obtain the corresponding values of v by replacing u in Eq. (7) by each of the above values, as indicated below.

$$9 + 2v^2 + 6 = 17 \qquad \blacktriangleleft \text{replacing } u \text{ by 3 in (7)}$$
$$2v^2 = 2$$
$$v^2 = 1$$
$$v = \pm 1$$

Therefore, if $u = 3$, $v = \pm 1$. Now we replace (u, v) by $(3, 1)$ and $(3, -1)$ in $x = u + v$ and $y = u - v$ and get

$$\begin{array}{lll} & x = 3 + 1 = 4 & y = 3 - 1 = 2 \\ \text{and} & x = 3 - 1 = 2 & y = 3 - (-1) = 4 \end{array}$$

Hence, two pairs in the solution set are $(4, 2)$ and $(2, 4)$. If we replace u by $-\tfrac{1}{5}$ in (7), we have

$$\tfrac{1}{25} + 2v^2 - \tfrac{2}{5} = 17$$
$$1 + 50v^2 - 10 = 425$$
$$50v^2 = 434$$
$$v^2 = \tfrac{434}{50} = \tfrac{217}{25}$$
$$v = \pm \frac{\sqrt{217}}{5}$$

Finally,

$$x = -\frac{1}{5} + \frac{\sqrt{217}}{5} \qquad y = -\frac{1}{5} - \frac{\sqrt{217}}{5}$$

and

$$x = -\frac{1}{5} - \frac{\sqrt{217}}{5} \qquad y = -\frac{1}{5} - \left(-\frac{\sqrt{217}}{5} \right)$$

Therefore the simultaneous solution set is

$$\left\{ (4, 2), (2, 4), \left(\frac{-1 + \sqrt{217}}{5}, \frac{-1 - \sqrt{217}}{5} \right), \left(\frac{-1 - \sqrt{217}}{5}, \frac{-1 + \sqrt{217}}{5} \right) \right\}$$

Exercise 12.5 Elimination by Addition or Subtraction and by Substitution; Symmetric Equations

Find the solution set of the pair of equations in each of the following problems.

1 $x^2 + xy - 2y^2 = 0$
$11x^2 + 3xy - 11y^2 = 27$

2 $x^2 + 4xy - 17y^2 = -20$
$x^2 + xy - 6y^2 = 0$

3 $2x^2 - xy - 3y^2 = 0$
$x^2 - 2xy + 2y^2 = 5$

4 $x^2 - xy - 2y^2 = 0$
$5x^2 + 3xy + y^2 = 27$

5 $4x^2 + 3xy + 2y^2 = 3$
$2x^2 + 6xy + y^2 = -3$

6 $5x^2 - 2xy - 2y^2 = 1$
$7x^2 - 3xy - 3y^2 = 1$

7 $x^2 + 2xy - 5y^2 = -5$
$2x^2 - xy - 6y^2 = 4$

8 $4x^2 + 6xy - y^2 = 1$
$3x^2 + 4xy - 2y^2 = -3$

9 $2x^2 + 4xy - 3y^2 = 3$
$x^2 + xy - 2y^2 = 4$

10 $x^2 - 2xy - 2y^2 = 2$
$2x^2 - 4xy - 3y^2 = 3$

11 $3x^2 - xy - 3y^2 = 4$
$5x^2 + xy - 2y^2 = 8$

12 $4x^2 + 6xy - y^2 = 3$
$2x^2 - 5xy + 3y^2 = 1$

13 $5x^2 - 3xy - 9y^2 = 3$
$3x^2 + xy - 3y^2 = 1$

14 $x^2 - 6xy + 4y^2 = 1$
$x^2 - 10xy + 8y^2 = -4$

15 $6x^2 - 5xy + 2y^2 = 72$
$5x^2 + 4xy - 3y^2 = -45$

16 $3x^2 + 4xy - 5y^2 = 6$
$5x^2 + 3xy - 7y^2 = 3$

17 $x^2 + y^2 - 3x = 1$
$x^2 + y^2 + 3y = 7$

18 $4x^2 + 4y^2 - 11x = 19$
$3x^2 + 3y^2 - 11y = 17$

19 $x^2 + y^2 + 6x = 52$
$x^2 + y^2 - 6y = 4$

20 $x^2 + y^2 - 3x = 7$
$2x^2 + 2y^2 + y = 23$

21   $x^2 + y^2 - 5x + y = -4$
       $x^2 + y^2 - 3x + 2y = 1$

22   $x^2 + y^2 - 2x - 3y = 1$
       $3x^2 + 3y^2 - 6x - 4y = 13$

23   $x^2 + y^2 + 2x + 2y = 6$
       $2x^2 + 2y^2 + 3x + 3y = 10$

24   $x^2 + y^2 + x + y = 12$
       $3x^2 + 3y^2 + 2x + 2y = 32$

25   $x^2 + y^2 + 2x + y = 6$
       $x^2 + y^2 + 3x + 2y = 1$

26   $x^2 + y^2 + 12x - 24y = 1$
       $x^2 + y^2 - 11x - y = -22$

27   $3x^2 + 3y^2 - 13x - y = -2$
       $5x^2 + 5y^2 - 16x + 4y = 25$

28   $2x^2 + 2y^2 - 15x + 4y = 23$
       $3x^2 + 3y^2 + x + 53y = 340$

29   $x^2 - 3xy + y^2 + 3x + 3y = 20$
       $3x^2 - 4xy + 3y^2 - 6x - 6y = 5$

30   $x^2 + xy + y^2 - x - y = 1$
       $x^2 - xy + y^2 - x - y = -1$

31   $2x^2 + 2xy + 2y^2 - x - y = 22$
       $3x^2 + 2xy + 3y^2 + 4x + 4y = 52$

32   $2x^2 - xy + 2y^2 - 3x - 3y = 5$
       $x^2 - 2xy + y^2 + 3x + 3y = 16$

33   $4x^2 - 5xy + 4y^2 - 6x - 6y = -8$
       $2x^2 - 3xy + 2y^2 - 2x - 2y = -2$

34   $2x^2 + xy + 2y^2 - 5x - 5y = 57$
       $x^2 - xy + y^2 + 7x + 7y = 84$

35   $x^2 - 2xy + y^2 + 3x + 3y = -6$
       $x^2 - 2xy + y^2 + x + y = -2$

36   $5x^2 + 6xy + 5y^2 - 32x - 32y = -32$
       $4x^2 - 4xy + 4y^2 - 8x - 8y = 36$

37   $xy + x + y = 11$
       $2xy - 3x - 3y = -3$

38   $3xy + 4x + 4y = -2$
       $xy + x + y = -1$

39   $xy + 2x + 2y = 21$
       $2xy - x - y = 12$

40   $3xy - 2x - 2y = 4$
       $xy - x - y = 0$

12.10   PROBLEMS LEADING TO SYSTEMS OF EQUATIONS

In the introduction to this chapter we stated that a worded problem can frequently be solved more easily if more than one unknown is introduced and more than one equation is employed in the solution. The method of setting up the equations mainly involves letting the unknowns represent quantities which are asked for in the problem. The general rule is *the number of equations formed must be equal to the number of unknowns introduced.* The problem must be examined to see if the necessary number of equations can be formed before a decision is made on the number of unknowns to use.

EXAMPLE 1   A real estate dealer received $1,200 in rents on two dwellings last year, and one of the dwellings brought $10 per month more than the other. Find the monthly rental on each if the more expensive house was vacant for 2 months.

Solution On inspecting the problem we see that there are two basic relations involved, one between the separate rentals and the other between the monthly rentals and the income per year. The monthly rentals differ by $10; hence, if we let

$x =$ monthly rental of the more expensive house in dollars
$y =$ monthly rental of the less expensive house in dollars

then

$$x - y = 10 \tag{1}$$

Furthermore, since the first of the two houses was rented for 10 months and the other was rented for 12 months, we know that $10x + 12y$ is the total annual income. Therefore,

$$10x + 12y = 1{,}200 \tag{2}$$

We now solve Eqs. (1) and (2) simultaneously by eliminating y:

$$
\begin{array}{ll}
12x - 12y = 120 & \blacktriangleleft \text{ multiplying (1) by 12} \qquad (3)\\
\underline{10x + 12y = 1{,}200} & \blacktriangleleft \text{ (2) recopied} \qquad\qquad\quad (2)\\
22x = 1{,}320 & \blacktriangleleft \text{ equating the sums of corresponding members}\\
x = 60 & \blacktriangleleft \text{ solving for } x\\
60 - y = 10 & \blacktriangleleft \text{ replacing } x \text{ by 60 in (1)}\\
\phantom{60 - {}}y = 50 & \blacktriangleleft \text{ solving for } y
\end{array}
$$

Therefore the solution set $\{(x, y)\} = \{(60, 50)\}$, and it follows that the monthly rentals are $60 and $50, respectively.

Check $$60 - 50 = 10 \quad \text{and} \quad 10(60) + 12(50) = 1{,}200$$

EXAMPLE 2 A tobacco dealer mixed 12 pounds of one grade of tobacco with 10 pounds of another grade to obtain a blend worth $54. He then made a second blend worth $61 by mixing 8 pounds of the first grade with 15 pounds of the second grade. Find the price per pound of each grade.

Solution In this problem we have two basic relations that we can use to form two equations. We let

$x =$ price per pound of the first grade, in dollars
$y =$ price per pound of the second grade, in dollars

and then

$$12x + 10y = 54 \qquad \blacktriangleleft \text{ using the numbers of pounds as coefficients} \tag{4}$$
and the values of the blends as constant terms

$$8x + 15y = 61 \tag{5}$$

We eliminate y by subtraction.

$$36x + 30y = 162 \qquad \blacktriangleleft \text{multiplying (4) by 3} \qquad (6)$$
$$\underline{16x + 30y = 122} \qquad \blacktriangleleft \text{multiplying (5) by 2} \qquad (7)$$
$$20x \qquad\quad = 40 \qquad \blacktriangleleft \text{equating the differences of the members}$$
$$x = 2 \qquad \blacktriangleleft \text{solving for } x$$
$$16 + 15y = 61 \qquad \blacktriangleleft \text{replacing } x \text{ by 2 in (5)}$$
$$15y = 45$$
$$y = 3$$

Therefore the solution set of Eqs. (4) and (5) is $\{(2, 3)\}$, and it follows that the prices of the two grades are \$2 and \$3 per pound.

Check

$$12(2) + 10(3) = 24 + 30 = 54 \qquad \blacktriangleleft \text{from (4)}$$
$$8(2) + 15(3) = 16 + 45 = 61 \qquad \blacktriangleleft \text{from (5)}$$

EXAMPLE 3

Two airfields A and B are 720 miles apart, and B is due east of A. A plane flew from A to B in $1\frac{4}{5}$ hours and then returned to A in 2 hours. If the wind blew with a constant velocity from the west during the entire trip, find the speed of the plane in still air and the speed of the wind.

Solution

The problem is illustrated in Fig. 12.7. The essential point in solving this problem is that the wind helps the plane in flying from A to B and hinders it in flying from B to A. We therefore have the basis for two equations that involve the speed of the plane, the speed of the wind, and the time for each trip. If we let

$$x = \text{speed of plane in still air, in miles per hour}$$
$$y = \text{speed of wind, in miles per hour}$$

then, since the wind blew constantly from the west,

$$x + y = \text{speed of plane from } A \text{ to } B \text{ (wind helping)}$$
$$x - y = \text{speed of plane from } B \text{ to } A \text{ (wind hindering)}$$

The distance traveled each way was 720 miles, and we have the following equations based on the formula distance/rate $=$ time:

$$\frac{720}{x + y} = 1\tfrac{4}{5} = \text{time required for eastward trip} \qquad (8)$$

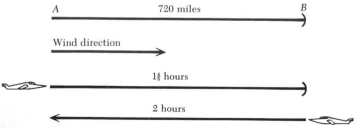

Figure 12.7

$$\frac{720}{x - y} = 2 = \text{time required for westward trip} \tag{9}$$

We solve these equations simultaneously for x and y.

$3{,}600 = 9x + 9y$	◀ multiplying (8) by $5(x + y)$ (10)
$720 = 2x - 2y$	◀ multiplying (9) by $x - y$ (11)
$7{,}200 = 18x + 18y$	◀ multiplying (10) by 2 (12)
$\underline{6{,}480 = 18x - 18y}$	◀ multiplying (11) by 9 (13)
$13{,}680 = 36x$	◀ equating the sums of corresponding members of (12) and (13)
$x = 380$	◀ solving for x
$3{,}600 = 3{,}420 + 9y$	◀ replacing x by 380 in (10)
$y = 20$	

Therefore the solution set of Eqs. (8) and (9) is $\{(380, 20)\}$, and it follows that the speed of the plane in still air is 380 miles per hour and the speed of the wind is 20 miles per hour.

Check

$$\frac{720}{380 + 20} = \frac{720}{400} = 1.8 \qquad ◀ \text{ from (8)}$$

$$\frac{720}{380 - 20} = \frac{720}{360} = 2 \qquad ◀ \text{ from (9)}$$

EXAMPLE 4

A circular flower garden is surrounded by a walk. The area of the walk is $\frac{11}{25}$ of the area of the garden, and the sum of the inner and outer circumferences of the walk is 88π feet. Find the radius of the garden and the width of the walk.

Solution

The walk is bounded by two concentric circles, and the radius of the inner circle is the radius of the garden. We let

$$x = \text{radius of inner circle, in feet}$$
$$y = \text{width of walk, in feet}$$

as indicated in Fig. 12.8. Then $x + y$ is the radius of the outer circle.

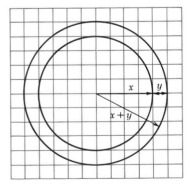

Figure 12.8

Consequently

$$\pi(x + y)^2 = \text{area of outer circle}$$
$$\pi x^2 = \text{area of inner circle}$$

and $$\pi(x + y)^2 - \pi x^2 = \text{area of walk}$$

Consequently, by the statement in the problem,

$$\pi(x + y)^2 - \pi x^2 = \tfrac{11}{25}(\pi x^2)$$

This is one of the required equations, and we simplify it as follows:

$$25x^2 + 50xy + 25y^2 - 25x^2 = 11x^2 \qquad \blacktriangleleft \text{ multiplying each member}$$
by $25/\pi$ and squaring $x + y$

$$11x^2 - 50xy - 25y^2 = 0 \qquad \blacktriangleleft \text{ adding } -11x^2 \text{ to each} \qquad (14)$$
member, combining terms,
and multiplying by -1

Furthermore

$$2\pi(x + y) = \text{circumference of outer circle}$$
$$2\pi x = \text{circumference of inner circle}$$

Therefore,

$$2\pi(x + y) + 2\pi x = 88\pi \qquad \blacktriangleleft \text{ since the sum of the inner and}$$
outer circumferences is 88π feet

This may be simplified to

$$2x + y = 44 \qquad \blacktriangleleft \text{ dividing each member by } 2\pi \text{ and combining terms} \quad (15)$$

Hence we have the equations

$$11x^2 - 50xy - 25y^2 = 0 \qquad\qquad\qquad (14)$$
$$2x + y = 44 \qquad\qquad\qquad (15)$$

The simultaneous solution set of (14) and (15) can be obtained by substitution since (15) is readily solved for y in terms of x. This method, however, leads to tedious computation. Therefore, since the right member of (14) is 0, we shall solve (14) for x in terms of y by the quadratic formula and get

$$x = \frac{50y \pm \sqrt{2{,}500y^2 + 1{,}100y^2}}{22} = \frac{50y \pm 60y}{22}$$

$$= 5y \text{ and } -\tfrac{10}{22}y$$

We discard the negative root, since x and y stand for positive numbers, and complete the solution by replacing x by $5y$ in (15). Thus,

$$10y + y = 44$$
$$11y = 44$$
$$y = 4$$

Hence, $x = 5y = 20$ and the solution set of (14) and (15) is $\{(20, 4)\}$. Thus the radius of the garden is 20 feet, and the width of the walk is 4 feet.

Check

Since the radius of the outer circle is 24 feet, the area of the walk is $24^2\pi - 20^2\pi = 176\pi$, the area of the garden is $20^2\pi = 400\pi$, and $176\pi/400\pi = \frac{11}{25}$. Furthermore, the sum of the circumferences is $2\pi(24) + 2\pi(20) = 88\pi$.

Exercise 12.6   Stated Problems Solvable by Use of Simultaneous Equations

1   One winter Lloyd and Brad earned $77.50 a month from their paper routes. When summer came, Lloyd lost one-fifteenth of his customers, Brad lost one-tenth of his, and their total monthly earnings were then $71. How much did each boy earn per month in the winter?

2   Brian and Bruce had saved a total of $205. After Brian bought a bicycle for $80, and Bruce spent $60 for a camera, Brian had $5 less than Bruce. How much did each boy have before his purchase?

3   An engineer found working expenses averaged $12 a day on days spent with customers and $2.50 a day on days spent in the office. If the expenses totalled $183 in 20 days, how many days were spent in the office?

4   A certain company employs 57 people, including 22 college graduates, in its two branch offices. If two-fifths of those in the first office and three-eighths of those in the second office are college graduates, how many people are employed in each office?

5   The girls in a certain district collected $1,536 from their annual candy sale. If there were 80 girls in all, each Girl Scout collected $12, and each Campfire Girl collected $24, how many girls in each organization sold candy?

6   A beach apartment rented for $250 per month during the 3 summer months and for $90 per month during the remainder of the year. During 1 year, it was occupied for only 9 months, and the rentals amounted to $1,130. How many months was it occupied for each portion of the year?

7   Mr. Black could normally drive from his home to the airport in 36 minutes. One holiday, lack of parking space at the airport forced him to drive to a hotel a few miles from the airport and go the rest of the way on an airport bus. This did not affect the distance he traveled to the airport but, including 15 minutes he waited for the bus, increased his travel time by 24 minutes. If he averaged 50 miles per hour driving his own car and the bus averaged 20 miles per hour, how far did he ride on the bus?

8   A contestant in a certain television quiz program received $100 at the

start of the program. He received $75 for each question he answered correctly and was penalized $40 for each question he missed. If he attempted 12 questions and received $425, how many questions did he answer correctly and how many did he miss?

9   Mrs. Johnson estimated that it cost her 26 cents to drive to work each day if she took three passengers each of whom paid an identical fee for the ride. When two additional passengers joined the car pool, she reduced the fee 10 cents a person and found she earned 24 cents a day from her drive. What was the total cost of driving to work and the fee paid by each of the first three passengers?

10   Susan earned $1,800 by working part time for 9 months and full time for 2 months. At the same monthly wages, Anne earned $1,560 by working part time for 10 months and full time for 1 month. Find the monthly wage for each type of work.

11   An individual receives $309.70 annually on two investments, with the first earning 6 percent and the second earning 5 percent. The next year the rates were interchanged and $316.20 was earned. Find the amount of each investment.

12   A tour bus and a private car left a resort hotel at the same time, but traveled in opposite directions, for a drive around a scenic loop. When they met, the bus had traveled 32 miles and the car 48 miles. Find the average speed of each if the car returned to the hotel 1 hour and 20 minutes before the bus.

13   A salesperson made a trip to a nearby town, visited customers on the way, and returned home by a direct route. The average speed to the town was 8 miles per hour, and the average speed returning home was 60 miles per hour. Find the length of the trips to and from the town if the return trip took $6\frac{2}{3}$ hours less time than the trip to the town and the total time gone was $8\frac{1}{3}$ hours.

14   A person built a cement-block wall along one longer side and two shorter sides of a rectangular backyard. The wall for the longer side required a retaining wall and cost $6.50 a foot, while the wall for the shorter sides cost $3.50 a foot. If the total cost of the wall was $1,037.50 and the wall for the longer side cost $197.50 more than the rest, what were the dimensions of the yard?

15   A pilot flew 1,680 miles due north and returned to the starting point. The wind was blowing from the north at a constant velocity during the entire trip. Find the groundspeed of the plane and the velocity of the wind if the trip north required 7 hours and the return trip required 6 hours.

16   A landscape gardener planned a flower bed to contain some shrubs that cost $6.50 apiece and some flowers that cost $1.80. Plants for the bed as planned would cost $69.30. To reduce the cost by $14.10, it was decided to use $\frac{3}{2}$ as many flowers and $\frac{2}{3}$ as many shrubs. How many plants of each kind were first planned for the bed?

17 A shipper sent 800 pounds of merchandise from New York to Boston and 950 pounds from New York to Philadelphia. The freight rate was 2 cents per 100 pounds per mile. The bill was for a total of 320 miles, and the amount of the bill was $53.93. How far is it from New York to Boston and from New York to Philadelphia?

18 Mrs. Green went from her home to the airport in her car at an average rate of 40 miles per hour. She then boarded a plane that averaged 300 miles an hour to fly to another city, where she rode to a hotel in a limousine that traveled 30 miles per hour. The entire trip of 625 miles required 2 hours 45 minutes of travel time, and she spent twice as long in the limousine as in her own car. How long did she spend in each mode of travel?

19 One month an automobile agency received 45 cars including sedans, sports cars, and station wagons. There were twice as many sports cars as station wagons. The dealer expected to gross $137,000 from these cars by pricing the sedans at $3,200 apiece, the sports cars at $2,700 apiece, and the station wagons at $3,500 apiece. How many of each style car were received?

20 A family on vacation traveled an average of 240 miles a day at a cost of 12 cents per mile. Their meal costs averaged $24 per day, and their motel costs averaged $27 per night. The total cost of the vacation was $948, and the motel costs were $12 more than the mileage costs. How many miles did they travel, and how many nights did they spend in motels?

21 The product of two positive numbers is 45, and the sum of their squares is 106. What are they?

22 What are the dimensions of a rectangle if its area is 540 square feet and its diagonal is 39 feet long?

23 A small rectangular park that has an area of 15,000 square feet is surrounded by a concrete walk 4 feet wide. Find the dimensions of the park if the walk has an area of 2,064 feet.

24 The combined areas of two externally tangent circles is 25π square inches. Find the radius of each circle if the centers are 7 inches apart.

25 A rectangle of area 60 square feet is inscribed in a circle of area $169\pi/4$ square feet. Find the dimensions of the rectangle.

26 Mrs. Simmons had a square carpet and a rectangular carpet of which the length was $\frac{3}{2}$ the width. The areas of the two carpets combined were 375 square feet. The square carpet cost $10 a square yard, and the rectangular one $12 a square yard. If she paid $50 more for the square carpet than for the rectangular one, what were the dimensions of each?

27 A rectangle is constructed in a right triangle by selecting a point on the hypotenuse and drawing perpendiculars to the legs. Find the dimensions of the rectangle if the area is 36 square inches and the legs of the triangle are 16 and 12 inches.

28 A florist expected to earn \$1,950 from the sale of corsages at a football game. At half time she had sold 500 corsages. She then reduced the price by 50 cents apiece and sold the rest by the end of the game. If her earnings during the second half were \$375, how many corsages did she have, and what was her first price?

29 A block of stock was bought for \$7,800. After holding it for a year, the buyer received a cash dividend of \$1.50 per share and a stock dividend of 10 shares. The stock was then sold for \$3 per share more than it cost. Find the number of shares bought and the cost per share if the profit on the transaction was \$1,220.

30 Two circles are tangent to each other, and the smaller circle is inside the larger. The lengths of the radii differ by 3 inches, and the area between the circles is 66 square inches. Find the lengths of the radii. (Use $\frac{22}{7}$ for π.)

31 A piece of wire 76 inches in length is cut into two pieces. One piece is bent into a square and the other into a circle. If the sum of the areas of the two figures is 218 square inches, find the length of the side of the square and the radius of the circle.

32 A person made two investments, the first being \$500 more than the second. If, during the first year, the rate of the second investment was 2 percent more than that of the first and the income from each was \$120, find the amount of the first investment and the rate that it earned.

12.11 GRAPHICAL SOLUTION OF A SYSTEM OF INEQUALITIES

In this section we shall explain the use of the graphical method for finding the solution set of a linear inequality in two variables and of a system of linear inequalities in two variables. We shall consider only $x > 0$, $x < 0$, $y > ax + b$, and $y < ax + b$ since any linear inequality in two variables is equivalent to one of these types.

In most of our discussion we shall be concerned with finding the solution set of a statement such as $y \geq ax + b$. This is a combination of an inequality and an equation, but we shall refer to it as an inequality. We shall use the notation $P(x, y)$ to stand for the point whose coordinates are (x, y).

In order to find the solution set of $y \geq ax + b$, we begin by constructing the graph of $y = ax + b$. Now if $P(x, y)$ is on the graph of $y = ax + b$ and if $y' > y$, then $P(x, y')$ is above the graph. Furthermore, since $y' > y$, it follows that $y' > ax + b$. Therefore, the solution set of $y \geq ax + b$ is $\{(x, y) | P(x, y)$ is on or above the graph of $y = ax + b\}$. Furthermore, for similar reasons, the solution set of $y \leq ax + b$ is $\{(x, y) | P(x, y)$ is on or below the graph of $y = ax + b\}$.

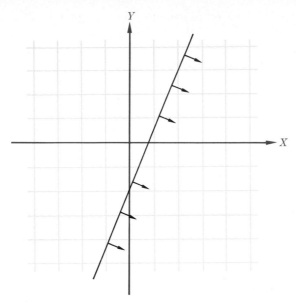

Figure 12.9

In keeping with the above discussion, the solution set of $y \leq 3x - 2$ is $\{(x, y) | P(x, y) \text{ is on or below the graph of } y = 3x - 2\}$. For example, $(1, -1)$ is below the lines since $-1 < 3(1) - 2 = 1$. Hence P is in the region indicated by the arrows in Fig. 12.9.

In the remainder of this section we shall be concerned with determining the region in which $P(x, y)$ lies if (x, y) belongs to the simultaneous solution set of two or more inequalities. The method consists of the following steps: first, draw the graph of each related equation; second, for each line, indicate by arrows the region in which $P(x, y)$ lies if (x, y) is a solution pair of the related inequality; third, determine the intersection of the regions obtained in the second step. We shall refer to this region as the *region determined by the inequalities*.

EXAMPLE Find the region determined by $x \geq 0$, $y \geq 0$, $y \geq 3x - 3$, and $y \leq 0.5x + 2$.

Solution We begin by sketching the graphs of $x = 0$, $y = 0$, $y = 3x - 3$, and $y = 0.5x + 2$. The first one is the Y axis, the second is the X axis, the third is through $(0, -3)$ and $(1, 0)$, and the last is through $(0, 2)$ and $(4, 4)$. (The graphs are shown in Fig. 12.10.) We now indicate the half plane determined by each inequality by use of arrows. We thus find that the region determined by the given set of inequalities is the boundary and interior of the quadrilateral with vertices at the origin, $Q(1, 0)$, $I(2, 3)$, and $R(0, 2)$.

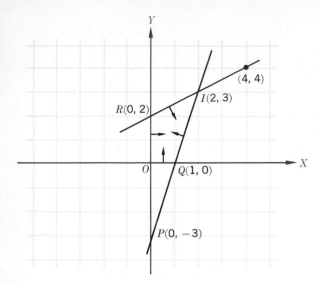

Figure 12.10

If all points of the line segment PQ are in a set S of the xy plane whenever P and Q are in it, we say that the set S is a *convex set of points* or a *convex region*.

Convex region

According to this definition, a circle and the quadrilateral $OQIR$ of Fig. 12.10 are convex sets, as is a half plane.

Polygonal set

The intersection of two or more closed half planes is called a *polygonal set of points*.

Finite polygonal region

If a polygonal set of points has a finite area, then it is called a *finite polygonal region* and its boundary is called a *convex polygon*.

Convex polygon

The quadrilateral $OQIR$ in Fig. 12.10 is a convex polygon.

Exercise 12.7   Convex Polygons

Indicate the region determined by the inequality in each of problems 1 to 4 by use of a figure.

1 $y \geq -3$ 2 $x \leq 2$
3 $x - y + 1 \leq 0$ 4 $2x + y - 4 \geq 0$

By use of a figure, show the convex region determined by the given set of inequalities in each of problems 5 to 12.

5 $x \geq 0,\ y \geq 0,\ x + y - 1 \leq 0$
6 $x \geq 0,\ y \leq 0,\ x - y - 2 \leq 0$
7 $y \leq 0,\ 3x + 2y + 4 \geq 0,\ 2x - y - 2 \leq 0$

276

8    $x - y + 2 \geq 0, \; x + 2y - 2 \geq 0, \; 5x + y - 8 \leq 0$
9    $x + y - 1 \leq 0, \; x - y - 1 \geq 0, \; x + y + 1 \geq 0, \; -x + y - 1 \leq 0$
10    $x \geq 0, \; y \geq 0, \; x \leq 1, \; y \leq 2$
11    $-x + y - 1 \leq 0, \; x - 2y + 2 \geq 0, \; x \leq 2, \; x + y + 1 \geq 0$
12    $x - 2y + 2 \geq 0, \; x + 3y + 2 \geq 0, \; 2x - y - 5 \leq 0, \; x + 2y - 2 \geq 0$

Show that the lines determined by the equations in each of problems 13 to 16 do not bound a convex region by selecting two points P and Q in the region and showing that not all points on the segment PQ are in the region.

13    $x = 0, \; 2y - x + 8 = 0, \; x + y = 2, \; 3x + y - 6 = 0$
14    $x = 0, \; y = 2x, \; y = 2, \; x = 5, \; y = 4$
15    $2y = x + 2, \; 4x + y = 16, \; x = 2, \; y = 0$
16    $2x - y + 4 = 0, \; x + y + 2 = 0, \; x = -4, \; x = 1, \; y = -3$

Show that the inequalities in each of problems 17 to 20 do not determine a finite polygonal region.

17    $x \leq 0, \; y \geq 0, \; x + y - 1 \leq 0$
18    $x \geq 0, \; y \leq -1, \; 3x + y - 5 \geq 0$
19    $x \geq 1, \; y \geq -1, \; y - x - 1 \geq 0$
20    $x \geq -2, \; y \geq -3, \; y \leq 1, \; x - y - 3 \geq 0$

Show that the quadrilateral with the points given in each of problems 21 to 24 as vertices is not a convex set.

21    $(-1, 2), (2, 3), (5, 3), (1, 5)$ 22 $(1, 2), (4, 7), (-1, 6), (1, 5)$
23    $(-1, 1), (3, 1), (1, 2), (0, 4)$ 24 $(1, 3), (3, 0), (2, 2), (6, 4)$

Find the vertices of the polygonal region determined by the inequalities in each of problems 25 to 28.

25    $4x - y - 7 \leq 0, \; x + 5y \leq 7, \; 5x + 4y + 7 \geq 0$
26    $4x - y \leq 16, \; 3x + 2y \leq 12, \; 5x - 3y - 1 \geq 0, \; x + 2y + 5 \geq 0$
27    $x + y - 4 \leq 0, \; x - y - 6 \leq 0, \; x + y + 2 \geq 0, \; 3x + y \geq 0, \; y \leq 0$
28    $2x - 3y - 10 \leq 0, \; 2x + y - 2 \geq 0, \; x - 3y + 6 \geq 0, \; 3x + 2y - 15 \leq 0$

12.12   PARAMETRIC EQUATIONS OF A LINE

We are interested in expressing the variables x and y that occur in the equation of a line in terms of a third variable since we can then find the largest and smallest values of a linear function in an interval.

THEOREM 1   If l is a line segment through two distinct points $P_1(x_1, y_1)$ and $P_2(x_2, y_2)$, if t is a variable whose domain is all real numbers, and if $P(x, y)$ is an unspecified point on l, then x and y can be expressed as

Parametric form
$$x = x_1 + t(x_2 - x_1) \tag{1}$$

and
$$y = y_1 + t(y_2 - y_1) \tag{2}$$

Proof ▶   If l is a directed line segment with P_1P_2 positive, as in Fig. 12.11, and if $P(x, y)$ is any point on l, then a number t is defined by $P_1P = tP_1P_2$. A unique value of t is determined for each position of P on P_1P_2, and conversely; hence, there is a one-to-one correspondence between the values of t and the points on l. If $P = P_1$, then $t = 0$. If $t \neq 0$ and l is not vertical, then from Fig. 12.11 we can see that

$$\frac{TP}{SP_2} = \frac{P_1P}{P_1P_2} = \frac{tP_1P_2}{P_1P_2} = t$$

Therefore,

$$\frac{x - x_1}{x_2 - x_1} = t \tag{3}$$

Note that if P is between P_1 and P_2, then $x_1 < x < x_2$ and $0 < t < 1$. Consequently, solving for x, we find $x = x_1 + t(x_2 - x_1)$, as stated in (1). Equation (2) may be proved similarly.

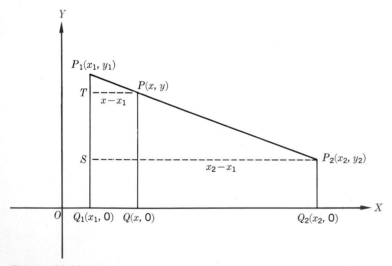

Figure 12.11

EXAMPLE 1   Express the equation of the segment between $P_1(1, 4)$ and $P_2(6, -2)$ in parametric form.

Solution   If we use (1) and (2), we find that $x = 1 + t(6 - 1) = 1 + 5t$ and $y = 4 + t(-2 - 4) = 4 - 6t$.   Therefore, $x = 1 + 5t$ and $y = 4 - 6t$ are parametric equations of the segment P_1P_2 if $0 \le t \le 1$.

12.13   EXTREMA OF A LINEAR FUNCTION

Maximum
Minimum
Extrema

If there is a largest value M of a function f and a smallest value m of it, then M and m are called the *maximum* and *minimum*, respectively, of f and are often referred to as the *extrema*.

THEOREM 2   If $f(t) = ct + d$, $c \ne 0$ and d constants, and if $\alpha \le t \le \beta$, then extrema M and m of f occur for $t = \alpha$ and $t = \beta$; furthermore,

Values of extrema

$$\text{if } c > 0, \text{ then } m - f(\alpha) \text{ and } M - f(\beta) \tag{1}$$

and $$\text{if } c < 0, \text{ then } m = f(\beta) \text{ and } M = f(\alpha) \tag{5}$$

Proof ▶   If $c > 0$ and $t_2 > t_1$, then $f(t_2) > f(t_1)$; hence $f(t)$ increases with t, and the graph, as shown in Fig. 12.12 rises as it moves to the right.   Therefore, the least value m of f is $f(\alpha)$ and the greatest value M is $f(\beta)$.   Statement (5) can be proved similarly.

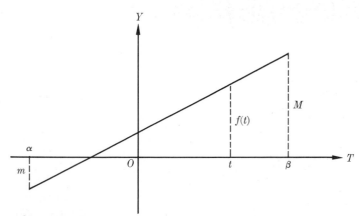

Figure 12.12

EXAMPLE 2 Find the maximum M and minimum m values of $f(t) = 2t - 3$ for the domain $\{t \mid 1 \leq t \leq 4\}$.

Solution Since the coefficient of t is positive, we know by use of (4) that the minimum occurs for $t = 1$ and the maximum for $t = 4$. They are

$$m = f(1) = -1 \quad \text{and} \quad M = f(4) = 5$$

THEOREM 3 If $P_1(x_1, y_1)$ and $P_2(x_2, y_2)$ are two points in the xy plane and if $P(x, y)$ is the set of points in the segment that connects P_1 and P_2, then the extrema of $f(x, y) = ax + by + c$ are $f(x_1, y_1)$ and $f(x_2, y_2)$.

Proof ▶ If the values of x and y as given by (1) and (2) are used, we have

$$\begin{aligned}
f(x, y) &= f[x_1 + t(x_2 - x_1), y_1 + t(y_2 - y_1)] \\
&= a[x_1 + t(x_2 - x_1)] + b[y_1 + t(y_2 - y_1)] + c \\
&= ax_1 + by_1 + c + [a(x_2 - x_1) + b(y_2 - y_1)]t \\
&= g(t)
\end{aligned} \tag{6}$$

Thus, $g(t)$ is a constant or a first-degree equation in t and is of the form

$$g(t) = Ct + D$$

where C and D are constants and $0 \leq t \leq 1$ since P is in the segment P_1P_2.

We notice that if $C = 0$, then $g(t) = f(x, y) = D$ and both the maximum and the minimum are D; furthermore, if $C \neq 0$, then by Theorem 2, g assumes its extrema for $t = 0$ and $t = 1$. We now point out that for $C > 0$, $g(t)$ increases as t does; hence, the minimum is $g(0) = f(x_1, y_1)$ and the maximum is $g(1) = f(x_2, y_2)$. Furthermore, if $C < 0$, then the maximum is $g(0) = f(x_1, y_1)$ and the minimum is $g(1) = f(x_2, y_2)$.

EXAMPLE 3 If P_1 and P_2 have the coordinates $(1, 7)$ and $(5, -1)$ and $f(x, y) = 2x - y + 3$, find the extrema.

Solution By use of Theorem 3, we know that the extrema are $f(1, 7) = (2 \times 1) - 7 + 3 = -2$ and $f(5, -1) = (2 \times 5) - (-1) + 3 = 14$. Hence, the minimum is -2 and the maximum is 14.

12.14 LINEAR PROGRAMMING

If a problem involves two variables x and y, if the conditions of the problem restrict the domain of (x, y) such that (x, y) must satisfy a specified system of linear inequalities, and if a given linear combination $f(x, y)$ of *x* and *y* is to be a maximum or a minimum subject to the restricting inequalities, then the determination of (x, y) is called *linear programming*.

We shall now state and prove a theorem about linear programming and then give a strictly mathematical and a business application.

THEOREM 4   If *S* is a convex polygon, if $f(x, y) = ax + by + c$, if the domain of *f* is $\{(x, y) \mid (x, y)$ are coordinates of a point of $S\}$, and if *B* is the polygonal boundary of *S*, then *f* has a maximum *M* and a minimum *m* in the domain *S* and it is assumed at some vertex of *B*.

Proof ▶   If $P(x, y)$ is restricted to a side P_1P_2 of *B*, then by Theorem 3 of Sec. 12.13 the maximum M_b and minimum m_b values of *f* are attained at the end points P_1 and P_2 and these points are vertices of the polygonal boundary, as indicated in Fig. 12.13. If $P_3(x_3, y_3)$ is a point of *S* but is not on *B*, we shall continue the proof of the theorem by letting $P_1(x_1, y_1)$ be a vertex of the boundary *B* such that $f(x_1, y_1) = M_b$ and drawing a segment between P_1 and P_3 and extending it until it meets a side, say at $P_4(x_4, y_4)$. Now, by Theorem 3 of Sec. 12.13, if $P(x, y)$ is any point on P_1P_4, then *f* attains its maximum at P_1 or P_4, but it must occur at P_1 since the maximum M_b of *f* on the boundary *B* occurs at P_1. Consequently, the minimum of *f* for $P(x, y)$ on P_1P_4 occurs at P_4 and is $f(x_4, y_4)$. Therefore,

$$m_b \le f(x_4, y_4) \le f(x_3, y_3) \le M_b \tag{1}$$

We see from (1) that the function value $f(x_3, y_3)$ for any interior point P_3 is greater than or equal to the minimum m_b of *f* on the boundary *B* and less than or equal to the maximum M_b of *f* on *B*. Hence, M_b and m_b are the maximum and minimum, respectively, of *f* for all points of *S* and each is attained at a vertex of the boundary since M_b is at the vertex P_1 and m_b is at P_2 or P_5 if P_4 is on P_2P_5. Similarly, m_b is at P_6 or P_5 if P_3 is so

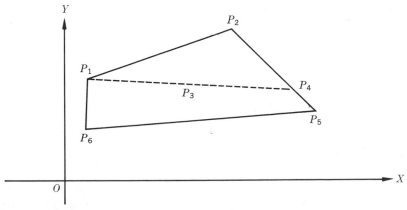

Figure 12.13

chosen that P_4 is on P_6P_5. Thus, m_b is always at a vertex if S is a quadrilateral. A similar argument can be used if S is any convex polygon.

The inequalities that determine the region S of the theorem are called the *restraints,* the region S is called the set of *feasible solutions,* and the linear function f is called the *objective function.*

EXAMPLE 1

Find the maximum and minimum values of $f(x, y) = 2x - 3y + 4$ in the region S with vertices at $(0, 5)$, $(6, 4)$, $(7, 1)$, and $(-3, -2)$.

Solution

Since the region S, shown in Fig. 12.14 with the four given points as vertices, is convex and f is a linear function in two variables, we know by use of Theorem 4 that the extrema of f are attained at vertices of the boundary. Therefore, we shall evaluate f at each vertex. The values are $f(0, 5) = (2 \times 0) - (3 \times 5) + 4 = -11$, $f(6, 4) = (2 \times 6) - (3 \times 4) + 4 = 4$, $f(7, 1) = (2 \times 7) - (3 \times 1) + 4 = 15$, and $f(-3, -2) = [2 \times (-3)] - [3 \times (-2)] + 4 = 4$. Consequently, the maximum value of f in S is $f(7, 1) = 15$ and the minimum value is $f(0, 5) = -11$.

EXAMPLE 2

A maker of animal shoes specializes in horseshoes, mule shoes, and oxen shoes and can produce 200 sets of shoes per unit of time. He has standing orders for 60 sets of horseshoes and 20 sets of oxen shoes and can sell at most 150 sets of horseshoes and 50 sets of mule shoes. How many sets of each type should he produce to make a maximum profit provided his profit on a set of shoes is \$0.40 for horseshoes, \$0.50 for mule shoes, and \$0.30 for oxen shoes?

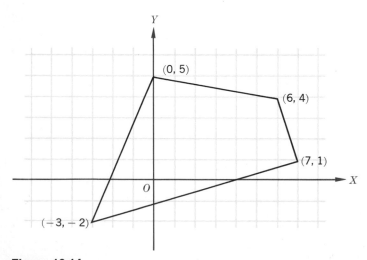

Figure 12.14

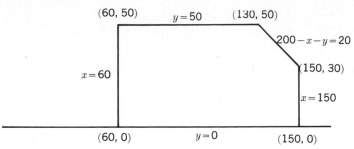

Figure 12.15

Solution

If we represent the number of sets of horseshoes produced by x and of mule shoes by y, then $200 - x - y$ is the number of sets of oxen shoes. Consequently, his profit in terms of dollars is

$$f(x, y) = 0.4x + 0.5y + 0.3(200 - x - y)$$
$$= 0.1x + 0.2y + 60$$

This objective function is subject to the restraints imposed by the problem. They are:

$$x \geq 60 \qquad \blacktriangleleft \text{since he has a standing order for 60 sets of horseshoes}$$
$$200 - x - y \geq 20 \qquad \blacktriangleleft \text{since he has a standing order for 20 sets of oxen shoes}$$
$$x \leq 150 \qquad \blacktriangleleft \text{since he cannot sell more than 150 sets of horseshoes}$$
$$y \leq 50 \qquad \blacktriangleleft \text{since he cannot sell more than 50 sets of mule shoes}$$
$$x \geq 0 \quad y \geq 0 \quad 200 - x - y \geq 0 \qquad \blacktriangleleft \text{since all three types are for sale}$$

Figure 12.15 shows these restraints graphically, along with the vertices of the polygonal boundary of the set of feasible solutions as obtained by solving the equations of the pairs of bounding lines that meet in a vertex. We now calculate $f(x, y) = 0.1x + 0.2y + 60$ at each vertex and find that $f(60, 0) = 66$, $f(150, 0) = 75$, $f(150, 30) = 81$, $f(130, 50) = 83$, and $f(60, 50) = 76$. Consequently, the shoe maker has the greatest profit if he makes 130 sets of horseshoes, 50 sets of mule shoes, and 20 sets of oxen shoes.

Exercise 12.8   Extrema, Linear Programming

Find a parametric representation for the line segment $P_1 P_2$ in each of problems 1 to 4.

1 $P_1(2, 3), P_2(5, 8)$
2 $P_1(3, 5), P_2(-1, 2)$

283

3 $P_1(-4, 0)$, $P_2(2, -5)$
4 $P_1(7, 3)$, $P_2(-2, -3)$

Find the maximum and minimum values of $f(t)$ for the given range of values of t in each of problems 5 to 8.

5 $f(t) = 2t + 5$, $0 \leq t \leq 4$
6 $f(t) = 3t - 2$, $-2 \leq t \leq 3$
7 $f(t) = -3t + 2$, $1 \leq t \leq 6$
8 $f(t) = -t - 4$, $-5 \leq t \leq 2$

In each of problems 9 to 12, express $f(x, y)$ as $g(t)$ for the segment between P_1 and P_2 and then find its extrema.

9 $f(x, y) = 3x + 2y - 5$, $P_1(2, 1)$, $P_2(8, 6)$
10 $f(x, y) = 2x - y + 3$, $P_1(-3, 0)$, $P_2(2, 3)$
11 $f(x, y) = -x + 4y + 2$, $P_1(4, -3)$, $P_2(-1, 4)$
12 $f(x, y) = -4x - 3y + 1$, $P_1(0, 5)$, $P_2(4, -2)$

Find the extrema of $f(x, y)$ for $P(x, y)$ on P_1P_2 in each of problems 13 to 16.

13 $f(x, y) = 5x + 2y - 3$, $P_1(2, 3)$, $P_2(5, -1)$
14 $f(x, y) = -2x + 3y - 1$, $P_1(4, -1)$, $P_2(-1, 3)$
15 $f(x, y) = -4x - 2y + 2$, $P_1(3, 2)$, $P_2(-2, -4)$
16 $f(x, y) = x - 5y + 4$, $P_1(-5, 0)$, $P_2(1, 7)$

For each system of restraints or linear inequalities in problems 17 to 28, find the set S of feasible solutions, select a point of S, and show that each inequality is satisfied by its coordinates. Select a point not in S, and show that at least one of the inequalities is not satisfied. Find the vertices of the convex polygon. Finally find the maximum and minimum values of the objective function.

17 $-2x + 3y - 6 \leq 0$, $y \geq 0$, $x \leq 0$, $f(x, y) = 2x + y - 1$
18 $x + y + 3 \geq 0$, $x \leq 0$, $y \leq 0$, $f(x, y) = 3x - 2y + 4$
19 $2x - y + 2 \geq 0$, $x + y - 2 \leq 0$, $y \geq 0$, $f(x, y) = -x + 3y - 5$
20 $3x - 4y + 12 \geq 0$, $3x + 2y - 6 \leq 0$, $y \geq -2$, $f(x, y) = -4x + 5y + 3$
21 $3x - 2y + 6 \geq 0$, $x + y + 2 \geq 0$, $x - y - 3 \leq 0$, $x + y - 3 \leq 0$, $f(x, y) = x + y - 8$
22 $-x + y - 2 \leq 0$, $3x + 2y + 6 \geq 0$, $3x - 4y - 12 \leq 0$, $x + 2y - 4 \leq 0$, $f(x, y) = -3x - 2y + 7$
23 $x - y + 1 \geq 0$, $x + y + 1 \geq 0$, $-x + y + 1 \geq 0$, $-x - y + 1 \geq 0$, $f(x, y) = -4x + 3y + 8$

24 $2x - y + 2 \geq 0,\ 4x + y + 4 \geq 0,\ -4x + 3y + 12 \geq 0,\ -2x - 3y + 6 \geq 0,$
$f(x, y) = 5x - 4y - 2$

25 $3x + y + 4 \geq 0,\ x + 4y + 5 \geq 0,\ 5x + 2y - 11 \leq 0,\ -x + 3y - 8 \leq 0,$
$f(x, y) = 4x + y + 6$

26 $-2x + 3y - 8 \leq 0,\ x + 5y + 4 \geq 0,\ -x + y + 2 \geq 0,\ x + 4y - 7 \leq 0,$
$f(x, y) = -3x - 2y + 9$

27 $-x + 3y - 9 \leq 0,\ -2x + y - 8 \leq 0,\ -x - y - 4 \leq 0,\ 2x - y - 4 \leq 0,\ 3x + 2y$
$- 6 \leq 0,\ f(x, y) = -5x + 6y + 2$

28 $3x - 4y + 12 \geq 0,\ x + y + 4 \geq 0,\ -x + 5y + 8 \geq 0,\ -4x - y + 11 \geq 0,$
$y \leq 3,\ f(x, y) = 6x + 5y - 4$

29 During the time that two machines are not otherwise needed, they are used to make fire pokers and tongs. It is anticipated that machines A and B will be free for 10 and 7.5 hours, respectively, during a certain period. Pokers require 1 hour of time from machine A and 1 hour from machine B and sell for a profit of $3, whereas tongs require 2 hours of time from machine A and 1.5 hours from machine B and sell for a profit of $5.50. How many of each should be made so as to make the profit a maximum?

30 A truck farmer has 20 acres available for planting with pepper, rhubarb, and tomatoes. He has reason to think that he can make a profit of $300/acre on peppers, $225/acre on rhubarb, and $250/acre on tomatoes. He cannot take care of more than 10 acres of pepper, more than 12 acres of rhubarb, or more than 7 acres of tomatoes. In order to make a maximum profit, how many acres of each should he grow?

31 An animal food is to be a mixture of product A and B. The content and cost of 1 pound of each is:

Product	Protein, g	Fat, g	Carbohydrates, g	Cost
A	180	2	240	$0.50
B	36	8	200	$0.24

How much of each product should be used to minimize the cost if each bag must contain at least 612 grams of protein and 22 grams of fat and at most 1,880 grams of carbohydrates?

32 A manufacturer produces three types of products in 500 hours. Product A requires 4 hours per unit, product B requires 6 hours per unit, and product C requires 2 hours per unit. The profits per unit of output are $50 on A, $80 on B, and $30 on C. He has an order for 25 A units, 30 B units, and 70 C units. If he can sell all he can produce, what combination should he produce to maximize profits, assuming that he cannot produce more than 120 A units and 40 B units?

Exercise 12.9   Review

Construct the graph of the function defined by the equation in each of problems 1 to 6.

1 $3x + 4y = 12$ **2** $2x - 3y = 8$
3 $4x + y^2 = 8$ **4** $9x^2 + 4y^2 = 72$
5 $xy = -6$ **6** $4x^2 - 9y^2 = 36$

Obtain the solution set of each pair of equations in problems 7 to 12 by graphing and by substitution.

7 $2x - y = 3$
     $3x + 2y = 8$
8 $x + y = 5$
     $2x^2 - y^2 = 14$
9 $3x + 2y = 1$
     $xy = -1$
10 $\dfrac{1}{x} + \dfrac{1}{y} = 5$

     $\dfrac{3}{x} + \dfrac{2}{y} = 12$
11 $x^2 + y^2 = 5$
     $2x^2 - 5y^2 = 3$
12 $bx + ay = ab$
     $x^2 - ay = -ab$

Solve the systems of equations in problems 13 to 16 by any method.

13 $2x^2 - xy + y^2 = 4$
     $x^2 + 2xy - y^2 = 1$
14 $x^2 + y^2 + 2xy = 1$
     $2x^2 + 2y^2 + 3xy = 8$
15 $2xy + x + y = -5$
     $xy - 2x - 2y = 0$
16 $3x^2 + 3y^2 + x + y = 6$
     $x^2 - xy + y^2 + 2x + 2y = 3$

Show the convex region determined by the set of inequalities in each of problems 17 and 18; find the vertices and extrema of $f(x, y) = 3x - 2y + 7$ in the region.

17 $x - 4y + 7 \geq 0$, $3x + 7y + 2 \geq 0$, $4x + 3y - 10 \leq 0$
18 $5x + 2y + 7 \geq 0$, $x - 7y - 6 \leq 0$, $2x + y - 12 \leq 0$, $x + 4y - 13 \leq 0$

19 A farmer has 1,800 acres available for planting wheat, cotton, and maize. He thinks that he can clear $110 per acre on wheat, $90 on cotton, and $80 on maize. He cannot take care of more than 1,100 acres of wheat, more than 400 acres of cotton, or more than 700 acres of maize. How much of each should he plant in order to make the largest possible profit?

Matrices
and
Determinants

The concepts of a matrix and of a determinant are important ones in mathematics. Determinants were invented or devised independently by Kiowa, a Japanese, in 1683 and Leibnitz, a German, in 1693 and were rediscovered in 1750 by Cramer, a Swiss, who used them for solving systems of linear equations. Matrices were invented by Cayley, an Englishman, during the nineteenth century and are used in solving systems of linear equations and in connection with computers.

13.1 MATRICES AND THEIR BASIC PROPERTIES

A rectangular array of numbers of the type

$$\begin{bmatrix} a_{11} & a_{12} & \cdots & a_{1g} & \cdots & a_{1n} \\ a_{21} & a_{22} & \cdots & a_{2g} & \cdots & a_{2n} \\ \cdots & \cdots & \cdots & \cdots & \cdots & \cdots \\ a_{i1} & a_{i2} & \cdots & a_{ij} & \cdots & a_{in} \\ \cdots & \cdots & \cdots & \cdots & \cdots & \cdots \\ a_{m1} & a_{m2} & \cdots & a_{mj} & \cdots & a_{mn} \end{bmatrix}$$

$m \times n$ matrix

Element

Equality of matrices

Sum of two matrices

The product $A_{m \times p}\ B_{p \times n}$

is called an $m \times n$ matrix. If $m = n$, it is called a *square matrix*. As seen from the array, an $m \times n$ matrix consists of m rows (horizontal) and n columns (vertical). Each number in the matrix is called an *element* of the matrix. In the above array, a_{ij} is the element in the ith row and jth column. Sometimes $A = (a_{ij})$ is used to designate the matrix. The number i belongs to $\{1, 2, 3, \ldots, m\}$, and j is an element of $\{1, 2, 3, \ldots, n\}$. The symbol $A_{m \times n}$ is often used to indicate that A is an $m \times n$ matrix.

Two m by n matrices are equal if and only if each element of one is equal to the corresponding element of the other. For example,

$$\begin{bmatrix} 3 & 2 & 1 \\ 2 & \sqrt{9} & 7 \end{bmatrix} = \begin{bmatrix} 3 & \sqrt{4} & 1 \\ \sqrt[3]{8} & 3 & 7 \end{bmatrix} \quad \text{and} \quad \begin{bmatrix} 5 & -3 \\ 4 & -2 \end{bmatrix} = \begin{bmatrix} 2x+1 & y \\ z+2 & w-1 \end{bmatrix}$$

if and only if $5 = 2x + 1, -3 = y, 4 = z + 2$, and $-2 = w - 1$; hence, if and only if $x = 2, y = -3, z = 2$, and $w = -1$.

The sum of two $m \times n$ matrices A and B is the $m \times n$ matrix obtained by adding corresponding elements of A and B. The sum is defined if and only if both matrices are $m \times n$. Consequently the sum of

$$A = \begin{bmatrix} 1 & 3 & 5 \\ 2 & 4 & 6 \end{bmatrix} \quad \text{and} \quad B = \begin{bmatrix} -2 & 7 & 8 \\ 9 & -5 & -6 \end{bmatrix}$$

is $$A + B = \begin{bmatrix} 1 + (-2) & 3 + 7 & 5 + 8 \\ 2 + 9 & 4 + (-5) & 6 + (-6) \end{bmatrix} = \begin{bmatrix} -1 & 10 & 13 \\ 11 & -1 & 0 \end{bmatrix}$$

The product AB of two matrices is defined if and only if the number of columns in A is equal to the numbers of rows in B. The product $(AB)_{m \times n}$ of the matrices $A_{m \times p}$ and $B_{p \times n}$ is the matrix with the element $c_{ij} = \sum_{k=1}^{p} a_{ik}\, b_{kj}$ in the ith row and jth column. The element c_{ij} is found by adding the product of the first element a_{i1} in the ith row of A and the first element b_{1j} in the jth column of B, the product of the second element a_{i2} in the ith row of A and the second element b_{2j} in the jth column of B, . . . , and the product of the pth element a_{ip} in the ith row of A and the pth element a_{pj} in the jth column of B. If AB and BA are both defined, they may or may not be equal.

EXAMPLE 1

Find the product $(AB)_{2\times2}$ if

$$A_{2\times3} = \begin{bmatrix} a_{11} & a_{12} & a_{13} \\ a_{21} & a_{22} & a_{23} \end{bmatrix} \quad \text{and} \quad B_{3\times2} = \begin{bmatrix} b_{11} & b_{12} \\ b_{21} & b_{22} \\ b_{31} & b_{32} \end{bmatrix}$$

Solution

If we follow the procedure outlined above, we get

$$(AB)_{2\times2} = \begin{bmatrix} a_{11}\,b_{11} + a_{12}\,b_{21} + a_{13}\,b_{31} & a_{11}\,b_{12} + a_{12}\,b_{22} + a_{13}\,b_{32} \\ a_{21}\,b_{11} + a_{22}\,b_{21} + a_{23}\,b_{31} & a_{21}\,b_{12} + a_{22}\,b_{22} + a_{23}\,b_{32} \end{bmatrix}$$

EXAMPLE 2

Find the product AB and BA if $A = \begin{bmatrix} 3 & -2 \\ 1 & 4 \end{bmatrix}$ and $B = \begin{bmatrix} 5 & 1 \\ -6 & 3 \end{bmatrix}$.

Solution

$$AB = \begin{bmatrix} (3)(5) + (-2)(-6) & (3)(1) + (-2)(3) \\ (1)(5) + (4)(-6) & (1)(1) + (4)(3) \end{bmatrix}$$

$$= \begin{bmatrix} 27 & -3 \\ -19 & 13 \end{bmatrix}$$

$$BA = \begin{bmatrix} 16 & -6 \\ -15 & 24 \end{bmatrix}$$

EXAMPLE 3

If $A = \begin{bmatrix} a_1 & a_2 & a_3 \\ b_1 & b_2 & b_3 \end{bmatrix}$ and $B = \begin{bmatrix} x \\ y \\ z \end{bmatrix}$, find AB.

Solution

By the usual procedure, we have

$$AB = \begin{bmatrix} a_1x + a_2y + a_3z \\ b_1x + b_2y + b_3z \end{bmatrix}$$

The product kA

The product of a scalar k and the matrix A is denoted by kA and is the matrix obtained by multiplying each element of A by the constant k. Symbolically $kA = k(a_{ij}) = (ka_{ij})$. For example, if

$$A = \begin{bmatrix} 1 & 3 & 6 \\ 0 & -2 & 5 \end{bmatrix} \quad \text{then} \quad 4A = \begin{bmatrix} 4 & 12 & 24 \\ 0 & -8 & 20 \end{bmatrix}$$

Identity matrix

If A is an $n \times n$ matrix and if each entry in the main diagonal (the diagonal from upper left to lower right) is one and all other elements are zero, the matrix is called an identity *matrix* since it acts as an identity for matrix multiplication. It is designated by I. Thus

$$I_2 = \begin{bmatrix} 1 & 0 \\ 0 & 1 \end{bmatrix} \quad \text{and} \quad I_3 = \begin{bmatrix} 1 & 0 & 0 \\ 0 & 1 & 0 \\ 0 & 0 & 1 \end{bmatrix}$$

are identity matrices.

Zero matrix

If each element of a matrix is zero, the matrix is called a zero matrix and is designated by O.

Transpose of a matrix

The matrix obtained by interchanging the rows and columns of a matrix A is called the transpose of A and is designated by A^T. Consequently, the transpose of

$$A = \begin{bmatrix} 3 & 1 & 2 \\ 4 & 0 & -5 \end{bmatrix} \quad \text{is} \quad A^T = \begin{bmatrix} 3 & 4 \\ 1 & 0 \\ 2 & -5 \end{bmatrix}$$

Exercise 13.1 Properties of Matrices

Find the values of a, b, c, and d in problems 1 to 4.

1 $[a \quad b \quad c \quad d] = [2 \quad 3 \quad -1 \quad 0]$

2 $\begin{bmatrix} a & b \\ c & d \end{bmatrix} = \begin{bmatrix} 1 & 5 \\ -2 & 3 \end{bmatrix}$

3 $\begin{bmatrix} a+2 & 3c-1 \\ 2b+3 & d-2 \end{bmatrix} = \begin{bmatrix} 5 & 8 \\ 7 & 0 \end{bmatrix}$

4 $\begin{bmatrix} a+2b & 2a-b & c+2 \\ a & 3d-4 & b+1 \end{bmatrix} = \begin{bmatrix} 5 & 0 & 6 \\ 1 & 2 & 3 \end{bmatrix}$

Use $A = \begin{bmatrix} 2 & 3 & 1 \\ 0 & -4 & 5 \end{bmatrix}$ and $B = \begin{bmatrix} -5 & 2 & 4 \\ 3 & 0 & -1 \end{bmatrix}$ in evaluating the matrices in problems 5 to 16.

5 $-A$	**6** $2A$	**7** $0B$	**8** $3B$
9 $A+B$	**10** $A-B$	**11** $2A-3B$	**12** $3A+2B$
13 A^T	**14** B^T	**15** $A+B^T$	**16** A^T+B

If
$$A_i = \begin{bmatrix} a_i & b_i & c_i \\ d_i & e_i & f_i \end{bmatrix} \qquad \text{where } i = 1, 2, 3$$

are any three 2×3 matrices and h and k are real numbers, prove the statements in problems 17 to 24.

17 $A_1 + A_2 = A_2 + A_1$
18 $A_1 + (A_2 + A_3) = (A_1 + A_2) + A_3$
19 If $A_1 + A_3 = A_2 + A_3$, then $A_1 = A_2$
20 If $A_1 - A_2 = A_1 - A_3$, then $A_3 = A_2$
21 $h(A_1 + A_2) = hA_1 + hA_2$
22 $(h + k)A_1 = hA_1 + kA_1$
23 $A_1 + (-A_1) = 0$
24 $A_1 + 0 = A_1$

Find the matrix X in each of problems 25 to 28.

25 $X - \begin{bmatrix} 2 & 3 \\ -1 & 4 \end{bmatrix} = \begin{bmatrix} 5 & 8 \\ 2 & 1 \end{bmatrix}$

26 $X + \begin{bmatrix} 3 & 2 \\ 0 & 1 \end{bmatrix} = 2\begin{bmatrix} -1 & 3 \\ 2 & 4 \end{bmatrix}$

27 $2X + \begin{bmatrix} 2 & 4 \\ 0 & 6 \end{bmatrix} = 3\begin{bmatrix} -2 & 0 \\ 2 & 4 \end{bmatrix}$

28 $3X - 2\begin{bmatrix} 3 & -1 \\ 2 & 4 \end{bmatrix} = \begin{bmatrix} -3 & -1 \\ 2 & 1 \end{bmatrix}$

Find the products called for in problems 29 to 36 where

$$A = \begin{bmatrix} 2 & 3 & -2 \\ 1 & 0 & 4 \\ -1 & 2 & -3 \end{bmatrix} \quad \text{and} \quad B = \begin{bmatrix} 0 & -2 & 1 \\ 3 & -5 & 2 \\ 1 & 2 & -3 \end{bmatrix}$$

29 $A(B^T)$
30 $B^T A$
31 $A^T B$
32 $B(A^T)$
33 Show that $A^T B = (B^T A)^T$
34 Show that $A(B^T) \neq A^T B$.
35 Show that $(AB)^T \neq A^T B^T$.
36 Show that $(AB)^T = B^T A^T$.

Find the indicated product in each of problems 37 to 40.

37 $\begin{bmatrix} 1 & 3 \\ 2 & 5 \end{bmatrix}\begin{bmatrix} 2 & 1 & -3 \\ 3 & 0 & 4 \end{bmatrix}$

38 $\begin{bmatrix} 0 & 2 \\ 5 & -1 \\ -2 & 6 \end{bmatrix}\begin{bmatrix} 1 & 0 \\ 0 & 1 \end{bmatrix}$

39 $\begin{bmatrix} 2 & 1 & 3 \\ 4 & 0 & 1 \\ -1 & 2 & 5 \end{bmatrix}\begin{bmatrix} -2 & 0 & 1 \\ 3 & 4 & 0 \\ 0 & -1 & 2 \end{bmatrix}$

40 $\begin{bmatrix} 3 & 4 & -2 \\ -1 & 2 & 5 \end{bmatrix}\begin{bmatrix} 1 \\ 2 \\ 0 \end{bmatrix}$

If A, B, and C are any 2×2 matrices, prove the statements in problems 41 to 44.

41 $A(BC) = (AB)C$	**42** $A(B + C) = AB + AC$
43 $(AB)^T = B^T A^T$	**44** $AB \neq BA$

13.2 MATRICES AND SYSTEMS OF LINEAR EQUATIONS

If we have the system of linear equations

$$
\begin{aligned}
a_{11}x_1 + a_{12}x_2 + \cdots + a_{1n}x_n &= k_1 \\
a_{21}x_2 + a_{22}x_2 + \cdots + a_{2n}x_n &= k_2 \\
&\vdots \\
a_{m1}x_n + a_{m2}x_2 + \cdots + a_{mn}x_n &= k_m
\end{aligned}
\tag{13.1}
$$

we can (1) interchange any two equations; (2) multiply any equation by a nonzero constant; (3) add a nonzero multiple of any equation to any other equation and have a system that is equivalent to the given system; i.e., we can have a system of equations with the same solution as the given system.

The matrix whose elements are the coefficients in the system of equations (13.1) and whose elements occur in the relative position in the system is called the *coefficient matrix*. If the constant terms are adjoined on the right of the coefficient matrix, as an $(n + 1)$st column, the new matrix is called the *augmented matrix*. Therefore, if the system of equations is

Coefficient matrix

Augmented matrix

$$
\begin{aligned}
x + y + z &= 2 \\
2x + 5y + 3z &= 1 \\
3x - y - 2z &= -1
\end{aligned}
\tag{1}
$$

the matrix of the coefficients is

$$
A = \begin{bmatrix} 1 & 1 & 1 \\ 2 & 5 & 3 \\ 3 & -1 & -2 \end{bmatrix}
$$

and

$$
B = \begin{bmatrix} 1 & 1 & 1 & 2 \\ 2 & 5 & 3 & 1 \\ 3 & -1 & -2 & -1 \end{bmatrix}
$$

is the augmented matrix.

If we know the augmented matrix, we have each equation of the system just as clearly as if the variables and equality signs were written in.

In keeping with the relation between the system of equations and the augmented matrix and statements (1), (2), and (3) just after equations (13.1), we can save time and space by performing the following *row operations* on the augmented matrix of the system of equations:

1 *Interchange any two rows of the augmented matrix.*

2 *Multiply any row of the augmented matrix by a nonzero constant.*

3 *Add a nonzero multiple of any row to any other row.*

We shall make use of these row operations to change the augmented matrix into one that represents a set of equations equivalent to the given set and which becomes the identity matrix after deleting the right-hand column. This is desirable since such a matrix is immediately translated into a system of equations with only one variable in each equation.

EXAMPLE

We shall illustrate the procedure by working with the system (1) above.

Solution

The matrix B is the augmented matrix for the system (1). We want to perform row operations on it so as to replace each element of the main diagonal by a 1 and the elements below and above the main diagonal by zeros.

We normally begin by getting a 1 as the first element in the main diagonal, but it is already 1; hence we write

$$B_1 = \begin{bmatrix} 1 & 1 & 1 & 2 \\ 2 & 5 & 3 & 1 \\ 3 & -1 & -2 & -1 \end{bmatrix}$$

We now get zero as the second element in the first column by subtracting twice each element of row one from the corresponding element of row two. This operation is indicated by $R_2' = R_2 - 2R_1$. We get a zero in the third row of column one by performing $R_3' = R_3 - 3R_1$. Thus,

$$B_2 = \begin{bmatrix} 1 & 1 & 1 & 2 \\ 0 & 3 & 1 & -3 \\ 0 & -4 & -5 & -7 \end{bmatrix} \qquad \begin{array}{l} R_2' = R_2 - 2R_1 \\ R_3' = R_3 - 3R_1 \end{array}$$

$$C_1 = \begin{bmatrix} 1 & 1 & 1 & 2 \\ 0 & 1 & \frac{1}{3} & -1 \\ 0 & -4 & -5 & -7 \end{bmatrix} \qquad R_2' = \tfrac{1}{3}R_2$$

$$C_2 = \begin{bmatrix} 1 & 0 & \frac{2}{3} & 3 \\ 0 & 1 & \frac{1}{3} & -1 \\ 0 & 0 & -\frac{11}{3} & -11 \end{bmatrix} \qquad \begin{array}{l} R_1' = R_1 - R_2 \\[4pt] R_3' = R_3 + 4R_2 \end{array}$$

$$D_1 = \begin{bmatrix} 1 & 0 & \frac{2}{3} & 3 \\ 0 & 1 & \frac{1}{3} & -1 \\ 0 & 0 & 1 & 3 \end{bmatrix} \qquad R_3' = -\tfrac{3}{11}R_3$$

$$D_2 = \begin{bmatrix} 1 & 0 & 0 & 1 \\ 0 & 1 & 0 & -2 \\ 0 & 0 & 1 & 3 \end{bmatrix} \qquad \begin{array}{l} R_1' = R_1 - \tfrac{2}{3}R_3 \\ R_2' = R_2 - \tfrac{1}{3}R_3 \end{array}$$

The matrix D_2 represents the system of equations $x = 1, y = -2, z = 3$. The solution set $\{1, -2, 3\}$ can be checked by substituting in the given system.

If we had cared to do so, we could have stopped with matrix D_1 since the system of equations represented by it is readily solved. The last row in D_1 represents the equation $z = 3$. The other two equations from D_1 are $y + z/3 = -1$ and $x + 2z/3 = 3$. From the first of these, we get $y = -1 - 1 = -2$, and from the second one, we find that $x = 3 - 2 = 1$. Consequently, as found earlier, the solution set of the given system is $\{1, -2, 3\}$. In most situations the equation obtained from the first row of the matrix corresponding to D_1 will contain more than two variables, but the value of all except one of them can be found from earlier equations.

13.3 THE INVERSE OF A MATRIX

Inverse of a matrix

The inverse of a matrix M is represented by M^{-1} and is the matrix such that $MM^{-1} = I$. We shall not do so, but it can be shown that if the matrix

$$\begin{bmatrix} a_{11} & a_{12} & \cdots & a_{1n} & 1 & 0 & \cdots & 0 \\ a_{21} & a_{22} & \cdots & a_{2n} & 0 & 1 & \cdots & 0 \\ \cdots\cdots\cdots\cdots\cdots\cdots\cdots\cdots\cdots \\ a_{n1} & a_{n2} & \cdots & a_{nn} & 0 & 0 & \cdots & 1 \end{bmatrix}$$

is transformed, by the use of row operations, into

$$\begin{bmatrix} 1 & 0 & \cdots & 0 & A_{11} & A_{12} & \cdots & A_{1n} \\ 0 & 1 & \cdots & 0 & A_{21} & A_{22} & \cdots & A_{2n} \\ \cdots\cdots\cdots\cdots\cdots\cdots\cdots\cdots\cdots \\ 0 & 0 & \cdots & 1 & A_{n1} & A_{n2} & \cdots & A_{nn} \end{bmatrix}$$

then, the inverse of

$$M = \begin{bmatrix} a_{11} & a_{12} & \cdots & a_{1n} \\ a_{21} & a_{22} & \cdots & a_{2n} \\ \cdots\cdots\cdots\cdots\cdots \\ a_{n1} & a_{n2} & \cdots & a_{nn} \end{bmatrix}$$

is

$$M^{-1} = \begin{bmatrix} A_{11} & A_{12} & \cdots & A_{1n} \\ A_{21} & E_{22} & \cdots & A_{2n} \\ \cdots\cdots\cdots\cdots\cdots \\ A_{n1} & A_{n2} & \cdots & A_{nn} \end{bmatrix}$$

EXAMPLE

Find the inverse of

$$M = \begin{bmatrix} 1 & 2 & 5 \\ 2 & 3 & 8 \\ -1 & 1 & 2 \end{bmatrix}$$

and prove it is the inverse by showing that $MM^{-1} = I$.

Solution   We begin by augmenting M on the right with a 3 by 3 identity matrix. Thus we have

$$\begin{bmatrix} 1 & 2 & 5 & 1 & 0 & 0 \\ 2 & 3 & 8 & 0 & 1 & 0 \\ -1 & 1 & 2 & 0 & 0 & 1 \end{bmatrix}$$

We now must use row operations in order to get a 3 by 3 identity matrix on the left. For this, we perform the operations indicated below.

$$B_1 = A$$

$$B_2 = \begin{bmatrix} 1 & 2 & 5 & 1 & 0 & 0 \\ 0 & -1 & -2 & -2 & 1 & 0 \\ 0 & 3 & 7 & 1 & 0 & 1 \end{bmatrix} \qquad \begin{array}{l} R_2' = R_2 - 2R_1 \\ R_3' = R_3 + R_1 \end{array}$$

$$C_1 = \begin{bmatrix} 1 & 2 & 5 & 1 & 0 & 0 \\ 0 & 1 & 2 & 2 & -1 & 0 \\ 0 & 3 & 7 & 1 & 0 & 1 \end{bmatrix} \qquad R_2' = -R_2$$

$$C_2 = \begin{bmatrix} 1 & 0 & 1 & -3 & 2 & 0 \\ 0 & 1 & 2 & 2 & -1 & 0 \\ 0 & 0 & 1 & -5 & 3 & 1 \end{bmatrix} \qquad \begin{array}{l} R_1' = R_1 - 2R_2 \\ \\ R_3' = R_3 - 3R_2 \end{array}$$

$$D_1 = C_2$$

$$D_2 = \begin{bmatrix} 1 & 0 & 0 & 2 & -1 & -1 \\ 0 & 1 & 0 & 12 & -7 & -2 \\ 0 & 0 & 1 & -5 & 3 & 1 \end{bmatrix} \qquad \begin{array}{l} R_1' = R_1 - R_3 \\ R_2' = R_2 - 2R_3 \end{array}$$

Consequently, the inverse of the given matrix M is

$$M^{-1} = \begin{bmatrix} 2 & -1 & -1 \\ 12 & -7 & -2 \\ -5 & 3 & 1 \end{bmatrix}$$

and it can be verified by showing that $MM^{-1} = I$.

Exercise 13.2   Solution Sets, Inverses

Find the solution set of the system of equations in each of problems 1 to 12 by use of row operations on matrices.

1 $2x + 3y = 7$
    $x - 4y = -2$

2 $x - 3y = 9$
    $3x + y = 7$

3 $4x + 3y = 1$
    $3x - 4y = 7$

4 $3x - 2y = 0$
    $x + 3y = 11$

5 $x + y + z = 6$
    $x - y + z = 2$
    $x + y - z = 0$

6 $2x + y - z = 6$
    $3x - y + 2z = 3$
    $x + 2y + 3z = 1$

7 $2x - 3y + 2z = 4$
$x - 2y - 2z = 0$
$3x - 7y + 4z = 4$

8 $x + y + 3z = 2$
$2x + 3y = 5$
$-x + 6z = 4$

9 $x + y + z + w = 2$
$x + y + z = 3$
$y - z - w = 3$
$x + z + w = 0$

10 $x + z + w = 0$
$x + y - w = -2$
$y + z + w = 0$
$x + y + z = -1$

11 $x + y + z - w = 1$
$y - z + w = -5$
$x + y + z = 0$
$x - w = 3$

12 $x + y - w = 0$
$x + 2y + w = 0$
$y + z + 3w = 1$
$2x + 3z - 2w = -2$

13 to 20 Find the inverse of the matrix of the coefficients of the system of equations in each of problems 1 to 8.

13.4 INVERSIONS

In any permutation of integers, an *inversion* occurs if an integer precedes a smaller integer. For example, in the permutation 5 4 3 1 2 7 6: 7 precedes 6; 5 precedes 4, 3, 1, and 2; 4 precedes 3, 1, and 2; 3 precedes 1 and 2. Hence, the permutation contains $1 + 4 + 3 + 2 = 10$ inversions.

The following theorem is necessary for the development of the theory of determinants of order greater than 2.

Theorem on inversions

If two adjacent integers in any permutation are interchanged, the number of inversions in the permutation is increased or decreased by 1.

Proof ▶

If s and t are the two integers of the permutation that are interchanged, an inversion is introduced if $s < t$, and one is removed if $s > t$.

The following theorem will be employed in deriving the properties of determinants:

If
$$e_1, e_2, e_3 \ldots , e_n \tag{1}$$

is a permutation of the integers

$$1, 2, 3, \ldots , n \tag{2}$$

and if k inversions occur in (1)*, then* (1) *can be transformed into* (2) *by k interchanges of consecutive terms.*

The meaning of this theorem and the method for proving it are illustrated below.

In the permutation 4 1 5 3 2: 5 precedes 3 and 2; 4 precedes 1, 3, 2; and 3 precedes 2. Hence, we have $2 + 3 + 1 = 6$ inversions. To arrange these integers in numerical order, we proceed as follows:

1   Start with the largest integer, 5, that is out of place. Move it over 3 and 2. Thus, by *two interchanges*, get 4 1 3 2 5.

2   Move the next largest integer, 4, over 1, 3, and 2 to get 1 3 2 4 5. This involves *three interchanges*.

3   Finally, interchange 3 and 2 and obtain 1 2 3 4 5.

Consequently, the desired arrangement is accomplished by $3 + 2 + 1 = 6$ *interchanges*.

13.5 DETERMINANTS OF ORDER n

We represent an element of a determinant of order n by a_{ij}. In the double subscript, i denotes the row in which the element occurs, and j denotes the column. Thus a_{13} is the element in the first row and third column. In terms of this notation

Determinant of order n

$$D_n = \begin{vmatrix} a_{11} & a_{12} & \cdots & a_{1j} & \cdots & a_{1n} \\ a_{21} & a_{22} & \cdots & a_{2j} & \cdots & a_{2n} \\ \cdots & \cdots & \cdots & \cdots & \cdots & \cdots \\ a_{i1} & a_{i2} & \cdots & a_{ij} & \cdots & a_{in} \\ \cdots & \cdots & \cdots & \cdots & \cdots & \cdots \\ a_{n1} & a_{n2} & \cdots & a_{nj} & \cdots & a_{nn} \end{vmatrix} \tag{13.2}$$

is a number called a determinant of order n. The number is represented by a homogeneous polynomial obtained by:

Procedure for expanding a determinant

1   Forming every possible product by taking one and only one factor from each row and each column of D_n.

2   Arranging the factors in each product so that the first, or row, subscripts are in order of magnitude.

3   Prefixing each product by a plus or minus sign according as the number of inversions in the second subscript is even or odd.

4   Taking the algebraic sum of the products obtained in steps 1, 2, and 3.

The polynomial obtained by the above procedure is called the *expansion* of the determinant D_n.

The following theorem is a direct consequence of the above definition.

THEOREM   *The expansion of a determinant of order n contains $n!$ terms.*

Proof ▶   Each term in the expansion of D_n will be of the form $a_{1e_1} a_{2e_2} a_{3e_3} \cdots a_{ne_n}$, where $e_1 e_2 e_3 \cdots e_n$ is a permutation of the integers $1 \cdot 2 \cdot 3 \cdots n$, and there will be one term in the expansion for each permutation of these integers. Since $n!$ permutations can be formed from n different elements taken n at a time, there will be $n!$ terms in the expansion of D_n.

We shall illustrate the application of the above four steps by using them to obtain the expansion of

$$D_3 = \begin{vmatrix} a_{11} & a_{12} & a_{13} \\ a_{21} & a_{22} & a_{23} \\ a_{31} & a_{32} & a_{33} \end{vmatrix}$$

We first form 3! permutations of the integers 1, 2, 3. Thus, we get 123, 231, and 312 and also 321, 132, and 213. Each of the first three has an even number of inversions, and each of the last three has an odd number. Now we form the products of three factors using 1, 2, and 3, respectively, as the first subscripts in each product and one of the above permutations as the second subscripts. Thus we obtain

$$D = a_{11}a_{22}a_{33} + a_{12}a_{23}a_{31} + a_{13}a_{21}a_{32} - a_{13}a_{22}a_{31} - a_{11}a_{23}a_{32} - a_{12}a_{21}a_{33}$$

It should be noticed that the number of inversions in the second subscripts of the first three terms is even and in the second three terms, odd. Hence, the first three terms are preceded by a positive sign and the second three terms by a negative sign.

The above method for obtaining the expansion of a determinant is rather tedious and would be much more so for determinants of order greater than 3. Fortunately, there is another method which can be applied to determinants of order greater than 3, and we discuss this method in the next section.

13.6 EXPANSION OF DETERMINANTS OF ORDER n

We shall give some definitions and notation for use in connection with determinants.

Minor　　The *minor* of the element a_{ij} of a determinant D_n is the determinant that remains after deleting the ith row and the jth column of D_n. The minor of a_{ij} is designated by $m(a_{ij})$.

Cofactor　　The *cofactor* of a_{ij} is the minor of a_{ij} if $i+j$ is an even number and the negative of the minor if $i+j$ is an odd number. We designate the cofactor of a_{ij} by A_{ij}.

According to this definition,

$$A_{ij} = (-1)^{i+j}m(a_{ij}) \tag{13.3}$$

Expansion of a determinant by minors　　The expansion of a determinant D_n of order n can be expressed in terms of the minors of the elements of the ith row in the following way:

$$D_n = (-1)^{i+1}a_{i1}m(a_{i1}) + (-1)^{i+2}a_{i2}m(a_{i2}) + (-1)^{i+3}a_{i3}m(a_{i3})$$
$$+ \cdots + (-1)^{i+j}a_{ij}m(a_{ij}) + \cdots + (-1)^{i+n}a_{in}m(a_{in}) \tag{13.4}$$

By use of (13.3) we can express (13.4) in terms of cofactors as follows:

$$D_n = a_{i1}A_{i1} + a_{i2}A_{i2} + a_{i3}A_{i3} + \cdots + a_{ij}A_{ij} + \cdots + a_{in}A_{in} \qquad (13.5)$$

Similarly, the expansion of D_n in terms of the minors and cofactors of the elements of the jth column are, respectively,

$$D_n = (-1)^{1+j}a_{1j}m(a_{1j}) + (-1)^{2+j}a_{2j}m(a_{2j}) + (-1)^{3+j}a_{3j}m(a_{3j})$$
$$+ \cdots + (-1)^{i+j}a_{ij}m(a_{ij}) + \cdots + (-1)^{n+j}a_{nj}m(a_{nj}) \qquad (13.6)$$

and

$$D_n = a_{1j}A_{1j} + a_{2j}A_{2j} + a_{3j}A_{3j} + \cdots + a_{ij}A_{ij} + \cdots + a_{nj}A_{nj} \qquad (13.7)$$

The proof for a determinant of order n is very abstract and will not be given here.

We shall now expand

$$D_3 = \begin{vmatrix} a_{11} & a_{12} & a_{13} \\ a_{21} & a_{22} & a_{23} \\ a_{31} & a_{32} & a_{33} \end{vmatrix}$$

in terms of the elements of the first row by use of (13.5) and see that we get the same expansion as was obtained in Sec. 13.5. We shall write out each minor instead of merely indicating it. Thus, we have

$$D_3 = (-1)^{1+1}\,a_{11}\begin{vmatrix} a_{22} & a_{23} \\ a_{32} & a_{33} \end{vmatrix} + (-1)^{1+2}\,a_{12}\begin{vmatrix} a_{21} & a_{23} \\ a_{31} & a_{33} \end{vmatrix} + (-1)^{1+3}\,a_{13}\begin{vmatrix} a_{21} & a_{22} \\ a_{31} & a_{32} \end{vmatrix}$$

$$= a_{11}(a_{22}a_{33} - a_{23}a_{32}) - a_{12}(a_{21}a_{33} - a_{23}a_{31}) + a_{13}(a_{21}a_{32} - a_{22}a_{31})$$
$$= a_{11}A_{11} + a_{12}A_{12} + a_{13}A_{13}$$

EXAMPLE

Obtain the value of

$$D = \begin{vmatrix} 2 & 5 & 4 & 3 \\ 3 & 2 & 5 & 1 \\ 4 & 0 & 2 & 1 \\ 3 & 0 & 3 & 2 \end{vmatrix}$$

Solution

We expand in terms of the minors of the second column and obtain

$$D = (-1)^{1+2}(5)\begin{vmatrix} 3 & 5 & 1 \\ 4 & 2 & 1 \\ 3 & 3 & 2 \end{vmatrix} + (-1)^{2+2}(2)\begin{vmatrix} 2 & 4 & 3 \\ 4 & 2 & 1 \\ 3 & 3 & 2 \end{vmatrix} + 0 + 0$$

We next expand each of the third-order determinants in terms of the minors of the first row and obtain

$$D = -5[3(4-3) - 5(8-3) + 1(12-6)]$$
$$+ 2[2(4-3) - 4(8-3) + 3(12-6)]$$
$$= -5(3 - 25 + 6) + 2(2 - 20 + 18)$$
$$= -5(-16) + 2(0) = 80$$

Exercise 13.3 Expansion of Second- and Third-Order Determinants

Find the value of the determinant in each of problems 1 to 32.

1 $\begin{vmatrix} 2 & 1 \\ 5 & 4 \end{vmatrix}$

2 $\begin{vmatrix} 3 & 2 \\ 5 & 7 \end{vmatrix}$

3 $\begin{vmatrix} 4 & 2 \\ 3 & -1 \end{vmatrix}$

4 $\begin{vmatrix} 7 & 3 \\ -2 & 1 \end{vmatrix}$

5 $\begin{vmatrix} 3 & 5 \\ -1 & -2 \end{vmatrix}$

6 $\begin{vmatrix} 3 & -1 \\ 2 & -7 \end{vmatrix}$

7 $\begin{vmatrix} 4 & -2 \\ -3 & 5 \end{vmatrix}$

8 $\begin{vmatrix} -5 & 2 \\ -1 & -3 \end{vmatrix}$

9 $\begin{vmatrix} -2 & 1 \\ -7 & -3 \end{vmatrix}$

10 $\begin{vmatrix} -6 & 3 \\ -1 & -2 \end{vmatrix}$

11 $\begin{vmatrix} 2 & -3 \\ 5 & 4 \end{vmatrix}$

12 $\begin{vmatrix} 17 & -1 \\ 2 & 0 \end{vmatrix}$

13 $\begin{vmatrix} a & c \\ e & s \end{vmatrix}$

14 $\begin{vmatrix} c & r \\ t & a \end{vmatrix}$

15 $\begin{vmatrix} m & t \\ h & -3a \end{vmatrix}$

16 $\begin{vmatrix} 3s & -2t \\ 4s & e \end{vmatrix}$

17 $\begin{vmatrix} 0 & 2 & 3 \\ 4 & 1 & 0 \\ 2 & 1 & 6 \end{vmatrix}$

18 $\begin{vmatrix} 3 & -1 & 0 \\ 0 & 2 & 4 \\ -2 & 3 & -5 \end{vmatrix}$

19 $\begin{vmatrix} 2 & 0 & 1 \\ -1 & 0 & 3 \\ 2 & 4 & 0 \end{vmatrix}$

20 $\begin{vmatrix} 3 & 5 & -3 \\ 0 & -2 & 0 \\ 2 & 0 & -1 \end{vmatrix}$

21 $\begin{vmatrix} 0 & 1 & 5 \\ 4 & 0 & -3 \\ -3 & 1 & 2 \end{vmatrix}$

22 $\begin{vmatrix} 1 & 2 & -3 \\ -2 & 3 & 1 \\ 0 & -1 & 2 \end{vmatrix}$

23 $\begin{vmatrix} 2 & 0 & 1 \\ 0 & 1 & 2 \\ 1 & 2 & 0 \end{vmatrix}$

24 $\begin{vmatrix} 3 & 0 & 1 \\ 0 & 5 & -2 \\ -1 & 3 & 0 \end{vmatrix}$

25 $\begin{vmatrix} 2 & -1 & 3 \\ -2 & 1 & -3 \\ 0 & -2 & 0 \end{vmatrix}$

26 $\begin{vmatrix} 1 & 2 & 1 \\ 0 & 3 & 0 \\ 2 & 4 & 2 \end{vmatrix}$

27 $\begin{vmatrix} -1 & 5 & 2 \\ 0 & 0 & 3 \\ 2 & 7 & -4 \end{vmatrix}$

28 $\begin{vmatrix} 2 & -1 & 0 \\ -2 & 1 & 3 \\ -1 & 0 & 3 \end{vmatrix}$

29 $\begin{vmatrix} 0 & 1 & 7 \\ 2 & 0 & -1 \\ 0 & -3 & 2 \end{vmatrix}$

30 $\begin{vmatrix} 0 & 2 & 3 \\ 1 & 0 & -2 \\ 2 & -3 & 0 \end{vmatrix}$

31 $\begin{vmatrix} 2 & -4 & 5 \\ 3 & 0 & -1 \\ 0 & 5 & 0 \end{vmatrix}$

32 $\begin{vmatrix} 3 & 0 & -5 \\ 0 & 1 & 0 \\ -2 & 0 & 3 \end{vmatrix}$

13.7 PROPERTIES OF DETERMINANTS

Although the expansion of a determinant by minors enables us to express the determinant in terms of determinants of lower order, the computation in calculating the value of a determinant of order 4 or more by use of this method would be very tedious. The computation can be greatly simplified if we use the following properties:

1 *If the rows of one determinant are the same as the columns of another, the two determinants are equal.*

Proof ▶

We shall prove the theorem for determinants of the second and third order. If we expand the two determinants

$$D_2 = \begin{vmatrix} a_{11} & a_{12} \\ a_{21} & a_{22} \end{vmatrix} \quad \text{and} \quad D_2' = \begin{vmatrix} a_{11} & a_{21} \\ a_{12} & a_{22} \end{vmatrix}$$

we get $a_{11}a_{22} - a_{12}a_{21}$ in both cases since multiplication is commutative. Hence, the theorem is true for $n = 2$. We shall now prove it for $n = 3$ by considering

$$D_3 = \begin{vmatrix} a_{11} & a_{12} & a_{13} \\ a_{21} & a_{22} & a_{23} \\ a_{31} & a_{32} & a_{33} \end{vmatrix} \quad \text{and} \quad D_3' = \begin{vmatrix} a_{11} & a_{21} & a_{31} \\ a_{12} & a_{22} & a_{32} \\ a_{13} & a_{23} & a_{33} \end{vmatrix}$$

If we expand D_3 in terms of the elements of the first row and D_3' in terms of the elements of the first column, we obtain $a_{11}A_{11} + a_{12}A_{12} + a_{13}A_{13}$ in each case. Hence the theorem is true for $n = 3$.

We shall prove that the properties stated in the remainder of this section hold for determinants of order 4. The arguments used, however, are general, and can be applied to a determinant of any given order.

2 *If two columns (or rows) of a determinant are interchanged, the value of the determinant is equal to the negative of the value of the given determinant.*

Proof ▶

The proof of this statement is as follows. By (13.6), with $n = 4$, the expansion of the determinant D in terms of the minors of the elements of the jth column is

$$D = (-1)^{1+j}a_{1j}m(a_{1j}) + (-1)^{2+j}a_{2j}m(a_{2j}) + (-1)^{3+j}a_{3j}m(a_{3j}) \\ + (-1)^{4+j}a_{4j}m(a_{4j})$$

where the exponent of -1 is the sum of the numbers of the row and column in which the element a_{ij} appears.

Now if we interchange the jth column with the column immediately to the left, we obtain a new determinant D'. This operation changes neither the elements of the jth column nor the minors of the elements. It does, however, decrease the number of the column by 1; hence, the jth column of D becomes the $(j-1)$st column of D'. Therefore, the expansion of D' will be the same as the expansion of D except that the exponent of -1 in each term will be decreased by 1. Hence, $D = -D'$.

If the two columns interchanged are not adjacent, we can prove that this interchange can be accomplished by an odd number of interchanges of adjacent columns. For example, to interchange the first and fourth columns of D, we interchange the fourth column successively with the third, the second, and the first, placing it immediately at the left of the first column. Next we interchange the first column successively with the second and third, placing it in the position vacated by the fourth. Hence, we have made $3 + 2 = 5$ interchanges, and therefore have five changes in sign. Consequently, if two nonadjacent columns of a determinant are interchanged, the value of the determinant obtained will be the negative of the value of the original determinant.

3 *If the corresponding elements of two columns (or rows) of a determinant are identical, the value of the determinant is zero.*

Proof ▶

The proof of this statement is as follows. If any two columns of the determinant D are identical, and if we obtain the determinant D' by interchanging these two columns, then $D = -D'$. On the other hand, since the two columns interchanged are identical, $D = D'$. Therefore $D = -D$, and it follows that $D = 0$.

4 *If the elements of a column (or row) of a determinant are multiplied by k, the value of the determinant is multiplied by k.*

Proof ▶

To prove this statement, we shall multiply the elements of the jth column of the determinant D by k and obtain the determinant D''. If we expand D'' in terms of the minors of the elements of the jth columns, we obtain

$$D'' = \text{sum of products } (-1)^{i+j} k a_{ij} m(a_{ij}), \quad i = 1, 2, 3, 4, \ldots, n$$
$$= k\,[\text{sum of products } (-1)^{i+j} a_{ij} m(a_{ij}), i = 1, 2, 3, 4, \ldots, n]$$
$$= kD$$

5 *If the elements of the jth column of a determinant D are of the form $a_{ij} + b_{ij}$, then D is the sum of the determinants D' and D'' in which all the columns of D, D', and D'' are the same except the jth; furthermore, the jth column of D' is a_{ij}, $i = 1, 2, 3, 4, \ldots, n$, and the jth column of D'' is b_{ij}, $i = 1, 2, 3, 4, \ldots, n$.*

Proof ▶

We shall prove that this property is valid for a determinant of the fourth order in which the third column is of the form $a_{ij} + b_{ij}$. If

$$D = \begin{vmatrix} a_{11} & a_{12} & a_{13} + b_{13} & a_{14} \\ a_{21} & a_{22} & a_{23} + b_{23} & a_{24} \\ a_{31} & a_{32} & a_{33} + b_{33} & a_{34} \\ a_{41} & a_{42} & a_{43} + b_{43} & a_{44} \end{vmatrix}$$

and we expand D in terms of the minors of the elements of the third column, we obtain

$$D = (a_{13} + b_{13})m(a_{13}) - (a_{23} + b_{23})m(a_{23}) + (a_{33} + b_{33})m(a_{33}) -$$
$$(a_{43} + b_{43})m(a_{43})$$
$$= [a_{13}m(a_{13}) - a_{23}m(a_{23}) + a_{33}m(a_{33}) - a_{43}m(a_{43})] +$$
$$[b_{13}m(a_{13}) - b_{23}m(a_{23}) + b_{33}m(a_{33}) - b_{43}m(a_{43})]$$

By (13.6) the first and second bracketed expressions are respectively the expansions of

$$D' = \begin{vmatrix} a_{11} & a_{12} & a_{13} & a_{14} \\ a_{21} & a_{22} & a_{23} & a_{24} \\ a_{31} & a_{32} & a_{33} & a_{34} \\ a_{41} & a_{42} & a_{43} & a_{44} \end{vmatrix} \quad \text{and} \quad D'' = \begin{vmatrix} a_{11} & a_{12} & b_{13} & a_{14} \\ a_{21} & a_{22} & b_{23} & a_{24} \\ a_{31} & a_{32} & b_{33} & a_{34} \\ a_{41} & a_{42} & b_{43} & a_{44} \end{vmatrix}$$

Therefore, $D = D' + D''$.

6 *If in a given determinant D the elements of the kth column a_{ik}, $i = 1, 2, 3, \ldots, n$, are replaced by $a_{ik} + ta_{ij}$, where a_{ij}, $i = 1, 2, 3, \ldots, n$, are the elements of the jth column, the determinant obtained is equal to D.*

Proof ▶

We shall prove that this property is true for a determinant of order 4 in which we multiply each element of the fourth column by t and add the product to the corresponding element of the second column. The same method can be used to prove that the property holds for any two determinants of any order. If

$$D = \begin{vmatrix} a_{11} & a_{12} & a_{13} & a_{14} \\ a_{21} & a_{22} & a_{23} & a_{24} \\ a_{31} & a_{32} & a_{33} & a_{34} \\ a_{41} & a_{42} & a_{43} & a_{44} \end{vmatrix} \quad \text{and} \quad D' = \begin{vmatrix} a_{11} & a_{12} + ta_{14} & a_{13} & a_{14} \\ a_{21} & a_{22} + ta_{24} & a_{23} & a_{24} \\ a_{31} & a_{32} + ta_{34} & a_{33} & a_{34} \\ a_{41} & a_{42} + ta_{44} & a_{43} & a_{44} \end{vmatrix}$$

then, by properties 4 and 5,

$$D' = \begin{vmatrix} a_{11} & a_{12} & a_{13} & a_{14} \\ a_{21} & a_{22} & a_{23} & a_{24} \\ a_{31} & a_{32} & a_{33} & a_{34} \\ a_{41} & a_{42} & a_{43} & a_{44} \end{vmatrix} + t \begin{vmatrix} a_{11} & a_{14} & a_{13} & a_{14} \\ a_{21} & a_{24} & a_{23} & a_{24} \\ a_{31} & a_{34} & a_{33} & a_{34} \\ a_{41} & a_{44} & a_{43} & a_{44} \end{vmatrix}$$

Now the first determinant in the above pair is equal to D, and the second is equal to zero, since two columns are identical. Therefore, $D' = D$.

By property 1, the above property is true if the word "column" is replaced by "row."

By a repeated application of property 6 to a determinant D of order n, we can obtain a determinant in which all of the elements except one of some row or column are zeros, and the determinant thus obtained will be equal to D. We may then expand the determinant in terms of the minors of the row or column that contains the zeros and thus have the given determinant equal to the product of a constant and a determinant of order $n - 1$. We shall illustrate the method with two examples. We shall use the notation $C_k + tC_j$ to indicate that each element in the jth column of D is multiplied by t and the product is added to the corresponding element of the kth column, and we shall use a similar notation for rows.

7 *If each element of a row is multiplied by the cofactor of the corresponding element of another row and the products added, the sum is zero.*

Proof ►   We shall prove this theorem for a third-order determinant, but the procedure can be applied to one of any order. We shall consider the elements of the first row of

$$D = \begin{vmatrix} a_{11} & a_{12} & a_{13} \\ a_{21} & a_{22} & a_{23} \\ a_{31} & a_{32} & a_{33} \end{vmatrix}$$

and the cofactors of the elements of the third row and prove that

$$a_{11}A_{31} + a_{12}A_{32} + a_{13}A_{33} = 0 \qquad (1)$$

Since $A_{31} = \begin{vmatrix} a_{12} & a_{13} \\ a_{22} & a_{23} \end{vmatrix}$ $A_{32} = -\begin{vmatrix} a_{11} & a_{13} \\ a_{21} & a_{23} \end{vmatrix}$ $A_{33} = \begin{vmatrix} a_{11} & a_{12} \\ a_{21} & a_{22} \end{vmatrix}$

the left member of (1) is the expansion of

$$\begin{vmatrix} a_{11} & a_{12} & a_{13} \\ a_{11} & a_{12} & a_{13} \\ a_{21} & a_{22} & a_{23} \end{vmatrix}$$

in terms of the elements of the first row. The value of the determinant is zero since two rows are identical.

EXAMPLE 1   Find the value of

$$D = \begin{vmatrix} 2 & 1 & 4 & 1 \\ 4 & 2 & 6 & 2 \\ 3 & 5 & 2 & 3 \\ 7 & 3 & 1 & 3 \end{vmatrix}$$

Solution   We first notice that the second and fourth columns have three elements in common; hence, we perform $C_2 - C_4$ and get

$$D = \begin{vmatrix} 2 & 0 & 4 & 1 \\ 4 & 0 & 6 & 2 \\ 3 & 2 & 2 & 3 \\ 7 & 0 & 1 & 3 \end{vmatrix}$$

Now we expand in terms of the minors of the elements of the second column and get

$$D = -2 \begin{vmatrix} 2 & 4 & 1 \\ 4 & 6 & 2 \\ 7 & 1 & 3 \end{vmatrix}$$

Finally, we perform the operation $R_2 - 2R_1$, expand the determinant thus obtained in terms of the minors of the elements of the second row, and get

$$D = -2 \begin{vmatrix} 2 & 4 & 1 \\ 0 & -2 & 0 \\ 7 & 1 & 3 \end{vmatrix} = (-2)(-2) \begin{vmatrix} 2 & 1 \\ 7 & 3 \end{vmatrix} = 4(6-7) = -4$$

EXAMPLE 2　　Obtain the value of the determinant

$$D = \begin{vmatrix} 2 & 3 & 5 & 1 \\ 4 & 2 & 3 & 5 \\ 3 & 1 & 4 & 2 \\ 5 & 4 & 2 & 3 \end{vmatrix}$$

Solution　　If we examine the above determinant, we see that we cannot obtain a determinant with three zeros in either one row or one column by adding or subtracting corresponding terms in either rows or columns. We can, however, obtain a determinant in which the elements in the first row are 0, 0, 0, 1 by performing successively the operations $C_1 - 2C_4$, $C_2 - 3C_4$, and $C_3 - 5C_4$ and then writing column 4 unchanged. Thus we get

$$D = \begin{vmatrix} 0 & 0 & 0 & 1 \\ -6 & -13 & -22 & 5 \\ -1 & -5 & -6 & 2 \\ -1 & -5 & -13 & 3 \end{vmatrix}$$

If we now expand in terms of the minors of the elements of the first row, we get

$$D = -1 \begin{vmatrix} -6 & -13 & -22 \\ -1 & -5 & -6 \\ -1 & -5 & -13 \end{vmatrix}$$

Now we notice that the first two terms in the second and third rows are the same, so we perform the operation $R_2 - R_3$ and get

$$D = -1 \begin{vmatrix} -6 & -13 & -22 \\ 0 & 0 & 7 \\ -1 & -5 & -13 \end{vmatrix} = (-1)(-7) \begin{vmatrix} -6 & -13 \\ -1 & -5 \end{vmatrix} = 7(30 - 13) = 119$$

Exercise 13.4　Properties of Determinants

Prove by use of properties 1 to 7 and without expanding that the statement in each of problems 1 to 16 is true.

1　$\begin{vmatrix} a & b & e \\ n & e & w \\ d & t & e \end{vmatrix} = \begin{vmatrix} a & n & d \\ b & e & t \\ e & w & e \end{vmatrix}$

2　$\begin{vmatrix} 2 & a & m \\ a & c & t \\ t & e & n \end{vmatrix} = \begin{vmatrix} 2 & a & t \\ a & c & e \\ m & t & n \end{vmatrix}$

3　$3\begin{vmatrix} a & d & g \\ b & e & h \\ c & f & i \end{vmatrix} = \begin{vmatrix} a & b & c \\ 3d & 3e & 3f \\ g & h & i \end{vmatrix}$

4　$\begin{vmatrix} 2 & 6 & 8 \\ 1 & 3 & 4 \\ 5 & 7 & 3 \end{vmatrix} = \begin{vmatrix} 8 & 5 & 4 \\ 6 & 1 & 3 \\ 4 & 7 & 2 \end{vmatrix}$

5 $\begin{vmatrix} 3 & 2 & 1 \\ 9 & 0 & 3 \\ 15 & 6 & 5 \end{vmatrix} = \begin{vmatrix} 1 & 1 & 4 \\ 0 & 3 & 12 \\ 3 & 5 & 20 \end{vmatrix}$ **6** $\begin{vmatrix} 5 & 5 & 1 \\ 20 & 5 & 4 \\ 35 & 5 & 7 \end{vmatrix} = 0$

7 $\begin{vmatrix} 10 & 2 & 5 \\ 8 & 4 & 4 \\ 4 & 5 & 2 \end{vmatrix} = \begin{vmatrix} 2 & 7 & 4 \\ 3 & 3 & 6 \\ 7 & 2 & 14 \end{vmatrix}$ **8** $\begin{vmatrix} 8 & 0 & 2 \\ 0 & 5 & 0 \\ 4 & 2 & 1 \end{vmatrix} = \begin{vmatrix} 3 & 5 & 6 \\ 2 & 2 & 4 \\ 5 & 3 & 10 \end{vmatrix}$

9 $\begin{vmatrix} x_1 & y_1 & z_1 \\ x_2 & y_2 & z_2 \\ x_3 & y_3 & z_3 \end{vmatrix} = \begin{vmatrix} x_1 + ax_3 & y_1 + ay_3 & z_1 + az_3 \\ x_2 & y_2 & z_2 \\ x_3 & y_3 & z_3 \end{vmatrix}$

10 $\begin{vmatrix} a & b & c \\ d & e & f \\ g & h & i \end{vmatrix} = \begin{vmatrix} a+b-c & b & c \\ d+e-f & e & f \\ g+h-i & h & i \end{vmatrix}$ **11** $\begin{vmatrix} a & b \\ c & d \end{vmatrix} = \begin{vmatrix} a+2b & b \\ c+2d & d \end{vmatrix} = \begin{vmatrix} a & b-3a \\ c & d-3c \end{vmatrix}$

12 $\begin{vmatrix} a & b & c \\ d & e & f \\ g & h & i \end{vmatrix} = -\begin{vmatrix} a & c & b \\ d & f & e \\ g & i & h \end{vmatrix}$ **13** $\begin{vmatrix} 5 & 6 & 1 \\ 2 & 9 & 4 \\ 4 & 15 & 3 \end{vmatrix} = -3\begin{vmatrix} 2 & 3 & 5 \\ 5 & 2 & 4 \\ 1 & 4 & 3 \end{vmatrix}$

14 $\begin{vmatrix} 1 & 3 & 6 \\ 3 & 7 & 4 \\ 11 & -1 & 2 \end{vmatrix} = 4\begin{vmatrix} 1 & 2 & 3 \\ 3 & 5 & 2 \\ 11 & 5 & 1 \end{vmatrix}$ **15** $\begin{vmatrix} 2 & 3 & 1 \\ 6 & 1 & 7 \\ 10 & 5 & 3 \end{vmatrix} = 8\begin{vmatrix} 1 & 3 & 5 \\ 1 & 2 & 2 \\ 1 & 7 & 3 \end{vmatrix}$

16 $\begin{vmatrix} 1 & 4 & 3 \\ 2 & 5 & 2 \\ 3 & 4 & 1 \end{vmatrix} = \begin{vmatrix} -2 & 1 & 3 \\ 0 & 3 & 2 \\ 2 & 3 & 1 \end{vmatrix}$

Find the value of the determinant in each of problems 17 to 20 by use of property 6.

17 $\begin{vmatrix} -11 & 13 & -4 \\ 9 & -1 & 3 \\ -5 & 3 & -2 \end{vmatrix}$ **18** $\begin{vmatrix} 1 & 1 & 8 \\ 8 & 3 & 3 \\ 1 & 3 & 24 \end{vmatrix}$

19 $\begin{vmatrix} 6 & -4 & 4 \\ 3 & -3 & 2 \\ -3 & 6 & -1 \end{vmatrix}$ **20** $\begin{vmatrix} 1 & -4 & -2 \\ -1 & -4 & -3 \\ 1 & 7 & 4 \end{vmatrix}$

Find the replacement set for x so that the statement in each of problems 21 to 28 is true.

21 $\begin{vmatrix} 2 & 1 \\ -6 & x \end{vmatrix} = 0$ **22** $\begin{vmatrix} x & 3 \\ 4 & 2 \end{vmatrix} = 4$

23 $\begin{vmatrix} 1 & x \\ 2-x & -3 \end{vmatrix} = 0$ **24** $\begin{vmatrix} 2x-3 & 2 \\ -2 & x \end{vmatrix} = 9$

25 $\begin{vmatrix} x & 1 & 3 \\ 1 & x & 2 \\ 1 & 1 & 2 \end{vmatrix} = 0$

26 $\begin{vmatrix} x & 1 & 5 \\ 2 & 3 & 2 \\ 1 & -1 & x \end{vmatrix} = 4$

27 $\begin{vmatrix} 3 & 4 & 2 \\ 5 & x & 1 \\ -1 & 2 & 1 \end{vmatrix} = 0$

28 $\begin{vmatrix} x & 1 & 3 \\ 2 & 6 & x \\ 3 & 1 & 0 \end{vmatrix} = -50$

13.8 CRAMER'S RULE

In the eighteenth century, the Swiss mathematician Cramer devised a rule for obtaining the simultaneous solution set of a system of linear equations. We shall explain this rule and its use in this section. His rule states that if the determinant of the coefficients is not zero, the solution set of the system of equations

$$a_{11}x_1 + a_{12}x_2 + a_{13}x_3 + \cdots + a_{1j}x_j + \cdots + a_{1n}x_n = e_1$$
$$a_{21}x_1 + a_{22}x_2 + a_{23}x_3 + \cdots + a_{2j}x_j + \cdots + a_{2n}x_n = e_2$$
$$a_{31}x_1 + a_{32}x_2 + a_{33}x_3 + \cdots + a_{3j}x_j + \cdots + a_{3n}x_n = e_3$$
$$\cdots \cdots \cdots \cdots \cdots \cdots \cdots \cdots \cdots \cdots \cdots$$
$$a_{i1}x_1 + a_{i2}x_2 + a_{i3}x_3 + \cdots + a_{ij}x_j + \cdots + a_{in}x_n = e_i$$
$$\cdots \cdots \cdots \cdots \cdots \cdots \cdots \cdots \cdots \cdots \cdots$$
$$a_{n1}x_1 + a_{n2}x_2 + a_{n3}x_3 + \cdots + a_{nj}x_j + \cdots + a_{nn}x_n = e_n$$

is

$$\left\{ \frac{N(x_1)}{D}, \frac{N(x_2)}{D}, \frac{N(x_3)}{D}, \ldots, \frac{N(x_j)}{D}, \ldots, \frac{N(x_n)}{D} \right\}$$

where

$$D = \begin{vmatrix} a_{11} & a_{12} & a_{13} & \cdots & a_{1j} & \cdots & a_{1n} \\ a_{21} & a_{22} & a_{23} & \cdots & a_{2j} & \cdots & a_{2n} \\ a_{31} & a_{32} & a_{33} & \cdots & a_{3j} & \cdots & a_{3n} \\ \cdots & \cdots & \cdots & \cdots & \cdots & \cdots & \cdots \\ a_{i1} & a_{i2} & a_{i3} & \cdots & a_{ij} & \cdots & a_{in} \\ \cdots & \cdots & \cdots & \cdots & \cdots & \cdots & \cdots \\ a_{n1} & a_{n2} & a_{n3} & \cdots & a_{nj} & \cdots & a_{nn} \end{vmatrix}$$

and the determinant $N(x_j)$, for $j = 1, 2, 3, \ldots, n$, is obtained by replacing the jth column of D by the column of constant terms e_i, for $i = 1, 2, 3, \ldots, n$.

We shall prove Cramer's rule for a system of three linear equations. The method used is applicable regardless of the number of equations. We shall consider the system

$$a_1x + b_1y + c_1z = d_1 \tag{1}$$
$$a_2x + b_2y + c_2z = d_2 \tag{2}$$
$$a_3x + b_3y + c_3z = d_3 \tag{3}$$

The determinant of the coefficients is

$$D = \begin{vmatrix} a_1 & b_1 & c_1 \\ a_2 & b_2 & c_2 \\ a_3 & b_3 & c_3 \end{vmatrix}$$

We define the determinants formed when the constant terms are substituted for the coefficients of x, y, and z in D as

$$D_x = \begin{vmatrix} d_1 & b_1 & c_1 \\ d_2 & b_2 & c_2 \\ d_3 & b_3 & c_3 \end{vmatrix} \qquad D_y = \begin{vmatrix} a_1 & d_1 & c_1 \\ a_2 & d_2 & c_2 \\ a_3 & d_3 & c_3 \end{vmatrix} \qquad D_z = \begin{vmatrix} a_1 & b_1 & d_1 \\ a_2 & b_2 & d_2 \\ a_3 & b_3 & d_3 \end{vmatrix}$$

We shall show that, if $D \neq 0$, then the solution set $\{(x, y, z)\}$ of the system is $\{(D_x/D, D_y/D, D_z/D)\}$. For this purpose, we multiply the first column of D by x, and by property 4, we have

$$xD = \begin{vmatrix} a_1x & b_1 & c_1 \\ a_2x & b_2 & c_2 \\ a_3x & b_3 & c_3 \end{vmatrix} \tag{4}$$

Next, we multiply the elements of the second column of the above determinant by y, and the elements of the third column by z and add the products to the elements of the first column. Thus by property 6 we get

$$xD = \begin{vmatrix} a_1x + b_1y + c_1z & b_1 & c_1 \\ a_2x + b_2y + c_2z & b_2 & c_2 \\ a_3x + b_3y + c_3z & b_3 & c_3 \end{vmatrix}$$

$$= \begin{vmatrix} d_1 & b_1 & c_1 \\ d_2 & b_2 & c_2 \\ d_3 & b_3 & c_3 \end{vmatrix} \quad \blacktriangleleft \text{ since } a_ix + b_iy + c_iz = d_i,$$
$$\text{for } i = 1, 2, 3$$

$$xD = D_x$$

Consequently, $x = D_x/D$.

By a similar argument, we can show that $y = D_y/D$ and $z = D_z/D$.

We shall now verify that these values of x, y, and z satisfy Eq. (1). If we replace x, y, and z respectively in the left member of (1) by these values, we have

$$\frac{1}{D}\left(a_1D_x + b_1D_y + c_1D_z \right) \tag{5}$$

and we must show that this expression is equal to d_1. If we expand D_x in terms of the elements of the first column, we obtain

$$D_x = d_1 \begin{vmatrix} b_2 & c_2 \\ b_3 & c_3 \end{vmatrix} - d_2 \begin{vmatrix} b_1 & c_1 \\ b_3 & c_3 \end{vmatrix} + d_3 \begin{vmatrix} b_1 & c_1 \\ b_2 & c_2 \end{vmatrix} = d_1A_1 + d_2A_2 + d_3A_3$$

since the second-order determinants together with the sign that precedes each are respectively the cofactors of a_1, a_2, and a_3 in D. Therefore,

$$a_1 D_x = a_1 d_1 A_1 + a_1 d_2 A_2 + a_1 d_3 A_3 \tag{6}$$

Similarly,
$$b_1 D_y = b_1 d_1 B_1 + b_1 d_2 B_2 + b_1 d_3 B_3 \tag{7}$$

and
$$c_1 D_z = c_1 d_1 C_1 + c_1 d_2 C_2 + c_1 d_3 C_3 \tag{8}$$

We now equate the sums of the left and right members of (6), (7), and (8) and get

$$
\begin{aligned}
a_1 D_x + b_1 D_y + c_1 D_z &= d_1 (a_1 A_1 + b_1 B_1 + c_1 C_1) \\
&\quad + d_2 (a_1 A_2 + b_1 B_2 + c_1 C_2) \\
&\quad + d_3 (a_1 A_3 + b_1 B_3 + c_1 C_3) \\
&= d_1 D + d_2 (0) + d_3 (0) = d_1 D
\end{aligned}
$$

since the expression in the first parentheses is the expansion of D in terms of the elements of the first column and the expression in each of the second and third parentheses is the sum of the products of the elements of the first column of D and the cofactors of the elements of the second and third columns, respectively, of D.   Therefore by property 4 the two latter sums are zero.   Consequently, expression (5) is equal to $d_1 D/D = d_1$. Therefore, since (5) is the left member of Eq. (1) for $x = D_x/D$, $y = D_y/D$, $z = D_z/D$, these values satisfy the equation.   By a similar method we can verify that these values also satisfy Eqs. (2) and (3).   Hence the simultaneous solution set of Eqs. (1), (2), and (3) is $\{(D_x/D, D_y/D, D_z/D)\}$.

By use of a similar argument, we can show that the solution set of the equations

$$
\begin{aligned}
a_1 x + b_1 y &= d_1 \\
a_2 x + b_2 y &= d_2
\end{aligned}
$$

is $\{(D_x/D, D_y/D)\}$, provided $D \neq 0$ and

$$
D = \begin{vmatrix} a_1 & b_1 \\ a_2 & b_2 \end{vmatrix} \qquad
D_x = \begin{vmatrix} d_1 & b_1 \\ d_2 & b_2 \end{vmatrix} \qquad \text{and} \qquad
D_y = \begin{vmatrix} a_1 & d_1 \\ a_2 & d_2 \end{vmatrix}
$$

If the determinant of the coefficients D is zero, the system of equations is not independent, and the solution set may be the empty set $\varnothing$, or it may contain an infinitude of elements.

We illustrate the use of Cramer's rule with three examples.

EXAMPLE 1　　Find the simultaneous solution set of the equations

$$3x - 6y - 2 = 0 \tag{9}$$
$$4x + 7y + 3 = 0 \tag{10}$$

Solution　　by use of Cramer's rule.

We first add 2 to each member of (9) and -3 to each member of (10) and get

$$3x - 6y = 2$$
$$4x + 7y = -3$$

We now obtain the solution set by the following steps:

1   We form the determinant D whose elements are the coefficients of the unknowns in the order in which they appear, and get

$$D = \begin{vmatrix} 3 & -6 \\ 4 & 7 \end{vmatrix} = 21 + 24 = 45$$

2   Replace the column of coefficients of x in D by the constant terms and get

$$D_x = \begin{vmatrix} 2 & -6 \\ -3 & 7 \end{vmatrix} = 14 - 18 = -4$$

3   Replace the column of coefficients of y in D by the constant terms and get

$$D_y = \begin{vmatrix} 3 & 2 \\ 4 & -3 \end{vmatrix} = -9 - 8 = -17$$

4   By Cramer's rule

$$x = \frac{D_x}{D} = \frac{-4}{45} = -\frac{4}{45}$$
$$y = \frac{D_y}{D} = \frac{-17}{45} = -\frac{17}{45}$$

Hence the simultaneous solution set is $\{(-\frac{4}{45}, -\frac{17}{45})\}$.

Check   Replacing x and y in the given equations by the appropriate elements of the solution set, we have

$$3(-\tfrac{4}{45}) - 6(-\tfrac{17}{45}) - 2 = \frac{-12 + 102 - 90}{45} = 0 \qquad \blacktriangleleft \text{ from (9)}$$

$$4(-\tfrac{4}{45}) + 7(-\tfrac{17}{45}) + 3 = \frac{-16 - 119 + 135}{45} = 0 \qquad \blacktriangleleft \text{ from (10)}$$

EXAMPLE 2   Use Cramer's rule to solve the system of equations

$$3x + y - 2z = -3$$
$$2x + 7y + 3z = 9$$
$$4x - 3y - z = 7$$

Solution   The terms in the left members are arranged in the proper order and only the constant terms appear in the right members.   Hence, we proceed as follows:

Step 1:
$$D = \begin{vmatrix} 3 & 1 & -2 \\ 2 & 7 & 3 \\ 4 & -3 & -1 \end{vmatrix}$$

$$= 3 \begin{vmatrix} 7 & 3 \\ -3 & -1 \end{vmatrix} - 1 \begin{vmatrix} 2 & 3 \\ 4 & -1 \end{vmatrix} - 2 \begin{vmatrix} 2 & 7 \\ 4 & -3 \end{vmatrix}$$

$$= 3(-7 + 9) - 1(-2 - 12) - 2(-6 - 28)$$

$$D = 6 + 14 + 68 = 88$$

Step 2:
$$D_x = \begin{vmatrix} -3 & 1 & -2 \\ 9 & 7 & 3 \\ 7 & -3 & -1 \end{vmatrix}$$

$$= -3 \begin{vmatrix} 7 & 3 \\ -3 & -1 \end{vmatrix} - 1 \begin{vmatrix} 9 & 3 \\ 7 & -1 \end{vmatrix} - 2 \begin{vmatrix} 9 & 7 \\ 7 & -3 \end{vmatrix}$$

$$= -3(-7 + 9) - 1(-9 - 21) - 2(-27 - 49)$$

$$= -6 + 30 + 152 = 176$$

Step 3:
$$D_y = \begin{vmatrix} 3 & -3 & -2 \\ 2 & 9 & 3 \\ 4 & 7 & -1 \end{vmatrix}$$

$$= 3 \begin{vmatrix} 9 & 3 \\ 7 & -1 \end{vmatrix} + 3 \begin{vmatrix} 2 & 3 \\ 4 & -1 \end{vmatrix} - 2 \begin{vmatrix} 2 & 9 \\ 4 & 7 \end{vmatrix}$$

$$= 3(-9 - 21) + 3(-2 - 12) - 2(14 - 36)$$

$$= -90 - 42 + 44 = -88$$

Step 4:
$$D_z = \begin{vmatrix} 3 & 1 & -3 \\ 2 & 7 & 9 \\ 4 & -3 & 7 \end{vmatrix}$$

$$= 3 \begin{vmatrix} 7 & 9 \\ -3 & 7 \end{vmatrix} - 1 \begin{vmatrix} 2 & 9 \\ 4 & 7 \end{vmatrix} - 3 \begin{vmatrix} 2 & 7 \\ 4 & -3 \end{vmatrix}$$

$$= 3(49 + 27) - 1(14 - 36) - 3(-6 - 28)$$

$$= 228 + 22 + 102 = 352$$

Step 5:
$$x = \frac{D_x}{D} = \frac{176}{88} = 2$$

$$y = \frac{D_y}{D} = \frac{-88}{88} = -1$$

$$z = \frac{D_z}{D} = \frac{352}{88} = 4$$

by Cramer's rule. Hence the solution set is $\{(2, -1, 4)\}$; it can be checked by the usual method.

EXAMPLE 3 Show that the following equations are not independent:

$$5x + 4y + 11z = 3$$
$$6x - 4y + 2z = 1$$
$$x + 3y + 5z = 2$$

Solution

$$D = \begin{vmatrix} 5 & 4 & 11 \\ 6 & -4 & 2 \\ 1 & 3 & 5 \end{vmatrix} = 5\begin{vmatrix} -4 & 2 \\ 3 & 5 \end{vmatrix} - 4\begin{vmatrix} 6 & 2 \\ 1 & 5 \end{vmatrix} + 11\begin{vmatrix} 6 & -4 \\ 1 & 3 \end{vmatrix}$$

$$= 5(-20 - 6) - 4(30 - 2) + 11(18 + 4)$$
$$D = -130 - 112 + 242 = 0$$

Hence, since $D = 0$, the equations are not independent, and no unique solution set exists.

Exercise 13.5   Use of Cramer's Rule

Find the solution set of each of the following systems of equations by use of Cramer's rule.

1 $3x + y = 7$
  $x - 3y = -1$

2 $5x - y = 6$
  $2x + 5y = -3$

3 $6x - 7y - 3$
  $7x - 9y = -1$

4 $2x + 5y = 0$
  $3x - y = 17$

5 $5x + 3y = 0$
  $3x + 4y = -11$

6 $3x - y = -9$
  $5x + 9y - -1$

7 $6x + 5y = -1$
  $5x + 6y = -10$

8 $4x - 7y = -2$
  $3x + 4y = 17$

9 $3x + 2y = 2$
  $8x + 4y = 5$

10 $2x + 5y = 2$
  $-4x + 5y = 1$

11 $x + y = 1$
  $6x + y = 3$

12 $7x - 3y = 3$
  $x - 9y = -1$

13 $2x - 3y + z - 2$
  $3x + 2y - 3z = -4$
  $5x - 3y - z = -1$

14 $4x + 3y - 5z = 6$
  $3x + 2y + 4z = 14$
  $5x - 2y - 7z = 1$

15 $7x - 2y + z = 5$
  $6x + 3y - 5z = 2$
  $5x + 7y - 9z = 1$

16 $x - y - z = 0$
  $3x - 4y - 2z = -1$
  $8x - 11y + 3z = 5$

17 $x + y + z = -1$
  $2x + 3y + 2z = -5$
  $3x - 2y + 9z = 0$

18 $4x + 2y + 5z = 3$
  $5x - y + 9z = 19$
  $2x + 3y + 4z = 2$

19 $x + y - z = 4$
  $x - y + z = 6$
  $x + y + z = -2$

20 $2x + y + z = 0$
  $3x + y + 3z = 0$
  $x + y + z = -2$

21 $3x + 2y + 4z = 4$
  $2x + 3y - z = 2$
  $5x + 7y + z = 4$

22 $x + 2y - 6z = 1$
  $3x - y + z = 1$
  $5x - y - z = 2$

23 $2x - 2y + 2z = 1$
  $6x - y + 3z = 2$
  $5x + y - z = 1$

24 $6x + y + 10z = 2$
  $3x - y + 2z = 1$
  $4x + y + 4z = 1$

25 $x + y + z = 2a$
  $x - y - z = 0$
  $x + y - z = 2b$

26 $x - y - z = 0$
  $x + y + z = 4a$
  $x - y + z = 2a - 2b$

27 $x + y + z = 3a$
  $x - y + z = a + 2b$
  $2x - y - z = 0$

28 $x + y - z = 3a - b$
  $2x - 3y + z = -2a + 6b$
  $x - 2y - 3z = 2a$

29 $x + z = 4$
  $x - y = 1$
  $y + z = 3$

30 $x + 2y = 4$
  $x + z = 0$
  $2y + 3z = -2$

31 $3x + 2z = 0$ **32** $y + z = 1$
     $4x + 3y = 5$ $x - y = 2$
     $5y - z = -17$ $x + z = 3$

Exercise 13.6   Review

1   Find x and y so that $\begin{bmatrix} 2x + y & 9 \\ 13 & 3 \end{bmatrix} = \begin{bmatrix} 4 & x - 3y \\ 13 & 3 \end{bmatrix}$

2   Find x, y, and z so that $\begin{bmatrix} x + y & 2 & 7 \\ 5 & y + z & 4 \\ 1 & -1 & 6 \end{bmatrix} = \begin{bmatrix} 3 & 2 & 7 \\ 5 & 0 & 4 \\ x + z & -7 & 6 \end{bmatrix}$

Perform the indicated operations in problems 3 to 6.

3 $\begin{bmatrix} 3 & x - y & 4 \\ x & 5 & 0 \end{bmatrix} + \begin{bmatrix} -2 & y & y - 4 \\ -x & 1 & x \end{bmatrix}$ **4** $\begin{bmatrix} x & 3 & z \\ y & 4 & -3 \\ z & -5 & 1 \end{bmatrix} + \begin{bmatrix} 2 - x & 5 & -z \\ -2y & -1 & 3 \\ -2 & 7 & 2 \end{bmatrix}$

5 $\begin{bmatrix} 2 & 5 \\ -1 & 3 \\ 0 & 4 \end{bmatrix} \begin{bmatrix} 3 & 5 & -1 & 0 \\ 1 & -2 & 7 & 4 \end{bmatrix}$ **6** $\begin{bmatrix} 1 & 3 & -2 \\ 2 & 0 & 4 \end{bmatrix} \begin{bmatrix} 4 & 2 & -1 \\ -3 & 0 & 5 \\ 6 & -5 & 1 \end{bmatrix}$

Find the inverse of the matrix in each of problems 7 and 8 by use of row operations and show that $MM^{-1} = I$.

7 $\begin{bmatrix} 1 & -4 & 2 \\ -1 & 6 & -3 \\ -1 & 1 & 0 \end{bmatrix}$ **8** $\begin{bmatrix} 2 & 3 & -1 & 6 \\ 2 & 4 & -1 & 7 \\ -1 & -2 & 1 & -4 \\ 2 & 2 & -1 & 6 \end{bmatrix}$

Solve the system of equations in each of problems 9 and 10 by use of row operations and by use of determinants.

9 $2x + y = 5$ **10** $x + y + z = 0$
    $x - 3y = -8$ $2x - y = 3$
                           $4y + z = 1$

In each of problems 11, 12, and 13, show that if

$$M = \begin{bmatrix} a_{11} & a_{12} & a_{13} \\ a_{21} & a_{22} & a_{23} \\ a_{31} & a_{32} & a_{33} \end{bmatrix} \quad \text{then} \quad M^{-1} = \frac{1}{D} \begin{bmatrix} A_{11} & A_{21} & A_{31} \\ A_{12} & A_{22} & A_{32} \\ A_{13} & A_{23} & A_{33} \end{bmatrix}$$

by finding M^{-1} and showing that $MM^{-1} = I$, where D is the determinant of the matrix and A_{ij} is the cofactor of a_{ij}.

11 $\begin{bmatrix} 3 & 2 \\ 1 & 4 \end{bmatrix}$ **12** $\begin{bmatrix} 1 & 0 & 2 \\ 3 & -1 & -2 \\ 0 & 3 & -4 \end{bmatrix}$ **13** $\begin{bmatrix} 0 & 1 & -1 \\ 1 & -1 & 2 \\ 2 & 0 & -2 \end{bmatrix}$

14 Show that I is self-inverse.

Find the value of the determinant in each of problems 15 to 20.

15 $\begin{vmatrix} 3 & 2 \\ -1 & 5 \end{vmatrix}$ **16** $\begin{vmatrix} 0 & 3 \\ 4 & 2 \end{vmatrix}$ **17** $\begin{vmatrix} 4 & -7 \\ -5 & 9 \end{vmatrix}$

18 $\begin{vmatrix} 1 & 2 & 3 \\ 3 & 1 & 2 \\ 2 & 3 & 1 \end{vmatrix}$ **19** $\begin{vmatrix} 2 & 3 & 4 \\ 3 & 2 & 3 \\ 4 & 3 & 2 \end{vmatrix}$ **20** $\begin{vmatrix} 0 & 1 & 2 \\ -2 & 0 & 3 \\ -5 & 0 & 4 \end{vmatrix}$

Prove the statement in each of problems 21 to 24 by use of properties 1 to 7 without expanding.

21 $\begin{vmatrix} a & b & d \\ d & a & b \\ b & d & a \end{vmatrix} = \begin{vmatrix} a & d & b \\ b & a & d \\ d & b & a \end{vmatrix}$ **22** $\begin{vmatrix} 3 & 1 & 3 \\ 2 & -2 & 1 \\ 4 & 2 & -3 \end{vmatrix} = \begin{vmatrix} 3 & 2 & 3 \\ 6 & -12 & 3 \\ 4 & 4 & -3 \end{vmatrix}$

23 $\begin{vmatrix} x_1 & y_1 & z_1 \\ x_2 & y_2 & z_2 \\ x_3 & y_3 & z_3 \end{vmatrix} = \begin{vmatrix} x_1 - ay_1 & y_1 + bz_1 & z_1 \\ x_2 - ay_2 & y_2 + bz_2 & z_2 \\ x_3 - ay_3 & y_3 + bz_3 & z_3 \end{vmatrix}$

24 $\begin{vmatrix} 8 & 1 & 4 \\ 4 & 2 & 2 \\ 6 & 5 & 3 \end{vmatrix} = 2 \begin{vmatrix} 4 & 2 & 3 \\ 1 & 2 & 5 \\ 4 & 2 & 3 \end{vmatrix} = 0$

Find the replacement set for x so that the statements in problems 25 and 26 are true.

25 $\begin{vmatrix} x & 5 \\ 4 & 3 \end{vmatrix} = 7$ **26** $\begin{vmatrix} 1 & 2 & x \\ x & 0 & 3 \\ 5 & 4 & 5 \end{vmatrix} = 24$

Find the solution sets of the systems of equations in problems 27 to 30 by use of Cramer's rule.

27 $2x + y = 3$
$5x + 3y = 5$

28 $6x + 5y = 9$
$3x + y = 0$

29 $x - y + z = 0$
$3x + 2y - z = -3$
$4x + 3y - 2z = -5$

30 $x + y - z = -1$
$y + 2z = 6$
$x - 3z = -7$

14

Ratio, Proportion, and Variation

In education circles one frequently hears the term "IQ," or *intelligence quotient*. In a machine shop one may hear references to the *gear ratio*. Highway engineers are interested in the *grades* of highways, and carpenters continually deal with the *pitch* of a roof. Each of the italicized words is the quotient of two numbers and is the name of a measure of some particular thing. For example, a person's IQ is used as a measure of his mental ability, and both the grade of a highway and the pitch of a roof are measures of a slope. In this chapter we discuss ratios, and we shall be especially interested in pairs of numbers that vary in such a way that their ratio or product never changes.

14.1 RATIO

By definition, the *ratio* of any number a to any nonzero number b is the quotient obtained by dividing a by b, that is, $\frac{a}{b}$. Hence, the ratio of 10 feet to 2 feet is $\frac{10}{2} = 5$, and the ratio of 6 pounds to 15 pounds is $\frac{6}{15} = \frac{2}{5}$.

The ratio $\frac{a}{b}$ is commonly written, for convenience in writing and typesetting, as a/b; it is also written as $a : b$. The colon in $a : b$ should be read as a sign of division. Much of the difficulty often encountered in the study of ratios can be avoided if it is remembered that the expressions $\frac{a}{b}$, a/b, and $a : b$ are equivalent to each other and are the symbolic expressions of "the ratio of a to b."

If a and b are magnitudes of the same kind, then they must be expressed in the same unit if the ratio a/b is to have a meaning. Thus, in order to find the ratio of 3 inches to 2 feet, we reduce 2 feet to 24 inches; then the desired ratio is $\frac{3}{24} = \frac{1}{8}$. In such cases, the ratio a/b represents an abstract number and is the answer to the question, "The number a is what multiple of b, or what fractional part of b?"

Although we ordinarily think of a ratio as a relation that involves quantities of the same kind, we frequently see a ratio expressed between magnitudes that are entirely different in their nature. For example, if a body moves s feet in t seconds at a uniform speed, then the velocity v of the body is expressed as

$$v = \frac{s}{t}$$

The value of the ratio is the number of feet that a body moves in 1 second. Also, the price P per acre of a farm is equal to the ratio of the total cost C to the number of acres n or

317

$$P = \frac{C}{n}$$

Again, the value of the ratio is the portion of C that corresponds to 1 acre.

Thus if a and b do not represent magnitudes of the same kind, the ratio $a : b$ represents the portion of a that corresponds to one unit of b.

14.2 PROPORTION

If $a = kb$, $c = kd$, and b and d are not zero, we have

$$\frac{a}{b} = \frac{c}{d} = k$$

Proportion   ***The fractional equation*** $\dfrac{a}{b} = \dfrac{c}{d}$ ***is called a proportion***   (14.1)

The proportion (14.1) is frequently expressed in the form

$$a : b = c : d$$

and is read "a is to b as c is to d."

Extremes   The numbers a and d are called the *extremes* of a proportion, and c
Means   and b are the *means*.

Proportions are used extensively in geometry, trigonometry, physics, chemistry, and many other subjects that involve elementary algebra. We shall state below several theorems dealing with proportions. The proofs of these theorems depend upon the properties of fractions and, for the most part, will be left as exercises for the student.

$$\textit{If } \frac{a}{b} = \frac{c}{d}, \textit{ then } ad = bc \tag{14.2}$$

$$\textit{If } \frac{a}{b} = \frac{c}{d}, \textit{ then } \frac{b}{a} = \frac{d}{c} \textit{ and } \frac{a}{c} = \frac{b}{d} \tag{14.3}$$

$$\textit{If } \frac{a}{b} = \frac{c}{d}, \textit{ then } \frac{a+b}{b} = \frac{c+d}{d} \textit{ and } \frac{a-b}{b} = \frac{c-d}{d} \tag{14.4}$$

The first statement in (14.4) can be proved by adding 1 to each member of the given equation and simplifying. The second statement is proved by a similar method.

$$\textit{If } \frac{a}{b} = \frac{c}{d}, \textit{ then } \frac{a+b}{a-b} = \frac{c+d}{c-d} \tag{14.5}$$

Mean   In the proportion $a/c = c/d$, or $a : c = c : d$, c is called the *mean propor-*
proportional   *tional* to (or between) a and d. It follows from (14.2) that $c^2 = ad$. Hence $c = \pm\sqrt{ad}$. Therefore,

> ***The two mean proportionals to (or between) two numbers are***
> ***the positive and negative square roots of their product***   **(14.6)**

A statement resembling a proportion is frequently used to indicate that three ratios are equal. For example,

$$a : b : c = x : y : z$$

is a short way of stating that

$$a : b = x : y \qquad a : c = x : z \qquad \text{and} \qquad b : c = y : z$$

or that

$$\frac{a}{x} = \frac{b}{y} = \frac{c}{z}$$

If we set each of the above ratios equal to k, we have

$$a = kx \qquad b = ky \qquad \text{and} \qquad c = kz$$

and it follows that

$$\frac{a + b + c}{x + y + z} = k$$

Hence we have the theorem

$$\textbf{\textit{If } } a : b : c = x : y : z, \textbf{\textit{ then }} \frac{a + b + c}{x + y + z} = \frac{a}{x} = \frac{b}{y} = \frac{c}{z} \qquad (14.7)$$

EXAMPLE 1   If $a/b = c/d$, $a - b = 12$, $c - 6$, and $d = 2$, find b.

Solution
$$\frac{a}{b} = \frac{c}{d} \qquad \blacktriangleleft \text{ given}$$

$$\frac{a - b}{b} = \frac{c - d}{d} \qquad \blacktriangleleft \text{ by (14.4)}$$

$$\frac{12}{b} = \frac{6 - 2}{2} \qquad \blacktriangleleft \text{ since } a - b = 12,\ c = 6,\ \text{and } d = 2$$

$$4b = 24 \qquad \blacktriangleleft \text{ by (14.2)}$$

$$b = 6$$

EXAMPLE 2   Find c if c is the mean proportional to 6 and 24.

Solution
$$c = \pm\sqrt{6(24)} = \pm 12 \qquad \blacktriangleleft \text{ by (14.6)}$$

EXAMPLE 3   If the sides of two triangles are x, y, z, and 5, 18, and 21, respectively, and the perimeter of the first triangle is 176, find x, if $x : y : z = 5 : 18 : 21$.

Solution

$$x : y : z = 5 : 18 : 21 \qquad \blacktriangleleft \text{ given}$$

$$\frac{x + y + z}{5 + 18 + 21} = \frac{x}{5} \qquad \blacktriangleleft \text{ by (14.7)}$$

$$\frac{176}{44} = \frac{x}{5} \qquad \blacktriangleleft \text{ since } x + y + z = 176$$

$$44x = 880$$

$$x = 20$$

Exercise 14.1   Ratio and Proportion

Express the indicated ratio in each of problems 1 to 8 as a fraction and simplify.

1	4 weeks to 2 days	**2**	$2.20 to 7 nickels
3	3 months to 4 years	**4**	21° to 14′
5	5 gallons to 6 quarts	**6**	4 yards to 5 feet
7	1 mile to 330 yards	**8**	6 bushels to 7 pecks

Find the value of the indicated ratio in each of problems 9 to 16.

9	144 eggs to 8 hens	**10**	273 miles to 21 gallons
11	336 miles to 6 hours	**12**	234 students to 9 classes
13	$3,105 to 27 acres	**14**	$7.56 to 4 pounds of beef
15	$11,700 to 9 months	**16**	221 marbles to 13 boys

17   Find the density of a piece of wood if 50 cubic centimeters of it weighs 42 grams. The density of a body is defined to be the ratio of the weight of a body to its volume.

18   Find the specific gravity of a piece of metal if 1 cubic foot of it weighs 137.50 pounds and 1 cubic foot of water weighs 62.5 pounds. The specific gravity of a body is the ratio of the weight of the body to the weight of an equal body of water.

19   What is the heat of fusion of ice if 3,200 calories are required to melt 40 grams? The heat of fusion of a solid material is the number of calories required to melt one gram of the material.

20   Find the specific heat of aluminum if 50.4 calories are required to increase the temperature of 40 grams by 6° C. The specific heat of a substance is the number of calories required to raise the temperature of 1 gram by 1° C.

Find the value of x in each of problems 21 to 28.

21 $x : 7 = 9 : 20$
23 $34 : 17 = x : 5$
25 $(3 + x) : (1 + x) = 10 : 6$
27 $7 : (9 + x) = (x - 3) : 4$

22 $12 : x = 18 : 6$
24 $4 : 3 = 8 : x$
26 $5 : 4 = (x + 7) : (x + 5)$
28 $(x + 11) : 4 = 6 : (3 - x)$

Find the mean proportional between the pair of numbers in each of problems 29 to 32.

29 1, 16   **30** 3, 27   **31** 4, 25   **32** 12, 27

Find the third proportional to the pair of numbers in each of problems 33 to 36.

33 2, 6   **34** 1, 5   **35** 3, 12   **36** 5, 15

Find the fourth proportional to the triplet of numbers in each of problems 37 to 40.

37 7, 2, 14   **38** 8, 6, 12   **39** 12, 3, 28   **40** 10, 15, 6
41 If $x : y = 3 : 4$ and $x - y = -1$, find x and y.
42 If $x : 2 = y : 4$ and $x + y = 15$, find x and y.
43 If $x : y = 3 : 2$ and $x + y = 15$, find x and y.
44 If $x : 12 = y : 3$ and $x - y = 3$, find x and y.

14.3 VARIATION

The term *variation* is often used in dealing with ratios. The remainder of this chapter will be devoted to a discussion of the types and uses of variation. The four types are *direct variation, inverse variation, joint variation,* and *combined variation.*

Direct variation   If a number y *varies directly* as another number x, $x \neq 0$, then $y = kx$. The word "directly" is often omitted when discussing direct variation. If the weight w of a piece of pipe varies directly as its length L, then $w = kL$.

Inverse variation   If one number y *varies inversely* as another x, then $y = k/x$. Thus, if the volume V of a confined mass of gas at a constant temperature varies inversely as the pressure P, then $V = k/P$. This is known as Boyle's law.

Joint variation If one number *varies jointly* as two or more others, then it varies as their product. Thus, if the volume V of a box varies jointly as its length L, width W, and height H, then $V = kLWH$.

Combined variation If one number varies jointly as several others and inversely as still others, then the variation is referred to as a *combined variation*. Thus, if y varies jointly as x and z and inversely as w, then $y = kxz/w$.

Constant of variation In each of the four types of variations defined above, k is called the *constant of variation*. The constant can be determined if a set of values for the variables is known. Thus if in the example given for direct variation, $w = 90$ for $L = 15$, then $90 = k(15)$ and $k = 6$; hence, $w = 6L$.

A typical problem in variation involves a set of values for all the variables and a second set for all but one of the variables. After the variation has been expressed as an equation, the value of the constant of variation can be found by making use of the complete set of values of the variables as in the previous example. Finally, the value of the variable not included in the incomplete set of values can be determined by use of the incomplete set. If we want to find the value of w for $L = 17$, we need only substitute 17 for L in $w = 6L$ and thereby get $w = 6(17) = 102$.

EXAMPLE 1 The horsepower required to propel a ship varies as the cube of the speed. If the horsepower required for a speed of 15 knots is 10,125, find the horsepower required for a speed of 20 knots.

Solution If we let

$$P = \text{required horsepower}$$
$$s = \text{speed in knots}$$

then, since P varies as s^3, we have

$$P = ks^3 \qquad (1)$$

We are given that $P = 10,125$ for $s = 15$ knots. By substituting these values in (1), we get

$$10,125 = k(15^3)$$

and

$$k = \frac{10,125}{15^3} = \frac{10,125}{3,375} = 3$$

Now we substitute $k = 3$ and $s = 20$ in (1) and have

$$P = 3(20^3) = 3(8,000)$$
$$= 24,000 \text{ horsepower}$$

EXAMPLE 2 The weight of a rectangular block of metal varies jointly as the length, the width, and the thickness. If the weight of a 12- by 8- by 6-inch block of aluminum is 18.7 pounds, find the weight of a 16- by 10- by 4-inch block.

Solution

1 We let

$$W = \text{weight, pounds}$$
$$l = \text{length, inches}$$
$$w = \text{width, inches}$$
$$t = \text{thickness, inches}$$

Then, since the weight varies jointly as the length, width, and thickness, we have

$$W = klwt$$

2 When $l = 12$ inches, $w = 8$ inches, and $t = 6$ inches, $W = 18.7$ pounds. Therefore,

$$18.7 = k(12)(8)(6)$$
$$= 576k$$

and

$$k = \frac{18.7}{576}$$

3 On substituting $k = 18.7/576$, $l = 16$, $w = 10$, and $t = 4$ in the equation $W = klwt$, we obtain

$$W = \frac{18.7}{576}(16)(10)(4)$$
$$= 20.8 \text{ pounds}$$

as the weight of the 16- by 10- by 4-inch block. The reader should note that in the example k is the weight of 1 cubic inch of aluminum.

EXAMPLE 3

The safe load of a beam with a rectangular cross section that is supported at each end varies jointly as the product of the width and the square of the depth and inversely as the length of the beam between supports. If the safe load of a beam 3 inches wide and 6 inches deep with supports 8 feet apart is 2,700 pounds, find the safe load of a beam of the same material that is 4 inches wide and 10 inches deep with supports 12 feet apart.

Solution

1 We let

$$w = \text{width of beam, inches}$$
$$d = \text{depth of beam, inches}$$
$$l = \text{length between supports, feet}$$
$$L = \text{safe load, pounds}$$

Then

$$L = \frac{kwd^2}{l}$$

2 According to the first set of data, when $w = 3$, $d = 6$, and $l = 8$, then $L = 2,700$. Therefore,

$$2,700 = \frac{k(3)(6^2)}{8}$$

$$21,600 = 108k$$

and $\qquad\qquad k = 200$

3 Consequently, if $w = 4$, $d = 10$, $l = 12$, and $k = 200$, we have

$$L = \frac{200(4)(10^2)}{12}$$

$$= 6,666\tfrac{2}{3}$$

Exercise 14.2   Variation

Express the variation in each of problems 1 to 4 as an equation and find the constant of variation.

1 W varies directly as t and is 18 for $t = 6$.
2 s varies inversely as v and is 11 for $v = 3$.
3 m varies jointly as p and q and is 72 for $p = 4$ and $q = 2$.
4 v varies inversely as n and jointly as q and r and is 108 for $n = 5$, $q = 12$, and $r = 15$.

Find the constant of variation in each of problems 5 to 8.

5 y varies as x and $x : y = 5 : 3$.
6 y varies inversely as x and $x : 4 = 5 : y$.
7 y varies directly as x and inversely as z and $y : x = 7 : z$.
8 y varies jointly as p and q and inversely as r^2 and $2p : ry = 5r : q$.
9 If the resistance of the air is neglected, the velocity attained by a compact body falling from rest varies as the time of the fall. If the velocity of a falling body is 96.6 feet per second at the end of 3 seconds, find its velocity at the end of 5 seconds.
10 The mechanical advantage of a jackscrew varies directly as the length of the lever arm. If the mechanical advantage of a jackscrew is 192 when a 3-foot lever arm is used, what is the mechanical advantage for a 2-foot lever arm?
11 If the other variables are constant, the lift on a wing of a plane varies with the density of the air. If the density of the air at sea level is 0.08 pound per cubic foot and the lift on a wing is 2,500 pounds, find the lift at an altitude at which the density of the air is 0.06 pound per cubic foot.
12 If the temperature is constant, the density of a gas varies directly as the pressure. If the density of air at sea level is 0.078 pound per

cubic foot and the pressure is 2,160 pounds per square foot, what is the density where the pressure is 1,800 pounds per square foot?

13 If other factors are equal, the centrifugal force on a curve varies inversely as the radius of the turn. If the centrifugal force is 18,000 pounds for a curve of radius 50 feet, find the centrifugal force on a curve of radius 150 feet.

14 The weight of a body situated above the surface of the earth varies inversely as the square of the distance from the body to the center of the earth. If a boy weighs 121 pounds on the surface, how much would he weigh 400 miles above the surface? Assume the radius of the earth to be 4,000 miles.

15 For the same load, the amount a wire stretches varies inversely as the square of the diameter. If a wire with a diameter of 0.6 inch is stretched 0.006 inch by a given load, how much will a wire of the same material with a diameter of 0.2 inch be stretched by the same load?

16 Under the same load, the sag of beams of the same material, length, and width varies inversely as the cube of the thickness. If a beam 4 inches thick sags $\frac{1}{64}$ inch when a load is placed on it, find the sag of a beam 2 inches thick under the same load.

17 The force necessary to pull a wire through the surface of a liquid varies jointly as the length of the wire and the surface tension of the liquid. If 360 dynes will pull a wire 12 centimeters in length through a liquid with a surface tension of 200 dynes per centimeter, find the force necessary to pull a wire 8 centimeters long through a liquid with a surface tension of 40 dynes per centimeter.

18 The pressure on the bottom of a container holding a liquid varies jointly as the depth and the specific gravity of the liquid. The pressure on the bottom of a beaker filled to a depth of 10 inches with mercury is 4.90 pounds per square inch. The specific gravity of mercury is 13.6. Find the pressure on the bottom of an upright drum filled to a depth of 3 feet with oil of specific gravity 0.8.

19 The kinetic energy of a moving body varies jointly as its mass and the square of its velocity. If the kinetic energy of a body of mass 10 grams moving at 5 centimeters per second is 125 ergs, find the kinetic energy of a mass of 6 grams moving at 10 centimeters per second.

20 The number of pounds of steam that will flow through an orifice varies jointly as the steam pressure and the area of a cross section of the orifice. If 70 pounds of steam under a pressure of 210 pounds per square inch will flow through an orifice of cross-sectional area of 15 square inches, how much steam under a pressure of 180 pounds per square inch will flow through an orifice with cross-sectional area of 12 square inches in the same amount of time?

21 The velocity of a jet of liquid from a pressure tank varies as the square root of the gauge pressure and inversely as the square root of the

specific gravity of the liquid. If water (specific gravity 1) is discharged at the rate of 24 feet per second from a tank in which the gauge pressure is 500 pounds per square foot, find the velocity of discharge of gasoline (specific gravity 0.64) from a tank with gauge pressure 125 pounds per square inch.

22 The resistance of a wire to a current varies directly as the length and inversely as the square of the diameter. If the resistance of a 75-foot wire of diameter 0.01 inch is 21.6 ohms, find the resistance of a 100-foot wire of diameter 0.02 inch.

23 The crushing load of a circular pillar varies directly as the fourth power of the diameter and inversely as the square of the height of the pillar. If 256 tons is needed to crush a pillar 8 inches in diameter and 20 feet high, find the load that is needed to crush a pillar 6 inches in diameter and 15 feet high.

24 The current I that flows through a system varies directly as the electromotive force E and inversely as the resistance R. If, in a system, a current of 20 amperes flows through a resistance of 24 ohms with an electromotive force of 120 volts, find the current that 200 volts will send through a resistance of 24 ohms.

25 The centrifugal force on a plane varies jointly as the mass of the plane and the square of the velocity and inversely as the radius of the turn. If the centrifugal force is 12,000 pounds on a 3,200-pound plane making a turn of radius 100 feet at a velocity of 75 miles per hour, find the force on a 4,200-pound plane making a turn with a 200-foot radius at a velocity of 100 miles per hour.

26 The sag in a loaded beam varies directly as the cube of the length and inversely as the width and the cube of the thickness. If a loaded beam 6 inches wide, 1 foot thick, and 4 feet long sags $\frac{1}{8}$ inch, find the sag of a beam 6 feet long, 1 foot wide, and 1 foot thick under the same load.

27 The safe load of a beam with a rectangular cross section that is supported at each end varies jointly as the product of the width and the square of the depth and inversely as the length of the beam between supports. If the safe load of a beam 4 inches wide, 8 inches deep, and 8 feet long is 6,400 pounds, find the safe load on a beam 3 inches wide, 8 inches deep, and 12 feet long.

28 Find the safe load for the second beam in problem 27 if the beam is turned on its side.

Exercise 14.3   Review

1 Find the ratio of 4 dimes to 3 quarters.
2 Find the ratio of 75 cents to 15 apples.
3 Find the density of a cork of 60 cubic centimeters if it weighs 24 grams.
4 Find the value of x if $18 : x = 27 : 9$.

5 Find the value of x if $2 : (x + 1) = x : 3$.

6 Find the mean proportional between 4 and 36.

7 Find the third proportional to 2 and 5.

8 Find the fourth proportional to 5, 2, 15.

9 Find x and y if $x : 5 = y : 10$ and $x + y = 9$.

10 If a shot was heard 3 seconds after it was fired 3,315 feet away, when would a person 8,840 feet from the gun hear the shot?

11 How much meat is needed for a meat loaf for 31 people if 7 pounds are needed for 14 people?

Graphs
of the
Trigonometric
Functions

In this chapter, we shall study the way in which the trigonometric function values change as the angle varies through a set of values, then sketch the graphs of the functions, and finally determine the effect on the graph of multiplying the function value by a constant, of multiplying the angle by a constant, and of adding a constant to a multiple of an angle.

15.1 PERIODIC FUNCTIONS

We are able to get a complete sample of the graph of a trigonometric function by examining a limited portion of the graph because of a repeating property of the function that is known as *periodicity*.

If f is a function whose domain consists of all members θ in a specified set, and if p is a number such that

$$f(\theta + p) = f(\theta) \tag{15.1}$$

Periodic function
Period

for all admissible values of θ, then f is said to be a *periodic function* and to have the period p. Furthermore, if p is the smallest positive number for which (15.1) is true, then p is called the *period*.

We can show that the sine, cosine, and tangent are periodic functions and that the period of the sine and cosine is 2π whereas that of the tangent is π by use of the identities for the sum of two angles. Thus,

$$\sin (x + 2\pi) = \sin x \cos 2\pi + \cos x \sin 2\pi$$
$$= \sin x$$
$$\cos (x + 2\pi) = \cos x \cos 2\pi - \sin x \sin 2\pi$$
$$= \cos x$$

and

$$\tan (x + \pi) = \frac{\tan x + \tan \pi}{1 - \tan x \tan \pi}$$
$$= \tan x$$

since $\sin 2\pi = 0$, $\cos 2\pi = 1$, and $\tan \pi = 0$. The other three trigonometric ratios are the reciprocals of these three; consequently, we know that *each*

Periods of the
functions

of the trigonometric functions is periodic and has 2π as a period; furthermore, *the tangent and cotangent have π as a period.*

We now know that the function values of sine, cosine, secant, and cosecant are repeated at intervals of 2π; hence we can get a complete sample of the graph of each by sketching the graph over a horizontal interval of length 2π; furthermore, we can get a complete sample of the graphs of the tangent and cotangent functions by taking a horizontal strip of length π.

15.2 VARIATION OF THE SINE AND COSINE

It is the purpose of this section to determine the range of the sine and cosine functions for $0 \leq \theta \leq 2\pi$ and to investigate the manner in which the function values change as θ increases continuously in the interval. For this purpose we shall use Fig. 15.1, in which the center of the circle is the origin, the radius is unity, the angle θ is in standard position, and $P(x, y)$ is the point where the terminal side of θ intersects the circumference of the circle.

Variation of sine and cosine

By definition, $\sin \theta = y/OP = y/1 = y$ and $\cos \theta = x/1 = x$. For $\theta = 0$, P coincides with A, and so $y = 0$ and $x = 1$. As θ increases continuously from 0 to $\pi/2$, $y = \sin \theta$ increases continuously from 0 to 1 and $x = \cos \theta$ decreases continuously from 1 to 0. As θ continues to increase continuously to π and then to $3\pi/2$ and finally to 2π, $y = \sin \theta$ varies continuously through the intervals 1 to 0, 0 to -1, and -1 to 0, and $x = \cos \theta$ varies through the intervals 0 to -1, -1 to 0, and 0 to 1. We summarize this information by arranging it in tabular form:

Variation of θ	Variation of $P(x, y)$	Variation of $y = \sin \theta$	Variation of $x = \cos \theta$
0 to $\dfrac{\pi}{2}$	$A(1, 0)$ to $B(0, 1)$	0 to 1	1 to 0
$\dfrac{\pi}{2}$ to π	$B(0, 1)$ to $C(-1, 0)$	1 to 0	0 to -1
π to $\dfrac{3\pi}{2}$	$C(-1, 0)$ to $D(0, -1)$	0 to -1	-1 to 0
$\dfrac{3\pi}{2}$ to 2π	$D(0, -1)$ to $A(1, 0)$	-1 to 0	0 to 1

15.3 BOUNDS AND AMPLITUDE

If we refer again to Fig. 15.1, we see that $x = \cos \theta$ and $y = \sin \theta$ are never greater than 1 or less than -1; furthermore, $\tan \theta$ is larger than any number that can be selected if θ is acute and near enough to $\pi/2$ and is negative and numerically larger than any chosen number if θ is a second-quadrant angle and sufficiently near $\pi/2$. Thus, $\sin \theta$ and $\cos \theta$ have a property of being hemmed in that $\tan \theta$ does not possess. A function that has this property is said to be *bounded* and is more precisely defined by the statement: A function f is *bounded above* if a number M exists such that $f(x) \leq M$ for all x in the domain of f and is *bounded below* if a number m exists such that $f(x) \geq m$ for all x in the domain of f.

Bounded function

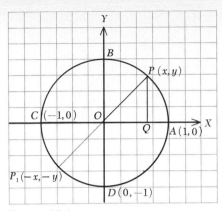

Figure 15.1

If a function f is bounded above, the smallest M for which $f(x) \leq M$ for all x in the domain of f is called the *least upper bound;* furthermore, if f is bounded below, the largest m for which $f(x) \geq m$ for all x in the domain of f is called the *greatest lower bound.* The least upper and greatest lower bounds of $\sin \theta$ and $\cos \theta$ are 1 and -1.

If M and m are the least upper and greatest lower bounds of a periodic function f, then $\frac{1}{2}(M - m)$ is called the *amplitude* of f. Thus, the amplitude of $\{(\theta, \sin \theta)\}$ is $\frac{1}{2}[1 - (-1)] = 1$, and so is that of $\{(\theta, \cos \theta)\}$.

15.4   THE GRAPH OF $y = \sin x$

Since the equation $y = \sin x$ defines a function for all real values of x, it has a graph in the usual rectangular coordinate system. This graph consists of the points $(x, \sin x)$ that constitute the function.

We shall make use of the periodicity and bounds of $\sin x$ as developed in this chapter along with a table of corresponding values of x and $y = \sin x$ for selected values of x. We shall assign the multiples of $\pi/6$ and $\pi/4$ between 0 and 2π to x and enter each of them and the corresponding values of $\sin x$ in the following table.

x	0	$\pi/6$	$\pi/4$	$\pi/3$	$\pi/2$	$2\pi/3$	$3\pi/4$	$5\pi/6$	π
$\sin x = y$	0	.5	.7	.9	1	.9	.7	.5	0

x	$7\pi/6$	$5\pi/4$	$4\pi/3$	$3\pi/2$	$5\pi/3$	$7\pi/4$	$11\pi/6$	2π
$\sin x = y$	$-.5$	$-.7$	$-.9$	-1	$-.9$	$-.7$	$-.5$	0

We now draw a pair of coordinate axes, choose any convenient unit on the Y axis, and use approximately 3.14 times this unit distance to represent π on the X axis. If we locate the points (x, y) from the table and draw

331

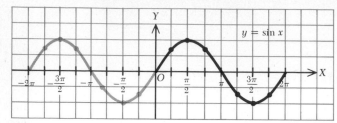

Figure 15.2

a smooth curve through them, we obtain the dark part of the curve in Fig. 15.2. It is of horizontal length 2π. If we use the fact that $y = \sin x$ is periodic, with 2π as a period, we can extend the curve as far as we wish in each direction. The light portion was obtained in this way. The bounds are used so that the curve is never above $y = 1$ or below $y = -1$.

15.5   THE GRAPH OF $y = \cos x$

The graph of $y = \cos x$ can be drawn by making use of the bounds and periodicity along with a table of values. We shall use the same values of x as in the preceding section and make the following table.

x	0	$\pi/6$	$\pi/4$	$\pi/3$	$\pi/2$	$2\pi/3$	$3\pi/4$	$5\pi/6$	π	$7\pi/6$	$5\pi/4$
$\cos x = y$	1	.9	.7	.5	0	−.5	−.7	−.9	−1	−.9	−.7

x	$4\pi/3$	$3\pi/2$	$5\pi/3$	$7\pi/4$	$11\pi/6$	2π
$\cos x = y$	−.5	0	.5	.7	.9	1

If we locate the points determined by these number pairs and draw a smooth curve through them, we obtain the dark part of the curve in Fig. 15.3. Since it is of horizontal length 2π, we can extend the graph by making use of the periodicity. The light portion was so obtained.

A glance at Figs. 15.2 and 15.3 or a comparison of the tables in Secs. 15.4 and 15.5 shows that the graphs of $y = \sin x$ and $y = \cos x$ have the

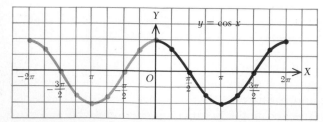

Figure 15.3

same shape and that the latter can be obtained from the former by moving the graph of $y = \sin x$ exactly $\pi/2$ units to the left.

15.6 VARIATION OF THE TANGENT

We shall now consider $\tan \theta = y/x$ in which both x and y assume all values between -1 and 1, inclusive. The value of $\tan 0 = \frac{0}{1} = 0$, but, as θ increases and becomes nearer and nearer to $\pi/2$, y becomes as near to 1 as one wishes and x becomes arbitrarily near zero; hence, $\tan \theta$ remains positive and is larger than any selected number. This state of affairs is described in mathematical language by saying that if the angle θ increases toward $\pi/2^-$, then $\tan \theta$ increases without bound. This is put in symbols as

$$\lim_{\theta \to \pi/2^-} \tan \theta = \infty$$

Variation of the tangent

If, however, θ is in the second quadrant and approaches $\pi/2^+$, then $\tan \theta$ is negative and numerically larger than any chosen number. We then write

$$\lim_{\theta \to \pi/2^+} \tan \theta = -\infty$$

If θ approaches π from either direction, then $\tan \theta$ approaches zero and $\tan \pi = 0$.

A similar argument can be used to show that $\tan \theta$ approaches infinity as θ increases toward $3\pi/2$ and $\tan \theta$ approaches minus infinity as θ decreases toward $3\pi/2$. Consequently, the variation of $\tan \theta$ is as shown in the following table.

θ varies from	$P(x, y)$ varies from	$\tan \theta$ varies from
0 to $\pi/2$	(1, 0) to (0, 1)	0 to ∞
$\pi/2$ to π	(0, 1) to (-1, 0)	$-\infty$ to 0
π to $3\pi/2$	(-1, 0) to (0, -1)	0 to ∞
$3\pi/2$ to 2π	(0, -1) to (1, 0)	$-\infty$ to 0

15.7 THE GRAPH OF $y = \tan x$

We shall assign the multiples of $\pi/6$ and $\pi/4$ between $-\pi/2$ and $\pi/2$ to x and enter them and the corresponding values of $\tan \theta$ in a table. Thus, we have

x	$-\pi/2$	$-\pi/3$	$-\pi/4$	$-\pi/6$	0	$\pi/6$	$\pi/4$	$\pi/3$	$\pi/2$
$\tan x = y$	$-\infty$	-1.7	-1	$-.6$	0	$.6$	1	1.7	∞

In drawing the graph, we make use of the fact that the symbol ∞ under $\pi/2$ indicates that if x is sufficiently near $\pi/2$, then $\tan x$ is greater than any pre-

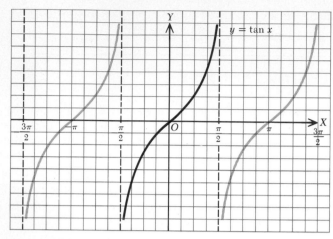

Figure 15.4

viously selected number and therefore the curve continues to rise as it approaches the line $x = \pi/2$. Now, by using this fact, locating the points (x, y) determined by the table, and drawing a smooth curve through them, we obtain the darker part of the graph given in Fig. 15.4. The lighter colored part is obtained by making use of the fact that π is the period of $\tan x$.

15.8 THE GRAPHS OF THE OTHER THREE FUNCTIONS

The graphs of $y = \cot x$, $y = \sec x$, and $y = \csc x$ can be drawn by making use of the periodicity and a table of values. The following table gives several values of x from 0 to 2π and most corresponding function values of the

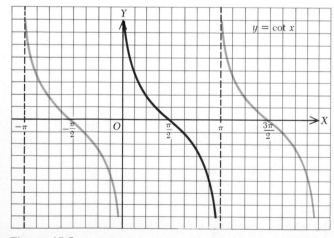

Figure 15.5

x	0	$\dfrac{\pi}{6}$	$\dfrac{\pi}{3}$	$\dfrac{\pi}{2}$	$\dfrac{2\pi}{3}$	$\dfrac{5\pi}{6}$	π	$\dfrac{7\pi}{6}$	$\dfrac{4\pi}{3}$	$\dfrac{3\pi}{2}$	$\dfrac{5\pi}{3}$	$\dfrac{11\pi}{6}$	2π
$\cot x$	∞	$\sqrt{3}$	$\dfrac{\sqrt{3}}{3}$	0	$\dfrac{-\sqrt{3}}{3}$	$-\sqrt{3}$	$-\infty$						
$\sec x$	1	$\dfrac{2}{\sqrt{3}}$	2	∞	-2	$\dfrac{-2}{\sqrt{3}}$	-1	$\dfrac{-2}{\sqrt{3}}$	-2	$-\infty$	2	$\dfrac{2}{\sqrt{3}}$	1
$\csc x$	∞	2	$\dfrac{2}{\sqrt{3}}$	1	$\dfrac{2}{\sqrt{3}}$	2	∞	-2	$\dfrac{-2}{\sqrt{3}}$	-1	$\dfrac{-2}{\sqrt{3}}$	-2	∞

three functions. Values have not been entered for $\cot x$ for angles greater than π, since $y = \cot x$ is periodic with period π. The darker part of the graph in each of Figs. 15.5 to 15.7 was obtained by use of the appropriate entries in the table of values and the lighter part by use of periodicity. A total of three periods is shown in Fig. 15.5 and two periods in each of Figs. 15.6 and 15.7.

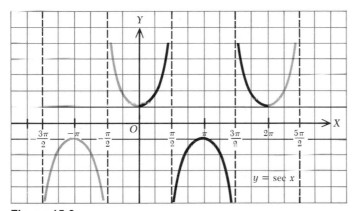

Figure 15.6

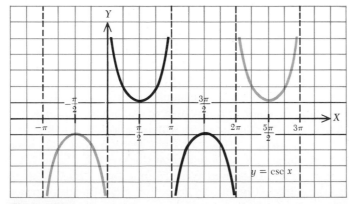

Figure 15.7

Exercise 15.1   Graphs of the Trigonometric Functions

Show the variation of the function value in each of problems 1 to 4 for the specific interval by making a table similar to the one in Sec. 15.4.

1 $\cos x, -3\pi/2 \le x \le \pi/2$ **2** $\sec x, -\pi/2 \le x \le 3\pi/2$
3 $\csc x, -\pi \le x \le \pi$ **4** $\tan x, 0 \le x \le \pi$

Make a table of corresponding values of the angle and the function value in each of problems 5 to 16 and then sketch the graph.

5 $y = \sin x, -\pi \le x \le \pi$ **6** $y = \cos x, -2\pi \le x \le 0$
7 $y = \cos x, -\pi/2 \le x \le 5\pi/2$ **8** $y = \sin x, -\pi/2 \le x \le 5\pi/2$
9 $y = \cot x, -\pi \le x \le 0$ **10** $y = \tan x, \pi/2 \le x \le 2\pi$
11 $y = \tan x, -\pi/2 \le x \le 3/2$ **12** $y = \cot x, \pi/2 \le x \le 5\pi/2$
13 $y = \sec x, \pi/2 \le x \le 3\pi$ **14** $y = \csc x, -\pi \le x \le 2\pi$
15 $y = \csc x, 0 \le x \le 3\pi$ **16** $y = \sec x, \pi \le x \le 7\pi/2$

15.9   GRAPHS OF FUNCTIONS OF THE TYPE $\{(x, y) \,|\, y = af(bx + c)\}$

The graphs of the trigonometric functions are used in most branches of mathematics as well as in physics, engineering, and other branches of science. We found in Secs. 15.4, 15.5, 15.7, and 15.8 how to sketch the graph of the function defined by $y = f(x)$, where f is a trigonometric function, and we shall see, in this section, how to sketch the graph of $\{(x, y) \,|\, y = af(bx + c)\}$, where a, b, and c are constants and f is a trigonometric function.

THE FUNCTION $\{(x, y) \,|\, y = f(bx)\}$ All of the values of the function will be taken on by $y = \sin bx$ as the argument bx changes by 2π. Thus

The period of f(bx)   for $0 \le bx < 2\pi$; hence if $b > 0$ for $0 \le x < 2\pi/b$. Thus, the period of $\{(x, y) \,|\, y = f(bx)\}$ where f is a trigonometric function is $2\pi/b$. Consequently, the period of a trigonometric function of bx is $1/b$ times the period of that function of x. Therefore, we know that multiplying the angle by a constant divides the period by that constant. For example, the period of $\tan x$ is known to be π; hence the period of $\tan 5x$ is $\pi/5$.

THE FUNCTION $\{(x, y) \,|\, y = f(bx + c)\}$ We shall now investigate the effect on the graph of adding a constant to a multiple of the angle. In order to do this, we shall compare the values of x for which bx and $bx + c$ have the

Relative positions of graphs   same value. If $bx + c = x_1$, then $x = (x_1 - c)/b = x_1/b - c/b$, and if $bx = x_1$, then $x = x_1/b$. The first value of x is c/b less than the second. Therefore, we know that the graph of a trigonometric function of $bx + c$ is c/b units to the left of that of the same function of bx. For example, the graph of the function determined by $y = \sin(2x + 3)$ is $\frac{3}{2}$ units to the left of the graph of the function determined by $y = \sin 2x$; furthermore, the graph of $y = \cos(3x - 4) = \cos[3x + (-4)]$ is $-\frac{4}{3}$ units to the left of that of $y = \cos 3x$. Some readers may prefer to say $\frac{4}{3}$ of a unit to the right instead of $-\frac{4}{3}$ of a unit to the left.

Some physicists and engineers would say that two graphs differ by b/c in phase instead of saying that one is to the left of the other.

THE FUNCTION $\{(x, y)|y = af(bx + c)\}$ We shall investigate the effect on the graph of multiplying the function by a constant.   If $y_1 = f(bx + c)$ and $y_2 = af(bx + c)$, then $y_2 = ay$. Therefore the ordinate of each point on the graph of the function determined by $y = af(bx + c)$ is a times the ordinate of the corresponding point on the graph of the function determined by $y = f(bx + c)$.   Consequently, multiplying a trigonometric function value by a constant multiplies the ordinate of each point on the graph by that constant.   For example, if a point on the graph of the function determined by $y = 3\sin(2x - 1)$ and a point on the graph of $\{(x, y)|y = \sin(2x - 1)\}$ have the same abscissa, then the ordinate of the former is 3 times that of the latter.

Ordinate of
$y = af(bx + c)$

EXAMPLE 1

Sketch the graphs of the functions determined by $y = \cos x$, $y = \cos 2x$, $y = \cos(2x + \pi)$, and $y = 3\cos(2x + \pi)$ about the same axes.

Solution

The graph of the function determined by $y = \cos x$ is shown in Fig. 15.3 and the part of it from $x = 0$ to $x = 2\pi$ is reproduced as a part of Fig. 15.8. The graph of $y = \cos 2x$ is also shown, and points on it are determined from the graph of $y = \cos x$ by making use of the fact that multiplying the

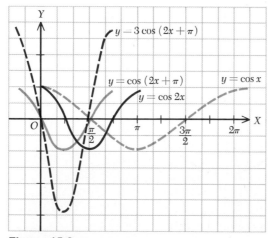

Figure 15.8

angle by a constant is the same as dividing the period by that constant. The graph of $y = \cos (2x + \pi)$ is shown in the figure, and each point on it is $\pi/2$ units to the left of the corresponding point on $y = \cos 2x$. Finally, the graph of the function determined by $y = 3 \cos (2x + \pi)$ is shown, and the ordinate of each point on it is 3 times the ordinate of the corresponding point on the graph of the function determined by $y = \cos (2x + \pi)$.

15.10   COMPOSITION OF ORDINATES

Quite often a function value is made up of the sum of two or more simpler ones. Under such circumstances, points on the graph can be obtained in the usual manner by assigning values in the domain to the independent variable and calculating each corresponding function value. The graph can then be obtained by drawing a smooth curve through the points thus determined. It is often a simpler task, however, to sketch the graph if we first sketch the graphs of the component parts and then add the ordinates of the components for all x in both domains.

EXAMPLE   Sketch the graph of $y = \frac{1}{2}x + \sin x$.

Solution   Since the given function is composed of the sum of two functions whose graphs we can draw readily, we shall begin by letting $y_1 = \frac{1}{2}x$ and $y_2 = \sin x$ and sketching their graphs as shown in Fig. 15.9. Then $y = y_1 + y_2$ for all x

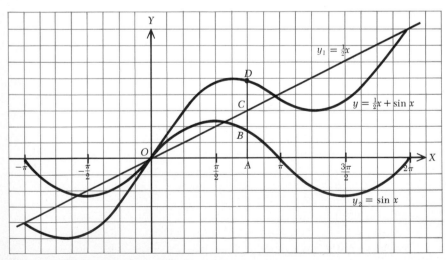

Figure 15.9

in the domain of both y_1 and y_2. For example, if in Fig. 15.9 $x = OA$, then $y_1 = AC$, $y_2 = AB$, and $y = y_1 + y_2 = AC + AB = AD$; hence, D is a point on the required graph. As many other points as desired can be obtained in this manner. The composite graph from $x = -\pi$ to $x = 2\pi$ is shown.

Exercise 15.2 Period, Amplitude, Phase, Composition of Ordinates

Find the period and amplitude of the function defined by the equation in each of problems 1 to 8.

1 $y = 2 \sin 3x$ **2** $y = 4 \cos 2x$ **3** $y = 3 \tan 5x$
4 $y = 5 \sec 4x$ **5** $y = 2 \tan \frac{1}{2}x$ **6** $y = 4 \sec \frac{1}{4}x$
7 $y = 7 \cos \frac{1}{3}x$ **8** $y = 6 \sin \frac{1}{5}x$

Determine the value of a so that the period of the function in each of problems 9 to 16 is the number given after the comma.

9 $\{(x, y) \mid y = \cos ax\}$, π **10** $\{(x, y) \mid y = \sin ax\}$, $\pi/2$
11 $\{(x, y) \mid y = \tan ax\}$, $\pi/3$ **12** $\{(x, y) \mid y = \sec ax\}$, $\pi/4$
13 $\{(x, y) \mid y = \csc ax\}$, 4π **14** $\{(x, y) \mid y = \cot ax\}$, 2π
15 $\{(x, y) \mid y = \sin ax\}$, 2π **16** $\{(x, y) \mid y = \cos ax\}$, 6π

Find the period and amplitude of the function defined by the first equation in each of problems 17 to 24 and the position of its graph relative to that of the function defined by the second equation.

17 $y = \tan (2x + \pi)$, $y = \tan 2x$ **18** $y = \sec (3x + 2\pi)$, $y = \sec 3x$
19 $y = \sin (4x + 3\pi)$, $y = \sin 4x$ **20** $y = \cos (3x + 4\pi)$, $y = \cos 3x$
21 $y = \cos (\frac{1}{2}x + \pi)$, $y = \cos \frac{1}{2}x$ **22** $y = \sin (\frac{1}{4}x + 2\pi)$, $y = \sin \frac{1}{4}x$
23 $y = \csc (\frac{1}{2}x - \pi)$, $y = \csc \frac{1}{2}x$ **24** $y = \cot (\frac{1}{3}x - 3\pi)$, $y = \cot \frac{1}{3}x$

Make use of the facts developed in Sec. 15.8 and sketch the graphs of the four functions defined by the equations in each of problems 25 to 32.

25 $y = \sin x$, $y = \sin 2x$, $y = \sin (2x + \pi)$, $y = 2 \sin (2x + \pi)$
26 $y = \cos x$, $y = \cos 3x$, $y = \cos (3x + 2\pi)$, $y = 2 \cos (3x + 2\pi)$
27 $y = \tan x$, $y = \tan 2x$, $y = \tan (2x - \pi)$, $y = 3 \tan (2x - \pi)$
28 $y = \cot x$, $y = \cot 2x$, $y = \cot (2x - 3\pi)$, $y = 2 \cot (2x - 3\pi)$
29 $y = \csc x$, $y = \csc \frac{1}{2}x$, $y = \csc (\frac{1}{2}x + \frac{1}{4}\pi)$, $y = 4 \csc (\frac{1}{2}x + \frac{1}{4}\pi)$
30 $y = \sec x$, $y = \sec \frac{1}{3}x$, $y = \sec (\frac{1}{3}x - \pi)$, $y = 2 \sec (\frac{1}{3}x - \pi)$

31 $y = \cos x$, $y = \cos \frac{2}{3}x$, $y = \cos \left(\frac{2}{3}x - 2\pi \right)$, $y = 4 \cos \left(\frac{2}{3}x - 2\pi \right)$
32 $y = \sin x$, $y = \sin \frac{3}{4}x$, $y = \sin \left(\frac{3}{4}x + \pi/2 \right)$, $y = 3 \sin \left(\frac{3}{4}x + \pi/2 \right)$

Sketch the graph of the function given in each of problems 33 to 40 by composition of ordinates.

33 $\{(x, y) \mid y = x + \cos x\}$ **34** $\{(x, y) \mid y = x - \sin x\}$
35 $\{(x, y) \mid y = \tan x - x\}$ **36** $\{(x, y) \mid y = x + \cot x\}$
37 $\{(x, y) \mid y = \sin x + \cos x\}$ **38** $\{(x, y) \mid y = \cos x - \sin x\}$
39 $\{(x, y) \mid y = \tan x + \sec x\}$ **40** $\{(x, y) \mid y = \sec x - \cot x\}$

Exercise 15.3 Review

Sketch the graph of the function in the specified interval in each of problems 1 to 6.

1 $y = \sin x$, $-\pi/2 \leq x \leq 3\pi/2$ **2** $y = \csc x$, $0 < x < \pi$
3 $y = \cos x$, $-\pi \leq x \leq \pi$ **4** $y = \sec x$, $\pi/2 < x < 3\pi/2$
5 $y = \tan x$, $-3\pi/2 < x < \pi/2$ **6** $y = \cot x$, $-\pi < x < \pi$
7 Find the period and amplitude of $y = 5 \sin 2x$.
8 Find the period and amplitude of $y = 3 \cos 5x$.
9 Find the period and amplitude of $y = 2 \tan 3x$.

Determine the value of a and of b so that the function in each of problems 10, 11, and 12 has the first number after the function as amplitude and the second as period.

10 $\{(x, y) \mid y = a \cos bx\}$, 3, π
11 $\{(x, y) \mid y = a \sin bx\}$, 0.5, 3π
12 $\{(x, y) \mid y = a \tan bx\}$, ∞, 4π

Find the period and amplitude of the function defined by the first equation in each of problems 13, 14, and 15 and the position of its graph relative to that of the function defined by the second equation.

13 $y = \tan (3x + 2\pi)$, $y = \tan 3x$
14 $y = 3 \cos (2x - \pi)$, $y = 3 \cos 2x$
15 $y = 2 \sin (\frac{1}{2}x + \pi/4)$, $y = 2 \sin \frac{1}{2}x$

Sketch about the same axes the graphs of the four functions defined by the equations in each of problems 16, 17, and 18.

16 $y = \sin x$, $y = \sin 3x$, $y = \sin (3x + \pi)$, $y = 4 \sin (3x + \pi)$

17 $y = \cos x$, $y = \cos 2x$, $y = \cos (2x + 3\pi)$, $y = 2 \cos (2x + 3\pi)$

18 $y = \tan x$, $y = \tan \frac{1}{2} x$, $y = \tan (\frac{1}{2} x + \pi/3)$, $y = 3 \tan (\frac{1}{2} x + \pi/3)$

Sketch the graph of the function given in each of problems 19 and 20.

19 $\{(x, y) \mid y = 2x + \sin x\}$ **20** $\{(x, y) \mid y = x - \cos x\}$

Complex Numbers

In Sec. 11.3, we stated that a complex number is a number of the form $a + bi$, where a and b are real and i represents $\sqrt{-1}$. We also discussed quadratic equations whose roots are complex numbers. The work of Gauss and Argand in the eighteenth century aroused an interest in complex numbers, which have since become of considerable importance in mathematics, physics, electronics, and electrical engineering. In this chapter we shall give another definition of a complex number and make some use of a third form for such numbers.

16.1 COMPLEX NUMBERS AS ORDERED PAIRS

We found in Chap. 2 that the set of real numbers is closed under several operations, including addition. The set of real numbers is not closed if the operation is taking square roots. This is clear since the square of zero is zero and the square of a positive or a negative number is positive. Consequently, the square root of a negative number cannot be a real number. Therefore, we shall introduce a new type of number that permits taking the square root of a negative number.

Complex number   An ordered pair (a, b) of real numbers is called a *complex number*. It can be associated with the point in the plane whose coordinates are (a, b).

In terms of this notation, the complex number that would have been written as $2 + 3i$ in Sec. 11.3 is written as $(2, 3)$. It may be thought of as a vector with real component 2 and imaginary component 3.

Equal complex numbers   Two complex numbers are equal if and only if they are associated with the same point. Consequently, we state as a matter of definition that

$$(a, b) = (c, d), \text{ if and only if } a = c \text{ and } b = d \qquad (16.1)$$

Therefore, $(x, y) = (3, 2)$ if and only if $x = 3$ and $y = 2$.

We shall now define the sum and product of two complex numbers by stating that

Sum
$$(a, b) + (c, d) = (a + c, b + d) \qquad (16.2)$$

Product   and
$$(a, b) \cdot (c, d) = (ac - bd, ad + bc) \qquad (16.3)$$

EXAMPLE 1   By use of Eq. (16.2), we have $(3, 2) + (4, -1) = (3 + 4, 2 - 1) = (7, 1)$.

EXAMPLE 2   By use of Eq. (16.3), we see that

$$(3, 2) \cdot (4, -1) = [3(4) - 2(-1), 3(-1) + 2(4)]$$
$$= (12 + 2, -3 + 8) = (14, 5)$$

We shall now consider a special case of Eqs. (16.2) and (16.3). If $b = d = 0$, then Eq. (16.2) becomes $(a, 0) + (c, 0) = (a+c, 0)$ and Eq. (16.3) becomes $(a, 0) \cdot (c, 0) = (ac, 0)$. Hence, complex numbers with the second of the ordered pairs equal to zero behave as real numbers do relative to addition and multiplication. Consequently, we define $(a, 0)$ as follows:

The complex number $(a, 0)$ is equal to the real number a
and at times will be written as a $\qquad$ **(16.4)**

We shall designate the number $(0, 1)$ by i and apply Eq. (16.3). Thus, we have

$$i^2 = (0, 1) \cdot (0, 1) = (0 \cdot 0 - 1 \cdot 1, 0 \cdot 1 + 1 \cdot 0) = (-1, 0) = -1$$

Hence, i is a square root of -1.

We now obtain a relation between complex numbers in the number pair form (a, b) and in the binomial form $a + bi$ of Sec. 11.3. If a and b are real numbers and $i = (0, 1)$, then

$$\begin{aligned} a + bi &= (a, 0) + (b, 0)(0, 1) \\ &= (a, 0) + (b \cdot 0 - 0 \cdot 1, b \cdot 1 + 0 \cdot 0) \\ &= (a, 0) + (0, b) = (a, b) \end{aligned}$$

We now know that

(a, b) and $a + bi$ are two forms of the same number $\qquad$ **(16.5)**

Real part
Imaginary part

The number a is called the *real part* and b the *coefficient of the imaginary part* of the number $(a, b) = a + bi$.

We shall now find the form taken by Eq. (16.3) if the second factor is a real number. If $(c, d) = (c, 0) = c$, then Eq. (16.3) becomes $(a, b)(c, 0) = (ac, bc)$. Consequently,

If c is a real number, then $(a, b)c = (ac, bc)$ $\qquad$ **(16.3a)**

The reader should show that $(a, b)c = c(a, b)$.

Identity

The complex number $(1, 0)$ is the *identity element for multiplication* since, by use of Eq. (16.3a), we have $(a, b)(1, 0) = (a, b)1 = (a, b)$.

We have defined and used addition and multiplication of complex numbers, and we shall now consider subtraction and division. We shall define the difference $(a, b) - (c, d)$ as the complex number (x, y) such that $(a, b) = (c, d) + (x, y)$. If we make use of the sum of two complex numbers as given by Eq. (16.2), we see that

$$(a, b) = (c + x, d + y)$$

Consequently,

$$a = c + x \qquad \text{and} \qquad b = d + y \qquad \blacktriangleleft \text{ by use of Eq. (16.1)}$$

Therefore, $x = a - c$ and $y = b - d$, and $(a, b) - (c, d) = (x, y)$ becomes

Subtraction $\qquad\qquad$ **$(a, b) - (c, d) = (a - c, b - d)$** $\qquad$ **(16.6)**

EXAMPLE 3   By use of Eq. (16.6) we get

$$(5, 7) - (-3, 1) = [5 - (-3), 7 - 1]$$
$$= (8, 6)$$

We shall define the quotient $(a, b)/(c, d)$, $(c, d) \neq (0, 0)$, as the number (x, y) such that $(a, b) = (c, d)(x, y)$. Now, applying the definition of a product as given by Eq. (16.3), we have

$$(a, b) = (cx - dy, cy + dx)$$

Consequently, by use of Eq. (16.1), we find that

$$cx - dy = a$$
$$dx + cy = b$$

This pair of simultaneous equations may be solved by either of several methods, and the solution is

$$\left(\frac{ac + bd}{c^2 + d^2}, \frac{bc - ad}{c^2 + d^2} \right)$$

Therefore, for $(c, d) \neq (0, 0)$, we know that

Quotient
$$\frac{(a, b)}{(c, d)} = \left(\frac{ac + bd}{c^2 + d^2}, \frac{bc - ad}{c^2 + d^2} \right) \tag{16.7}$$

EXAMPLE 4   By use of Eq. (16.7), we have

$$\frac{(3, 4)}{(5, 6)} = \left(\frac{3 \cdot 5 + 4 \cdot 6}{5^2 + 6^2}, \frac{4 \cdot 5 - 3 \cdot 6}{5^2 + 6^2} \right)$$

$$= \left(\frac{15 + 24}{25 + 36}, \frac{20 - 18}{25 + 36} \right)$$

$$= \left(\tfrac{39}{61}, \tfrac{2}{61} \right)$$

We shall now consider a special case of Eqs. (16.6) and (16.7). If $b = d = 0$, Eq. (16.6) becomes $(a, 0) - (c, 0) = (a - c, 0)$ and Eq. (16.7) becomes

$$\frac{(a, 0)}{(c, 0)} = \left(\frac{ac + 0 \cdot 0}{c^2 + 0^2}, \frac{0 \cdot c - a \cdot 0}{c^2 + 0^2} \right)$$

$$= \left(\frac{a}{c}, 0 \right)$$

Consequently, with $b = d = 0$, complex numbers behave as real numbers do relative to subtraction and division. This should have been expected, since subtraction and division are the inverses of addition and multiplication, respectively.

We have defined the four fundamental operations in the set of complex numbers $\{(a, b)\}$ and have shown that (a, b) and $a + bi$ are the same number. Consequently, applying Eqs. (16.2), (16.3), (16.6), and (16.7) to $a + bi$ and $c + di$, we have

$$(a + bi) + (c + di) = (a + c) + (b + d)i \qquad (16.2)$$
$$(a + bi)(c + di) = (ac - bd) + (ad + bc)i \qquad (16.3)$$
$$(a + bi) - (c + di) = (a - c) + (b - d)i \qquad (16.6)$$
$$\frac{(a + bi)}{(c + di)} = \left(\frac{ac + bd}{c^2 + d^2}, \frac{bc - ad}{c^2 + d^2}\right) \qquad (16.7)$$

Consequently, the four fundamental operations can be applied to complex numbers just as they are to other binomials. Furthermore, the quotient $(a + bi)/(c + di)$ can be obtained readily by multiplying the numerator and denominator by $c - di$.

Conjugate

The complex number $(c, -d) = c - di$ is called the *conjugate* of $(c, d) = c + di$ and is often indicated by placing a line above the number. Thus $\overline{3 + 4i} = 3 - 4i$ is the conjugate of $3 + 4i$.

EXAMPLE 5

Find the product of $2 - 3i$ and the conjugate of $4 - 5i$.

Solution

The conjugate of $4 - 5i$ is $4 - (-5i) = 4 + 5i$. Hence, the desired product is

$$(2 - 3i)\overline{(4 - 5i)} = (2 - 3i)(4 + 5i)$$
$$= 8 + 10i - 12i - 15i^2$$
$$= 23 - 2i \qquad \blacktriangleleft \text{ since } i^2 = -1$$

Exercise 16.1 Fundamental Operations

Determine the replacement for x and for y so that the two complex numbers in each of problems 1 to 8 are equal.

1 $(x, 2), (5, y)$
2 $(x + 1, y - 2), (4, -3)$
3 $(x - 3, y - 5), (4, 5 - y)$
4 $(x - y, 4), (2, x + y)$
5 $x + 7i, 2 - yi$
6 $x - 2i, y + xi$
7 $y + 6i, 2x + yi$
8 $y + 2xi, x - 3 + (y + 5)i$

Perform the indicated operation in each of problems 9 to 48 and leave each result in the same form as the problem.

9   $(2, 3) + (4, -5)$
10 $(-1, 3) + (0, 2)$
11 $(-7, 1) + (7, -1)$
12 $(3, -2) + (-5, 4)$

13 $(3 + 2i) + (1 - 4i)$ **14** $(-6 + i) + (6 - 2i)$

15 $(4 - 3i) + (3 + 5i)$ **16** $(2 - 6i) + (1 + 3i)$

17 $(4, 5) - (3, 1)$ **18** $(-2, 1) - (3, 0)$

19 $(-2, 3) - (-2, 3)$ **20** $(3, -2) - (4, -5)$

21 $(3 + 2i) - (2 + i)$ **22** $(4 - 5i) - (3 + 2i)$

23 $(-7 + 5i) - (3 + 5i)$ **24** $(6 - 3i) - (6 - 4i)$

25 $(3, 1)(2, 4)$ **26** $(2, -3)(4, 1)$

27 $(5, 3)(0, 1)$ **28** $(1, 0)(0, 1)$

29 $(2 + 5i)(3 + 2i)$ **30** $(3 - 2i)(2 + 5i)$

31 $(-4 + 3i)(2 - 3i)$ **32** $(5 - i)(3i)$

33 $\dfrac{(4, 5)}{(2, 1)}$ **34** $\dfrac{(2, -3)}{(5, 1)}$

35 $\dfrac{(3, 5)}{(5, 3)}$ **36** $\dfrac{(-2, 1)}{(3, -4)}$

37 $\dfrac{3 + i}{2 + 5i}$ **38** $\dfrac{2 + 3i}{5 - 2i}$

39 $\dfrac{-4 + 7i}{7 - 4i}$ **40** $\dfrac{5 - 3i}{2 - 7i}$

41 $\overline{(3, 4)}\,\overline{(3, 2)}$ **42** $\overline{(2, -1)(4, 3)}$

43 $\overline{(5 + 2i)}(3 - 2i)$ **44** $\overline{(3 + 5i)(2 - 3i)}$

45 $(3 - 4i) \div \overline{(2 - i)}$ **46** $\overline{(2 - 3i)} \div (2 + 5i)$

47 $\overline{(2, 1)} \div (3, -2)$ **48** $\overline{(3, 4)} \div (2, -3)$

If $Z = (x, y)$ and $W = (u, v)$, prove each of statements 49 to 52.

49 $\overline{Z} + \overline{W} = \overline{Z + W}$ **50** $\overline{Z} - \overline{W} = \overline{Z - W}$

51 $\overline{Z}\,\overline{W} = \overline{ZW}$ **52** $\overline{Z}/\overline{W} = \overline{Z/W}$

16.2 COMPLEX NUMBERS AS VECTORS

Vector quantity — Any quantity that has magnitude and direction is called a *vector quantity*. It can be represented in both magnitude and direction by a directed line Vector — segment, as shown in Fig. 16.1. A *vector is a directed line segment.*

We shall identify a vector by a boldface letter such as **v**. Since such a letter cannot be made in writing, it is customary to use a letter with an arrow above it. Thus, if **v** were used in print, $\vec{v}$ would be used in writing. The vector **v** in Fig. 16.1 is determined by the ordered number pair (x, y), Components — where **x** and **y** are vectors. In fact, **x** and **y** are called the *components* of **v**, Resultant — and **v** is their *resultant.*

We have seen in this section that an ordered number pair determines a vector, and we found in Sec. 16.1 that an ordered number pair is a complex number. Consequently, we conclude that *a complex number is a vector.*

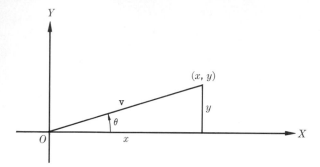

Figure 16.1

Absolute value
If $\mathbf{v} = (x, y)$ *is a vector or complex number, then its absolute value or modulus is designated by* $|\mathbf{v}|$ *or* r *and is defined by*

$$|\mathbf{v}| = r = \sqrt{x^2 + y^2} \qquad (16.8)$$

Amplitude
The angle from the positive real axis to the vector is called the *amplitude* or *argument* of the complex number. It is designated by θ in Fig. 16.1 and is

$$\theta = \arctan \frac{y}{x} \qquad (16.9)$$

since $\tan \theta = y/x$.

We defined the sum of two complex numbers or vectors in (16.2) in algebraic terms, and we shall now define the sum or resultant in geometric terms.

The sum or resultant of two vectors $\mathbf{v}_1$ and $\mathbf{v}_2$ is the vector $\mathbf{v}_1 + \mathbf{v}_2$, which begins at the origin as do $\mathbf{v}_1$ and $\mathbf{v}_2$ and terminates at the opposite vertex of a parallelogram that has $\mathbf{v}_1$ and $\mathbf{v}_2$ as two of its adjacent sides.

This definition is illustrated in Fig. 16.2.

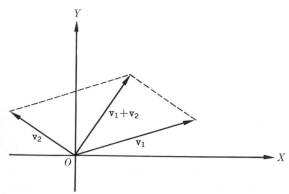

Figure 16.2

348

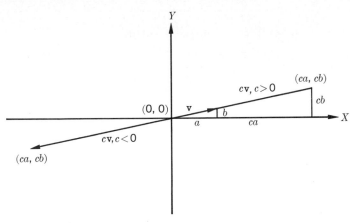

Figure 16.3

We showed in Eq. (16.3a) how to multiply a complex number by a real number, and we had $c(a, b) = (ca, cb)$. We shall now see how to show this product as a vector. Each component of (ca, cb) is c times the corresponding component of (a, b); this is shown in Fig. 16.3.

EXAMPLE   Find the absolute value and argument of $(3, -4)$.

Solution   The absolute value of $\mathbf{v} = (3, -4)$ by use of Eq. (16.8) is

$$|\mathbf{v}| = r = \sqrt{3^2 + (-4)^2} = 5$$

and
$$\theta = \arctan\frac{-4}{3}$$
$$= 306°50'$$

16.3   POLAR REPRESENTATION

By use of Fig. 16.1, the definitions of $\cos\theta$ and $\sin\theta$, and the fact that $|\mathbf{v}| = r$, we see that $\cos\theta = x/r$ and $\sin\theta = y/r$. Hence, $x = r\cos\theta$ and $y = r\sin\theta$. Therefore, the complex number

$$z = x + iy$$
$$= r\cos\theta + ir\sin\theta$$

Consequently, $$z = r(\cos\theta + i\sin\theta) \qquad (16.10)$$

Trigonometric, or polar, form of z   We shall frequently use the notation "cis θ" to stand for $\cos\theta + i\sin\theta$. The right member of Eq. (16.10) is called the *trigonometric*, or *polar*, form of z, and it is very useful in the process of multiplication and division.

349

EXAMPLE 1   Express the complex number $1 + i\sqrt{3}$ in polar form.

Solution   We shall first obtain the values of r and θ:

$$r = \sqrt{1^2 + (\sqrt{3})^2} = \sqrt{1 + 3} = 2 \qquad \blacktriangleleft \text{ by Eq. (16.8)}$$

$$\theta = \arctan\frac{\sqrt{3}}{1} \qquad\qquad \blacktriangleleft \text{ by Eq. (16.9)}$$

$$= 60°$$

Consequently, $1 + i\sqrt{3} = 2(\cos 60° + i \sin 60°) = 2 \text{ cis } 60°$.

EXAMPLE 2   Express $4 - 5i$ in polar form.

Solution   The absolute value of $4 - 5i$ is

$$r = \sqrt{4^2 + (-5)^2} = \sqrt{16 + 25} = \sqrt{41} \qquad \blacktriangleleft \text{ by Eq. (16.8)}$$

The argument is

$$\theta = \arctan\left(-\tfrac{5}{4}\right) \qquad \blacktriangleleft \text{ by Eq. (16.9)}$$

Consequently,

$$4 - 5i = \sqrt{41}\left[\cos\left(\arctan\frac{-5}{4}\right) + i \sin\left(\arctan\frac{-5}{4}\right)\right]$$

$$= \sqrt{41} \text{ cis }\left(\arctan\frac{-5}{4}\right)$$

Using a table of trigonometric functions, we find that $\arctan\frac{5}{4} = \arctan 1.25 = 51°20'$. Then, since the point representing $4 - 5i$ is in the fourth quadrant, we have

$$\theta = 360° - 51°20' = 308°40'$$

Hence, $4 - 5i = \sqrt{41} \text{ cis } 308°40'$.

Exercise 16.2   Vectors, Polar Representation

Locate the points that represent the complex number and its conjugate in each of problems 1 to 12 and draw the vectors.

1 $(4, 3)$	**2** $(5, -\sqrt{11})$	**3** $(-8, 15)$
4 $(-5, -12)$	**5** $(7, -24)$	**6** $(-5, 0)$
7 $(3, 2)$	**8** $(-4, 5)$	**9** $-7 + 24i$
10 $-5 + 7i$	**11** $3 - 5i$	**12** $6 + 7i$

Perform the operation indicated in each of problems 13 to 20 both algebraically and vectorially.

13 $(3, 2) + (4, -1)$	**14** $(-7, 3) + (3, 2)$
15 $(5, 4) + (-4, -2)$	**16** $(6, -1) + (-4, 3)$
17 $(4, 9) - (3, 7)$	**18** $(2, 5) - (3, 2)$
19 $(-1, 2) - (-5, 4)$	**20** $(8, -3) - (5, -3)$

If $\mathbf{u} = (2, -3)$, $\mathbf{v} = (-4, 1)$, $c = 2$, and $d = -3$, find the vector in each of problems 21 to 28 as an ordered number pair and find the absolute value and argument.

21 $c\mathbf{v}$	**22** $d\mathbf{u}$
23 $c\mathbf{u} + d\mathbf{v}$	**24** $(c + d)(\mathbf{u} + \mathbf{v})$
25 $cd(\mathbf{u} - \mathbf{v})$	**26** $(c - d)(\mathbf{u} - \mathbf{v})$
27 $c\mathbf{u}\mathbf{v}$	**28** $17d(\mathbf{u}/\mathbf{v})$

Express the complex numbers in each of problems 29 to 48 in polar form. Give the angle in problems 29 to 36 exactly and in the others to the nearest 10 minutes.

29 $(1, \sqrt{3})$	**30** $(1, -1)$	**31** $(-\sqrt{3}, 1)$
32 $(-2, 2)$	**33** $-1 + \sqrt{3}i$	**34** $\sqrt{3} - 3i$
35 $-i$	**36** -5	**37** $3 + 4i$
38 $24 - 7i$	**39** $-5 + 12i$	**40** $-15 - 8i$
41 $3 + 5i$	**42** $7 - 6i$	**43** $-2i + 3i$
44 $-11 - 3i$	**45** $(5, 4)$	**46** $(7, -1)$
47 $(-2, 5)$	**48** $(-3, -7)$	

16.4 THE PRODUCT OF TWO COMPLEX NUMBERS IN POLAR FORM

In this section and in Sec. 16.5, we shall show that the product and the quotient of two complex numbers are obtained very readily if the numbers are expressed first in polar form.

We shall consider the two complex numbers

$$z = (x, y) = x + iy = r(\cos \theta + i \sin \theta) = r \operatorname{cis} \theta$$

and

$$w = (u, v) = u + w = R(\cos \phi + i \sin \phi) = R \operatorname{cis} \phi$$

By Eq. (16.3), we have

$$zw = (x, y)(u, v) = (xu - yu, xv + yu) = (xu - yu) + i(xv + yu)$$

Now, if we replace x, y, u, and v by $r \cos \theta$, $r \sin \theta$, $R \cos \phi$, and $R \sin \phi$, respectively, and simplify, we have

$$zw = rR(\cos \theta \cos \phi - \sin \theta \sin \phi, \cos \theta \sin \phi + \sin \theta \cos \phi)$$

Furthermore,

$$\cos \theta \cos \phi - \sin \theta \sin \phi = \cos (\theta + \phi)$$

and

$$\cos \theta \sin \phi + \sin \theta \cos \phi = \sin \theta \cos \phi + \cos \theta \sin \phi$$
$$= \sin (\theta + \phi)$$

Therefore, since $zw = (r \text{ cis } \theta)(R \text{ cis } \phi)$, we have

$$(\boldsymbol{r} \textbf{ cis } \boldsymbol{\theta})(\boldsymbol{R} \textbf{ cis } \boldsymbol{\phi}) = \boldsymbol{rR} \textbf{ cis } (\boldsymbol{\theta} + \boldsymbol{\phi}) \qquad (16.11)$$

Consequently, we have the following theorem:

Product of two complex numbers The absolute value of the product of two complex numbers is the product of their absolute values. An argument of the product of two complex numbers is the sum of their arguments.

Since the product of two complex numbers is itself a complex number, we can obtain the product of three or more complex numbers by a repeated application of this theorem.

EXAMPLE Find the product of $1 + i$, $1 + \sqrt{3}i$, and $\sqrt{3} - i$.

Solution We first express each of these numbers in polar form and get

$$1 + i = \sqrt{2}(\cos \arctan 1 + i \sin \arctan 1)$$
$$= \sqrt{2}(\cos 45° + i \sin 45°)$$
$$= \sqrt{2} \text{ cis } 45°$$
$$1 + \sqrt{3}i = 2(\cos \arctan \sqrt{3} + i \sin \arctan \sqrt{3})$$
$$= 2 \text{ cis } 60°$$

Since the point representing $\sqrt{3} - i$ is in the fourth quadrant, the argument of $\sqrt{3} - i$ is 360° minus the acute angle whose tangent is $1/\sqrt{3}$; hence, $360° - 30° = 330°$. Therefore,

$$\sqrt{3} - 1 = 2 \text{ cis } (360° - 30°) = 2 \text{ cis } 330°$$

Therefore,

$$(1 + i)(1 + \sqrt{3}i)(\sqrt{3} - i) = (\sqrt{2} \text{ cis } 45°)(2 \text{ cis } 60°)(2 \text{ cis } 330°)$$
$$= 4\sqrt{2} \text{ cis } (45° + 60° + 330°)$$
$$= 4\sqrt{2} \text{ cis } 435°$$
$$= 4\sqrt{2} \text{ cis } (360° + 75°)$$
$$= 4\sqrt{2} \text{ cis } 75°$$

16.5 THE QUOTIENT OF TWO COMPLEX NUMBERS IN POLAR FORM

By Eq. (16.7), the quotient of the two complex numbers z and w is

$$\frac{z}{w} = \frac{(x, y)}{(u, v)} = \left(\frac{xu + yv}{u^2 + v^2}, \frac{yu - xv}{u^2 + v^2}\right)$$

If, in this equation, by use of Eq. (16.10) we replace x, y, u, and v by $r \cos \theta$, $r \sin \theta$, $R \cos \phi$, and $R \sin \phi$, respectively, and also replace $u^2 + v^2$ by R^2, we see, after simplifying, that

$$\frac{z}{w} = \frac{r \text{ cis } \theta}{R \text{ cis } \phi} = \left(\frac{rR(\cos \theta \cos \phi + \sin \theta \sin \phi)}{R^2}, i \frac{rR(\sin \theta \cos \phi - \cos \theta \sin \phi)}{R^2}\right)$$

Consequently, since

$$\cos \theta \cos \phi + \sin \theta \sin \phi = \cos (\theta - \phi)$$
$$\sin \theta \cos \phi - \cos \theta \sin \phi = \sin (\theta - \phi)$$

and $rR/R^2 = r/R$, we have

$$\frac{r \text{ cis } \theta}{R \text{ cis } \phi} = \frac{r}{R} \text{ cis } (\theta - \phi) \qquad (16.12)$$

Therefore, we have the following theorem:

Quotient of two complex numbers

The absolute value of the quotient of two complex numbers is equal to the absolute value of the dividend divided by the absolute value of the divisor. An argument of the quotient is equal to the argument of the dividend minus the argument of the divisor.

EXAMPLE

Express $1 + \sqrt{3}i$ and $1 + i$ in polar form and obtain their quotient.

Solution

If we express each of the given complex numbers in polar form, we have

$$\frac{1 + \sqrt{3}i}{1 + i} = \frac{2 \text{ cis arctan } \sqrt{3}}{\sqrt{2} \text{ cis arctan } 1}$$

$$= \frac{2 \text{ cis } 60°}{\sqrt{2} \text{ cis } 45°} \qquad \blacktriangleleft \text{ since arctan } \sqrt{3} = 60°$$
$$\text{and arctan } 1 = 45°$$

$$= \frac{2}{\sqrt{2}} \text{ cis } (60° - 45°) \qquad \blacktriangleleft \text{ by Eq. (16.12)}$$

$$= \sqrt{2} \text{ cis } 15°$$

Exercise 16.3   Products and Quotients in Polar Form

Perform indicated operations in problems 1 to 28 and express each result in the form r cis θ and $a + bi$.

1 $[2\ (\cos 31° + i \sin 31°)][5\ (\cos 14° + i \sin 14°)]$

2 $[3\ (\cos 23° + i \sin 23°)][4\ (\cos 7° + i \sin 7°)]$

3 $[7\ (\cos 38° + i \sin 38°)][3\ (\cos 22° + i \sin 22°)]$

4 $[5\ (\cos 74° + i \sin 74°)][11\ (\cos 16° + i \sin 16°)]$

5 $(2 \text{ cis } 38°)(3 \text{ cis } 82°)$ **6** $(5 \text{ cis } 92°)(3 \text{ cis } 43°)$

7 $(5 \text{ cis } 71°)(7 \text{ cis } 79°)$ **8** $(7 \text{ cis } 73°)(2 \text{ cis } 107°)$

9 $(2 \text{ cis } 139°)(3 \text{ cis } 71°)$ **10** $(5 \text{ cis } 178°)(2 \text{ cis } 47°)$

11 $(7 \text{ cis } 197°)(5 \text{ cis } 43°)$ **12** $(6 \text{ cis } 119°)(3 \text{ cis } 151°)$

13 $\dfrac{12(\cos 64° + i \sin 64°)}{3(\cos 34° + i \sin 34°)}$ **14** $\dfrac{15(\cos 72° + i \sin 72°)}{3(\cos 27° + i \sin 27°)}$

15 $\dfrac{22(\cos 85° + i \sin 85°)}{2(\cos 25° + i \sin 25°)}$ **16** $\dfrac{18(\cos 103° + i \sin 103°)}{9(\cos 13° + i \sin 13°)}$

17 $\dfrac{25(\cos 169° + i \sin 169°)}{5(\cos 49° + i \sin 49°)}$ **18** $\dfrac{35(\cos 146° + i \sin 146°)}{7(\cos 11° + i \sin 11°)}$

19 $\dfrac{26(\cos 177° + i \sin 177°)}{2(\cos 27° + i \sin 27°)}$ **20** $\dfrac{39(\cos 206° + i \sin 206°)}{3(\cos 26° + i \sin 26°)}$

21 $\dfrac{(18 \text{ cis } 187°)(2 \text{ cis } 58°)}{9 \text{ cis } 20°}$ **22** $\dfrac{(4 \text{ cis } 237°)(3 \text{ cis } 44°)}{6 \text{ cis } 41°}$

23 $\dfrac{(6 \text{ cis } 198°)(4 \text{ cis } 146°)}{12 \text{ cis } 44°}$ **24** $\dfrac{(10 \text{ cis } 249°)(21 \text{ cis } 170°)}{42 \text{ cis } 104°}$

25 $\dfrac{30 \text{ cis } 151°}{(5 \text{ cis } 37°)(3 \text{ cis } 84°)}$ **26** $\dfrac{48 \text{ cis } 183°}{(12 \text{ cis } 77°)(2 \text{ cis } 61°)}$

27 $\dfrac{126 \text{ cis } 147°}{(6 \text{ cis } 23°)(7 \text{ cis } 34°)}$ **28** $\dfrac{60 \text{ cis } 258°}{(3 \text{ cis } 72°)(4 \text{ cis } 36°)}$

Express the complex numbers in each of problems 29 to 44 in polar form, then perform the indicated operations, and leave the result in polar form.

29 $(1, 1)(1, -\sqrt{3})$ **30** $(-\sqrt{3}, 1)(1, \sqrt{3})$

31 $(\sqrt{3}, 1)(-1, \sqrt{3})$ **32** $(-2, 2)(3, -\sqrt{3})$

33 $(1, -1)^3$ **34** $(0, 1)^5$

35 $(-\sqrt{7}, \sqrt{7})^4$ **36** $(2, 0)^3$

37 $\dfrac{(-3, \sqrt{3})}{(1, -\sqrt{3})}$ **38** $\dfrac{(3, -\sqrt{3})}{(-1, 1)}$

39 $\dfrac{(-5, -5)}{(\sqrt{3}, -1)}$ **40** $\dfrac{(-3, \sqrt{3})}{(2, -2)}$

41 $\dfrac{(\sqrt{3}, -1)(0, 1)}{(-2, 0)}$ **42** $\dfrac{(4, 0)(0, -2)}{(-\sqrt{3}, 1)}$

43 $\dfrac{(-5, 5)\,(1, -\sqrt{3})}{(-\sqrt{3}, -1)}$ **44** $\dfrac{(2, -2)\,(1, \sqrt{3})}{(-3, 3)}$

16.6 | DEMOIVRE'S THEOREM

In this section we shall discuss a theorem proved by Abraham Demoivre (1667–1754) that is very useful for finding a power or a root of a complex number.

If we square the complex number $z = r \operatorname{cis} \theta$, we get

$$z^2 = (r \operatorname{cis} \theta)(r \operatorname{cis} \theta)$$
$$= r^2 \operatorname{cis}(\theta + \theta) = r^2 \operatorname{cis} 2\theta \qquad \blacktriangleleft \text{ by Eq. (16.11)}$$

Furthermore,

$$z^3 = z^2(z) = (r^2 \operatorname{cis} 2\theta)(r \operatorname{cis} \theta)$$
$$= r^3 \operatorname{cis}(2\theta + \theta) = r^3 \operatorname{cis} 3\theta \qquad \blacktriangleleft \text{ by Eq. (16.11)}$$

A repeated application of this procedure leads to the statement that

$$[r(\cos\theta + i\sin\theta)]^n - r^n(\cos n\theta + i\sin n\theta) \qquad \textbf{(16.13)}$$

In abbreviated form, the above theorem becomes

$$(r \operatorname{cis} \theta)^n - r^n \operatorname{cis} n\theta$$

Demoivre's theorem

This statement is known as Demoivre's theorem.

The proof of this theorem for integral values of n requires the use of mathematical induction and will be given as Example 2 in Sec. 22.1. At this point, we shall assume that the theorem holds for integral values of n, and we shall prove that it holds for any rational number p/q. For this purpose, we let

$$r^{p/q} = R \qquad \text{and} \qquad \frac{p}{q}\theta = \phi \qquad \qquad (1)$$

Then, $$r = R^{q/p} \qquad \text{and} \qquad \theta = \frac{q}{p}\phi \qquad \qquad (2)$$

Furthermore, $$r^p = R^q \qquad \text{and} \qquad p\theta = q\phi \qquad \qquad (3)$$

By use of the fact that $a^{s/r} = (a^s)^{1/r}$, we have

$$(r \operatorname{cis} \theta)^{p/q} = [(r \operatorname{cis} \theta)^p]^{1/q}$$
$$= (r^p \operatorname{cis} p\theta)^{1/q} \qquad \blacktriangleleft \text{ by Eq. (16.13)}$$
$$= (R^q \operatorname{cis} q\phi)^{1/q} \qquad \blacktriangleleft \text{ by (3) of this section}$$
$$= [(R \operatorname{cis} \phi)^q]^{1/q} \qquad \blacktriangleleft \text{ by Eq. (16.13)}$$
$$= (R \operatorname{cis} \phi)^{q/q} \qquad \blacktriangleleft \text{ by Eq. (5.2)}$$
$$= R \operatorname{cis} \phi$$
$$= r^{p/q}\left(\operatorname{cis}\frac{p}{q}\theta\right) \qquad \blacktriangleleft \text{ by (1) of this section}$$

EXAMPLE   Find the fifth power of $\sqrt{3} + i$.

Solution
$$(\sqrt{3} + i)^5 = \left(2 \text{ cis arctan} \frac{1}{\sqrt{3}}\right)^5$$

$$= (2 \text{ cis } 30°)^5 \qquad \blacktriangleleft \text{ since arctan} \frac{1}{\sqrt{3}} = 30°$$

$$= 2^5 \text{ cis } 5(30°) \qquad \blacktriangleleft \text{ by Eq. (16.13)}$$

$$= 32 \text{ cis } 150°$$

$$= 32 \text{ cis } (180° - 30°)$$

$$= 32(-\cos 30° + i \sin 30°)$$

$$= 32\left(-\frac{\sqrt{3}}{2} + i\frac{1}{2}\right)$$

$$= -16\sqrt{3} + 16i$$

16.7 ROOTS OF COMPLEX NUMBERS

In the set of real numbers, there is no square root of -9, no fourth root of -16, or in fact, no even root of any negative number; furthermore, there is only one odd root of a negative number. If, however, we employ complex numbers, we can obtain n nth roots of any given number by use of Demoivre's theorem. We shall illustrate the procedure to be followed with two examples.

EXAMPLE 1   Obtain the three cube roots of 64.

Solution   We first express $64 = 64 + 0i$ in polar form and get

$$64 = 64(\cos 0° + i \sin 0°)$$
$$= 64[\cos (0° + n360°) + i \sin (0° + n360°)] \qquad (1)$$

for any integral value of n, since $\sin (n360° + \theta) = \sin \theta$ and $\cos (n360° + \theta) = \cos \theta$. Now we let

$$64^{1/3} = r(\cos \theta + i \sin \theta) \qquad (2)$$

and raising each member of this equation to the third power, we get

$$64 = [r(\cos \theta + i \sin \theta)]^3$$
$$= r^3(\cos 3\theta + i \sin 3\theta) \qquad \blacktriangleleft \text{ by Eq. (16.13)} \qquad (3)$$

Now, by equating the expressions for 64 given by the right members of (1) and (3), we have

$$r^3(\cos 3\theta + i \sin 3\theta) = 64[\cos (0° + n360°) + i \sin (0° + n360°)] \qquad (4)$$

We next make use of the fact that if two complex numbers are equal, their absolute values are equal and their arguments are equal, and we obtain

$$r^3 = 64$$
$$r = 4 \qquad \blacktriangleleft \text{ since } 4^3 = 64$$

Here we use the real cube root of 64 since the absolute value of a complex number is a real number. Furthermore,

$$3\theta = 0° + n360°$$
$$\theta = n120°$$
$$= 0° \qquad \text{for } n = 0$$
$$= 120° \qquad \text{for } n = 1$$
$$= 240° \qquad \text{for } n = 2$$

Consequently, by referring to (2), we see that the three cube roots of 64 are

$$4(\cos 0° + i \sin 0°) = 4 \qquad \text{for } n = 0$$
$$4(\cos 120° + i \sin 120°) = -2 + 2\sqrt{3}i \qquad \text{for } n = 1$$
$$4(\cos 240° + i \sin 240°) = -2 - 2\sqrt{3}i \qquad \text{for } n = 2$$

There is no need to assign additional values to n, since to do so would yield angles that differ from those already obtained by integral multiples of 360°; consequently, the root determined by using one of these angles would be one of those already obtained.

EXAMPLE 2   Find the four fourth roots of $1 + \sqrt{3}i$.

Solution   If we express $1 + \sqrt{3}i$ in polar form, we have

$$1 + \sqrt{3}i = 2 \text{ cis arctan } \sqrt{3}$$
$$= 2 \text{ cis}(60° + n360°) \tag{5}$$

for all integral values of n. We next let

$$(1 + \sqrt{3}i)^{1/4} = r \text{ cis } \theta$$

then raise each member of this equation to the fourth power, and get

$$1 + \sqrt{3}i = r^4 \text{ cis } 4\theta \qquad \blacktriangleleft \text{ by Eq. (16.13)} \tag{6}$$

Now, by equating the expressions for $1 + \sqrt{3}i$ given in (6) and (5), we have

$$r^4 \text{ cis } 4\theta = 2 \text{ cis } (60° + n360°) \tag{7}$$

Consequently, by equating the values of the absolute values and also of the arguments in (7), we get

$$r^4 = 2 \qquad \text{and} \qquad 4\theta = 60° + n360°$$

Therefore, $r = 2^{1/4} \qquad$ and $\qquad \theta = 15° + n90°$

Hence we have

$$\begin{aligned} \theta &= 15° & \text{for } n = 0 \\ &= 105° & \text{for } n = 1 \\ &= 195° & \text{for } n = 2 \\ &= 285° & \text{for } n = 3 \end{aligned}$$

The corresponding values of the fourth roots of $1 + \sqrt{3}i$ are $2^{1/4}$ cis $15°$, $2^{1/4}$ cis $105°$, $2^{1/4}$ cis $195°$, $2^{1/4}$ cis $285°$.

If a root of a complex number is known in polar form, the approximate value in algebraic form can be obtained by using an approximation to each function value of each angle and calculating the real nth root of r by use of logarithms.

It is an interesting fact that the four fourth roots of $1 + \sqrt{3}i$ are equally spaced about the circumference of a circle of radius $r^{1/4} = 2^{1/4}$, the smallest argument being $15°$. This is a special case of the following statement: *The n nth roots of a + bi are equally spaced about the circumference of a circle of radius* $r^{1/n} = (\sqrt{a^2 + b^2})^{1/n}$, *the smallest argument being θ/n.*

Exercise 16.4 Powers and Roots

Use Demoivre's theorem to find the indicated power in each of problems 1 to 20. If needed, use Table II in the Appendix to get the amplitude of the given number to the nearest 10 minutes.

1	$(2i)^5$	**2**	$(-3i)^4$
3	$(-5)^3$	**4**	7^3
5	$(1 + \sqrt{3}i)^7$	**6**	$(\sqrt{3} + i)^5$
7	$(1 - i)^4$	**8**	$(\sqrt{3} - i)^4$
9	$(-1 - \sqrt{3}i)^5$	**10**	$(-1 + i)^6$
11	$(1 + i)^7$	**12**	$(-\sqrt{3} - i)^8$
13	$[(-5, 12)]^4$	**14**	$[(8, -15)]^3$
15	$[(-3, -4)]^5$	**16**	$[(24, 7)]^2$
17	$(3 - 5i)^3$	**18**	$(-4 + 7i)^4$
19	$(5 + 7i)^6$	**20**	$(-3 - 8i)^3$

Obtain the indicated roots in each of problems 21 to 36 by use of Demoivre's theorem.

21 The cube roots of $1 + \sqrt{3}i$
22 The cube roots of -1
23 The square roots of $4i$
24 The square roots of $-1 + \sqrt{3}i$

25 The fourth roots of 81
26 The fourth roots of $\sqrt{3} - i$
27 The fifth roots of $-1 - i$
28 The fifth roots of $1 - \sqrt{3}i$
29 The sixth roots of $(64, 0)$
30 The sixth roots of $(0, -1)$
31 The seventh roots of $(1, -1)$
32 The seventh roots of $-\sqrt{3} - i$
33 The eighth roots of $-1 - \sqrt{3}i$
34 The eighth roots of $(-1, 0)$
35 The ninth roots of $(1, -1)$
36 The tenth roots of $1 - \sqrt{3}i$

Find the solution set of the equation in each of problems 37 to 44.

37 $x^2 + 9 = 0$ **38** $x^2 + 4i = 0$
39 $x^3 + 27i = 0$ **40** $x^3 - 8 = 0$
41 $x^4 + 16 = 0$ **42** $x^4 + i = 0$
43 $x^5 - 32i = 0$ **44** $x^5 + 1 = 0$

Exercise 16.5   Review

1 For what values of x and y is $(x - 3, y + 4) = (1, 3)$?
2 Add $(3, -4)$ and $(2, 5)$.
3 Subtract $(4, 5)$ from $(6, 3)$.
4 Multiply $(2, 3)$ by $(1, -4)$.
5 Divide $5 - i$ by $1 + i$.
6 Multiply $3 + 2i$ by $\overline{4 - 5i}$.
7 If $\mathbf{u} = 3 - 2i$, $\mathbf{v} = 4 + 3i$, $c = -2$, and $d = 3$, find $c\mathbf{u} + d\mathbf{uv}$.
8 Express $5 - 5i$ in polar form.
9 Express $(3, 2)$ in polar form.

Perform the indicated operations in problems 10 to 13 by use of polar forms.

10 $(1 + \sqrt{3}i)(-1 + i)$ **11** $(-\sqrt{3} + i)(-1 - \sqrt{3}i)$
12 $(1 - \sqrt{3}i)/(-1 - i)$ **13** $(1 - i)/(-1 - \sqrt{3}i)$
14 Find $(1 + i)^4$ by use of polar forms.
15 Find $(1 - \sqrt{3}i)^5$ by use of polar forms.
16 Find the cube roots of $1 + i$.
17 Find the fourth roots of $-1 - \sqrt{3}i$.
18 Solve: $x^2 + 1 = 0$
19 Solve: $x^5 + 32i = 0$

Higher-Degree Equations

In this chapter we shall discuss equations of the type $f(x) = 0$, where $f(x)$ is a polynomial of degree greater than 2. Unfortunately, no direct method exists for obtaining the solution set of such equations if the degree of the polynomial is greater than 4, and the direct methods for solving equations in which the polynomial is of degree 3 or 4 are long and tedious. However, we shall present several theorems that will enable us to obtain pertinent information about the roots. We shall also discuss methods for obtaining an approximation of any irrational root to the desired degree of accuracy.

17.1 POLYNOMIAL EQUATIONS

Polynomial equation

An equation of the type

$$a_0x^n + a_1x^{n-1} + a_2x^{n-2} + \cdots + a_{n-1}x + a_n = 0 \qquad (17.1)$$

where n is a positive integer and $a_0, a_1, a_2, \ldots, a_n$ are constants, is a *polynomial* equation.

We shall make extensive use of the notation introduced in Sec. 6.4. For example, if $f(x) = 2x^3 + x^2 - 2x + 4$, then $f(2) = 2(2^3) + 2^2 - 2(2) + 4$ and $f(r) = 2r^3 + r^2 - 2r + 4$.

17.2 THE REMAINDER THEOREM

The computation involved in finding the solution set of (17.1) is greatly simplified if the left member is factored into linear factors or if one or more linear factors are found. The remainder theorem stated and proved below is useful for this purpose and is essential in finding the solution set of (17.1) when the left member is not readily factorable.

Remainder theorem

If a polynomial $f(x)$ is divided by $x - r$ until a remainder independent of x is obtained, then the remainder is equal to $f(r)$.

Before proving the remainder theorem, we shall illustrate its meaning in the following example.

If we divide $x^3 - 2x^2 - 4x + 5$ by $x - 3$, using the method of Sec. 2.12, we obtain $x^2 + x - 1$ as the quotient and 2 as the remainder. Note that the remainder is a constant and does not depend on x. Hence, since $x - r = x - 3$, it follows that $r = 3$, and, according to the remainder theorem, the remainder must be $f(r) = f(3)$. Since

$$f(x) = x^3 - 2x^2 - 4x + 5$$

we have
$$f(3) = 3^3 - 2(3^2) - 4(3) + 5$$
$$= 27 - 18 - 12 + 5 = 2$$

and this is indeed equal to the remainder obtained by dividing $f(x)$ by $x - 3$.

Proof ▶   In the division process, we have the following relation between the dividend, divisor, quotient, and remainder:

$$\text{Dividend} = (\text{quotient})(\text{divisor}) + \text{remainder}$$

Hence, if the quotient obtained by dividing $f(x)$ by $x - r$ is $Q(x)$ and the remainder is R, we have

$$f(x) = Q(x)(x - r) + R \qquad\qquad (17.2)$$

Equation (17.2) is true for all values of x including $x = r$. Hence, we may substitute r for x in (17.2) and thereby obtain

$$f(r) = Q(r)(r - r) + R$$

Therefore, $f(r) = R$ ◀ since $r - r = 0$

This proves the remainder theorem.

17.3 THE FACTOR THEOREM AND ITS CONVERSE

If r is a root of $f(x) = 0$, then $f(r) = 0$. Hence, by the remainder theorem, R in (17.2) is zero, and (17.2) becomes

$$f(x) = Q(x)(x - r)$$

Therefore, $x - r$ is a factor of $f(x)$, and we have the factor theorem.

Factor theorem   *If r is a root of the polynomial equation $f(x) = 0$, then $x - r$ is a factor of $f(x)$.*

Conversely, if $x - r$ is a factor of $f(x)$, then the remainder obtained by dividing $f(x)$ by $x - r$ is equal to zero. Hence, by the remainder theorem, $f(r) = 0$. Therefore, r is a root of $f(x) = 0$. Hence we have the converse of the factor theorem.

Converse of
factor theorem   *If $x - r$ is a factor of the polynomial $f(x)$, then r is a root of $f(x) = 0$.*

17.4 SYNTHETIC DIVISION

Synthetic division   In the application of the theorems in the two preceding sections it will frequently be necessary to divide a polynomial by $x - r$. When the dividend is a polynomial and the divisor is of the form $x - r$, most of the steps in long division can be eliminated and the process can be reduced to a very short form called *synthetic division*. We illustrate the process by first dividing $2x^3 - 10x^2 + 7x - 9$ by $x - 3$ by long division and then indicating the steps that can be eliminated.

$$
\begin{array}{l}
\quad\quad\dfrac{2x^2 -\ 4x\ -5}{2x^3 - 10x^2 + 7x - 9}\big(x - 3 \\[4pt]
\quad\quad \underline{(2x^3) - 6x^2} \\[2pt]
\quad\quad\quad\quad -4x^2\ \ + [7x] \\[2pt]
\quad\quad\quad\quad \underline{(-4x^2) + 12x} \\[2pt]
\quad\quad\quad\quad\quad\quad\ -\ 5x\ -[9] \\[2pt]
\quad\quad\quad\quad\quad\quad\ \underline{(-\ 5x) + 15} \\[2pt]
\quad\quad\quad\quad\quad\quad\quad\quad\quad -24
\end{array}
$$

We shall now examine the above process to ascertain the portions that can be eliminated without interfering with the essential steps. The division process requires that each term enclosed in parentheses in the above computation be the same as the term above it. Furthermore, the terms enclosed in brackets are terms in the dividend written in a new position. If these two sets of terms are eliminated, we have

$$
\begin{array}{l}
\quad\quad\dfrac{2x^2 -\ 4x\ -5}{2x^3 - 10x^2 + 7x - 9}\big(x - 3 \\[4pt]
\quad\quad\quad\ \underline{-\ 6x^2} \\[2pt]
\quad\quad\quad\ -\ 4x^2 \\[2pt]
\quad\quad\quad\quad\quad\quad +12x \\[2pt]
\quad\quad\quad\quad\quad\quad -\ 5x \\[2pt]
\quad\quad\quad\quad\quad\quad\quad\quad \underline{+15} \\[2pt]
\quad\quad\quad\quad\quad\quad\quad\quad -24
\end{array}
$$

We can save space by writing $12x$ and 15 on the same line with $-6x^2$. Also, it is not necessary to write the various powers of x, because the problem tells us what they should be. Then a shorter form for the work is

$$
\begin{array}{l}
\quad\quad\dfrac{2 -\ 4-\ 5}{2 - 10 +\ 7 -\ 9}\big(1 - 3 \\[4pt]
\quad\quad\ \underline{-\ 6 + 12 + 15} \\[2pt]
\quad\quad\ -\ 4 -\ 5 - 24
\end{array}
$$

This method is applicable only if the divisor is $x - r$. Therefore, it is not necessary for the coefficient of x to appear in the divisor. Furthermore, in subtraction, we change the signs in the subtrahend and add. This change becomes automatic if we replace -3 in the divisor by 3. When these changes are made, the problem becomes

$$
\begin{array}{l}
\quad\quad\dfrac{2 -\ 4-\ 5}{2 - 10 +\ 7 -\ 9}\big(3 \\[4pt]
\quad\quad\quad\ \underline{6 - 12 - 15} \\[2pt]
\quad\quad\ -\ 4 -\ 5 - 24
\end{array}
$$

Finally, we write the 2 in the dividend as the first term in the last line.

Then the numbers in the quotient are the same as the first three numbers in the last line. Therefore, the quotient can be omitted and we have

$$
\begin{array}{r}
2 - 10 + 7 - 9\, \underline{(3} \\
6 - 12 - 15 \\
\hline
2 - 4 - 5 - 24
\end{array}
$$

Steps in synthetic
division

Now we see that the essential steps in the process can be carried out mechanically as follows: Write the first 2 in the third line; multiply by the divisor 3 and place the product, 6, under -10; add and obtain -4; multiply -4 by the divisor 3 and obtain -12; write -12 under 7; add and obtain -5; multiply -5 by 3; place the product under -9; add and obtain -24. Then the coefficients in the quotient are 2, -4, and -5, so the quotient is $2x^2 - 4x - 5$ and the remainder is -24.

Rule for synthetic
division

This problem illustrates the following rule for synthetic division. In order to divide $f(x)$ by $x - r$ synthetically,

1 Arrange the coefficients of $f(x)$ in order of descending powers of x, supplying zero as the coefficient of each missing power.

2 Replace the divisor $x - r$ by r.

3 Bring down the coefficient of the largest power of x, multiply it by r, place the product beneath the coefficient of the second largest power of x and add the product to that coefficient. Multiply the sum just obtained by r, place the product beneath the coefficient of the next largest power of x, and add the product to the coefficient. Continue this process until there is a product added to the constant term.

4 The last number in the third row is the remainder, and the others, reading from left to right, are the coefficients of the quotient, which is of degree one less than $f(x)$.

EXAMPLE

Determine the quotient and remainder obtained by dividing $2x^4 + x^3 - 16x^2 + 18$ by $x + 2$.

Solution

Since $x - r = x + 2$, it follows that $r = -2$. We now write the coefficients of the dividend in a line, supply zero as the coefficient of the missing power of x, carry out the steps in the synthetic division, and get

$$
\begin{array}{r}
2 + 1 - 16 0 + 18\, \underline{(-2} \\
\downarrow\, - 4 + 6 + 20 - 40 \\
\hline
2 - 3 - 10 + 20 - 22
\end{array}
$$

Exercise 17.1   Remainder and Factor Theorems

Find the remainder if the first polynomial is divided by the second in each of problems 1 to 8.

1 $x^3 + 7x^2 - 6x - 5,\ x - 1$
2 $x^3 - 9x^2 + 8x + 4,\ x - 3$
3 $x^4 + 3x^3 - 7x^2 - 8x - 2,\ x - 2$
4 $x^4 - 4x^3 + 2x^2 - 9x + 5,\ x - 4$
5 $3x^3 - 2x^2 - 4x + 4,\ x + 1$
6 $5x^3 + 9x^2 - 3x + 3,\ x + 2$
7 $2x^4 + 7x^3 + 4x^2 - 2x + 7,\ x + 3$
8 $3x^4 + 16x^3 + 6x^2 - 2x - 13,\ x + 5$

Show, by use of the factor theorem, that the binomial in each of problems 9 to 20 is a factor of the polynomial.

9 $2x^3 + 3x^2 - 6x + 1,\ x - 1$
10 $2x^3 - 4x^2 + 4x - 8,\ x - 2$
11 $3x^3 - 9x^2 - 4x + 12,\ x - 3$
12 $-2x^3 + 6x^2 + 2x + 24,\ x - 4$
13 $5x^4 + 8x^3 + x^2 + 2x + 4,\ x + 1$
14 $3x^4 + 9x^3 - 4x^2 - 9x + 9,\ x + 3$
15 $2x^4 + 8x^3 - 9x^2 + 3x - 10,\ x + 5$
16 $5x^4 + 17x^3 + 6x^2 + 9x + 27,\ x + 3$
17 $x^n - a^n,\ x - a$
18 $x^n - a^{2n},\ x - a^2$
19 $x^n - a^n,\ x + a,\ n$ even
20 $x^n + a^n,\ x + a,\ n$ odd

Find the solution set of the equation in each of problems 21 to 28 by use of the converse of the factor theorem.

21 $(x - 1)(x - 3)(x + 4) = 0$
22 $(x - 6)(x + 5)(x - 2) = 0$
23 $(x - 2)(2x - 1)(3x + 2) = 0$
24 $(x + 5)(3x - 4)(5x + 7) = 0$
25 $(2x^2 - 5x + 3)(x + 1) = 0$
26 $(3x^2 - 2x - 5)(2x - 3) = 0$
27 $(3x^2 + 5ax - 2a^2)(x^2 + x - 6) = 0$
28 $(2x^2 + bx - b^2)(2x^2 - x - 3) = 0$

By use of synthetic division, find the quotient and remainder if the polynomial in each of problems 29 to 40 is divided by the binomial.

29 $x^3 + 3x^2 - 5x + 2, \ x - 1$
30 $x^3 - 3x^2 + 4x - 2, \ x - 2$
31 $2x^3 + 5x^2 + 3x + 7, \ x + 2$
32 $3x^3 + 10x^2 + 4x - 2, \ x + 3$
33 $5x^4 + 7x^3 - 9x^2 - 2x + 3, \ x + 1$
34 $7x^4 + 15x^2 + 3x^2 - 2x + 9, \ x + 2$
35 $2x^4 - 6x^3 - 5x^2 - 7x - 8, \ x - 4$
36 $2x^4 - 8x^3 - 9x^2 - 4x - 11, \ x - 5$
37 $x^5 - 6x^4 + 5x^3 + 5x^2 - 4x + 7, \ x - 2$
38 $x^5 - 3x^4 - 2x^3 + 4x^2 + x - 9, \ x - 3$
39 $2x^6 + 2x^5 + 7x^4 + 8x^3 - 2x^2 + x + 3, \ x + 1$
40 $3x^6 + 5x^5 - 3x^4 - 4x^3 - 11x^2 + 3x - 6, \ x + 2$

17.5 THE GRAPH OF A POLYNOMIAL

We shall make extensive use of the graph of a polynomial in investigating the roots of a polynomial equation, and for that reason an understanding of such graphs is essential. The method for obtaining the graph of $y = f(x)$, where $f(x)$ is a polynomial, is the same as that used in preceding chapters except that synthetic division may be employed for obtaining corresponding values of x and y. With the exception of values of x in a comparatively short range, the values of y change very rapidly and become very large numerically, and it is advisable to use different scales on the X and Y axes. For example, in Fig. 17.1 the unit length on the X axis is 5 times the unit length on the Y axis, and in Fig. 17.2 the unit length on the X axis is 10 times the unit length on the Y axis. We shall illustrate the graphing of a polynomial with two examples.

EXAMPLE 1   Construct the graph of $y = 2x^3 - 3x^2 - 12x + 6$.

Solution   To obtain the graph of $y = 2x^3 - 3x^2 - 12x + 6$, we assign the values $-3, -2, -1, 0, 1, 2, 3, 4$ to x and compute each corresponding value of y either by direct substitution or by use of synthetic division and the remainder theorem. For example, if $x = 4, f(4)$ is the remainder when $f(x)$ is divided by $x - 4$. By the synthetic division below, we see that $f(4) = 38$.

$$
\begin{array}{r}
2 - 3 - 12 + \ 6 \lfloor 4 \\
\underline{+ 8 + 20 + 32} \\
2 + 5 + \ 8 + 38
\end{array}
$$

The following table of values shows the value of y that corresponds to each of the above values of x:

x	-3	-2	-1	0	1	2	3	4
y	-39	2	13	6	-7	-14	-3	38

The next problem is to choose a scale such that the points determined by the above table will lie in a convenient area. Here we shall let the unit length on the X axis be 5 times the unit length on the Y axis, as indicated in Fig. 17.1. If we plot the points in the above table and join them with a smooth curve, we obtain the graph in Fig. 17.1.

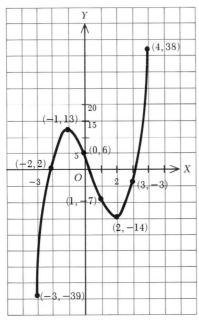

Figure 17.1

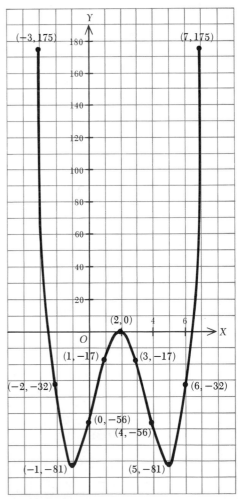

Figure 17.2

EXAMPLE 2   Construct the graph of $y = x^4 - 8x^3 + 6x^2 + 40x - 56$.

Solution   The following is a table of corresponding values of x and of y for the given equation.

x	-3	-2	-1	0	1	2	3	4	5	6	7
y	175	-32	-81	-56	-17	0	-17	-56	-81	-32	175

Since, in this table, x varies from -3 to 7 and y varies from -81 to 175, we shall choose a scale such that the length of the unit on the X axis is 10 times the length of the unit on the Y axis. The graph determined by the table and scale is shown in Fig. 17.2.

17.6   THE NUMBER OF ROOTS OF A POLYNOMIAL EQUATION

Fundamental theorem of algebra

The method for determining the number of roots of a polynomial equation depends upon the fundamental theorem of algebra, which states that *every polynomial equation has at least one root.*† By this theorem the equation

$$f(x) = a_0x^n + a_1x^{n-1} + \cdots + a_{n-1}x + a_n = 0 \tag{17.3}$$

has at least one root. If r_1 is this root, then $x - r_1$ is a factor of $f(x)$ and we can write $f(x) = (x - r_1)Q_1(x)$, where $Q_1(x)$ is a polynomial of degree $n - 1$. Again by the above theorem, $Q_1(x) = 0$ has at least one root, r_2, and therefore $Q_1(x) = (x - r_2)Q_2(x)$, where $Q_2(x)$ is a polynomial of degree $n - 2$. Consequently,

$$f(x) = (x - r_1)(x - r_2)Q_2(x)$$

By continuing this process, we obtain

$$f(x) = (x - r_1)(x - r_2) \cdots (x - r_n)Q_n(x) \tag{17.4}$$

where Q_n is of degree $n - n = 0$ and is therefore a constant. If we perform the indicated multiplication in (17.4), we obtain a polynomial of the type

$$Q_n x^n + \text{terms of degree less than } n \text{ in } x$$

Hence, Q_n is the coefficient of x^n in $f(x)$ and, consequently, is equal to a_0. Hence the factored form of $f(x)$ is

$$f(x) = a_0(x - r_1)(x - r_2)(x - r_3) \cdots (x - r_n) \tag{17.5}$$

† For proof see Lars V. Ahlfors, "Complex Analysis," 2d ed., p. 99, McGraw-Hill Book Company, New York, 1966.

By the converse of the factor theorem, $r_1, r_2, r_3, \ldots, r_n$ are roots of $f(x) = 0$. Furthermore, $f(x)$ has no other roots, since no one of the factors in (17.5) is zero for $x = r_{n+1}$ unless r_{n+1} is one of $r_1, r_2, r_3, \ldots, r_n$. Hence, $f(r_{n+1}) \neq 0$, and r_{n+1} is not a root. If $r_1 = r_2 = r_3 = \cdots = r_s$, where $s < n$, then

$$f(x) = (x - r_s)^s (x - r_{s+1}) \cdots (x - r_n)$$

Multiplicity of a root

and r_s is called a root of *multiplicity s.* Therefore we have the following theorem:

Theorem on the number of roots of a polynomial equation

A polynomial equation of degree n has exactly n roots, where a root of multiplicity s is counted as s roots.

17.7 BOUNDS OF THE REAL ROOTS

The theorem given in this section enables us to find a number that is greater than or equal to the largest real root of an equation and another number that is smaller than or equal to the least root of the equation. Thus we can restrict the range in which the real roots are known to lie. Any number that is larger than or equal to the greatest root of an equation is called *an upper bound of the roots;* any number that is smaller than or equal to the least root of an equation is called a *lower bound of the roots.*

Upper bound
Lower bound

We shall now state and then prove a theorem that enables us to determine upper and lower bounds.

Theorem on bounds

If the coefficient of x^n in the polynomial equation $f(x) = 0$ is positive and if there are no negative terms in the third line of the synthetic division of $f(x)$ by $x - k$, $k > 0$, then k is an upper bound of the real roots of $f(x) = 0$. Furthermore, if the signs in the third line of the synthetic division of $f(x)$ by $x - (-k) = x + k$ are alternately plus and minus,† then $-k$ is a lower bound of the real roots.

EXAMPLE

Find an upper and a lower bound of the real roots of the equation $x^3 - 2x^2 + 3x + 3 = 0$.

Solution

The synthetic-division processes of the left member of $x^3 - 2x^2 + 3x + 3 = 0$ by $x - 3$ and also by $x + 1$ are given below.

$$
\begin{array}{r}
1 - 2 + 3 + 3 \underline{|3} \\
3 + 3 + 18 \\
\hline
1 + 1 + 6 + 21
\end{array}
\qquad
\begin{array}{r}
1 - 2 + 3 + 3\underline{|-1} \\
-1 + 3 - 6 \\
\hline
1 - 3 + 6 - 3
\end{array}
$$

---

† When one or more zeros occur in the third line of the synthetic division of $f(x)$ by $x + k$, $-k$ is a lower bound of the real roots if, after each zero is replaced by either a plus or a minus sign, the signs in the third line of the synthetic division are alternately plus and minus.

In the first case, the terms in the third row are all positive. Hence, 3 is an upper bound. In the second case, the terms in the third row are alternately plus and minus. Therefore, -1 is a lower bound of the roots.

In order to prove the first part of the theorem, we shall divide $f(x)$ by $x - k$ and get the quotient $Q(x)$ and the remainder R. Then

$$f(x) = Q(x)(x - k) + R \tag{1}$$

The coefficients in $Q(x)$ and the value of R are the numbers in the third row of the division of $f(x)$ by $x - k$. If these numbers are positive or zero and if x is greater than k, then $Q(x)(x - k) + R > 0$ since $x > k > 0$ and $x - k > 0$. Hence there are no real roots of $f(x) = 0$ that are greater than k; that is, k is an upper bound of the real roots, as stated in the first part of the theorem.

If the quotient obtained by dividing $f(x)$ by $x + k$ is $q(x)$ and if the remainder is R', then

$$f(x) = q(x)(x + k) + R' \tag{2}$$

Furthermore, if as stated by the theorem the terms in the third line of the synthetic division are alternately plus and minus, then the coefficients in $q(x)$ and R' are alternately plus and minus. The expression $q(x)$ is of degree $n - 1$, and its first coefficient is the same as the coefficient of x^n in $f(x)$ and therefore is positive by the statement of the theorem. Furthermore, since R' is the last term of the third line of the synthetic division and since there are $n + 1$ terms in this line, R' is negative when n is odd and positive when n is even.

We now choose a positive number h such that $-h < -k$, or $-h + k < 0$, substitute $-h$ for x in (2), and get

$$f(-h) = q(-h)(-h + k) + R' \tag{3}$$

The substitution of $-h$ for x in $q(x)$ will change the signs of the terms of odd degree only. Since the terms in $q(x)$ are alternately plus and minus, then the terms in $q(-h)$ are all plus or all minus. If the degree n of $f(x)$ is odd, then the degree of $q(x)$ is $n - 1$ and is therefore even. Hence, the first term in $q(-h)$, as well as all the other terms, is plus. Thus, since $-h + k$ is negative, $q(-h)(-h + k)$ is negative. Furthermore, when n is odd, R' is negative. Consequently, $f(-h)$ is the sum of two negative quantities and hence is not zero, and $-h$ is not a root of $f(x)$.

If n is even, $n - 1$ is odd, and then all terms in $q(-h)$ are negative and R' is positive. Therefore, since $q(-h)(-h + k)$ is positive, $f(-h)$ is the sum of two positive numbers and is not zero; in other words, $-h$ is not a root of $f(x)$.

Thus, in either case, $-h$ is not a root of $f(x)$, and $-k$ is therefore a lower bound of the real roots. This completes the proof of the theorem.

17.8 LOCATING THE REAL ROOTS OF A POLYNOMIAL EQUATION

Prior to the middle of the nineteenth century a great deal of research was devoted to the problem of solving a polynomial equation in terms of the coefficients. Cardan and Tartaglia published methods for solving such equations of degrees 3 and 4, but the methods were long and tedious. Near the middle of the nineteenth century Galois and Abel proved that no general method exists if the degree of the equation is greater than four. Therefore, we resort to a method of trial and error and of successive approximation to obtain the real roots of polynomial equations.

The first step in our method is to determine the interval in which each of the real roots lies, and we use the graph of $f(x)$ for this purpose. It can be proved that if $f(x)$ is a polynomial, the graph of $y = f(x)$ is a continuous curve. This means that it contains no gaps and is not made up of separate or disjointed parts. Since the graph is continuous, if $f(a)$ and $f(b)$ have different signs, the curve crosses the X axis an odd number of times between $x = a$ and $x = b$. Since $y = 0$ at each of these points of crossing, the abscissa of each such point is a root of $f(x) = 0$. Therefore, we have the following theorem:

Location theorem *If $f(x)$ is a polynomial, and if $f(a)$ and $f(b)$ have different signs, there is an odd number of real roots of $f(x) = 0$ between $x = a$ and $x = b$.*

EXAMPLE Locate the roots of $x^3 - 3x^2 - 6x + 9 = 0$.

Solution To locate the roots of $x^3 - 3x^2 - 6x + 9 = 0$, we consider the function $y = x^3 - 3x^2 - 6x + 9$, assign consecutive integers from -3 to 5 to x, compute each corresponding value of y, and record the results.

x	-3	-2	-1	0	1	2	3	4	5
y	-27	1	11	9	1	-7	-9	1	29

Since $f(-3) = -27$ and $f(-2) = 1$, there is an odd number of roots between $x = -3$ and $x = -2$. Similarly, there is an odd number of roots between $x = 1$ and $x = 2$, and between $x = 3$ and $x = 4$. Furthermore, since the equation is of degree 3, it has exactly three roots. Therefore, exactly one root lies in each of the above intervals.

Exercise 17.2   Graphs, Bounds, Location Theorem

Sketch the graph of the function defined in each of problems 1 to 12 for the indicated domain.   Estimate the zeros of each function to one decimal place.

1 $y = x^3 - 7x + 6, -4 \leq x \leq 3$
2 $y = x^3 + 4x^2 + x - 6, -4 \leq x \leq 2$
3 $y = x^3 - 5x^2 + 2x + 8, -2 \leq x \leq 5$
4 $y = x^3 - 6x^2 + 5x + 12, -2 \leq x \leq 5$
5 $y = 2x^3 - 5x^2 + 4, -2 \leq x \leq 4$
6 $y = 2x^3 + 7x^2 + 3x - 6, -3 \leq x \leq 2$
7 $y = 6x^3 - 13x^2 + 2x + 6, -2 \leq x \leq 3$
8 $y = 9x^3 + 9x^2 - 7x - 4, -2 \leq x \leq 2$
9 $y = x^4 - x^3 - 4x^2 + x + 1, -2 \leq x \leq 3$
10 $y = x^4 + 5x^3 + 3x^2 - 7x + 2, -4 \leq x \leq 2$
11 $y = 2x^4 + 3x^3 - 4x^2 - 4x + 3, -3 \leq x \leq 2$
12 $y = 3x^4 - x^3 - 13x^2 - 4x + 6, -2 \leq x \leq 3$

In each of problems 13 to 20, state the degree of the equation, find all roots and the multiplicity of each.   Verify the fact that the sum of the multiplicities is equal to the degree of the equation.

13 $(x - 2)^2(3x + 1)^3 = 0$
14 $(2x + 3)^4(3x + 4)^2 = 0$
15 $(x + 3)^3(2x - 1)^2(4x + 3)^2 = 0$
16 $(3x + 7)^5(2x + 3)^3(5x - 2) = 0$
17 $(2x - 3)^6(3x + 1)^4(2x + 5)^3 = 0$
18 $(7x + 8)(8x + 7)^7(3x - 1)^3 = 0$
19 $(3x + 11)^3(5x - 4)^2(8x - 9)^7 = 0$
20 $(2x + 3)^6(3x - 1)(7x + 2) = 0$

In each of problems 21 to 32, find the greatest integral lower bound and the smallest integral upper bound for the roots by use of the theorem of Sec. 17.7 and also locate each root between two consecutive integers.

21 $2x^3 - 5x^2 - 6x + 4 = 0$
22 $2x^3 - 3x^2 - 14x - 6 = 0$
23 $2x^3 - 11x^2 + 12x + 7 = 0$
24 $3x^3 - 8x^2 - 8x + 8 = 0$
25 $3x^3 + x^2 - 16x + 10 = 0$
26 $5x^3 + 19x^2 + x - 1 = 0$
27 $4x^3 + 19x^2 + 4x - 6 = 0$
28 $7x^3 + 13x^2 - 44x + 6 = 0$
29 $x^4 + 2x^3 - 10x^2 - 6x + 1 = 0$
30 $x^4 - 2x^3 - 11x^2 + 6x + 2 = 0$
31 $x^4 + 8x^3 + 17x^2 + 4x - 2 = 0$
32 $x^4 + 8x^3 + 12x^2 - 16x - 16 = 0$

17.9 RATIONAL ROOTS OF A POLYNOMIAL EQUATION

In Sec. 17.8 we stated that we shall use a method of trial and error for obtaining the roots of a general polynomial of degree greater than 2. Fortunately, we have methods that enable us to limit the field of investigation; two methods were discussed in Secs. 17.7 and 17.8. As a first step in solving a polynomial equation, it is advantageous to decide whether or not the equation has *rational* roots. In this section we develop a theorem that enables us to make this decision, provided the coefficients in the equation are integers. The theorem also enables us to select a set of numbers that contains all rational roots of the equation. For example, the theorem will enable us to say that if $2x^3 - x^2 + 3x - 3 = 0$ has a rational root, it must be one of the numbers $\pm\frac{1}{2}, \pm\frac{3}{2}, \pm1$, or ±3 and that if no one of these numbers is a root, the equation has no rational roots. We shall next state and prove the theorem and then illustrate its use.

If q and p are integers that have no common factor greater than 1, and if q/p is a root of the equation

Theorem on rational roots

$$a_0x^n + a_1x^{n-1} + \cdots + a_{n-1}x + a_n = 0 \tag{1}$$

where the a_i, for $i = 0, 1, 2, 3, \ldots, n$, are integers, then q is a factor of a_n and p is a factor of a_0.

Proof ▶

By hypothesis q/p is a rational root of (1). Consequently, if we substitute q/p for x in (1) and then multiply by p^n, we have

$$a_0q^n + a_1pq^{n-1} + \cdots + a_{n-1}p^{n-1}q + a_np^n = 0 \tag{2}$$

Now we add $-a_np^n$ to each member and divide each member of the resulting equation by q and obtain

$$a_0q^{n-1} + a_1pq^{n-2} + \cdots + a_{n-1}p^{n-1} = -\frac{a_np^n}{q} \tag{3}$$

The left member of (3) is the sum of the products of integral powers of integers and is therefore an integer. Consequently, $-a_np^n/q$ is an integer. Hence, since p and q have no common factors greater than 1, q is a factor of a_n.

If in (2) we add $-a_0q^n$ to each member and divide by p, we get

$$a_1q^{n-1} + \cdots + a_{n-1}p^{n-2}q + a_np^{n-1} = -\frac{a_0q^n}{p} \tag{4}$$

Therefore, since the left member of (4) is an integer, a_0q^n/p is an integer and, since p and q have no common factor greater than 1, p is a factor of a_0.

EXAMPLE 1   Determine the set of rational numbers that contains the rational roots of

$$2x^3 - 3x^2 - 11x + 6 = 0 \tag{5}$$

and then find all rational roots.

Solution   Since the numerator of each rational root must be a factor of 6 and the denominator a factor of 2, the possible numerators are $\pm 6, \pm 3, \pm 2$, and ± 1, and the possible denominators are ± 2 and ± 1. Hence, the set of possibilities for the rational roots is $\pm 6, \pm 3, \pm 2, \pm 1, \pm\frac{6}{2}, \pm\frac{3}{2}, \pm\frac{2}{2}, \pm\frac{1}{2}$. If we eliminate the duplications and arrange the remaining numbers in order of magnitude, we have

$$-6, -3, -2, -\tfrac{3}{2}, -1, -\tfrac{1}{2}, \tfrac{1}{2}, 1, \tfrac{3}{2}, 2, 3, 6$$

To determine which of these numbers are roots, we must test them one by one by synthetic division until three roots are found or until all have been tested. It is usually advisable to test the positive integers first, starting with the smallest. The test applied to the possible roots 1 and 2 reveals that neither is a root nor a bound of the roots. However, when we test 3, we get

$$
\begin{array}{r}
2 - 3 - 11 + 6\underline{|3} \\
6 + 9 - 6 \\
\hline
2 + 3 - 2 \quad0
\end{array}
$$

Therefore, 3 is a root. Furthermore, the left member of (1) can now be expressed in the factored form $(x-3)(2x^2 + 3x - 2)$, and the equation becomes

$$(x-3)(2x^2 + 3x - 2) = 0$$

Since the other two roots of (5) are also roots of $2x^2 + 3x - 2 = 0$, we complete the solution by solving that quadratic equation by factoring:

$$(2x-1)(x+2) = 0$$

By setting each factor equal to zero and solving for x we get $x = \frac{1}{2}$ and $x = -2$. Therefore the roots of (5) are -2, $\frac{1}{2}$, and 3.

Depressed   Example 1 illustrates the meaning and use of the *depressed equation,*
equation   which we now define. If $f(x) = 0$ is a polynomial equation and if $f(x) = (x-r)q(x)$, then $q(x) = 0$ is called the depressed equation of $f(x) = 0$ that corresponds to the root r. Obviously, since $f(x) = (x-r)q(x) = 0$, all roots of $f(x)$ except possibly r are roots of $q(x) = 0$. Furthermore, all roots of $q(x) = 0$ are also roots of $f(x) = 0$. It is possible that r is also a root of $q(x) = 0$. If so, it is a multiple root of $f(x) = 0$.

Since the degree of the depressed equation is one less than the degree of the given equation, the computation involved in obtaining the rational roots is less complicated if, after a root is found, the depressed equation is used in searching for the remaining roots.

EXAMPLE 2 Find the set of numbers that includes all rational roots of

$$6x^4 - 5x^3 - 27x^2 - 2x + 8 = 0 \qquad (6)$$

and then find the rational roots.

Solution Since the possible numerators are factors of 8 and the possible denominators are factors of 6, the set of numbers that includes all rational roots of (6) is

$$\{-8, -4, -\tfrac{8}{3}, -2, -\tfrac{4}{3}, -1, -\tfrac{2}{3}, -\tfrac{1}{2}, -\tfrac{1}{3}, -\tfrac{1}{6}, \tfrac{1}{6}, \tfrac{1}{3}, \tfrac{1}{2}, \tfrac{2}{3}, 1, \tfrac{4}{3}, 2, \tfrac{8}{3}, 4, 8\}$$

If we test the positive integers first, starting with 1, we find that neither 1, 2, nor 4 is a root and that 4 is an upper bound. For the negative integers, we find that neither -1 nor -2 is a root but -2 is a lower bound. Hence we discard $-8, -4, -\tfrac{8}{3}, -2, -1, 1, 2, 4$, and 8 as possible roots and start testing the positive fractions. When we try $x = \tfrac{1}{2}$, we get

$$
\begin{array}{r}
6 - 5 - 27 - 2 + 8 \,\lfloor \tfrac{1}{2} \\
 \; \, 3 - 1 - 14 - 8 \\
\hline
6 - 2 - 28 - 16 \quad 0
\end{array}
$$

Hence $\tfrac{1}{2}$ is a root, and $6x^3 - 2x^2 - 28x - 16 = 0$ is the depressed equation. We divide each member by 2 and get

$$3x^3 - x^2 - 14x - 8 = 0 \qquad (7)$$

The remaining roots of (6) are also roots of (7), and each root of (7) is a root of (6). The set of numbers that includes the rational roots of (7) is

$$\{\pm 8, \pm 4, \pm 2, \pm 1, \pm \tfrac{8}{3}, \pm \tfrac{4}{3}, \pm \tfrac{2}{3}, \pm \tfrac{1}{3}\}$$

However, since no number is a root of (7) unless it is also a root of (6), we discard all numbers in the above set that were tested on (6) or ruled out because of the bounds. Therefore, we eliminate all integers, all numbers greater than 4, and all numbers less than -2. Hence the set is reduced to

$$\{-\tfrac{4}{3}, -\tfrac{2}{3}, -\tfrac{1}{3}, \tfrac{1}{3}, \tfrac{2}{3}, \tfrac{4}{3}, \tfrac{8}{3}\}$$

and we must test each of these one by one until we find a root or discover that no one of the set is a root. The test applied to the positive numbers reveals that no one of them is a root. However, when we try $-\tfrac{2}{3}$, we have

$$
\begin{array}{r}
3 - 1 - 14 - 8 \,\lfloor -\tfrac{2}{3} \\
 -2 + 2 + 8 \\
\hline
3 - 3 - 12 \quad 0
\end{array}
$$

Hence, $-\tfrac{2}{3}$ is a root, and $3x^2 - 3x - 12 = 0$, or $x^2 - x - 4 = 0$, is the further depressed equation. Since this equation is quadratic, we solve it with the quadratic formula and get

$$x = \frac{1 \pm \sqrt{1 + 16}}{2} = \frac{1 \pm \sqrt{17}}{2}$$

Therefore, the four roots of Eq. (6) are $\frac{1}{2}, -\frac{2}{3}, \frac{1}{2}(1 + \sqrt{17})$, and $\frac{1}{2}(1 - \sqrt{17})$.
Since $\frac{1}{2}(1 + \sqrt{17})$ and $\frac{1}{2}(1 - \sqrt{17})$ are irrational, these roots did not appear in the set of possible rational roots obtained as a first step in the solution.

17.10 THE PROCESS OF OBTAINING ALL RATIONAL ROOTS

We suggest the following procedure as a systematic and efficient method for investigating the rational roots of a polynomial equation.

Steps in the process of obtaining all rational roots

1 List the set of numbers that are possible rational roots.

2 Test the positive integers in the set, starting with the smallest. If an upper bound that is not a root is found, discard this bound and all larger numbers in the set. If a root is found, use the depressed equation for the remainder of the investigation.

3 Repeat the above procedure for negative integers.

4 Test the fractions that remain after considering any bound that has been found.

Exercise 17.3 Solution Sets of Polynomial Equations

Find the solution set of each of the following equations. Make use of the depressed equation.

1 $x^3 - 7x + 6 = 0$

2 $x^3 + 2x^2 - 5x - 6 = 0$

3 $x^3 - x^2 - 10x - 8 = 0$

4 $x^3 + 5x^2 + 2x - 8 = 0$

5 $2x^3 + x^2 - 13x + 6 = 0$

6 $3x^3 + 19x^2 + 16x - 20 = 0$

7 $2x^3 - 13x^2 + 17x + 12 = 0$

8 $3x^3 - x^2 - 20x - 12 = 0$

9 $6x^3 + 7x^2 - x - 2 = 0$

10 $6x^3 - x^2 - 31x - 10 = 0$

11 $4x^3 - 8x^2 - 15x + 9 = 0$

12 $10x^3 + 11x^2 - 31x + 10 = 0$

13 $x^3 - 4x^2 + 3x + 2 = 0$

14 $x^3 + x^2 - 8x - 6 = 0$

15 $x^3 - 2x - 1 = 0$

16 $2x^3 + 6x^2 + 3x - 2 = 0$

17 $2x^3 - 3x^2 + 2x - 3 = 0$

18 $2x^3 + x^2 + 8x + 4 = 0$

19 $3x^3 + x^2 + x - 2 = 0$

20 $3x^3 + 2x^2 - 2x + 5 = 0$

21 $6x^4 + x^3 - 22x^2 - 11x + 6 = 0$

22 $6x^4 + 17x^3 - 14x^2 - 27x + 18 = 0$

23 $8x^4 + 42x^3 + 57x^2 + 7x - 6 = 0$
24 $12x^4 - 59x^3 + 37x^2 + 34x - 24 = 0$
25 $x^4 - x^3 - 11x^2 + 5x + 30 = 0$
26 $x^4 - 4x^2 - x + 2 = 0$
27 $2x^4 - x^3 - 21x^2 + 39x - 18 = 0$
28 $3x^4 - 8x^3 - 4x^2 + 24x - 15 = 0$
29 $4x^4 + 4x^3 + x^2 + 4x - 3 = 0$
30 $6x^4 - 11x^3 + 14x^2 - 44x - 40 = 0$
31 $4x^4 - 8x^3 + x^2 - x - 2 = 0$
32 $3x^4 - 5x^3 - 7x^2 - 21x + 18 = 0$

17.11   APPROXIMATION OF IRRATIONAL ROOTS

We stated in Sec. 17.8 that the general method for solving polynomial equations of degrees 3 and 4 is long and tedious and that general methods for solving polynomial equations of degree greater than 4 do not exist. Therefore, we must resort to some method of approximation to obtain the irrational roots of such equations.   Several methods exist, and we shall

Basis for approximating use one that depends upon the following fact.   If $f(x)$ is a polynomial and $y_1 = f(x_1)$ and $y_2 = f(x_2)$, then if x_1 and x_2 are sufficiently near each other, the portion of the graph of $y = f(x)$ between (x_1, y_1) and (x_2, y_2) will lie very near the straight line, or secant, that connects these two points. Consequently, in Fig. 17.3a, if the graph crosses the X axis at $(a, 0)$ and the straight line crosses it at $(b, 0)$, the root a of $f(x) = 0$ will be very near to b.

We can calculate the value of b by use of Fig. 17.3a, where the lines CD and EC are parallel to the X and Y axes, respectively, and the coordi-

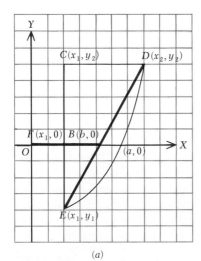

(a)

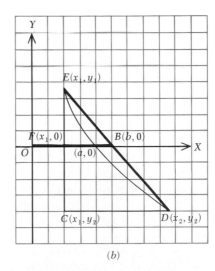

(b)

Figure 17.3

nates of C, D, E, F, and B are as indicated. The triangles EFB and ECD are similar. Hence

$$\frac{FB}{CD} = \frac{EF}{EC} \tag{1}$$

By the distance formula,

$$FB = b - x_1 \qquad CD = x_2 - x_1 \qquad EF = 0 - y_1 = -y_1 \qquad \text{and} \qquad EC = y_2 - y_1$$

Substituting these values in (1), we obtain

$$\frac{b - x_1}{x_2 - x_1} = \frac{-y_1}{y_2 - y_1} \tag{2}$$

Now we solve (2) for b and get

$$\boldsymbol{b = x_1 - \frac{y_1(x_2 - x_1)}{y_2 - y_1}} \tag{17.6}$$

We obtain the same result using Fig. 17.3b. However, in this case, using positive directions, we have

$$\frac{FB}{CD} = \frac{FE}{CE}$$

or in terms of coordinates,

$$\frac{b - x_1}{x_2 - x_1} = \frac{y_1 - 0}{y_1 - y_2}$$

Nevertheless, if we solve this equation for b, we obtain (17.6).

If the graph of $y = f(x)$ is to the right of the secant, as in Fig. 17.3a, then $b < a$, but if the graph is to the left of the secant, as in Fig. 17.3b, then $b > a$. Furthermore, if $f(x_1)$ and $f(b)$ have the same sign, then $b < a$, but if $f(x_1)$ and $f(b)$ have different signs, then $b > a$.

The steps in obtaining successive approximations to the root follow.

1 Determine integral values of x_1 and x_2 so that $x_2 - x_1 = 1$ and $f(x_1)$ and $f(x_2)$ have different signs.

2 Substitute these values in (17.6), calculate b, and round the result off to one decimal place.

3 Label this value b_1. Then we say that b_1 is the first approximation to a. Usually b_1 will differ from a by less than 0.1. That is, usually $b_1 < a < b_1 + 0.1$ or $b_1 - 0.1 < a < b_1$. This, however, is not always true.

4 Determine to one decimal place values for x_1 and x_2 such that $x_2 - x_1 = 0.1$ and $f(x_1)$ and $f(x_2)$ differ in sign. Usually these values will be either b_1 and $b_1 + 0.1$ or $b_1 - 0.1$ and b_1. If, however, neither of these two pairs of values satisfies the conditions, other values near b_1 must be tried.

5   After x_1 and x_2 have been found and the corresponding values of y_1 and y_2 calculated, substitute them in (17.6), compute b_1, and round the result off to three decimal places. Label the result b_2 and call it the second approximation to a. Usually b_2 will differ from a by less than 0.001. However, if $f(b_2)$ and $f(x_1)$ have the same signs, then $b_2 < a$ and $b_2 + 0.001$ may be a better approximation to a, so that the latter value should be tested. Likewise, if $f(b_2)$ and $f(x_1)$ have opposite signs, then $b_2 > a$, and $b_2 - 0.001$ should be tested.

6   The third approximation is obtained by the above procedure using x_1 and x_2 so that $x_2 - x_1 = 0.01$. Then b_3 will usually differ from a by less than 0.00001. The calculation, however, is long and tedious. Methods that depend on calculus are usually employed if approximations correct to more than three decimal places are desired.

EXAMPLE

Calculate to three decimal places the largest positive root of

$$2x^3 - 3x^2 - 12x + 6 = 0 \qquad (3)$$

Solution

The graph of

$$y = f(x) = 2x^3 - 3x^2 - 12x + 6 \qquad (4)$$

is shown in Fig. 17.1, and from it we see that the largest positive root of (3) is between 3 and 4, that $f(3) = -3$, and $f(4) = 38$. Hence with $(x_1, y_1) = (3, -3)$ and $(x_2, y_2) = (4, 38)$, Eq. (17.6) yields

$$b = 3 - \frac{-3(4-3)}{38-(-3)}$$
$$= 3 + \tfrac{3}{41}$$

Therefore, $b_1 = 3.1$

By synthetic division

$$
\begin{array}{rrrr|l}
2 - 3 & -12 & +6 & \underline{3.1} \\
 +6.2 + & 9.92 & -6.448 & \\
\hline
2 + 3.2 - & 2.08 & -0.448 &
\end{array}
$$

and $f(3.1) = -0.448$. Hence $f(b_1) = f(3.1)$ and $f(x_1) = f(3)$ have the same sign, so that 3.1 is less than a. Therefore we find $f(3.2)$ by synthetic division as below:

$$
\begin{array}{rrrr|l}
2 - 3 & -12 & +6 & \underline{3.2} \\
 6.4 + & 10.88 & -3.584 & \\
\hline
2 + 3.4 - & 1.12 & +2.416 &
\end{array}
$$

Thus, $f(3.2) = 2.416$.

Now, using Eq. (17.6) with $(x_1, y_1) = (3.1, -0.488)$ and $(x_2, y_2) = (3.2, 2.416)$, we obtain

$$b_1 = 3.1 - \frac{-0.448(3.2 - 3.1)}{2.416 - (-0.448)}$$

$$= 3.1 + \frac{0.0448}{2.864}$$

$$= 3.1 + 0.0156 \cdots$$

Hence, $b_2 = 3.116$ to three decimal places

By synthetic division, we find that $f(3.116) = -0.0110$, and since $f(3.1)$ and $f(3.116)$ have the same sign, 3.116 is less than the root. Furthermore, by synthetic division, $f(3.117) = 0.0165$. Hence we conclude that the root of (3) to three decimal places is 3.116.

To enable the reader to check his understanding of the method, we give below the results in each step of the approximation to three decimal places of the root of (3) that is between 0 and 1.

1   $f(0) = 6$, $f(1) = -7$. If $(x_1,\, y_1) = (0,\, 6)$, $(x_2,\, y_2) = (1,\, -7)$, then (17.6) yields $b_1 = 0.5$.

2   $f(0.5) = -0.5$, $f(x_1) = f(0) = 6$. Hence $f(x_1)$ and $f(0.5)$ have different signs, so that 0.5 is greater than the root. However, $f(0.4) = 0.848$. Hence, for the second approximation, we use $(x_1,\, y_1) = (0.4,\, 0.848)$ and $(x_2,\, y_2) = (0.5,\, -0.5)$, substitute in (17.6), and get

$$b_2 = 0.463 \qquad \text{to three decimal places}$$

3   $f(0.463) = -0.0006$ to four decimal places. Since $f(x_1) = f(0.4) = 0.848$, 0.463 is greater than the root. Hence we try 0.462 and find that $f(0.462) = 0.0218$. Therefore, since $f(0.463)$ is nearer to zero than $f(0.462)$, we conclude that the root to three decimal places is 0.463.

Exercise 17.4   Irrational Roots

Find to two decimal places the least positive root of the equation in each of problems 1 to 12.

1	$x^3 + 3x^2 + 3x - 1 = 0$	2	$x^3 + 3x^2 + 4x - 1 = 0$
3	$x^3 + 3x^2 - 45x + 17 = 0$	4	$x^3 + 6x^2 + 9x - 1 = 0$
5	$x^3 + 6x^2 + 12x - 5 = 0$	6	$x^3 + 6x^2 + 11x - 5 = 0$
7	$x^3 + 6x^2 + 16x - 8 = 0$	8	$x^3 - 21x + 7 = 0$
9	$x^4 - 2x^3 - 2x - 1 = 0$	10	$4x^4 - 4x^3 - 3x^2 - 7x + 1 = 0$
11	$2x^4 + x^3 - x^2 - 3x - 2 = 0$	12	$x^4 - x^3 - 2x^2 - 3x - 1 = 0$

Find to three decimal places the numerically largest negative root of the equation in each of problems 13 to 20.

13	$x^3 + x^2 - x + 1 = 0$	14	$x^3 - 4x^2 - 4x + 8 = 0$

15 $8x^3 + 8x^2 - 14x + 1 = 0$

16 $4x^3 + 6x^2 + 3x + 5 = 0$

17 $x^4 + 2x^3 - 4x - 4 = 0$

18 $4x^4 + 8x^3 + 5x^2 - 6x - 6 = 0$

19 $x^4 - 3x^3 - x^2 + 2x - 4 = 0$

20 $x^4 - 3x^3 - 2x^2 + 2x - 12 = 0$

Find to two decimal places each root of the equations given in problems 21 to 28.

21 $x^3 + 3x^2 - 3x - 7 = 0$

22 $x^3 + 6x^2 - 28 = 0$

23 $x^3 - 6x^2 + 6x + 2 = 0$

24 $3x^3 - 9x^2 + 2 = 0$

25 $2x^4 - 6x^3 + x^2 + 4x + 1 = 0$

26 $4x^4 - 28x^3 + 63x^2 - 52x + 14 = 0$

27 $2x^4 + 2x^3 - 9x^2 + 2x + 2 = 0$

28 $9x^4 + 12x^3 - 49x^2 + 22x + 4 = 0$

Exercise 17.5 Review

Find the quotient and remainder obtained by dividing the first polynomial by the second in each of problems 1 to 3.

1 $x^3 - 5x^2 + 7x - 6,\ x - 1$

2 $2x^4 - 7x^3 - 3x^2 + 17x + 3,\ x - 2$

3 $4x^5 + 5x^4 + 2x^3 - 3x^2 - 5x - 3,\ x + 1$

4 Show that $x + 2$ is a factor of $x^4 + 2x^3 - 5x^2 - 4x + 12$.

5 Find the roots of $(x - 1)^3(2x + 3)^4(3x - 2) = 0$ and the multiplicity of each.

6 Show that $q(x) = x^2 - x - 2$ is a factor of $p(x) = 2x^4 - 5x^3 + 4x^2 + x - 10$ by dividing $p(x)$ by one factor of $q(x)$ and then dividing the quotient thus obtained by the other factor of $q(x)$.

7 Sketch the graph of $y = 2x^3 - 9x^2 + 6x + 1$ for $-1 \leq x \leq 4$.

8 Sketch the graph of $y = 2x^4 - 14x^3 + 21x^2 - 4x - 5$ for $-1 \leq x \leq 5$.

Find an integral lower bound and an integral upper bound for the roots of the equation in each of problems 9 and 10. Locate each root between a pair of consecutive integers.

9 $3x^3 - 5x^2 - 7x + 6 = 0$

10 $6x^4 - x^3 - 59x^2 - 41x + 35 = 0$

Find the solution set of the equation in each of problems 11, 12, and 13. Make use of the depressed equation.

11 $4x^3 - 12x^2 + 11x - 3 = 0$

12 $3x^4 - x^3 - 6x^2 - 10x + 4 = 0$

13 $6x^5 - x^4 + 13x^3 + 2x^2 - 44x + 24 = 0$

Find to two decimal places the largest irrational root of each of the equations given in problems 14 and 15.

14 $2x^4 - 6x^3 - 9x^2 + 14x + 6 = 0$

15 $x^4 - 4x^3 - 3x^2 + 20x - 10 = 0$

Logarithms

Logarithms were invented in the seventeenth century. They are useful and efficient in arithmetical computation, important in the application of mathematics to chemistry, physics, and engineering, and indispensable for some parts of advanced mathematics. In this chapter we shall develop some properties of logarithms and show how they are used in numerical computation. We also use logarithms to obtain the solution sets of certain types of equations. The theory of logarithms is based on the laws of exponents, and the reader is advised to review these laws now.

18.1 APPROXIMATIONS

In this chapter and in Chaps. 19 and 20 we shall deal extensively with numbers that are approximations. In this section we discuss computations that involve such numbers and the degree of accuracy of the results of such computations. The problem of accuracy involves the concept of significant figures and the method of rounding off a number to the required degree of accuracy.

Any number that is obtained by measurement is an approximation. For example, if the thickness of a sheet of metal is measured with a micrometer calibrated in thousandths of an inch and found to be 0.053 inch, it is understood that the actual thickness is between 0.0525 inch and 0.0535 inch including the former number but not the latter, and we write $0.0525 \le$ thickness < 0.0535. Similarly, if the length and width of a rectangle as measured with a rule calibrated in tenths of a centimeter are 10.6 and 5.2 centimeters, then

$$10.55 \le \text{length} < 10.65 \quad \text{and} \quad 5.15 \le \text{width} < 5.25$$

Now in connection with the statement that the area is $10.6(5.2) = 55.12$ square centimeters, the question arises: How many digits in 55.12 are accurate? We answer this question by referring to the equation above and observing that

$$10.55(5.15) \le \text{area} < 10.65(5.25)$$
or
$$54.3325 \le \text{area} < 55.9125$$

Since the first number in this equation is only slightly more than 54 and the other is nearly 56, we cannot be certain of the area even to two digits. We shall, however, say that the area is probably nearer 55 square centimeters than any other two-digit number. This discussion illustrates the following definition:

In a number that represents an approximation the digits *known* to be correct are called *significant*.

Significant digits

The digits 1, 2, 3, 4, 5, 6, 7, 8, and 9 are always significant if used in connection with a measurement, as are any zeros between any two of these digits. Zeros on the right may or may not be significant. Zeros whose only function is as an aid in placing the decimal point are never significant.

Such zeros occur between the decimal and the first nonzero digit and only in positive numbers less than 1.

In computation involving approximate numbers we do not use a digit if its accuracy is doubtful. Eliminating such a digit or such digits is called *rounding off a number*. The procedure of rounding a number off to *n* significant digits consists of the following steps.

1 If the decimal point is to the right of the *n*th digit, replace all digits between the *n*th digit and the decimal point by zeros and discard all digits to the right of the decimal point.

2 If the decimal point is to the left of the *n*th digit, discard all digits to the right of the *n*th digit.

3 If the digit following the *n*th digit is 5, 6, 7, 8, or 9, increase the *n*th digit by 1.

4 If the digit following the *n*th digit is 0, 1, 2, 3, or 4, the *n*th digit is not changed.

We shall illustrate rounding off by observing that 48.257 rounded off to three digits is 48.3, since the decimal point is to the left of the third digit and the fourth digit is 5; 3.842 rounded off to three digits is 3.84, since the decimal point is to the left of the third digit and the fourth digit is 2; 3,842 rounded off to three digits is 3,840, since the decimal point is at the right of the third digit and the fourth digit is 2; and 0.0020651 rounded off to three digits is 0.00207 since the decimal point is to the left of the third digit and the first digit dropped is 5.

As illustrated in the discussion that precedes the definition of significant digits, the result of a computation depends upon the number of significant digits in the numbers involved. We shall now present three rules that are employed in computation.

1 *If the number of significant digits in M and N are equal or differ by 1, we round off the product MN and the quotient M/N to the number of significant digits in the less accurate one of M and N.*

2 *If the number of significant digits in M and N differ by 2 or more, we round off the one with the greater number of significant figures so that it contains one more significant digit than the other and then proceed as in rule 1.*

Before giving the third rule, we illustrate rules 1 and 2 by noticing that since the number of significant figures in $M = 27.3$ and $N = 145.8$ differs by only 1, we find the product by multiplying M and N together and then rounding off to three digits, because one of the numbers contains only three significant digits. Thus,

$$MN = 27.3(145.8)$$
$$= 3,980.34 = 3,980 \qquad \text{to three significant digits}$$

Furthermore, since the number of significant digits in $M = 583.27$ is two more than in $N = 34.7$, we round M off to one more digit than in N and have $M = 583.3$ to four figures. Then

$$\frac{M}{N} = \frac{583.3}{34.7} = 16.81 = 16.8 \qquad \text{to three digits}$$

When dealing with a sum, we are not primarily interested in the number of significant figures in the data. Rather, we are interested in the precision of the data. That is, we are interested in whether the data measure the integral number of units, the tenths of a unit, the hundredths of a unit, or smaller fractions of a unit. Consequently, we are interested in the number of significant figures in the *decimal portion* of the addends, since we add digits with the same place value. We therefore use the following rule in adding approximate numbers.

> **3** *Round off each addend to one more decimal place than there is in the addend which has the least number of decimal places; then add the resulting numbers and round off the last digit in the sum. A similar procedure is used in subtraction.*

EXAMPLE   Find the sum of 28.72, 3.683, 7.2, and 23.7864.

Solution   The third number contains only one decimal place; hence, the others should be rounded off to two decimal places before adding. If this is done, the addends and their sum are

$$\begin{array}{r} 28.72 \\ 3.68 \\ 7.2 \\ \underline{23.79} \\ 63.39 \end{array}$$

After rounding off to one decimal place, we find the sum to be 63.4.

18.2   SCIENTIFIC NOTATION

In the previous section we stated the conditions under which some zeros in an approximate number are significant and others are not. We did not discuss the significance of the final zero or zeros in a number. The fact is that in some cases the final zero or zeros are significant and in other cases they are not. For example, if the approximate number 1,269.8 is rounded off to four digits, we have 1,270 and the zero is significant. If, however, 1,269.8 is rounded off to two digits, we have 1,300 and neither zero is significant. This ambiguity with respect to the final zeros is removed if the number is expressed in *scientific notation*. A number is in scientific notation if it is expressed as the product of two factors with the first factor a number equal to or greater than 1 but less than 10 and with all its digits significant and with the other factor an integral power of 10. For example, the numbers 1,270 and 1,300 mentioned above expressed in scientific notation are $1.270(10^3)$ and $1.3(10^3)$, respectively. Note that in the first number, all four digits are significant and in the second only two digits are significant. Before explaining the procedure for expressing a number in scientific notation, we shall define the reference position for the decimal point.

The *reference position* for the decimal point in a number N is immediately to the right of the first nonzero digit in N.

For example, the reference position for the decimal point in each of the following numbers is indicated by a caret:

$$2_{\wedge}16.34 \qquad 3_{\wedge}021 \qquad 0.004_{\wedge}623$$

The scientific notation for the number N is $N'(10^c)$ with all digits in N appearing in N' and with the decimal point in N' in reference position. The exponent c is chosen so that $N'(10^c) = N$ and is equal to the number of digits between the reference position and the decimal point in N and is positive or negative according as the decimal point is to the right or to the left of the reference position.

Number	N'	c	Scientific notation
312	3.12	2	$3.12(10^2)$
1,235	1.235	3	$1.235(10^3)$
0.0621	6.21	-2	$6.21(10^{-2})$
0.00004326	4.326	-5	$4.326(10^{-5})$
1,250, zero significant	1.250	3	$1.250(10^3)$
3,650, zero not significant	3.65	3	$3.65(10^3)$

EXAMPLE

Solution

Convert $M = 3,270$ (zero not significant) and $N = 43$ to scientific notation and then find the product MN. Express the result in scientific notation.

In scientific notation $M = 3.27(10^3)$, $N = 4.3(10)$, and

$$
\begin{aligned}
MN &= (3.27)(10^3)(4.3)(10) \\
&= (3.27)(4.3)(10^4) \\
&= 14.061(10^4) \qquad \blacktriangleleft \text{ since } (3.27)(4.3) = 14.061 \\
&= 14(10^4) \qquad \blacktriangleleft \text{ rounding off to two digits} \\
&= 1.4(10)(10^4) \qquad \blacktriangleleft \text{ since } 14 = 1.4(10) \\
&= 1.4(10^5)
\end{aligned}
$$

Exercise 18.1 Approximations

Round off the number in each of problems 1 to 16 to four significant digits. Use scientific notation if needed.

1	372.413	**2**	83.2936	**3**	0.784208
4	0.0679924	**5**	0.00312261·	**6**	0.08439743
7	485.75	**8**	57.3492	**9**	12,345.2
10	372,094	**11**	87,538.4	**12**	458,993
13	2,359.3	**14**	5,794.28	**15**	382,412
16	5,555.5				

Express the number in each of problems 17 to 32 in scientific notation. Assume that all printed digits are significant.

17	3,782	**18**	37.06
19	80,596	**20**	7,623.42
21	0.0352	**22**	0.5968
23	0.000478	**24**	0.08306
25	3,780, zero significant	**26**	28,500, zeros significant
27	6,000, zeros significant	**28**	60,070, zero significant
29	3,780, zero not significant	**30**	28,500, zeros not significant
31	6,000, last zero not significant	**32**	60,070, zero not significant

Assume that all numbers in problems 33 to 52 are approximations, perform the indicated operations, round each result off to the proper number of digits, and express it in scientific notation.

33	(5.3)(37)	**34**	(0.39)(2.8)
35	(27)(16.4)	**36**	(5.13)(0.79)
37	(257.4)(30.7)	**38**	(0.597)(5.097)
39	(7.3)(7.859)	**40**	(85)(78.51)
41	$93 \div 26$	**42**	$97.3 \div 5.9$
43	$63.28 \div 0.88$	**44**	$57.931 \div 2.83$
45	$27.2 + 3.14 + 7.83$	**46**	$1.25 + 3.03 + 2.714$
47	$7.635 + 19.37 + 2.6$	**48**	$11.7 + 1.17 + 0.117$
49	$23.862 + 13.85 - 7.23$	**50**	$9.806 + 7.15 - 8.4$
51	$0.9857 + 0.018 - 0.32$	**52**	$5.38 + 83.5 - 0.385$

18.3 THE DEFINITION

In the statement

$$2^3 = 8$$

we use the term *exponent* to indicate the relationship that exists between 3 and 2. However, 3 is also related to 8 in that it expresses the power to which 2 must be raised to produce 8, and we indicate this relationship by the term *logarithm.* In other words, in the statement $2^3 = 8$, 3 is the exponent of 2 and is also the logarithm of 8, or more precisely, the logarithm to the base 2 of 8.

As a basis for our subsequent discussion, we present the following definition.

Logarithm

The *logarithm* to a given base of a positive number is the exponent that indicates the power to which the base must be raised in order to obtain the number.

The statement, "The logarithm to the base b of N is L" is written in abbreviated form as $\log_b N = L$. In terms of this notation, the definition becomes

$$\log_b N = L \text{ if and only if } b^L = N, N > 0, b > 0, b \neq 1 \qquad (18.1)$$

Note that the abbreviation "log" appears without a period and that the symbol for the base appears as a subscript. The following are examples of logarithms:

$$\log_8 64 = 2 \qquad \blacktriangleleft \text{ since } 8^2 = 64$$
$$\log_4 64 = 3 \qquad \blacktriangleleft \text{ since } 4^3 = 64$$
$$\log_{81} 9 = \tfrac{1}{2} \qquad \blacktriangleleft \text{ since } 81^{1/2} = 9$$
$$\log_a 1 = 0 \qquad \blacktriangleleft \text{ since } a^0 = 1, \text{ for } a \neq 0$$

We see from these examples that a logarithm may be integral or fractional. In fact, it can be and often is irrational. Furthermore, in many cases, if two of the three letters in (18.1) are known, the third can be found by inspection.

EXAMPLE 1 Find the replacement for N in $\log_7 N = 2$.

Solution By use of (18.1) we have $7^2 = N$; hence, $N = 49$.

EXAMPLE 2 Find the replacement for b if $\log_b 125 = 3$.

Solution By use of (18.1) we have

$$b^3 = 125 = 5^3$$

Hence, $\qquad\qquad b = 5$

EXAMPLE 3 Find the replacement for a if $\log_{27} 3 = a$.

Solution Again using (18.1), we have

$$27^a = 3$$

and, since $27^{1/3} = 3$, it follows that

$$a = \tfrac{1}{3}$$

18.4 | PROPERTIES OF LOGARITHMS

In this section we shall employ the laws of exponents (Chap. 5) and the definition of a logarithm to derive three important properties of logarithms. Later we shall show how to use these properties in numerical computations. The first two properties enable us to obtain the logarithm of the product and the quotient of two numbers in terms of the logarithms of the numbers, and the third enables us to express the logarithm of the power of a number in terms of the logarithm of the number.

If we are given

$$\log_b M = m \qquad \text{and} \qquad \log_b N = n \tag{1}$$

then, by (18.1), we have

$$M = b^m \qquad \text{and} \qquad N = b^n \tag{2}$$

Hence,
$$MN = (b^m)(b^n)$$
$$= b^{m+n} \qquad \blacktriangleleft \text{ by (2.38)}$$

Therefore,
$$\log_b MN = m + n \qquad \blacktriangleleft \text{ by (18.1)}$$

and it follows from (1) that

$$\log_b MN = \log_b M + \log_b N \tag{18.2}$$

Consequently, we have the following property:

Logarithm of
a product

The logarithm of the product of two numbers is equal to the sum of the logarithms of the numbers.

EXAMPLE 1

If $\log_{10} 3 = .4771$ and $\log_{10} 4 = .6021$, find $\log_{10} 12$.

Solution

$$\log_{10} 12 = \log_{10} (3)(4)$$
$$= \log_{10} 3 + \log_{10} 4 \qquad \blacktriangleleft \text{ by (18.2)}$$
$$= .4771 + .6021$$
$$= 1.0792$$

Property (18.2) can be extended to three or more numbers by the following process:

$$\log_b MNP = \log_b (MN)P$$
$$= \log_b (MN) + \log_b P$$
$$= \log_b M + \log_b N + \log_b P \qquad \blacktriangleleft \text{ by (18.2)}$$

Again, using (2), we have

$$\frac{M}{N} = \frac{b^m}{b^n}$$
$$= b^{m-n} \qquad \blacktriangleleft \text{ by (2.47)}$$

Hence, $\log_b \dfrac{M}{N} = m - n$ ◄ by (18.1)

and we have

$$\log_b \frac{M}{N} = \log_b M - \log_b N \tag{18.3}$$

Now we may state the following property:

Logarithm of
a quotient

The logarithm of the quotient of two numbers is equal to the logarithm of the dividend minus the logarithm of the divisor.

EXAMPLE 2   If $\log_{10} 3 = .4771$ and $\log_{10} 2 = .3010$, find $\log_{10} 1.5$.

Solution

$$\begin{aligned}
\log_{10} 1.5 &= \log_{10} \tfrac{3}{2} \\
&= \log_{10} 3 - \log_{10} 2 \qquad \text{◄ by (18.3)} \\
&= .4771 - .3010 \\
&= .1761
\end{aligned}$$

Finally, if we raise both members of $M = b^m$ to the kth power, we have

$$\begin{aligned}
M^k &= (b^m)^k \\
&= b^{km} \qquad \text{◄ by (2.40)}
\end{aligned}$$

Hence, by (18.1), $\log_b M^k = km$

and since $m = \log_b M$, it follows that

$$\log_b M^k = k \log_b M \tag{18.4}$$

and we have the third property of logarithms:

Logarithm of
a power

The logarithm of a power of a number is equal to the product of the exponent of the power and the logarithm of the number.

EXAMPLE 3   Find the logarithm of 3^2.

Solution

$$\begin{aligned}
\log_{10} 3^2 &= 2 \log_{10} 3 \\
&= 2(.4771) \\
&= .9542
\end{aligned}$$

Since a root of a number can be expressed as a fractional power, (18.4) can be employed to obtain the logarithm of a root of a number. Thus

$$\log_b \sqrt[k]{M} = \log_b (M)^{1/k} = \frac{1}{k} \log_b M$$

EXAMPLE 4   Find $\log_{10} \sqrt[3]{2}$.

Solution

$$\begin{aligned} \log_{10} \sqrt[3]{2} &= \log_{10} 2^{1/3} \\ &= \tfrac{1}{3} \log_{10} 2 \\ &= \frac{.3010}{3} \\ &= .1003 \end{aligned}$$

Exercise 18.2   Conversion of Exponential and Logarithmic Forms

Change the statement in each of problems 1 to 16 to logarithmic form and the statement in each of problems 17 to 28 to exponential form by making use of (18.1).

1 $5^2 = 25$ **2** $2^5 = 32$
3 $3^4 = 81$ **4** $6^3 = 216$
5 $3^{-2} = \frac{1}{9}$ **6** $4^{-3} = \frac{1}{64}$
7 $5^{-4} = \frac{1}{625}$ **8** $2^{-7} = \frac{1}{128}$
9 $(\frac{1}{2})^{-3} = 8$ **10** $(\frac{1}{3})^{-1} = 3$
11 $(\frac{1}{5})^{-5} = 3,125$ **12** $(\frac{1}{10})^{-4} = 10,000$
13 $49^{1/2} = 7$ **14** $64^{1/3} = 4$
15 $64^{5/6} = 32$ **16** $81^{3/4} = 27$
17 $\log_5 25 = 2$ **18** $\log_2 32 = 5$
19 $\log_7 343 = 3$ **20** $\log_3 243 = 5$
21 $\log_3 \frac{1}{9} = -2$ **22** $\log_2 \frac{1}{8} = -3$
23 $\log_5 \frac{1}{625} = -4$ **24** $\log_7 \frac{1}{7} = -1$
25 $\log_4 8 = \frac{3}{2}$ **26** $\log_{27} 9 = \frac{2}{3}$
27 $\log_{16} 32 = \frac{5}{4}$ **28** $\log_{25} 3,125 = 2.5$

In each of problems 29 to 52, find the replacement for b, n, or L so as to make the statement true.

29 $\log_3 81 = L$ **30** $\log_5 25 = L$
31 $\log_7 343 = L$ **32** $\log_2 32 = L$
33 $\log_{81} 27 = L$ **34** $\log_{16} 8 = L$
35 $\log_{125} 25 = L$ **36** $\log_{27} 9 = L$
37 $\log_2 n = 5$ **38** $\log_3 n = 3$
39 $\log_5 n = 2$ **40** $\log_2 n = 4$
41 $\log_{27} n = \frac{2}{3}$ **42** $\log_{64} n = \frac{5}{6}$
43 $\log_9 n = \frac{3}{2}$ **44** $\log_{32} n = \frac{3}{5}$
45 $\log_b 9 = 2$ **46** $\log_b 125 = 3$

47 $\log_b 64 = 6$	**48** $\log_b 64 = 2$
49 $\log_b 27 = \frac{3}{2}$	**50** $\log_b 4 = \frac{2}{3}$
51 $\log_b 27 = \frac{3}{4}$	**52** $\log_b 8 = \frac{3}{5}$

Given that $\log_2 8 = 3$, $\log_2 16 = 4$, $\log_2 64 = 6$, $\log_2 256 = 8$, $\log_2 1{,}024 = 10$, $\log_2 4{,}096 = 12$, $\log_2 16{,}384 = 14$, and $\log_2 65{,}536 = 16$, obtain the required logarithm in each of problems 53 to 64 by use of (18.2), (18.3), and (18.4).

53 $\log_2 (16)(1{,}024)$	**54** $\log_2 (8)(16{,}384)$
55 $\log_2 (256)(4{,}096)$	**56** $\log_2 (64)(65{,}536)$
57 $\log_2 \dfrac{4{,}096}{16}$	**58** $\log_2 \dfrac{16{,}384}{256}$
59 $\log_2 \dfrac{65{,}536}{1{,}024}$	**60** $\log_2 \dfrac{4{,}096}{64}$
61 $\log_2 (1{,}024)^3$	**62** $\log_2 (16{,}384)^2$
63 $\log_2 \sqrt[3]{4{,}096}$	**64** $\log_2 \sqrt[4]{65{,}536}$

18.5 COMMON, OR BRIGGS, LOGARITHMS

Common, or Briggs, logarithms

We stated previously that logarithms can be employed efficiently in numerical computation, and we shall use logarithms to the base 10 for this purpose. Logarithms to the base 10 are called the *common*, or *Briggs*, logarithms.

The common logarithm of a number that is not an integral power of 10 cannot be computed by elementary methods, but tables have been prepared that enable us to obtain a decimal approximation to the common logarithm of any positive number. These tables can also be used to find a number if its common logarithm is known. The use of these tables will be explained in the next three sections.

It is customary to omit the subscript that indicates the base in the notation for a common logarithm. Consequently, in the statement $\log N = L$, it will be understood that the base is 10.

If c is a real number, then 10^c is positive. Hence the common logarithm of zero or of a negative number does not exist as a real number.

18.6 CHARACTERISTIC AND MANTISSA

If we express a positive number $N \neq 1$ in scientific notation, we have

$$N = N'(10^c) \qquad (1)$$

where $\qquad 1 < N' < 10$

and $\qquad c$ is an integer

We now consider the following three situations:

1 If $N > 10$, then in (1), $c > 1$. For example, $231 = 2.31(10^2)$.

2 If $N < 1$, then $c < 0$. For example, $0.0231 = 2.31(10^{-2})$.

3 If $1 < N < 10$, then in (1), $N = N'$ and $c = 0$. For example, $2.31 = 2.31(10^0)$.

Now, if we equate the common logarithms of the members in (1), we have

$$\log N = \log N' + \log 10^c \qquad \blacktriangleleft \text{ by (18.2)}$$
$$= \log N' + c \log 10 \qquad \blacktriangleleft \text{ by (18.4)}$$
$$= \log N' + c \qquad \blacktriangleleft \text{ since log 10 = 1}$$

Thus, by the commutative property of addition, we have

$$\log N = c + \log N' \qquad (2)$$

Since $1 < N' < 10$, it follows that $10^0 < N' < 10^1$, and hence $0 < \log N' < 1$. From Table I in the Appendix, we can obtain a decimal approximation to the common logarithm of any number between 1 and 10, correct to four decimal places. Consequently, by referring to (2), we see that the common logarithm of any positive number not equal to 1 can be expressed approximately as an integer plus a nonnegative decimal fraction. Since $1 = 10^0$, $\log 1 = 0$. Thus, for log 1, the integer is zero, and the decimal fraction is zero. We are now in a position to state the following definition.

Characteristic and mantissa

If the common logarithm of a positive number is expressed as an integer plus a nonnegative decimal fraction, the integer is called the *characteristic* of the logarithm and the decimal fraction is called the *mantissa*.

In the expression for $\log N$ in (2), the characteristic is the integer c and the mantissa is $\log N'$, where c is the exponent of 10 in the scientific notation for N. In Sec. 18.2 we demonstrated that c is numerically equal to the number of digits between the reference position and the decimal point in N and that it is positive or negative according as the decimal point is to the right or to the left of the reference position. Therefore, we have the following rule for determining the characteristic of the common logarithm of a positive number:

Rule for determining the characteristic

The characteristic of the common logarithm of a positive number N is numerically equal to the number of digits between the reference position and the decimal point in N and is positive or negative according as the decimal point is to the right or to the left of the reference position.

EXAMPLE 1

The reference position in 236.78 is between 2 and 3. Hence there are two digits, 3 and 6, between the reference position and the decimal point. Since the decimal point is to the right of the reference position, the characteristic of the common logarithm of 236.78 is 2.

EXAMPLE 2   The characteristic of the common logarithm of 2.3678 is zero since the decimal point is in reference position.

EXAMPLE 3   The decimal point in 0.0023678 is three places to the left of the reference position. Therefore, the characteristic of log 0.0023678 is -3.

Since the position of the decimal point in the number N affects only the integer c in the scientific notation for N, and since the mantissa of log N is log N', the mantissa of log N depends only upon the sequence of digits in N.

If $N \geq 1$, the characteristic of log N is zero or a positive integer. Therefore, log N can be written as a single number. For example, if the mantissa of log 23,678 is .3743, then log $236.78 = 2 + .3743 = 2.3743$, and log $2.3678 = 0 + .3743 = 0.3743$. In finding the logarithm of a positive number less than 1, however, we have a different situation. For example, log $0.0023678 = -3 + .3743 = -2.6257 = -2 - .6257$. Now the decimal fraction $-.6257$ is negative and consequently is not a mantissa. We use the following device for dealing with such situations. If the characteristic of log N is $-c$, where $c > 0$, we express $-c$ in the form $(10 - c) - 10$; then we write the mantissa to the right of $10 - c$. For example, we express log 0.0023678 in the form $7.3743 - 10$ since the characteristic is -3, and $(10 - 3) - 10 = 7 - 10$. Similarly, log $0.23678 = 9.3743 - 10$, log $0.023678 = 8.3743 - 10$, and log $0.0000023678 = 4.3743 - 10$.

18.7 USE OF TABLES TO OBTAIN THE MANTISSA

In this section we shall explain the method of finding the mantissa of a logarithm by use of Table I of the Appendix. We shall first discuss numbers of three digits and as a specific example, shall consider 3.27. Since the mantissa is not affected by the decimal point in the number, we temporarily disregard the decimal point. We turn to Table I in the Appendix and look in the column headed by N on the left side of the page where we find the first two digits, 32. Then, in line with this and across the page in the column headed by the third digit, 7, we find the entry 5145. Except for the decimal point before it, this is the desired mantissa. Since the decimal point in 3.27 is in the reference position, the characteristic of log 3.27 is zero. Hence log $3.27 = 0.5145$.

As a second example, consider 0.00634. Again, we temporarily disregard the decimal point and find 63 in the column headed by N. In line with this and in the column headed by 4, we find the entry 8021. By the rule in Sec. 18.6, the characteristic of log 0.00634 is -3, which we write as $7 - 10$. Hence, log $0.00634 = 7.8021 - 10$.

If a number is composed of fewer than three digits, we mentally annex one or two zeros at the right and proceed as before. For example, to get the mantissa of log 72, we look up 720, and to get the mantissa of log 3, we look up 300. If all digits in a number after the third are zeros, we disregard them in the process of getting the mantissa.

In the discussion that follows, it will be necessary to use the expression "the mantissa of the logarithm to the base 10 of N" frequently. For the sake of brevity, we shall abbreviate the expression to ml N.

Linear interpolation

If a number is composed of four digits, we obtain the mantissa by use of *linear interpolation* which was discussed in Sec. 9.5 We illustrate with an example.

EXAMPLE

Find log 0.006324.

Solution

Since the position of the decimal point has no effect on the mantissa, it follows that

$$\text{ml } 0.006324 = \text{ml } 632.4$$

and we proceed as follows:

$$1 \left[.4 \begin{bmatrix} \text{ml } 633 \ = .8014 \\ \text{ml } 632.4 = \\ \text{ml } 632 \ = .8007 \end{bmatrix} .0007 \right]$$

$$(.4)(.0007) = .00028 = .0003 \quad \text{to one digit}$$

Consequently, ml $632.4 = .8007 + .0003 = .8010$

Since the decimal point in 0.006324 is three places to the left of the reference position, we have log $0.006324 = 7.8010 - 10$.

The interpolation process consists of simple operations which, after some practice, can be performed mentally, thus saving considerable time. We suggest the following steps, which can be carried out rapidly:

Steps in interpolation

1 Temporarily place the decimal point between the third and fourth digits of the given number.

2 Subtract the mantissas of the logarithms of the two three-digit numbers between which the given number lies.

3 Multiply the difference between the two mantissas by the fourth digit of the given number considered as a decimal fraction.

4 Add the product obtained in step 3 to the smaller of the mantissas in step 2.

If a number contains more than four digits, round it off to four places and proceed as above. For example, to get ml 17.6352, we find ml 17.64.

Exercise 18.3 Characteristic and Mantissa

Find the characteristic of the common logarithm of the number in each of problems 1 to 16.

1	307	**2**	5,607	**3**	82
4	982	**5**	49.6	**6**	8.27
7	62.9	**8**	2.97	**9**	0.238
10	0.0045	**11**	0.0501	**12**	0.000219
13	0.000007	**14**	0.0202	**15**	0.5903
16	0.00000601				

Find the common logarithm of the number in each of problems 17 to 64. Interpolate if necessary.

17	3.76	**18**	9.83	**19**	2.05
20	8.64	**21**	38.9	**22**	50.2
23	67.3	**24**	46.8	**25**	708
26	539	**27**	857	**28**	606
29	0.323	**30**	0.711	**31**	0.684
32	0.992	**33**	0.0307	**34**	0.0276
35	0.0596	**36**	0.0605	**37**	0.0052
38	0.000218	**39**	0.00923	**40**	0.0000606
41	2,762	**42**	5,839	**43**	7,667
44	9,893	**45**	77.81	**46**	187.7
47	6.604	**48**	594.8	**49**	0.5723
50	0.8702	**51**	0.05006	**52**	0.003007
53	78.253	**54**	78.255	**55**	807.06
56	2.4685	**57**	0.012472	**58**	0.37964
59	0.0021117	**60**	0.98768	**61**	49.8215
62	9,821.56	**63**	2.81574	**64**	54,321.2

18.8 USE OF TABLES TO FIND *N* WHEN log *N* IS GIVEN

The next problem in the use of the tables is to find N when log N is given. We shall illustrate the procedure with two examples.

EXAMPLE 1　Find N if log $N = 1.6191$.

Solution　We first look through the columns in the tables until we find the mantissa .6191. It is in the column headed by 6, and is in line with 41 in the column headed by N. Hence, the three digits in N are 416. The next step is to place the decimal point. Since the characteristic of $\log N$ is 1, the decimal point is one place to the right of the reference position, i.e., between 1 and 6. Therefore, $N = 41.6$.

If ml N is not listed in the table, we must resort to interpolation. By use of a four-place table, we cannot obtain accurately more than the first four digits in N. We obtain the first three from the table and determine the fourth by interpolation. If the characteristic of $\log N$ indicates that N contains more than four digits, then zeros are added after the fourth or scientific notation is used.

EXAMPLE 2

Find N if $\log N = 8.4978 - 10$.

Solution

We shall let T represent the number composed of the first four digits in N and determine T. Then by considering the characteristic of $\log N$ we place the decimal point in T and thus get N. The mantissa .4978 is not listed in the table, but the two nearest to it are .4969 and .4983. These two mantissas are ml 3,140 and ml 3,150, respectively. (We add the zero in each case in order to obtain four places for use in interpolation.) Now $4983 - 4969 = 14$, and $4978 - 4969 = 9$; therefore, 4978 is $\frac{9}{14}$ of the way from 4969 to 4983. Hence, T is approximately $\frac{9}{14}$ of the way from 3,140 to 3,150. Since

$$\frac{9}{14}(10) = 6.4 = 6 \qquad \text{to one digit}$$

we have $T = 3,140 + 6 = 3,146$. Since the characteristic of $\log N$ is $8 - 10 = -2$, the decimal point in N is two places to the left of the reference position. Hence, $N = 0.03146$.

The steps in the above process are shown in the condensed form given below.

$$\log N = 8.4978 - 10$$

$$14 \left[\begin{array}{l} .4983 = \text{ml } 3,150 \\ 9 \left[\begin{array}{l} .4978 = \text{ml } T \\ .4969 = \text{ml } 3,140 \end{array} \right. \end{array} \right] 10$$

$$\frac{9}{14}(10) = 6.4 = 6 \qquad \text{to one digit}$$

$$3,140 + 6 = 3,146 = T$$

Hence,
$$N = 0.03146$$

since the characteristic of $\log N$ is -2.

Exercise 18.4   Given log N; To Find N

If log N is the number given in each of problems 1 to 16, find N.   Express N in scientific notation in each of problems 9 to 16.

1	0.3284	**2**	1.4871	**3**	2.7627
4	10.9375	**5**	9.9661 − 10	**6**	8.2504 − 10
7	7.4683 − 10	**8**	8.5866 − 10	**9**	3.6646
10	4.5694	**11**	6.9106	**12**	5.7767
13	7.9576 − 10	**14**	6.0719 − 10	**15**	9.8215 − 10
16	8.7059 − 10				

In each of problems 17 to 28, log N is given, and the value of N to three significant digits is to be found by using the entry in the table that is at least as near the given value of log N as any other entry in the table.   Express N in scientific notation in problems 21 to 28.

17	1.3427	**18**	2.4973	**19**	0.8675
20	2.6384	**21**	3.7781	**22**	5.3142
23	4.1892	**24**	6.3847	**25**	9.8056 − 10
26	8.7186 − 10	**27**	7.8999 − 10	**28**	8.4599 − 10

By use of interpolation, find the value of N to four digits if log N is the number given in each of problems 29 to 44.   Use scientific notation as needed.

29	2.2933	**30**	1.4892	**31**	1.1891
32	0.1940	**33**	4.7764	**34**	6.8008
35	5.6789	**36**	3.0303	**37**	9.7013 − 10
38	7.7770 − 10	**39**	8.7654 − 10	**40**	6.2882 − 10
41	9.9036 − 10	**42**	7.6543 − 10	**43**	8.4591 − 10
44	6.5332 − 10				

18.9   LOGARITHMIC COMPUTATION

The main purpose of this section is to illustrate procedures to be followed in numerical computation, but before that we call attention to the following fact.   The results obtained by use of four-place tables as in this book are correct to at most four places but never to more places than the least number of places in any of the data used.   Thus, the product of a three- and a four-digit number should be calculated to only three places if four-place tables are used.

EXAMPLE 1   Use logarithms to evaluate $P = 2.36(0.358)(719)$.

Solution   By (18.2), log P is the sum of the logarithms of the separate factors. We suggest that an outline of the solution including each characteristic be made before looking up the mantissas. If this is done, we have

$$\begin{aligned}
\log 2.36 &= 0.\\
\log 0.358 &= 9. \qquad -10\\
\underline{\log 719 = 2.}\\
\log P &= \qquad\qquad \blacktriangleleft \text{ enter sum here}\\
P &=
\end{aligned}$$

Now we fill in the mantissas, add the logarithms and obtain log P. Thus, using Table I, we have

$$\begin{aligned}
\log 2.36 &= 0.3729\\
\log 0.358 &= 9.5539 - 10\\
\underline{\log 719 = 2.8567}\\
\log P &= 12.7835 - 10\\
&= 2.7835
\end{aligned}$$

Finally, from the table, we find that $P = 607$.

We have written the outline twice in this example to show how it looks after each step. In practice, this is not necessary.

EXAMPLE 2   Use logarithms to find the value of

$$N = \frac{38.9(27.4)}{563(1.35)}$$

Solution   In this problem, N is a quotient in which the dividend and divisor are both products; hence, we find the logarithm of the dividend and of the divisor by use of (18.2) and then obtain the logarithm of the quotient by subtracting the logarithm of the dividend from that of the divisor.

$$\begin{aligned}
\log 38.9 &= 1.5899\\
\underline{\log 27.4 = 1.4378}\\
\log \text{dividend} =& \qquad 3.0277 \qquad \blacktriangleleft \text{ sum}\\[6pt]
\log 563 &= 2.7505\\
\underline{\log 1.35 = 0.1303}\\
\log \text{divisor} =& \qquad 2.8808 \qquad \blacktriangleleft \text{ sum}\\
\log N =& \qquad 0.1469 \qquad \blacktriangleleft \text{ difference}\\
N =& \ 1.40
\end{aligned}$$

EXAMPLE 3   By use of logarithms, find the value of $N = 57.1/86.4$.

Solution   Since log 57.1 = 1.7566 and log 86.4 = 1.9365, we would obtain a negative number including a negative fraction by subtracting the latter from the former. This would be a correct value for log N, but we could not use it because, by definition, all mantissas are nonnegative. This situation can be avoided by adding 10 to and subtracting 10 from the logarithm of the numerator before subtracting the logarithm of the denominator. In this way we have

$$\log 57.1 = 11.7566 - 10$$
$$\log 86.4 = 1.9365$$
$$\overline{ \log N = 9.8201 - 10}$$
$$N = 0.661 \qquad \blacktriangleleft \text{ since the characteristic is } 9 - 10 = -1$$

EXAMPLE 4   Use logarithms to evaluate $N = \sqrt[7]{0.813}$.

Solution   We have
$$\log N = \log \sqrt[7]{0.813}$$
$$= \log 0.813^{1/7}$$
$$= \tfrac{1}{7} \log 0.813$$
$$= \tfrac{1}{7}(9.9101 - 10)$$

If we perform this indicated multiplication and subtraction, we have

$$\log N = \frac{9.9101}{7} - \frac{10}{7}$$
$$= 1.4157 - 1.4286$$
$$= -.0129$$

This is a negative fraction and cannot be used with our tables. The difficulty can be avoided by adding $60 - 60$ to log .813. Thus, we obtain $69.9101 - 70$, and the negative portion of the logarithm is a multiple of 7. Now
$$\log N = \tfrac{1}{7}(69.9101 - 70)$$
$$= 9.9872 - 10$$
$$N = 0.971$$

EXAMPLE 5   Use logarithms and interpolation to evaluate $N = 58.41(32.736)$.

Solution   We begin by rounding the five-figure number off to four figures before interpolating and have $N = 58.41(32.74)$. Then

$$\log 58.41 = 1.7665$$
$$\log 32.74 = 1.5151$$
$$\overline{ \log N = 3.2816}$$
$$N = 1,913$$

Exercise 18.5 Logarithmic Computation

Use the properties of logarithms to express the logarithm in each of problems 1 to 8 as the sum and difference of logarithms of first powers of numbers.

1 $\log \dfrac{y}{b-y}$

2 $\log \dfrac{x}{x+a}$

3 $\log a(x-a)$

4 $\log bc(x+y)$

5 $\log cx^2$

6 $\log \dfrac{\sqrt{x+a}}{x^3}$

7 $\log \dfrac{a^2x^3}{b\sqrt{x+b}}$

8 $\log \dfrac{a^2\sqrt[3]{x-a}}{b(x+a)^3}$

By use of the computation theorems, write the expression in each of problems 9 to 16 as the logarithm of a product, quotient, or power or a combination of them.

9 $\log a - \log b$

10 $\log x - \log y$

11 $\log a + \log (x+a)$

12 $\log a + \log b + \log c$

13 $\log a + 2 \log b$

14 $\frac{1}{2} \log a + 3 \log b$

15 $2 \log a + 3 \log b - \log c - \frac{1}{2} \log (x+a)$

16 $3 \log b + \frac{1}{3} \log (x-a) - \log a - 3 \log (x+a)$

By use of logarithms perform the computations indicated in problems 17 to 60. Obtain the answer in problems 17 to 44 to three digits and the answer in problems 45 to 60 to four.

17 $(3.86)(59.1)$

18 $(36.2)(2.78)$

19 $(807)(0.276)$

20 $(97.4)(4.62)$

21 $(6.23)(4.71)(32.8)$

22 $(80.3)(29.7)(0.316)$

23 $(71.1)(236)(0.0234)$

24 $(5.27)(7.52)(25.7)$

25 $\dfrac{948}{237}$

26 $\dfrac{836}{723}$

27 $\dfrac{38.1}{572}$

28 $\dfrac{2.91}{47.6}$

29 $\dfrac{(24.7)(3.28)}{47.6}$

30 $\dfrac{(5.14)(41.5)}{145}$

31 $\dfrac{(66.2)(0.573)}{98.2}$

32 $\dfrac{(0.528)(68.4)}{97.1}$

33 $\dfrac{976}{(23.5)(5.32)}$

34 $\dfrac{7.76}{(2.93)(3.87)}$

35 $\dfrac{503}{(38.2)(4.79)}$ **36** $\dfrac{606}{(71.9)(8.27)}$

37 7.25^2 **38** 2.68^3

39 $\sqrt[3]{9.84}$ **40** $\sqrt{596}$

41 $\sqrt{27.3}\ \sqrt[3]{0.629}$ **42** $\sqrt[3]{3.85}\ \sqrt[7]{0.581}$

43 $\sqrt[3]{76.3}\ \sqrt[4]{0.404}$ **44** $\sqrt[5]{2.83}\ \sqrt[3]{0.0346}$

45 $\dfrac{(29.63)(4.804)}{72.27}$ **46** $\dfrac{(802.3)(0.4719)}{382.3}$

47 $\dfrac{987.6}{(237.5)(57.32)}$ **48** $\dfrac{5,827}{(78.51)(73.44)}$

49 $\sqrt[3]{80.29}$ **50** $\sqrt[5]{776.2}$

51 $\sqrt[4]{271.3}$ **52** $\sqrt[7]{79.43}$

53 $\dfrac{17.32^{3/5}}{179.9}$ **54** $\dfrac{106.6^{4/3}}{874.1}$

55 $\dfrac{1,492}{1,976^{2/5}}$ **56** $\dfrac{1,939}{190.8^{5/4}}$

57 $\sqrt{\dfrac{(38.712)^3(2.384)}{98,753}}$ **58** $\sqrt[3]{\dfrac{(56.27)^2\ \sqrt{0.42503}}{417.6}}$

59 $\sqrt[4]{\dfrac{\sqrt{6.035}\ \sqrt[3]{0.2716}}{0.07097}}$ **60** $\sqrt{\dfrac{(4.2375)^3\ \sqrt{17.28}}{33.724}}$

18.10 LOGARITHMS TO BASES OTHER THAN 10

Natural logarithms

Although the common, or Briggs, logarithms are usually used in numerical computation, logarithms to bases other than 10 are more convenient in many situations.   In fact, the natural, or Naperian, logarithms are employed extensively in more advanced mathematics.   The base of this system is the irrational number $e = 2.718 \cdots$.   We shall next state and prove a theorem that establishes a relation between the logarithms of a number to two different bases.

If a and b are two positive numbers, then

Logarithm to two bases

$$\log_a N = \frac{\log_b N}{\log_b a} \tag{18.5}$$

To prove the theorem, we begin by letting $N = a^p$, then

$$\log_a N = \log_a a^p = p \tag{1}$$

Furthermore,

$$\log_b N = \log_b a^p = p \log_b a \qquad \blacktriangleleft \text{ by (18.4)}$$
$$= \log_a N \log_b a \qquad \blacktriangleleft \text{ by (1)}$$

By solving for $\log_a N$, we obtain (18.5).

We shall get a relation between the two most commonly used bases, 10 and $e = 2.718 \cdots$. If, in Eq. (18.5) we replace a by e and b by 10, we see that

$$\log_e N = \frac{\log_{10} N}{\log_{10} e} \qquad \textbf{(18.6a)}$$

The form of this equation can be altered by following the usual custom of writing $\ln N$ to designate the logarithm of N to the base e and replacing $\log_{10} e$ by its value 0.4343. Thus, we have

$$\ln N = 2.3026 \log_{10} N \qquad \textbf{(18.6b)}$$

since $1/\log_{10} e = 1/0.4343 = 2.3026$. A logarithm to the base e is called a *natural logarithm*.

If we use (18.6b), we see that

$$\ln 318 = 2.3026 \log 318$$
$$= 2.3026(2.5024)$$
$$= 5.7620$$

and that

$$\log_7 259 = \frac{\log 259}{\log 7} \qquad \blacktriangleleft \text{ by (18.5)}$$
$$= \frac{2.4133}{.8451}$$
$$= 2.8556$$

Exercise 18.6　Logarithms to Bases Other Than 10

Make use of the relation between $\log_a N$ and $\log_b N$ and a table of common logarithms to find the logarithms called for in problems 1 to 24.

1	$\log_e 7.63$	**2**	$\log_e 8.17$	**3**	$\log_e 1{,}350$
4	$\log_e 0.106$	**5**	$\log_2 862$	**6**	$\log_2 13.7$
7	$\log_2 5.43$	**8**	$\log_2 51.3$	**9**	$\log_4 63.2$
10	$\log_4 9.13$	**11**	$\log_4 6{,}240$	**12**	$\log_4 6.88$
13	$\log_5 1{,}492$	**14**	$\log_5 631$	**15**	$\log_5 1{,}066$
16	$\log_5 0.217$	**17**	$\log_6 64.31$	**18**	$\log_6 372.8$
19	$\log_7 1{,}776$	**20**	$\log_7 1{,}891$	**21**	$\log_{11} 7.13$
22	$\log_{11} 1{,}891$	**23**	$\log_{11} 119$	**24**	$\log_{11} 1{,}334$

18.11 EXPONENTIAL AND LOGARITHMIC EQUATIONS

An equation in which the unknown occurs in one or more exponents is called an *exponential equation.* *A logarithmic equation* is an equation in which the logarithm of the unknown or a function of it occurs. For example, $3^{x+2} = 7^{x-1}$ is an exponential equation, and $\log_2 (x - 2) + \log_2 (x + 1) = 2$ is a logarithmic equation.

Many exponential and logarithmic equations cannot be solved by methods heretofore discussed, but some can be solved by use of logarithms.

EXAMPLE 1   Solve $5^{2x} = 7^{x+1}$.

Solution   If we take the common logarithm of each member of the given equation, we have

$$\log 5^{2x} = \log 7^{x+1}$$
$$2x \log 5 = (x + 1) \log 7 \qquad \blacktriangleleft \text{ by (18.4)}$$
$$x(2 \log 5 - \log 7) = \log 7 \qquad \blacktriangleleft \text{ collecting terms}$$
$$x(\log 25 - \log 7) = \log 7 \qquad \blacktriangleleft \text{ by (18.4)}$$
$$x = \frac{\log 7}{\log 25 - \log 7} \qquad \blacktriangleleft \text{ solving for } x$$
$$= \frac{.8451}{1.3979 - .8451}$$
$$= \frac{.8451}{.5528} = 1.529$$

EXAMPLE 2   Solve $\log_2 (x - 1) + \log_2 (x + 1) = 3$.

Solution   If we apply (18.3) to the given equation, we get

$$\log_2 [(x - 1)(x + 1)] = 3$$

Hence, using each member of this equation as an exponent of 2, we have

$$(x - 1)(x + 1) = 2^3 = 8 \qquad \blacktriangleleft \text{ by (18.1)}$$
$$x^2 - 9 = 0 \qquad \blacktriangleleft \text{ adding } -8 \text{ to each number}$$

Therefore, $\{3, -3\}$ is the possible solution set, but we must check each in the given equation. This is necessary because we have not defined the logarithm of a negative number and, consequently, must rule out any value of x which would require the use of the logarithm of a negative number.

If $x = 3$, the left member of the given equation becomes

$$\log_2 2 + \log_2 4 = 1 + 2 = 3$$

as stated in the given equation. If $x = -3$, the left member of the given equation becomes $\log_2 (-4) + \log_2 (-2)$; hence, $x = -3$ cannot be admitted as a root.

EXAMPLE 3   Solve

$$5^{x+y} = 100 \tag{1}$$

and
$$2^{2x-y} = 10 \tag{2}$$

for x and y.

Solution   If we equate the common logarithms of the members of each of (1) and of (2), we get

$$(x + y) \log 5 = 2 \tag{3}$$
$$(2x - y) \log 2 = 1 \tag{4}$$

If we solve these equations for $x + y$ and $2x - y$ and then solve the resulting equations simultaneously, we obtain

$$x + y = \frac{2}{\log 5} = 2.86 \tag{5}$$

$$2x - y = \frac{1}{\log 2} = 3.32 \tag{6}$$

$$3x \qquad\quad = 6.18 \qquad \blacktriangleleft \ (5) + (6)$$
$$x = 2.06$$

Substituting 2.06 for x in (5) and solving for y gives $y = 0.80$. Therefore, the solution of the system is $\{(2.06, 0.80)\}$.

Exercise 18.7   Exponential and Logarithmic Equations

Solve each equation in problems 1 to 8 for x in terms of y.

1 $y = 10^x$ **2** $y = 10^{-x}$ **3** $y = 5(10^{-2x})$

4 $y = 3(10^{2x})$ **5** $y = \dfrac{e^x + e^{-x}}{2}$ **6** $y = \dfrac{e^x - e^{-x}}{2}$

7 $y = \log_e (\sqrt{x^2 + 1} + x)$ **8** $y = \log_e (x - \sqrt{x^2 - 1})$

Find the solution set of the equation in each of problems 9 to 32.

9 $5^x = 625$	**10** $3^x = 27$	**11** $2^x = 64$
12 $7^x = 343$	**13** $3^{x^2-1} = 27$	**14** $5^{x^2-3x} = 625$
15 $5^{x^2-4x+3} = 1$	**16** $4^{x^2+3x} = 256$	**17** $2^x = 41$
18 $5^{x+1} = 3.02$	**19** $195^x = 2.68$	**20** $3.02^x = 0.00739$

21 $\log_5 (x - 1) + \log_5 (x + 3) = 1$

22 $\log_5 (2x + 1) + \log_5 (3x - 1) = 2$

23 $\log_6 3 + \log_6 (x + 6) = 2$

24 $\log_3 (x + 2) + \log_3 (x + 4) = 1$

25 $\log_5 (2x + 4) - \log_5 (x - 1) = 1$

26 $\log_2 (3x + 1) - \log_2 (x - 3) = 3$

27 $\log_3 (x + 11) - \log_3 (x + 3) = 2$

28 $\log_5 (3x + 7) - \log_5 (x - 5) = 2$

29 $5^{x+2y} = 28$
   $4x + 2y = 3$

30 $2^{2x+3y} = 48$
   $x + y = 2$

31 $10^{x+2y} = 2$
   $\log 3x - \log 2y = 1$

32 $10^{x-3y} = 3$
   $\log 2x - \log y = 1$

18.12 GRAPHS OF THE LOGARITHMIC AND EXPONENTIAL FUNCTIONS

Several properties of logarithmic and exponential functions are revealed by a study of the graphs of

$$\{(x, y) \mid y = \log_a x\} \tag{1}$$

and
$$\{(x, y) \mid y = a^x\} \tag{2}$$

We shall sketch the graphs of (1) with $a = 10$ and $a = e$, using the following table of values of x and y:

x	0.2	0.4	0.6	0.8	1	2	4	6	8	10	20	30
$y = \log_{10} x$	$-.7$	$-.4$	$-.2$	$-.1$	0	.3	.6	.8	.9	1	1.3	1.5
$y = \ln x$	-1.6	$-.9$	$-.5$	$-.2$	0	.7	1.4	1.8	2.1	2.3	3.0	3.4

The graphs, shown in Fig. 18.1, indicate the following properties of the function (1):

1 $\log_a x$ is not defined for negative values of x.

2 $\log_a x$ increases as x increases.

3 $\log_a x$ is negative, zero, or positive according as x is less than, equal to, or greater than 1.

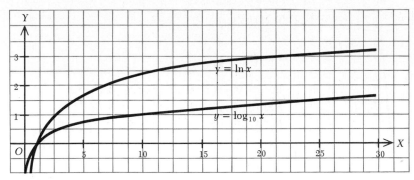

Figure 18.1

Next, we shall construct the graph of (2) with $a = e$ and $a = 2$. For this purpose, we must obtain a table of corresponding values of x and y, determined by $y = e^x$ and $y = 2^x$. To illustrate the method for computing the entries in this table, we shall obtain y if $x = 2.5$. We begin with

$$y = e^x$$

then
$$\log y = 2.5 \log e$$
$$= 2.5(.4343)$$
$$= 1.0858$$
$$y = 12.2 \qquad \blacktriangleleft \text{ from Table I}$$

By similar methods, we obtain the entries in the following table, which we use to construct the graphs in Fig. 18.2:

x	4	-2	-1	0	1	1.5	2	2.5	3
e^x	.02	.1	.4	1	2.7	4.5	7.4	12.2	20
2^x	.06	$\frac{1}{4}$	.5	1	2	2.8	4	5.7	8

The graphs indicate the following properties of the function (2):

1 a^x is never negative.

2 a^x increases as x increases.

3 a^x is less than 1, equal to 1, or greater than 1 according as x is negative, zero, or positive.

In the function $\{(x, y) \mid y = \log_a x\}$, the domain is $\{x \mid 0 < x < \infty\}$, and the range is $\{y \mid -\infty < y < \infty\}$. In the function $\{(x, y) \mid y = a^x\}$, the domain is $\{x \mid -\infty < x < \infty\}$, and the range is $\{y \mid 0 < y < \infty\}$. Hence the domain of $\log_a x$ is the range of a^x, and the range of $\log_a x$ is the domain of a^x. Furthermore, $\log_a a^x = x$. Hence a^x is the inverse of $\log_a x$. Also from Figs. 18.1 and 18.2, we see that the graph of $\ln x$ occupies the same

407

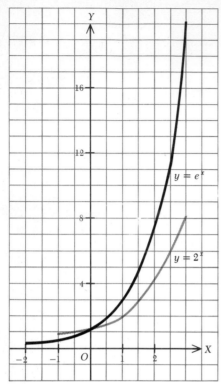

Figure 18.2

position relative to the X axis as the graph of e^x occupies with respect to the Y axis, as should be the case since each is the inverse of the other.

Exercise 18.8   Graphs of Exponential and Logarithmic Curves

Construct the graph of the function defined by the equation in each of problems 1 to 12.

1 $y = \log_2 x$	**2** $y = \log_3 x$	**3** $y = \log_7 x$
4 $y = \log_{11} x$	**5** $y = \log_{10} x^2$	**6** $y = \log_{10} x^3$
7 $y = \log_{10} 2x$	**8** $y = \log_{10} 3x$	**9** $y = 3^x$
10 $y = 3^{-x}$	**11** $y = 3^{2x}$	**12** $y = 3^{x^2}$

Exercise 18.9   Review

Round off the numbers in problems 1 to 6 to four digits and express each in scientific notation.

1 782.43	**2** 2,398.6	**3** 39.8752

408

4 0.28596 **5** 0.059855 **6** 0.0087684

Perform the indicated operations on the approximate numbers in problems 7 to 10. Carry each result to the appropriate number of digits. Use scientific notation if needed.

7 $(3.7)(58.96)$ **8** $389.2/783$
9 $2.85 + 3.715$ **10** $76.83 - 4.7146$
11 Change $3^5 = 243$ to logarithmic form.
12 Change $\log_2 128 = 7$ to exponential form.

In problems 13 to 15, solve for b, N, or L.

13 $\log_2 32 = L$ **14** $\log_b 81 = 4$ **15** $\log_6 N = 3$
16 If $\log N = 1.7176$, find N to three digits.
17 If $\log N = 0.3726$, find N to four digits.
18 If $N = 587.26$, find $\log N$.

Perform the indicated operations in problems 19 to 24 by use of logarithms.

19 $(39.7)(5.03)$ **20** $(8.627)(0.5964)$ **21** $785.9 \div 327.2$

22 $3.18 \div 0.0865$ **23** 8.25^3 **24** $\sqrt[3]{0.523}$

Find the logarithm called for in each of problems 25 to 27.

25 $\log_{11} 174$ **26** $\log_2 48.1$ **27** $\log_7 0.853$

28 Solve $e^x = \sqrt{\dfrac{1+y}{1-y}}$ for y.

29 Solve $\dfrac{e^x - e^{-x}}{e^x + e^{-x}} = y$ for x.

Find the solution set in problems 30 and 31.

30 $3^x = 37$ **31** $\log_2 (2x - 1) - \log_2 (x - 2) = 2$

Solve the pair of equations in each of problems 32 and 33 simultaneously.

32 $2^{x-y} = 5$ **33** $\log x + \log y = 4$
  $x + 2y = 3$ $\log 2x - \log 5y = 1$
34 Construct the graph of the function defined by $y = \log_7 3x$.
35 Construct the graph of the function defined by $y = 2^{3x}$.

Right Triangles

One of the more common applications of trigonometry is in determining the unknown sides and angles of a right triangle when certain sides and angles can be measured or are given. In order to determine unknown sides or angles, we must know the relations between the sides and angles of any right triangle. We must also know how many digits to use in the lengths of the sides if the angles are known to be correct to the nearest degree, 10 minutes, or minute. This chapter will be devoted to relations between the parts of right triangles, to the methods of solving right triangles, and to work with situations that involve right triangles.

19.1 CALCULATIONS WITH NUMBERS AND ANGLES

Rounding off was discussed in Sec. 18.1, and some discussion of procedures in calculation was presented. We shall now continue that discussion. If $\cos \theta$ is given as .94, we know that $\cos \theta$ is between .935 and .945 including .935 but not .945, since if it were greater than or equal to .945 or less than .935, it would not round off to .94. By use of Table II, we see that θ is between $19°10'$ and $20°40'$, and this does not establish θ closer than to the nearest degree. Since we obtain the function values of an angle from measurement of two sides of a triangle, the function values of an angle are ordinarily no more accurate than the lengths of the sides of the triangle in which the angles are located. We shall use the following table to determine

Accuracy of sides and angles

comparable degrees of accuracy in sides and angles of a triangle, even though the table might well be modified for angles near $0°$ and $90°$.

Significant digits in sides	Value of the angle accurate to the nearest
2	Degree
3	Multiple of 10 minutes
4	Minute

19.2 FUNCTION VALUES OF AN ACUTE ANGLE OF A RIGHT TRIANGLE

In order to avoid the inconvenience of having to put an acute angle of a right triangle in standard position when working with it, we shall define the sine, cosine, tangent, and cotangent of such an angle in terms of the sides and hypotenuse of a right triangle that contains the angle. We shall designate the right angle by C and the acute angles by A and B; furthermore, a is the side opposite A, b is the side opposite B, and c is the hypotenuse.

We shall construct a right triangle ABC as in Fig. 19.1 with the angle

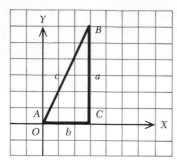

Figure 19.1

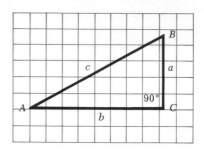

Figure 19.2

A in standard position and the sides and angles lettered as discussed. Consequently, the coordinates of the vertex B are (b, a) and the radius vector is c. Now, by applying the definition of the trigonometric ratios we have

Trigonometric
ratios of
acute angles

$$\sin A = \frac{a}{c} = \frac{\text{opposite side}}{\text{hypotenuse}}$$

$$\cos A = \frac{b}{c} = \frac{\text{adjacent side}}{\text{hypotenuse}}$$

$$\tan A = \frac{a}{b} = \frac{\text{opposite side}}{\text{adjacent side}}$$

$$\cot A = \frac{b}{a} = \frac{\text{adjacent side}}{\text{opposite side}}$$

These definitions are applicable only to acute angles of right triangles. Since a geometric figure can be moved without affecting its size or shape, the ratios of the sides remain the same regardless of the position of the triangle and can be used without reference to a coordinate system.

We shall now draw a right triangle as in Fig. 19.2 and find a relation between the function values of the acute angles.

$$\sin A = \frac{a}{c} = \cos B$$

$$\cos A = \frac{b}{c} = \sin B$$

$$\tan A = \frac{a}{b} = \cot B$$

$$\cot A = \frac{b}{a} = \tan B$$

Now if we compare the function values of A and B as listed above, it is clear that any trigonometric function value of either A or B is equal to the

412

corresponding cofunction value of the other; furthermore, *A* and *B* are complementary angles. Consequently, we have the following theorem:

Cofunction
complement
theorem

Any trigonometric function value of an acute angle is equal to the corresponding cofunction value of the complementary angle.

In keeping with this theorem, we have sin 42° = cos 48°, since sine and cosine are cofunctions and 42° and 48° are complementary angles, and also tan 31° = cot 59°, since tangent and cotangent are cofunctions and 31° and 59° are complementary angles.

It is because of the fact stated in the above theorem that the arrangement used in Table II is possible. The function names at the top and bottom of a column are cofunctions, and the angles directly across from one another are complementary.

19.3 LOGARITHMS OF FUNCTION VALUES OF ANGLES

Logarithmic
functions

We could find the value of a trigonometric function of an angle by use of Table II and then find its logarithm from Table I in order to obtain the logarithm of a function value of an angle. This is not necessary, however, since Table II gives the logarithms of the function values as well as the function values. The table is so constructed that 10 must be subtracted from each logarithm except the logarithmic tangent of angles larger than 45° and the logarithmic cotangent of angles less than 45°. Thus, if the function value of the angle is greater than 1, we do not subtract 10 from the entry, but if it is less than 1, we do subtract 10 from the entry. The characteristic and decimal point are not printed with each entry.

EXAMPLE 1

By use of tables, find log cot 5°20′.

Solution

To find log cot 5°20′, we look across from 5°20′ and under "Log." in the cotangents column and there find 0299. Then we look at the first entry in the block to get the characteristic 1. After placing the decimal point, we find that log cot 5°20′ = 1.0299.

EXAMPLE 2

Find log sin 56°10′.

Solution

We look across from 56°10′ and above "Log." in the sines column and find 9194. Then, after supplying the decimal point and the characteristic including −10, we see that log sin 56°10′ = 9.9194 − 10.

EXAMPLE 3   Find log tan $28°17'$ by use of interpolation.

Solution   The procedure is the same as in all linear interpolation, and we shall use the usual tabular form.

$$
10 \left[7 \begin{bmatrix} 28°10' & 9.7287 - 10 \\ 28°17' & ? \\ 28°20' & 9.7317 - 10 \end{bmatrix} c \right] .0030
$$

$$
c = \tfrac{7}{10}(.0030) = .0021
$$
$$
\log \tan 28°17' = 9.7287 - 10 + .0021 = 9.7308 - 10
$$

19.4 SOLUTION OF RIGHT TRIANGLES

Solution of right triangles   As we saw in Sec. 19.2, the definition of each function value of an acute angle of a right triangle includes an angle and two sides of the triangle. Therefore, if any two of these quantities are known, the third can be determined. If we want to find one of the sides of an acute angle, *we must select a function value of an angle such that two known parts and the desired unknown part are involved in the definition of that function value of that angle.* We can then find the value of the desired part by solving the resulting equation for it, or for a function value of it, and then performing the indicated calculations with or without the use of logarithms. If all unknown parts are determined, we say that the triangle is solved. It is desirable to draw a reasonably accurate figure approximately to scale before beginning the solution of the triangle. The table of comparable degrees of accuracy given in Sec. 19.1 should be kept in mind when solving a triangle.

EXAMPLE 1   Solve the right triangle in which $A = 27°30'$, zero not significant, and $b = 27.3$.

Solution without logarithms   Figure 19.3 shows the triangle determined by the given data. Since $A = 27°30'$, we know that $B = 90° - 27°30' = 62°30'$. If we decide to find a next, we can do so by use of $\tan A = a/b$. On solving for a, we have

$$
\begin{aligned}
a &= b \tan A \\
&= 27.3 \tan 27°30' \\
&= 27.3(.5206) \\
&= 14.21238 = 14.2 \qquad \text{to three figures}
\end{aligned} \tag{1}
$$

We shall now evaluate c. This can be done by means of the equation

$\cos A = b/c$; hence

$$c = \frac{b}{\cos A} = \frac{27.3}{\cos 27°30'}$$

$$= \frac{27.3}{.8870}$$

$$= 30.8 \qquad \text{to three figures}$$

Solution with logarithms

To solve the triangle by use of logarithms, we proceed as above down to and including Eq. (1). Then

$$\log a = \log 27.3 + \log \tan 27°30'$$
$$\log 27.3 = \;\;1.4362$$
$$\underline{\log \tan 27°30' = \;\;9.7165 - 10}$$
$$\log a = 11.1527 - 10 \qquad \blacktriangleleft \text{adding}$$
$$a = 14.2 \qquad\qquad \blacktriangleleft \text{since the characteristic is } 11 - 10 = 1$$

Now, as in solving without logarithms,

$$c = \frac{27.3}{\cos 27°30'}$$
$$\log c = \log 27.3 - \log \cos 27°30'$$
$$\log 27.3 = 11.4362 - 10 \qquad \blacktriangleleft \text{adding and subtracting 10}$$
$$\underline{\log \cos 27°30' = \;\;9.9479 - 10}$$
$$\log c = \;\;1.4883 \qquad\qquad \blacktriangleleft \text{subtracting}$$
$$c = 30.8$$

EXAMPLE 2

Solve the right triangle in which $a = 27.36$ and $c = 38.41$.

Solution

A sketch of the triangle is shown in Fig. 19.4. We shall find A by use of $\sin A$ and obtain

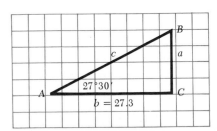

Figure 19.3

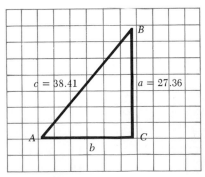

Figure 19.4

$$\sin A = \frac{a}{c}$$

$$= \frac{27.36}{38.41}$$

$$\log \sin A = \log 27.36 - \log 38.41$$

$$\log 27.36 = 11.4372 - 10$$

$$\underline{\log 38.41 = 1.5844}$$

$$\log \sin A = 9.8528 - 10 \qquad \blacktriangleleft \text{ subtracting}$$

$$A = 45°27' \qquad\qquad \blacktriangleleft \text{ interpolating}$$

Since A and B are the acute angles of a right triangle, we know that $B = 90° - 45°27' = 44°33'$. Furthermore, we can find b by use of $\tan B = b/a$. From this we have

$$b = a \tan B$$

$$= 27.36 \tan 44°33'$$

$$\log b = \log 27.36 + \log \tan 44°33'$$

$$\log 27.36 = 1.4372$$

$$\underline{\log \tan 44°33' = 9.9932 - 10}$$

$$\log b = 1.4304 \qquad \blacktriangleleft \text{ adding}$$

$$b = 26.94 \qquad \text{to four digits}$$

Exercise 19.1   Solution of Right Triangles

Solve the right triangle in problems 1 to 8 without the use of tables.

1	$B = 30°, c = 12$	**2**	$A = 60°, c = 36$
3	$A = 60°, b = 16$	**4**	$A = 30°, a = 20$
5	$B = 45°, a = 20$	**6**	$B = 45°, b = 7\sqrt{2}$
7	$b = 17, c = 17\sqrt{2}$	**8**	$a = 48, c = 48\sqrt{2}$

Solve the right triangle that has the parts given in each of problems 9 to 40. Obtain each result to the degree of accuracy justified by the given data.

9	$A = 41°10', c = 921$	**10**	$A = 32°30', c = 8.13$
11	$A = 73°40', b = 73.4$	**12**	$A = 68°50', b = 42.7$
13	$B = 37°30', a = 644$	**14**	$B = 25°20', a = 5.68$
15	$a = 3.19, c = 6.70$	**16**	$a = 538, c = 871$
17	$b = 0.415, c = 0.597$	**18**	$b = 1.78, c = 3.84$
19	$a = 52.7, b = 63.1$	**20**	$a = 0.419, b = 0.507$
21	$B = 37°12', c = 7.142$	**22**	$B = 57°34', c = 6.974$
23	$A = 81°14', c = 0.7891$	**24**	$A = 29°36', c = 8,055$

25	$B = 16°37'$, $a = 4,724$	**26**	$B = 71°46'$, $a = 13.46$
27	$A = 52°53'$, $b = 5.703$	**28**	$A = 63°17'$, $b = 0.4578$
29	$b = 7.117$, $c = 9.876$	**30**	$b = 2.013$, $c = 3.748$
31	$a = 56.78$, $b = 67.89$	**32**	$a = 0.3726$, $b = 0.5375$
33	$a = 3.24(10^3)$, $A = 39°10'$	**34**	$a = 2.73(10^2)$, $A = 63°20'$
35	$B = 27°40'$, $c = 8.18(10^3)$	**36**	$B = 58°50'$, $c = 8.13(10^2)$
37	$a = 3.724(10^4)$, $b = 2.637(10^4)$	**38**	$a = 4.523(10^5)$, $b = 5.127(10^5)$
39	$a = 8.976(10^4)$, $c = 1.234(10^5)$	**40**	$b = 9.147(10^4)$, $c = 1.329(10^5)$

19.5 | ANGLES OF ELEVATION AND DEPRESSION

Angles of elevation and depression

The ray from an observer to an object is called the *line of sight*. If the observer is at O and the object is at P, as in Fig. 19.5, and if H is on a horizontal ray through O and in the same vertical plane as OP, then the angle POH is called the *angle of depression* or *angle of elevation* of P according as P is below or above OH.

EXAMPLE

An observer on the ground finds that the angle of elevation of a balloon is $51°20'$. How high is the balloon if the point directly under it is 237 feet horizontally from the observer?

Solution

A sketch of the situation is shown in Fig. 19.6. To find the height, we solve $\tan 51°20' = h/237$ for h and have

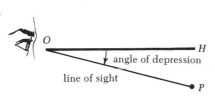

Figure 19.5

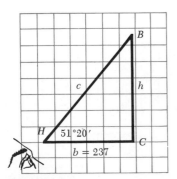

Figure 19.6

$$h = 237 \tan 51°20'$$
$$\log h = \log 237 + \log \tan 51°20'$$
$$\log 237 = 2.3747$$
$$\underline{\log \tan 51°20' = 0.0968}$$
$$\log h = 2.4715$$
$$h = 296$$

Therefore, the balloon is 296 feet high.

19.6 THE DIRECTION OF A RAY

Bearing of flight   The bearing of a ray or course of flight in air navigation is expressed as the clockwise angle less than 360° that the ray makes with the due north direction.   In Fig. 19.7, the directions of *OP*, *OQ*, *OR*, and *OS* are 70°, 150°, 230°, and 320°, respectively.   Each of these directions is measured clockwise from due north.

Bearing in marine   On the other hand, the direction or bearing of a ray in marine navi-
navigation   gation and in plane surveying is given by stating the acute angle the given ray makes with a due north or due south direction and whether it is to the east or west of that direction.   Thus, if the direction or bearing is 22° east of a due south direction, it is written as S22°E.   The directions N70°E, S30°E, S50°W, and N40°W shown in Fig. 19.8 correspond to the air navigation bearings given in Fig. 19.7.

19.7 VECTORS

Vector quantity   A quantity that has both magnitude and direction is called a *vector quantity*.   Consequently, forces, velocities, accelerations, and displacements are vector quantities.   Such a quantity can be represented by a directed line
Vector   segment, called a *vector*, whose length and direction represent the magnitude and direction, respectively, of the vector quantity.

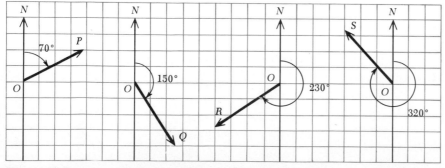

Figure 19.7

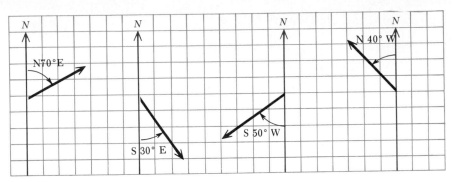

Figure 19.8

Resultant
Component

The single vector acting at a point that has the same effect as two other vectors acting at that point is called the *resultant*. The two other vectors are called the *components*. It has been verified experimentally that the resultant can be displayed as a diagonal of the parallelogram which has the two components as adjacent sides. In Fig. 19.9, the components are AB and AC, and AD is their resultant.

EXAMPLE

A ship is headed due north at a speed of 27.8 feet per second, and a man on the ship throws a ball due west at 36.4 feet per second. Find the speed and direction of the ball.

Solution

In constructing Fig. 19.10 to represent the situation, we first draw BC to represent the velocity of the ship and then draw CS to represent the velocity the ball would have had if the ship had not been moving. BS then represents the direction and velocity of the ball. Furthermore, the direction traveled by the ball is determined by angle B.

$$\tan B = \frac{36.4}{27.8}$$
$$\log \tan B = \log 36.4 - \log 27.8$$
$$\log 36.4 = 1.5611$$
$$\underline{\log 27.8 = 1.4440}$$
$$\log \tan B = 0.1171$$
$$B = 52°40' \qquad \text{to the nearest } 10'$$

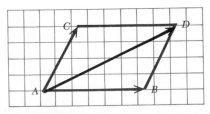

Figure 19.9

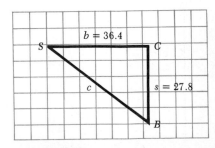

Figure 19.10

We can now find the magnitude of the resultant velocity c of the ball by use of

$$\sin B = \frac{b}{c}$$

$$c = \frac{b}{\sin B} = \frac{36.4}{\sin 52°40'}$$

$$\log c = \log 36.4 - \log \sin 52°40'$$

$$\log 36.4 = 11.5611 - 10$$

$$\underline{\log \sin 52°40' = \ \ 9.9004 - 10}$$

$$\log c = \ \ 1.6607$$

$$c = 45.8$$

Consequently, the ball travels N52°40'W at 45.8 feet per second.

Exercise 19.2   Angle of Elevation, Rays, and Vectors

1   Find the length of the diagonal of a rectangular tract of land if the sides are 130 meters and 100 meters.

2   A pilot flew his plane 505 miles at 22°10' in going from A to B. In returning by car, he followed roads due west and due south. How much further did he travel by car than by plane?

3   How tall is a tree that is on horizontal ground if the tip of its 6.40-meter shadow is 15.0 meters from the top of the tree?

4   An observer in the same horizontal plane as the base of a 475-foot tower finds the angle of elevation of its top to be 32°40'. How far is the observer from the base of the tower?

5   How tall is a tree if the angle of elevation of its top is 74°30' at a point 39.3 feet from its base?

6   The rafters of a roof are 21.5 feet long, and the roof rises 8.75 feet. What angle does the roof make with the horizontal?

7   A sewer pipe was to be placed under a level lot at an angle of 2°10' with the horizontal. At one point the trench for the pipe was 2.25 feet below the ground surface. How deep should the trench be at a point 250 feet away measured along the pipe?

8   A brace runs from a lower corner of a rectangular wall to the diagonally opposite corner. If the wall is 10.75 by 16.50 feet, find the length of the brace and the angle it makes with the longer side.

9   The entrance of a mine is 3,202 feet above sea level. The shaft is straight and has an angle of depression of 30°10'. Find the elevation of the lower end if the shaft is 200 feet long.

10   How long is a chord that subtends a central angle of 26°12' in a circle of radius 16.45 feet?

11 Find the height of a retaining wall that is to enable a man to level a lot that is 73.04 feet wide and slopes at 2°12′ with the horizontal.

12 Find the length of wire needed to reach from the top of the first pole to the top of the last pole if 12 telephone poles of uniform height are placed 50.0 feet apart on a uniform slope that makes 3°20′ with the horizontal.

13 Strings of lights are to reach from the top of a 25.0-foot pole and make an angle of elevation of 60°0′ with the horizontal ground. How far from the pole are they fastened to the ground?

14 The boom of a crane is 28.6 feet long, and the operators manual says that it is not to be lowered so as to make an angle of less than 45°50′ with the horizontal. What is the greatest horizontal distance from the crane that can be reached by the boom?

15 The gangplank of a ship is 28.6 feet long and makes an angle of 10°20′ with the horizontal. If the lower end of the gangplank is 3 feet from the dock's edge, how far is the ship from the dock?

16 An A-frame cabin is 25.69 feet high at the center and 40.52 feet wide at the base. What angle does the roof make with the floor?

17 A woman walked 510 yards N15°E and then S75°E for 630 yards. Find the direction and distance to her starting point.

18 A ranger wanted to know how far up a uniform slope a fire had traveled. An aerial photograph showed it had gone a horizontal distance of 165 feet. How far had the fire traveled if the angle of elevation of the hill was 47°20′?

19 A train traveled at 55 miles per hour on a straight track that headed N68°30′E from Center Point to Centerville. A car left Center Point when the train did and went 45.8 miles due east and then north to Centerville at 65 miles per hour. Which reached Centerville first and by how long?

20 During a fire in a refinery on Zumwalt Street, it was decided that everyone within 800 feet of the plant should be evacuated. Akron Avenue intersects Zumwalt Street at right angles and 250 feet from the plant. Was it necessary to evacuate a house on Akron Avenue that is 740 feet from the intersection?

21 What depth of cut is necessary in a ridge top that is 28.5 feet higher than a point 243.3 feet away horizontally in order to have a road from the point through the ridge with an angle of elevation of 4°3′?

22 A wall that runs S48°10′W for 135 feet is one side of a triangular lot. What is the area of the lot if the other sides run north-south and east-west?

23 Mason is due north of Dawson and due west of Clark. Dawson and Clark are connected by a highway that runs N29°20′E and is 59.8 miles long. How many miles out of her way did a woman go who went from Dawson to Clark by way of Mason?

24 A plane headed due north at 252 miles per hour in a west wind blow-

ing at 31.8 miles per hour. Find the ground speed and direction of its flight.

25 A plane headed due east at an air speed of 324 miles per hour in a south wind. If the ground speed of the plane was 325 miles per hour, find the direction of the flight and the speed of the wind.

26 A boy could have walked along a cutoff trail that heads N30°37′E, but he walked due east 153.9 feet to a bird nest and then due north to the cutoff. How far out of his way did he walk?

27 Two observers on an east-west highway find that the angles of elevation of a helicopter are 22°31′ and 24°42′. If the helicopter is 2,500 feet directly above a point on the highway east of the observers, how far apart are they?

28 A pilot flew 320 miles at 68°10′, then 540 miles at 158°10′, and back to the starting point. Find the length and direction of the last flight.

29 If one child pushed a toy due north with a force of 92 pounds and another pushed it due east with a force of 72 pounds, what direction did it move?

30 A pilot is flying at an air speed of 200 miles per hour in a west wind blowing at 30.6 miles per hour. In what direction must she head in order to fly due south? What is her speed relative to the ground?

31 At a point on a horizontal plane, the angle of elevation of a hill top is 23°43′. At a point 1,225 nearer the hill, the angle of elevation of the hill top is 30°22′. How high is the hill?

32 The angle of depression of a boat east of a helicopter is 34°6′, while the angle of depression of a boat south of it is 31°13′. How far apart are the boats if the helicopter is 1,200 feet above the harbor?

Exercise 19.3   Review

Solve the right triangles that are determined by the data in each of problems 1 to 8.

1   $A = 60°, b = 18$
2   $A = 38°10′, a = 11.3$
3   $A = 68°40′, c = 49.6$
4   $a = 2.87, b = 3.73$
5   $a = 39.5, c = 58.7$
6   $B = 40°20′, a = 2.34(10^3)$
7   $B = 23°30′, c = 4.28(10^4)$
8   $A = 28°31′, b = 45.67$
9   The sides of a rectangle are 162 and 128 meters. Find the length of the diagonal and the angle it makes with the shorter side.

10 Mr. Godfrey flew 473 miles at 140°20′ from *A* to *B* and then returned to *A* by going directly west and directly north. How far did he go in each direction?

11 Two observers on a horizontal north-south road find that the angles of elevation of a helicopter are 18°23′ and 41°36′. If the craft is 1,420 feet directly above the highway and between the observers, how far apart are they?

20

Oblique Triangles

We say that a triangle is solved if its sides, angles, and area are known. In this chapter we shall derive several formulas for use in solving triangles. A triangle can be solved if a side and two other parts are known. The required combination of parts is known if we are given two angles and a side, two sides and an angle opposite one of them, two sides and the included angle, or three sides. In order to determine a specified part of a triangle, we must select a formula that involves the desired part as the only unknown.

20.1 THE LAW OF SINES

The law of sines enables us to solve a triangle if certain parts are known. In order to derive it, we shall represent the sides of the triangle by a, b, and c and the angles opposite them by A, B, and C, respectively, as in Figs. 20.1 and 20.2. In each figure we have dropped a perpendicular from a vertex to the opposite side or to the opposite side produced and have called its length h and its foot D. From Fig. 20.1, we have $\sin A = h/b$ and $\sin B = h/a$. In Fig. 20.2, we have $\sin A = h/b$ and $\sin DBC = h/a$, but $\sin B = \sin DBC$, since B and angle DBC are supplements. Consequently $h = a \sin B = b \sin A$ from either figure. Now, if we divide each member of the last equation by $\sin A \sin B$, we have $a/(\sin A) = b/(\sin B)$. It can be proved in a similar manner that $a/(\sin A) = c/(\sin C)$. Consequently, it follows that

$$\frac{a}{\sin A} = \frac{b}{\sin B} = \frac{c}{\sin C} \qquad (20.1)$$

This can be stated in words as follows:

Law of sines
The three ratios obtained by dividing a side by the sine of the angle opposite it in any triangle are equal.

This statement is known as the *law of sines*, and it can be used if three of the four quantities involved in any two members of Eq. (20.1) are known. Thus, it can be used if we know two angles and a side or two sides and the angle opposite one of them. We shall consider the first of these two situations now and treat the second in Sec. 20.3.

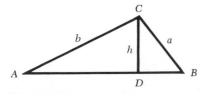

Figure 20.1

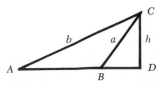

Figure 20.2

EXAMPLE

Find the other angle and the sides of a triangle in which $b = 309$, $A = 62°40'$, and $B = 73°20'$.

Solution

We shall begin by drawing a scale sketch of the situation, as shown in Fig. 20.3. Using the first equation of (20.1), we have

$$\frac{a}{\sin A} = \frac{b}{\sin B}$$

and by substituting in the given values, we get

$$\frac{a}{\sin 62°40'} = \frac{309}{\sin 73°20'}$$

We now solve for a and obtain

$$a = \frac{309}{\sin 73°20'} \sin 62°40'$$

We can now evaluate a with or without the use of logarithms, and we shall use them. Thus,

$$\log a = \log 309 + \log \sin 62°40' - \log \sin 73°20'$$

$$
\begin{array}{rl}
\log 309 = & 2.4900 \\
\log \sin 62°40' = & 9.9486 - 10 \\
\hline
\log 309 + \log \sin 62°40' = & 12.4386 - 10 \quad \blacktriangleleft \text{ adding} \\
\log \sin 73°20' = & 9.9814 - 10 \\
\hline
\log a = & 2.4572 \quad \blacktriangleleft \text{ subtracting} \\
a = & 287
\end{array}
$$

Since two angles are given, we can find the third by subtracting their sum from 180°; hence

$$C = 180° - (62°40' + 73°20')$$
$$= 44°0'$$

Now by using the second and third parts of (20.1) and substituting in the

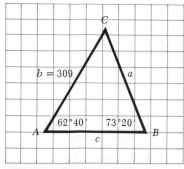

Figure 20.3

given and determined values, we have

$$\frac{c}{\sin 44°0'} = \frac{309}{\sin 73°20'}$$

$$c = \frac{309}{\sin 73°20'} \sin 44°0'$$

Consequently,

$$\log c = \log 309 + \log \sin 44°0' - \log \sin 73°20'$$

$$\begin{array}{ll}
\log 309 = & 2.4900 \\
\log \sin 44°0' = & 9.8418 - 10 \\
\hline
\log 309 + \log \sin 44°0' = & 12.3318 - 10 \qquad \blacktriangleleft \text{ adding} \\
\log \sin 73°20' = & 9.9814 - 10 \\
\hline
\log c = & 2.3504 \qquad \blacktriangleleft \text{ subtracting} \\
c = & 224 \\
\end{array}$$

20.2 THE AREA OF A TRIANGLE

We shall use the fact that the area of a triangle is equal to one-half the product of the base and the altitude as well as the value of the altitude found from Fig. 20.1 in order to obtain a new formula for the area. If, in connection with Figs. 20.1 and 20.2, we use $K = \frac{1}{2}ch$ and substitute $b \sin A$ for h, we see that

$$K = \tfrac{1}{2}bc \sin A \tag{20.2}$$

Area of a triangle
Since the lettering is immaterial, it follows that *the area of a triangle is one-half the product of any two sides and the sine of the included angle.*

We can get another formula for the area by replacing b in (20.2) by $(c \sin B)/(\sin C)$ as obtained by solving $b/(\sin B) = c/(\sin C)$ for b. Thus, we have

$$K = \frac{c^2 \sin A \sin B}{2 \sin C} \tag{20.3}$$

Another area formula
Since the lettering is immaterial, it follows that *the area of any triangle is the product of the square of any side and the sines of the adjacent angles divided by twice the sine of the opposite angle.*

EXAMPLE
Find the area of the triangle in which $b = 309$, $A = 62°40'$, and $B = 73°20'$.

Solution
This is the triangle used in the example of the last section. It is desirable to use a formula that includes only given parts, or as nearly so as possible, in order to minimize the possibility of carrying over any error that may

have been made. We shall use (20.3) with $C = 180° - (62°40' + 73°20') = 44°0'$ in addition to the given parts. Symbolically, we use

$$K = \frac{b^2 \sin A \sin C}{2 \sin B}$$

$$= \frac{309^2 \sin 62°40' \sin 44°0'}{2 \sin 73°20'}$$

Therefore,

$$
\begin{aligned}
2 \log 309 &= 4.9800 \\
\log \sin 62°40' &= 9.9486 - 10 \\
\log \sin 44°0' &= 9.8418 - 10 \\
\hline
\log \text{numerator} &= 24.7704 - 20 \\
\log \text{denominator} &= 10.2824 - 10 \\
\hline
\log K &= 4.4880 \\
K &= 3.08(10^4)
\end{aligned}
$$

$$
\begin{aligned}
\log 2 &= 0.3010 \\
\log \sin 73°20' &= 9.9814 - 10 \\
\hline
\log \text{denominator} &= 10.2824 - 10 \\
&\blacktriangleleft \text{recopied} \\
&\blacktriangleleft \log \text{numerator} - \log \text{denominator} \\
\text{to three digits}
\end{aligned}
$$

Exercise 20.1   The Law of Sines

Solve the following triangles. Interpolate if appropriate.

1   $A = 38°20'$, $B = 61°30'$, $a = 207$
2   $A = 71°10'$, $B = 23°40'$, $a = 429$
3   $A = 62°30'$, $B = 39°40°$, $b = 0.108$
4   $A = 54°40'$, $B = 43°20'$, $b = 2.16$
5   $A = 81°50'$, $C = 58°50'$, $b = 31.5$
6   $A = 47°30'$, $C = 54°30'$, $b = 0.0603$
7   $A = 63°10'$, $C = 29°40'$, $a = 3.77$
8   $A = 48°50'$, $C = 51°30'$, $a = 0.0444$
9   $B = 38°30'$, $C = 42°20'$, $c = 0.307$
10   $B = 54°20'$, $C = 30°30'$, $c = 109$
11   $B = 101°10'$, $C = 28°40'$, $c = 11.2$
12   $B = 34°40'$, $C = 41°50'$, $c = 3.24$
13   $A = 37°56'$, $B = 59°48'$, $a = 2,231$
14   $A = 72°37'$, $B = 39°39'$, $b = 39.39$
15   $A = 32°23'$, $B = 45°54'$, $c = 0.8228$
16   $A = 52°34'$, $B = 61°13'$, $a = 3,971$
17   $B = 73°32'$, $C = 41°29'$, $b = 0.02345$
18   $B = 47°47'$, $C = 53°35'$, $c = 702.8$
19   $B = 29°41'$, $C = 46°22'$, $a = 8.974$
20   $B = 45°14'$, $C = 37°23'$, $b = 43.21$
21   By how much do the other sides of a triangle differ if one side is 32.6 feet and the angles adjacent to that side are $29°40'$ and $70°30'$?
22   One diagonal of a parallelogram is 38.4 centimeters long and makes angles of $49°20'$ and $34°10'$ with the sides. How long is each side?

23 Two surveyors are known to be 1.76 kilometers apart and both observe a church spire. The spire is N41°20′E of one surveyor and N28°50′W of the other. How far is the church from each surveyor?

24 A man found the angle of elevation of the top of a building to be 23°20′ at one station and 50°10′ at another that is in the same horizontal plane as the first station and at the base of the 127-foot building. How far apart are the two stations?

20.3 THE AMBIGUOUS CASE

The situation in which two sides and an angle opposite one of them are given is the *ambiguous case* of the law of sines, since with such data there may be no solution, one solution, or two solutions. If we are given a, b, and A, we can use $a/(\sin A) = b/(\sin B)$, solve for $\sin b$, and obtain

$$\sin B = \frac{b \sin A}{a}$$

We can find the value of $\sin B$, since a, b, and A are known. The given data may be such that $\sin B$ is found to be greater than 1, equal to 1, or less than 1.

If $\sin B > 1$, then there is no angle B; hence, no triangle.

If $\sin B = 1$, then $B = 90°$ and B can or cannot be a part of a triangle that includes the given angle A according as $A + B < 180°$ or $A + B \geq 180°$. Therefore, there may be one or no triangle with the given data.

If $\sin B < 1$, we can find an acute angle B by use of a table of trigonometric function values. Furthermore, $180° - B = B'$ is another angle whose sine is equal to $\sin B$, since the sine of an angle and of its supplement are equal. Consequently, there are two triangles, one triangle, or no triangle with the given data according as $B + A < 180°$ and $B' + A < 180°$, $B + A < 180°$ and $B' + A \geq 180°$, or $B + A \geq 180°$ and $B' + A \geq 180°$.

There is no need to memorize the tests for the number of solutions in the various situations that may arise. Instead of memorizing the tests, *it is only necessary to evaluate* $\sin B$, *find all possible values of B that can be angles of a triangle, and use as many of them as fit into a triangle with the given angle.* The evaluation of $\sin B$ can be done with or without use of logarithms.

Solution of the ambiguous case

EXAMPLE 1

Find the number of triangles if $a = 176$, $b = 189$, and $A = 48°10′$.

Solution

If we substitute the given values in

$$\sin B = \frac{b \sin A}{a}$$

we get
$$\sin B = \frac{189 \sin 48°10′}{176}$$
$$= 0.8001$$

Consequently, $B = 53°10'$ and $B' = 180° - 53°10' = 126°50'$.   Therefore, $A + B = 48°10' + 53°10' = 101°20'$ and $A + B' = 48°10' + 126°50' = 175°0'$ are both less than 180°; hence, both B and B' can be used, and there are two solutions.   We shall not find the other parts in either solution, since to do so would require only the use of the law of sines in connection with triangle ABC and in connection with $AB'C'$.

EXAMPLE 2   How many solutions are there if $A = 45°30'$, $a = 723$, and $b = 578$?

Solution   If we use

$$\sin B = \frac{b \sin A}{a}$$

and substitute the given values, we have

$$\sin B = \frac{578 \sin 45°30'}{723}$$
$$= 0.5702$$

Therefore, $B = 34°50'$ and $B' = 180° - 34°50' = 145°10'$.   Now $A + B = 45°30' + 34°50' = 80°20' < 180°$ and $A + B' = 45°30' + 145°10' = 190°40' > 180°$; hence, B can be used in a triangle with A, and B' cannot be so used. Consequently, there is only one solution.

20.4 THE LAW OF COSINES

We shall now derive a second law for use in solving triangles.   It is called the law of cosines, and it can be used in the cases in which the law of sines is not applicable.   We shall consider any triangle ABC, as shown in Fig. 20.4, and drop a perpendicular from any vertex C to the opposite side or to the opposite side produced.

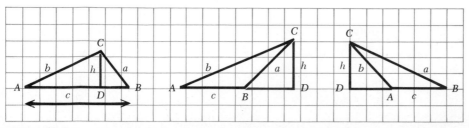

Figure 20.4

If we apply the pythagorean theorem to either triangle, we obtain

$$a^2 = h^2 + (DB)^2 \qquad (1)$$

We now express each term in the right member of this equation in terms of b, c, A, B, and C. Thus, since $\sin A = h/b$, we have $h = b \sin A$; furthermore,

$$
\begin{aligned}
DB &= c - AD \\
&= c - b \cos A \qquad \blacktriangleleft \text{ since } \cos A = AD/b
\end{aligned}
$$

Now, by substituting these expressions for h and DB in (1), we get

$$
\begin{aligned}
a^2 &= (b \sin A)^2 + (c - b \cos A)^2 \\
&= b^2 \sin^2 A + c^2 - 2bc \cos A + b^2 \cos^2 A \qquad \blacktriangleleft \text{ expanding} \\
&= b^2 (\sin^2 A + \cos^2 A) + c^2 - 2bc \cos A \qquad \blacktriangleleft \text{ combining similar terms}
\end{aligned}
$$

Since $\sin^2 A + \cos^2 A = 1$, we have

$$\boldsymbol{a^2 = b^2 + c^2 - 2bc \cos A} \qquad (20.4)$$

Equation (20.4), or those obtained from it by interchanging a with b or c, is called the *law of cosines*. It can be expressed in words as follows:

Law of cosines *The square of any side of a triangle is equal to the sum of the squares of the other two sides minus twice the product of those two sides and the cosine of the angle between them.*

The law of cosines involves all three sides and one angle and can be used if any three of these four quantities are known; hence, it can be used if the known parts are two sides and the angle opposite one of them, two sides and the included angle, or three sides. The data in the first of these situations are those of the ambiguous case that was discussed in the preceding section. The use of the other two sets of data will now be illustrated.

EXAMPLE 1   If in a triangle $a = 17$, $b = 22$, and $C = 47°$, find c.

Solution   Since the known angle is C, we must use the law of cosines in the form

$$
\begin{aligned}
c^2 &= a^2 + b^2 - 2ab \cos C \\
&= 17^2 + 22^2 - 2(17)(22) \cos 47° \\
&= 289 + 484 - 748(0.6820) \\
&= 262.864
\end{aligned}
$$

$$c = \sqrt{262.864} = 16 \qquad \text{to two digits}$$

If desired, the unknown angles could now be found by use of the law of sines.

EXAMPLE 2   Find the largest angle of the triangle in which $a = 36$, $b = 47$, and $c = 41$.

Solution   The largest angle is opposite the largest side; hence it is B.   Therefore, we use

$$b^2 = a^2 + c^2 - 2ac \cos B$$
$$47^2 = 36^2 + 41^2 - 2(36)(41) \cos B \quad \blacktriangleleft \text{ using the given values}$$
$$2{,}209 = 1{,}296 + 1{,}681 - 2{,}952 \cos B$$
$$\cos B = \frac{1{,}296 + 1{,}681 - 2{,}209}{2{,}952}$$
$$= 0.2602$$
$$B = 75° \quad \text{to the nearest degree}$$

Exercise 20.2   Ambiguous Case, Law of Cosines

In each of problems 1 to 12, determine the number of solutions.   In each of problems 9 to 12 find the unknown parts of the triangle by use of the law of sines.

1	$a = 193, b = 143, A = 55°50'$	**2**	$a = 54.1, b = 77.2, A = 40°30'$
3	$a = 725, b = 453, B = 23°10'$	**4**	$a = 10.7, b = 13.4, B = 24°20'$
5	$a = 0.247, b = 0.302, A = 69°40'$	**6**	$a = 6.47, c = 11.7, A = 36°50'$
7	$a = 2.36, c = 2.63, C = 32°40'$	**8**	$a = 70.3, c = 29.5, C = 47°30'$
9	$b = 0.777, c = 0.798, B = 50°10'$	**10**	$b = 413, c = 786, B = 71°40'$
11	$b = 12.3, c = 23.6, C = 28°20'$	**12**	$b = 0.214, c = 0.245, C = 41°50'$

Find the unknown side in each of problems 13 to 20 by use of the law of cosines.

13	$a = 15, b = 18, C = 30°$	**14**	$a = 28, b = 19, C = 60°$
15	$a = 13, b = 1.7, C = 45°$	**16**	$a = 0.21, b = 0.32, C = 45°$
17	$a = 3.1, c = 4.7, B = 135°$	**18**	$a = 53, c = 41, B = 150°$
19	$b = 0.67, c = 0.59, A = 120°$	**20**	$b = 81, c = 61, A = 135°$

Find the largest angle to the nearest $10'$ and the area in each of problems 21 to 24.

21	$a = 47, b = 43, c = 41$	**22**	$a = 19, b = 23, c = 17$
23	$a = 31, b = 41, c = 51$	**24**	$a = 17, b = 26, c = 29$

Find the smallest angle to the nearest $10'$ and the area in each of problems 25 to 28.

25	$a = 13, b = 16, c = 19$	**26**	$a = 19, b = 23, c = 29$

27 $a = 29, b = 31, c = 37$ **28** $a = 37, b = 41, c = 53$

29 Find the shortest diagonal of a parallelogram if its sides are 37 and 55 feet and meet at 58°.

30 Two forces are acting at the same point, and their directions make an angle of 42°40′. Find their resultant if they are 273 and 329 pounds.

31 Find the ground speed and flight direction of a plane whose heading is 225° at 270 miles per hour in a wind of 27 miles per hour from the east.

32 Find the speed and direction of a ball that is thrown N30°W at 34 miles per hour from a car that is moving due north at 41 miles per hour.

20.5   THE LAW OF TANGENTS

We shall use the law of sines as the basis for developing another formula for solving triangles that can be solved by the law of cosines. The new formula will be such that most of the computation in connection with it can be done by use of logarithms, whereas the law of cosines does not lend itself readily to logarithmic computation. If we begin with

$$\frac{\sin A}{a} = \frac{\sin B}{b} \tag{1}$$

and multiply each member by a and divide each member by $\sin B$, we get

$$\frac{\sin A}{\sin B} = \frac{a}{b} \tag{2}$$

Now, by subtracting 1 from each member of (2) and reducing to a common denominator, we have

$$\frac{\sin A - \sin B}{\sin B} = \frac{a - b}{b} \tag{3}$$

Furthermore, by adding 1 to each member of (2) and reducing to a common denominator, we obtain

$$\frac{\sin A + \sin B}{\sin B} = \frac{a + b}{b} \tag{4}$$

If we divide each member of (3) by the corresponding member of (4) and equate the quotients, we get

$$\frac{\sin A - \sin B}{\sin A + \sin B} = \frac{a - b}{a + b}$$

Hence, by use of problems 27 and 25 of Exercise 10.4 in the numerator and denominator respectively, we have

$$\frac{2 \cos\frac{1}{2}(A+B) \sin\frac{1}{2}(A-B)}{2 \sin\frac{1}{2}(A+B) \cos\frac{1}{2}(A-B)} = \frac{a-b}{a+b}$$

Now, by use of the ratio relations we see that

$$\frac{\tan\frac{1}{2}(A-B)}{\tan\frac{1}{2}(A+B)} = \frac{a-b}{a+b} \qquad (20.5)$$

Equation (20.5), or those obtained from it by replacing A or B by C, is known as the *law of tangents* and can be stated without symbols as follows:

Law of tangents

The tangent of half the difference of any two angles of a triangle divided by the tangent of half their sum is equal to the difference of the sides opposite them divided by the sum of the sides.

The law of tangents can be used if we know any two sides and the included angle, since if one angle is known, we can find half the sum of the other two angles.

EXAMPLE

Solve the triangle in which $a = 302$, $c = 376$, and $B = 71°20'$.

Solution

Since the given sides are a and c and c is the larger, we use

$$\frac{\tan\frac{1}{2}(C-A)}{\tan\frac{1}{2}(C+A)} = \frac{c-a}{c+a}$$

in order to have only positive numbers in the equation. We readily find that $c + a = 376 + 302 = 678$, $c - a = 376 - 302 = 74$, and $\frac{1}{2}(C + A) = \frac{1}{2}(180° - B) = \frac{1}{2}(180° - 71°20') = \frac{1}{2}(108°40') = 54°20'$. Therefore,

$$\frac{\tan\frac{1}{2}(C-A)}{\tan 54°20'} = \frac{74}{678}$$

$$\tan\tfrac{1}{2}(C-A) = \tfrac{74}{678} \tan 54°20'$$
$$\log \tan\tfrac{1}{2}(C-A) = \log 74 + \log \tan 54°20' - \log 678$$

$\log 74 =$	1.8692
$\log \tan 54°20' =$	0.1441

$\text{Sum} = 12.0133 - 10$ ◄ adding and subtracting 10
$\log 678 = \ \ 2.8312$

$\log \tan\tfrac{1}{2}(C-A) =$	$9.1821 - 10$	◄ subtracting
$\tfrac{1}{2}(C-A) =$	$8°40'$	◄ using Table II $\qquad (5)$
$\tfrac{1}{2}(C+A) =$	$54°20'$	◄ calculated above $\qquad (6)$
$C =$	$63°0'$	◄ (5) + (6)
$A =$	$45°40'$	◄ (6) − (5)

We can now find b by use of the law of sines. If we substitute the known values in $b/(\sin B) = a/(\sin A)$, we get

$$\frac{b}{\sin 71°20'} = \frac{302}{\sin 45°40'}$$

$$b = \frac{302 \sin 71°20'}{\sin 45°40'} \quad \blacktriangleleft \text{ solving for } b$$

$$\log b = \log 302 + \log \sin 71°20' - \log \sin 45°40'$$

$$\begin{aligned}
\log 302 &= 2.4800 \\
\log \sin 71°20' &= 9.9765 - 10 \\
\hline
\text{Sum} &= 12.4565 - 10 \\
\log \sin 45°40' &= 9.8545 - 10 \\
\hline
\log b &= 2.6020 \quad \blacktriangleleft \text{ subtracting} \\
b &= 400
\end{aligned}$$

To complete the solution we find the area by use of $K = \frac{1}{2}ac \sin B$.   By substituting therein we have

$$\begin{aligned}
K &= \ 0.5(302)(376) \sin 71°20' \\
\log 0.5 &= \ 9.6990 - 10 \\
\log 302 &= \ 2.4800 \\
\log 376 &= \ 2.5752 \\
\log \sin 71°20' &- \ 9.9765 \quad 10 \\
\log K &= \ 24.7307 - 20 \\
K &= \ 5.38(10^4)
\end{aligned}$$

Exercise 20.3   The Law of Tangents

Solve the following triangles.

1 $a = 17, b = 23, C = 48°$ **2** $a = 21, b = 37, C = 38°$

3 $a = 43, c = 57, B = 76°$ **4** $a = 71, c = 42, B = 96°$

5 $b = 203, c = 311, A = 38°40'$ **6** $b = 524, c = 607, A = 64°20'$

7 $b = 713, c = 535, A = 87°30'$ **8** $b = 478, c = 363, A = 106°20°$

9 $a = 331, b - 405, C = 48°40'$ **10** $a = 273, b = 364, C = 59°20'$

11 $a = 184, b = 237, C = 114°20'$ **12** $a = 923, b = 876, C = 37°40'$

13 $a = 2{,}359, c = 3{,}295, B = 72°36'$

14 $a = 4{,}783, c = 3{,}874, B = 101°14'$

15 $a = 5{,}964, c = 9{,}546, B = 83°22'$

16 $a = 7{,}117, c = 6{,}229, B = 26°48'$

17 Find the speed and direction of a boat that is headed S32°E across a stream that flows due north at 11 miles per hour provided the motor drives the boat at 17 miles per hour.

18 Find the ground speed of the plane and direction of flight if it has an air speed of 293 miles per hour and a heading of 208° in a wind of 35 miles per hour from the east.

19 If a point A is on the shore of a lake and is inaccessible to B and C, find BC provided $AB = 2.1(10)^3$ feet, $AC = 2.9(10^3)$ feet, and angle BAC is $38°$.

20 Two sides of a triangular tract of land are 146 and 117 feet. Find the angle between them if the area of the tract is 5.94 (10^3) square feet.

20.6 THE HALF-ANGLE FORMULAS

We shall now develop another formula for finding the angles of a triangle when the sides are known. For this purpose, we consider a triangle ABC of perimeter $2s$ with the inscribed circle of radius r. Figure 20.5 shows the triangle ABC, its inscribed circle, and three radii drawn to the points of tangency, D, E, and F. Therefore, the angles at D, E, and F are right angles. Furthermore, each angle of the triangle is bisected by the line segment which connects the vertex to the center O of the inscribed circle. Consequently, $\tan \frac{1}{2}A = r/AF$; we shall now express AF in terms of a and s. We know that

$$\begin{aligned}
2s = a + b + c &= (AF + AE) + (BF + BD) + (CD + CE) \\
&= 2AF + 2BD + 2CD \qquad \blacktriangleleft \text{ two tangents from a point} \\
&\qquad\qquad\qquad\qquad\qquad\qquad \text{to a circle are equal} \\
&= 2AF + 2(BD + CD) \\
&= 2AF + 2a
\end{aligned}$$

Therefore, by solving for $2AF$ and dividing by 2, we see that $AF = s - a$; hence, $\tan \frac{1}{2}A = r/AF$ becomes

Half-angle
formulas
$$\tan \tfrac{1}{2}A = \frac{r}{s - a} \tag{20.6a}$$

By means of a similar argument, we can show that

$$\tan \tfrac{1}{2}B = \frac{r}{s - b} \tag{20.6b}$$

$$\tan \tfrac{1}{2}C = \frac{r}{s - c} \tag{20.6c}$$

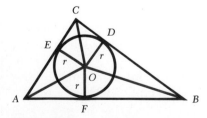

Figure 20.5

We shall now express r in terms of s, a, b, and c and, as an intermediate step, obtain another formula for the area of a triangle. Thus,

$$K = \text{area of } ABC = \text{area of } ABO + \text{area } BCO + \text{area } CAO$$
$$= \tfrac{1}{2}rc + \tfrac{1}{2}ra + \tfrac{1}{2}rb$$
$$= \tfrac{1}{2}r(a + b + c)$$
$$= \tfrac{1}{2}r(2s)$$

Therefore,

Third area
formula

$$\boldsymbol{K = rs} \tag{20.7}$$

Another expression for area is given by Heron's formula. It is

$$K = \sqrt{s(s - a)(s - b)(s - c)}$$

Consequently,

$$rs = \sqrt{s(s - a)(s - b)(s - c)}$$

$$r = \frac{\sqrt{s(s - a)(s - b)(s - c)}}{s}$$

$$= \sqrt{\frac{s(s - a)(s - b)(s - c)}{s^2}}$$

Now, dividing numerator and denominator by s, we have

$$\boldsymbol{r = \sqrt{\frac{(s - a)(s - b)(s - c)}{s}}} \tag{20.8}$$

EXAMPLE   Find the angles and area of a triangle in which $a = 372$, $b = 419$, and $c = 463$.

Solution   Since s, $s - a$, $s - b$, and $s - c$ are needed in computing r, K, and the tangents of the half angles, we shall now find them. Thus,

$$s = \tfrac{1}{2}(372 + 419 + 463) = 627$$
$$s - a = 627 - 372 = 255$$
$$s - b = 627 - 419 = 208$$
$$s - c = 627 - 463 = 164$$

Now, by using (20.8), we have

$$\log r = \tfrac{1}{2}[\log (s - a) + \log (s - b) + \log (s - c) - \log s]$$

$$\log (s - a) = \log 255 = 2.4065$$
$$\log (s - b) = \log 208 = 2.3181$$
$$\underline{\log (s - c) = \log 164 = 2.2148}$$
$$\text{Sum} = 6.9394$$
$$\underline{\log s = \log 627 = 2.7973}$$
$$4.1421 \qquad \blacktriangleleft \text{ subtracting}$$
$$\log r = 2.0711 \qquad \blacktriangleleft \text{ dividing by 2}$$

We shall not find r since only its logarithm is needed.

Now, by applying logarithms to

$$\tan \tfrac{1}{2}A = \frac{r}{s-a}$$

we get $\qquad \log \tan \tfrac{1}{2}A = \log r - \log (s-a)$

$$
\begin{aligned}
\log r &= 12.0711 - 10 \qquad \blacktriangleleft \text{ as calculated above} \\
\underline{\log (s-a)} &= \underline{2.4065} \qquad\; \blacktriangleleft \text{ as found above} \\
\log \tan \tfrac{1}{2}A &= 9.6646 - 10 \qquad \blacktriangleleft \text{ subtracting} \\
\tfrac{1}{2}A &= 24°50' \\
A &= 49°40'
\end{aligned}
$$

Similarly, $\qquad\qquad\qquad \tan \tfrac{1}{2}B = \dfrac{r}{s-b}$

$$
\begin{aligned}
\log \tan \tfrac{1}{2}B &= \log r - \log (s-b) \\
\log r &= 12.0711 - 10 \\
\underline{\log (s-b)} &= \underline{2.3181} \\
\log \tan \tfrac{1}{2}B &= 9.7530 - 10 \\
\tfrac{1}{2}B &= 29°30' \\
B &= 59°0'
\end{aligned}
$$

and $\qquad\qquad\qquad\qquad \tan \tfrac{1}{2}C = \dfrac{r}{s-c}$

$$
\begin{aligned}
\log \tan \tfrac{1}{2}C &= \log r - \log (s-c) \\
\log r &= 12.0711 - 10 \\
\underline{\log (s-c)} &= \underline{2.2148} \\
\log \tan \tfrac{1}{2}C &= 9.8563 - 10 \\
\tfrac{1}{2}C &= 35°40' \\
C &= 71°20'
\end{aligned}
$$

As a check, we shall find $A + B + C$; it is $49°40' + 59°0' + 71°20' = 180°$ as it should be, and the solution checks.

We shall now find the area by use of (20.7).   It is

$$
\begin{aligned}
K &= rs \\
\log K &= \log r + \log s \\
\log r &= 2.0711 \qquad\qquad \blacktriangleleft \text{ as above} \\
\underline{\log s} &= \underline{2.7973} \qquad\qquad \blacktriangleleft \text{ as above} \\
\log K &= 4.8684 \qquad\qquad \blacktriangleleft \text{ adding} \\
K &= 7.39(10^4)
\end{aligned}
$$

Exercise 20.4   Half-angle Formulas

Solve the following triangles.

1   $a = 23$, $b = 19$, $c = 18$
2   $a = 0.56$, $b = 0.71$, $c = 0.65$
3   $a = 8.7$, $b = 5.3$, $c = 4.2$
4   $a = 34$, $b = 23$, $c = 29$

5 $a = 371, b = 227, c = 342$
6 $a = 91.4, b = 78.6, c = 67.0$
7 $a = 2.97, b = 3.41, c = 2.58$
8 $a = 0.326, b = 0.274, c = 0.256$
9 $a = 0.0123, b = 0.0247, c = 0.0302$
10 $a = 34.6, b = 27.4, c = 28.8$
11 $a = 882, b = 664, c = 448$
12 $a = 5.97, b = 4.82, c = 3.73$
13 $a = 0.5789, b = 0.6481, c = 0.4444$
14 $a = 76.23, b = 59.80, c = 61.17$
15 $a = 8,998, b = 8,778, c = 8,118$
16 $a = 3.471, b = 2.997, c = 2.152$
17 What is the radius of the largest circular swimming pool that can be constructed on a triangular lot with sides 27, 32, and 41 feet?
18 How much sand is needed to fill a triangular pen to a depth of 9 inches if the sides are 11, 12, and 13 feet?
19 Find the angle between two forces of 334 pounds and 407 pounds if their resultant is 561 pounds.
20 The air speed and heading of a plane are 275 miles per hour and 230°, and the ground speed is 299 miles per hour in a wind of 23 miles per hour. Find the direction of the wind if the course of the plane is south of the heading.

Exercise 20.5   Review

1 $A = 29°50', B = 76°30', b = 572$
2 $A = 32°20', C = 44°50', b = 437$
3 $a = 15, b = 21, A = 37°$
4 $a = 146, c = 115, C = 53°40'$
5 $a = 247, b = 252, B = 47°30'$
6 If $a = 11, b = 14, c = 18$, find the smallest angle.
7 If $a = 43, b = 56, c = 77$, find the largest angle.
8 If $a = 19, c = 22, B = 60°$, find b.
9 If $b = 2.7, c = 3.5, A = 150°$, find a.
10 If $a = 4.23, b = 4.91, c = 5.52$, find C.
11 If $a = 31.7, b = 70.3, c = 42.6$, find B.
12 Find the longer diagonal of a parallelogram if its sides are 43 and 52 and meet at 127°.
13 Two sides of a triangular tract of land are 119 and 149 feet. Find the angle between these two sides if the area of the tract is $4.77(10^3)$ square feet.
14 A plane heads at 380 miles per hour in the direction of 170° but travels at 390 miles per hour due to a wind of 21 miles per hour. Find the direction of the wind if the plane travels south and west of the direction in which it headed.

Progressions

The inventor of chess, so it is said, asked that he be rewarded with one grain of wheat for the first square of the board, two grains for the second, four for the third, and so on for the 64 squares. Fortunately, this apparently modest request was examined before it was granted. By the twentieth square, the reward would have amounted to more than a million grains of wheat; by the sixty-fourth square the number called for would have been astronomical, and the amount would have far exceeded all the grain in the kingdom.

This story deals with a sequence of numbers. A sequence of numbers is a set arranged in an order determined by some rule or experiment. Such relationships have a great many important applications. Some of them are beyond the scope of this book, but we shall explore methods of dealing with a number of practical and interesting problems of this type.

21.1 ARITHMETIC PROGRESSIONS

Common
difference
Arithmetic
progression

A sequence of numbers such that each term after the first is obtained by adding a fixed number called the *common difference* to the preceding term is called an *arithmetic progression*. For example, the sequence 4, 9, 14, 19, 24 is an arithmetic progression with 5 as the common difference, and the sequence 10, 7, 4, 1, −2, −5 is an arithmetic progression with the common difference −3.

The following example uses an arithmetic progression in a problem involving an installment payment plan.

EXAMPLE 1

Mr. Brown bought a used piano priced at $600 and agreed to pay for it in six monthly installments. The first payment amounted to $103 and was paid at the end of the first month, and each subsequent payment was 50 cents less than the preceding one. Find the amount of the last payment and the total amount paid in the six installments.

Solution

Since Mr. Brown made five payments after the first and each was 50 cents less than the first, the last payment was 5($0.50) = $2.50 less than the first. Hence the last installment was $103 − $2.50 = $100.50. Furthermore, the total amount paid was $103 + $102.50 + $102 + $101.50 + $101 + $100.50 = $610.50.

Hereafter we shall let

a = the first term
l = the last term
d = the common difference
n = the number of terms
s = the sum of the terms

We shall also use the abbreviation AP for arithmetic progression.

If we apply the definition of an AP in which a is the first term and d is the common difference, we obtain

$$\text{First term} = a$$
$$\text{Second term} = a + d$$
$$\text{Third term} = a + d + d = a + 2d$$
$$\text{Fourth term} = a + 3d$$

As the above four terms illustrate, the coefficient of d increases by 1 as we progress from one term to the next and is 1 less than the number of the term. Therefore, it will be $n - 1$ in the nth or last term. Hence, the nth term of an AP is

Last term of an AP

$$l = a + (n - 1)d \qquad\qquad (21.1)$$

If three of l, a, n, and d are given, the fourth can be determined by substituting the given values in (21.1) and then solving the resulting equation.

EXAMPLE 2

If, in an AP, $a = 2$, $d = 0.01$, and $n = 100$, find the last term.

Solution

To obtain the last term in the AP, we substitute the given values in (21.1) and obtain

$$l = 2 + (100 - 1)(0.01) = 2.99$$

EXAMPLE 3

If, in an AP, $l = 130$, $d = 25$, and $a = 5$, find the number of terms in the progression.

Solution

To obtain the number of terms, we substitute the given values in (21.1), solve the resulting equation for n, and thus get

$$130 = 5 + (n - 1)25 \qquad \blacktriangleleft \text{ substituting given values in (21.1)}$$
$$(n - 1)25 = 130 - 5 \qquad \blacktriangleleft \text{ adding } -130 - (n - 1)25 \text{ to each}$$
$$\text{member and then dividing by } -1$$
$$(n - 1)25 = 125 \qquad \blacktriangleleft \text{ combining terms}$$
$$n - 1 = 5 \qquad \blacktriangleleft \text{ dividing by 25}$$
$$n = 6 \qquad \blacktriangleleft \text{ solving for } n$$

The second fundamental formula expresses s in terms of a, n, and l and is derived as follows:

By definition

$$s = a + (a + d) + (a + 2d) + (a + 3d) + \cdots + [a + (n - 1)d] \qquad (1)$$

Since there are n terms in the right member of (1) and each term involves only one a, we may combine the terms involving a and get

$$s = na + [d + 2d + 3d + \cdots + (n-1)d] \qquad (2)$$

We now reverse the order of the terms in the AP in (1) by writing l as the first term and using $-d$ as the common difference. Thus we obtain

$$s = l + (l - d) + (l - 2d) + (l - 3d) + \cdots + [l - (n-1)d] \qquad (3)$$

Therefore, after combining terms involving l, we have

$$s = nl - [d + 2d + 3d + \cdots + (n-l)d] \qquad (4)$$

Since the terms in the right member of (3) are those in the right member of (1) in reverse order, the values of s in (2) and (4) are the same; furthermore, the terms in the brackets in (2) and (4) are the same. Hence, if we add the corresponding members of (2) and (4), we get

$$2s = na + nl$$
$$= n(a + l)$$

Sum of AP   Consequently, $$s = \frac{n}{2}(a + l) \qquad (21.2)$$

A second formula for s is obtained by replacing l in (21.2) by its value $a + (n-1)d$. Therefore

$$s = \frac{n}{2}[a + a + (n-1)d]$$

or $$s = \frac{n}{2}[2a + (n-1)d] \qquad (21.3)$$

If any three of the numbers l, a, n, d, and s, other than l, d, and s, are given, each of the other two can be obtained by use of one of the formulas (21.1) to (21.3). If l, d, and s are given and the three values are substituted in the three formulas, the remaining unknowns a and n will appear in each of the equations obtained. Hence two of the resulting equations must be solved simultaneously for a and n.

EXAMPLE 4   Find the sum of the odd integers from 1 to 3,001 inclusive.

Solution   The consecutive odd integers from 1 to 3,001 inclusive form an AP with $a = 1$, $d = 2$, and $l = 3,001$. Since, however, both of the formulas for the sum s of the terms involve n, we must find n first. For this purpose, we shall use (21.1) and obtain

$3{,}001 = 1 + (n-1)2$	◀ substituting $a = 1$, $d = 2$, and $l = 3{,}001$
$(n-1)2 = 3{,}000$	◀ adding $-3{,}001 - (n-1)2$ to each member, combining terms, and then dividing by -1
$n - 1 = 1{,}500$	◀ dividing each member by 2
$n = 1{,}501$	◀ adding 1 to each member

Now, by using (21.2) with $a = 1$, $l = 3{,}001$, and $n = 1{,}501$, we get

$$s = \tfrac{1{,}501}{2}(1 + 3{,}001)$$
$$= (1{,}501)(1{,}501)$$
$$= 2{,}253{,}001$$

EXAMPLE 5 If $l = -2$, $n = 6$, and $s = 18$, find a and d.

Solution Since (21.2) involves a, s, n, and l, we can find a by substituting the given values in (21.2) and then solving for a. The steps in the solution follow.

$18 = \tfrac{6}{2}[a + (-2)]$ ◀ substituting the given values in (21.2)
$18 = 3(a - 2)$ ◀ simplifying
$18 = 3a - 6$ ◀ by the distributive law
$3a = 18 + 6$ ◀ adding $-3a - 18$ to each member and dividing by -1
$3a = 24$ ◀ combining terms
$a = 8$ ◀ dividing by 3

To obtain the value of d, we employ (21.1) with $l = -2$, $a = 8$, and $n = 6$, as shown below.

$-2 = 8 + (6 - 1)d$ ◀ substituting the given values in (21.2)
$-2 = 8 + 5d$ ◀ simplifying
$5d = -8 - 2$ ◀ adding $-5d + 2$ to each member and dividing by -1
$5d = -10$ ◀ combining terms
$d = -2$ ◀ dividing by 5

EXAMPLE 6 A man earns a salary of \$3,000 the first year and receives an increase of \$100 per year for the next 19 years. Find the total amount he earns.

Solution Here we use (21.3) with $a = \$3{,}000$, $n = 20$, and $d = \$100$ and get

$$s = \tfrac{20}{2}[\$6{,}000 + 19(\$100)]$$
$$= 10(\$6{,}000 + \$1{,}900)$$
$$= \$79{,}000$$

EXAMPLE 7 If in an AP, $d = 4$, $l = 31$, and $s = 136$, find a and n.

Solution If we substitute the given values of d, l, and s in (21.1), (21.2), or (21.3), we obtain an equation in two unknowns. Therefore, we must solve two of these equations simultaneously. We shall use (21.1) and (21.2) and solve the problem as follows:

$31 = a + (n - 1)4$ ◀ substituting the given values in (21.1)
$a + 4n = 35$ ◀ simplifying   (5)

$136 = \tfrac{n}{2}(a + 31)$ ◀ substituting the given values in (21.2)

$an + 31n = 272$ ◀ simplifying   (6)

Now we solve (5) and (6) simultaneously for a and n as follows.

$$a = 35 - 4n \qquad \blacktriangleleft \text{ solving (5) for } a \qquad\qquad (7)$$
$$(35 - 4n)n + 31n = 272 \qquad \blacktriangleleft \text{ substituting the value of } a$$
$$\text{from (7) in (6)}$$
$$4n^2 - 66n + 272 = 0 \qquad \blacktriangleleft \text{ simplifying}$$
$$2n^2 - 33n + 136 = 0 \qquad \blacktriangleleft \text{ dividing by 2}$$
$$n = 8 \text{ and } \tfrac{17}{2} \qquad \blacktriangleleft \text{ solving for } n$$

Since n is an integer, we must discard $\tfrac{17}{2}$ and use $n = 8$. Finally, by substituting $n = 8$ in (7) we obtain

$$a = 35 - 32 = 3$$

EXAMPLE 8

If in an AP, $a = 1$, $d = \tfrac{5}{2}$, and $s = 1\tfrac{19}{2}$, find n and l.

Solution

Since (21.3) involves a, d, n, and s, we use it and have

$$\frac{119}{2} = \frac{n}{2}[2 + (n - 1)\tfrac{5}{2}] \qquad \blacktriangleleft \text{ substituting in (21.3)}$$

$$238 = n[4 + (n - 1)5] \qquad \blacktriangleleft \text{ multiplying each member by 4}$$
$$5n^2 - n - 238 = 0 \qquad \blacktriangleleft \text{ simplifying}$$
$$n = 7 \text{ and } -6.8 \qquad \blacktriangleleft \text{ solving for } n$$

Since n must be a positive integer, we discard -6.8 and have $n = 7$.
We find l by substituting $a = 1$, $d = \tfrac{5}{2}$, and $n = 7$ in (21.1). Thus

$$l = 1 + (7 - 1)\tfrac{5}{2}$$
$$l = 1 + 15 = 16$$

Exercise 21.1   Arithmetic Progressions

Write the terms of the arithmetic progression that satisfy the conditions in each of problems 1 to 4.

1   $a = 3$, $d = 2$, $n = 6$ $\qquad\qquad$ 2 $a = 11$, $d = -3$, $n = 7$
3   $a = -4$, third term 2, $n = 5$
4   Second term 13, fourth term 9, $n = 8$

In each of problems 5 to 28, find the two of l, a, n, d, and s that are missing.

5   $a = 2$, $d = 3$, $n = 5$ $\qquad\qquad$ 6 $a = 17$, $d = -3$, $n = 7$
7   $a = 6$, $n = 6$, $l = 1$ $\qquad\qquad$ 8 $a = -3$, $x = 6$, $l = 12$
9   $n = 7$, $l = 5$, $s = 77$ $\qquad\qquad$ 10 $n = 6$, $l = 21$, $s = 66$
11   $a = -2$, $l = 13$, $s = 33$ $\qquad\qquad$ 12 $a = -7$, $l = 5$, $s = -7$
13   $a = 27$, $d = -6$, $l = -3$ $\qquad\qquad$ 14 $a = -8$, $d = 4$, $l = 12$

15 $a = 3, n = 7, s = 63$ **16** $a = 7, n = 7, s = -14$
17 $n = 7, d = -5, l = -18$ **18** $a = 11, d = -3, l = -7$
19 $n = 8, d = 1, s = 36$ **20** $n = 7, d = -3, s = 49$
21 $a = 11, d = -4, s = -7$ **22** $a = 11, d = -3, s = 14$
23 $l = -14, d = -6, s = 28$ **24** $l = 23, d = 5, s = 56$
25 $l = -23, d = -7, s = -14$ **26** $l = 17, d = 3, s = 58$
27 $l = -26, d = -9, s = 44$ **28** $l = 13, d = -5, s = 3$

29 Find the sum of all odd integers between 6 and 42.

30 What is the sum of all multiples of 7 between 5 and 50?

31 Assume that a compact body falls 16 feet during the first second, 48 feet during the next, 80 feet during the third, and so on. How far will it fall during the ninth second and during the first nine seconds?

32 A machine can be bought for $8,700. Find its value after 6 years if it depreciates 24 percent during the first year, 20 percent during the second, 16 percent during the third, and so on.

33 A student made 63 on the first of six algebra quizzes and seven more on each subsequent quiz than on the immediately preceding one. What was her grade on the last one, and what was her average on the six quizzes?

34 The three digits of a number form an AP with 15 as their sum. Find the number if it is increased by 792 by interchanging the units and hundred digits.

35 A bomb is dropped from an altitude of 4,624 feet. If air resistance is neglected, how much time elapses before the bomb hits the ground?

36 Mrs. Epstein made $6,300 in 1965, $6,600 in 1966, $7,000 in 1967, $7,500 in 1968, and so on including 1975. How much more did she earn in 1975 than in 1974?

37 Find the value of x if $x - 1$, $3x - 1$, and $4x + 1$ form an AP.

38 Find x and y so that $x, y,$ and $5x$ form an AP with sum 81.

39 Determine x so that $3x + 2$, $x^2 - x$, and $2x^2 - 6x + 1$ form an AP.

40 Determine x and y so that $x, 2y,$ and $3y + 1$ form an AP with the sum 18.

21.2 GEOMETRIC PROGRESSIONS

The second sequence that we shall study is the geometric progression, which we now define. A *geometric progression* (abbreviated as GP) is a sequence of numbers such that each term after the first is obtained by multiplying the immediately preceding term by a fixed number called the *common ratio*.

Geometric progression

Common ratio

The number of ancestors that a person had or has in each generation that precedes him is a geometric progression, since he had or has 2 parents, 4 grandparents, 8 great-grandparents, and so on. Geometric progressions have important applications in the fields of finance and biology.

The sequence 2, 6, 18, 54, 162 is a GP with the common ratio 3, and the sequence 32, -16, 8, -4, 2, -1, $\frac{1}{2}$ is a GP with the common ratio $-\frac{1}{2}$.

As in the case of arithmetic progressions, the problem of finding either the last term or the sum of the terms of a geometric progression containing a great many terms would be very laborious without the use of the formulas which we shall next develop.   We shall let

$$a = \text{the first term}$$
$$l = \text{the last term}$$
$$r = \text{the common ratio}$$
$$n = \text{the number of terms}$$
$$s = \text{the sum of the terms}$$

We shall first develop the formula for the last term of a GP.   For this purpose we write the first few terms of a progression, and from these we can see by inductive reasoning the form of the last, or nth, term.   Since r is the common ratio and a is the first term, we have, by definition

$$a = \text{the first term}$$
$$ar = \text{the second term}$$
$$ar^2 = \text{the third term}$$

and we multiply each term by r to get the next.   Therefore, it follows that the exponent of r increases by 1 as we move from any term to the next. Consequently, since the exponent of r is 1 in the second term, we conclude that the exponent of r in the nth term is $n - 1$.   Hence we have

Last term of a GP
$$l = ar^{n-1} \tag{21.4}$$

We shall obtain a compact formula for the sum of the terms in a GP by starting with the equation

$$s - a + ar + ar^2 + ar^3 + \cdots + ar^{n-2} + ar^{n-1} \tag{1}$$

Now we multiply each member of (1) by r and obtain

$$rs = ar + ar^2 + ar^3 + ar^4 + \cdots + ar^{n-1} + ar^n \tag{2}$$

If we compare (1) and (2), we see that except for a in (1) and ar^n in (2), the right-hand members are the same.   Hence if we subtract each member of (2) from the corresponding member of (1), we get

$$s - rs = a - ar^n$$

and, by solving for s, we have

$$s = \frac{a - ar^n}{1 - r} \tag{21.5}$$

We obtain a second formula for s by replacing ar^n by rl as justified by use of (21.4).   Thus

Sum of a GP $$s = \frac{a - rl}{1 - r}$$ **(21.6)**

We shall illustrate the use of these formulas with several examples.

EXAMPLE 1   If in a GP, $a = 243$, $r = \frac{1}{3}$, and $n = 5$, find l and s.

Solution   If we substitute $a = 243$, $r = \frac{1}{3}$, and $n = 5$ in (21.4) and (21.5), we get

$$l = 243(\tfrac{1}{3})^{5-1} = \tfrac{243}{81} = 3 \qquad \blacktriangleleft \text{ by using (21.4)}$$

$$s = \frac{243 - 243(\tfrac{1}{3})^5}{1 - \tfrac{1}{3}} = \frac{243 - 1}{\tfrac{2}{3}} = \frac{3(242)}{2} = 363 \qquad \blacktriangleleft \text{ by using (21.5)}$$

EXAMPLE 2   A man has 2 parents, 4 grandparents, 8 great-grandparents, and so on. Find the number of his ancestors during the eight generations preceding his own, provided there are no duplications.

Solution   In this problem, $a = 2$, $r = 2$, and $n = 8$. By substituting these values in (21.5), we get

$$s = \frac{2 - 2(2^8)}{1 - 2} = \frac{2 - 2(256)}{-1} = 510$$

EXAMPLE 3   A woman deposited \$100 in a savings bank that pays interest at the rate of 4 percent compounded semiannually. If she makes no withdrawals, find the amount that she will have to her credit at the end of 10 years.

Solution   Interest compounded semiannually is added to the principal at the end of each 6-month period, and this sum earns interest during the next 6 months. Since the interest rate is 4 percent per year, the rate is 2 percent per 6 months. Therefore, the amount of her investment at the end of the first 6 months is \$100 + (\$100)(0.02) = \$100(1.02). By continuing this argument, we reach the conclusion that the amount at the end of any interest period is the product of 1.02 and the amount at the end of the preceding period. Therefore, the amounts at the ends of the successive interest periods are \$100(1.02), \$100(1.02)^2, \$100(1.02)^3, and so on to 20 terms. This sequence is a GP with $a = \$100(1.02)$, $r = 1.02$, and $n = 20$. Consequently, the amount of the investment at the end of 10 years can be obtained by substituting the above values in (21.4). Thus we obtain

$$l = \$100(1.02)(1.02)^{20-1} = \$100(1.02)^{20} = \$100(1.4859) = \$148.59$$

If we are given any three of the elements s, n, a, r, and l, other than

l, n, and s, we can find each of the other two by use of one of the formulas (21.4) to (21.6). If, however, we are given l, n, and s, we find that the substitution of these values in any one of the three formulas leaves a and r as unknowns. Hence two of the resulting equations must be solved simultaneously for a and r.

EXAMPLE 4   If in a GP, $a = 2$, $l = 486$, and $n = 6$, find r and s.

Solution   If we substitute the given values in (21.4), we get

$$486 = 2r^5$$
$$r^5 = 243 \qquad \blacktriangleleft \text{ solving for } r^5$$
$$r = 3 \qquad \blacktriangleleft \text{ since } 3^5 = 243$$

Now we substitute $a = 2$, $n = 6$, and $r = 3$ in (21.5) and get

$$s = \frac{2 - 2(3^6)}{1 - 3} = \frac{2 - 1458}{-2} = 728$$

EXAMPLE 5   If in a GP, $l = -96$, $n = 6$, and $s = -63$, find a and r.

Solution   The substitution of the above values for l, n, and s in any one of (21.4) to (21.6) yields an equation in which a and r are unknown. We shall use (21.4) and (21.6), and after substituting shall solve the resulting equations simultaneously. Thus we obtain the following:

$$-96 = ar^{6-1} \qquad \blacktriangleleft \text{ substituting the given values in (21.4)}$$
$$ar^5 = -96 \qquad \blacktriangleleft \text{ simplifying} \tag{3}$$
$$-63 = \frac{a + 96r}{1 - r} \qquad \blacktriangleleft \text{ substituting the given values in (21.6)}$$
$$a = -63(1 - r) - 96r \qquad \blacktriangleleft \text{ multiplying each member by } 1 - r$$
$$\text{and solving for } a$$
$$a = -63 - 33r \qquad \blacktriangleleft \text{ simplifying} \tag{4}$$

Now we solve (3) and (4) simultaneously by substituting the value of a from (4) in (3). Thus,

$$(-63 - 33r)r^5 = -96$$
$$33r^6 + 63r^5 - 96 = 0 \qquad \blacktriangleleft \text{ simplifying}$$
$$11r^6 + 21r^5 - 32 = 0 \qquad \blacktriangleleft \text{ dividing by 3}$$

This is a polynomial equation of degree six, and, by using the method of Sec. 17.9, we find that the only rational roots are -2 and 1. We must discard 1 since the right side of (21.6) does not exist for that value. Hence, $r = -2$. Finally, we substitute -2 for r in (4) and get

$$a = -63 + 66 = 3$$

Exercise 21.2   Geometric Progressions

Write the terms of the geometric progression that satisfy the conditions in each of problems 1 to 4.

1 $a = 2, r = 3, n = 6$ **2** $a = 1, r = -3, n = 7$
3 $a = 2$, third term 8, $n = 5$
4 Second term 9, fourth term 1, $n = 6$

In each of problems 5 to 20, find the two of s, n, a, r, and l that are missing.

5 $a = 1, r = 3, n = 5$
6 $a = 1, r = -2, n = 7$
7 $a = 8, n = 5, l = \frac{1}{2}$
8 $a = 81, n = 6, l = \frac{1}{3}$
9 $l = -5, n = 6, s = 0$
10 $l = \frac{1}{49}, n = 6, s = \frac{19608}{49}$
11 $a = 12, l = \frac{3}{64}, s = \frac{1023}{64}$
12 $a = \frac{2}{3}, l = \frac{27}{8}, s = \frac{211}{24}$
13 $l = 162, n = 5, r = 3$
14 $l = \frac{3}{25}, n = 5, r = \frac{1}{3}$
15 $n = 6, r = -\frac{1}{4}, s = \frac{819}{16}$
16 $n = 5, r = \frac{1}{12}, s = \frac{22621}{12}$
17 $a = 1,296, r = \frac{1}{6}, s = 1,555$
18 $a = 1, r = -5, s = -2,604$
19 $l = 1, r = \frac{1}{2}, s = 255$
20 $l = -243, r = -3, s = -182$
21 If it were possible to save 1 cent on the first day of the month, 2 cents on the second day, 4 cents on the third day, and so on, on which day would the savings amount to a million dollars?
22 Ten people attended a sheriff's sale. The first had $10, the second had $20, the third $40, and so on. The sheriff offered a tract of land but said that $2,500 would be the lowest bid accepted. How many of the 10 people could make a bid? What was the largest bid a member of the group could make?
23 If a girl deposits $300 at the beginning of each year in a bank that pays 4.5 percent compounded annually, how much will be to her credit at the end of 5 years?
24 The number of bacteria in a culture triples every 2 hours. If there were n bacteria present at 2:55 A.M. one day, how many were present 24 hours later?
25 What is a $4,000 car worth after 5 years if, during each year, it depreciates 30 percent of the value at the beginning of the year?

26 If there are no duplicates, how many direct ancestors does a set of quadruplets have in the six generations beginning with their parents?

27 A man willed $\frac{1}{4}$ of his estate to his oldest child, $\frac{1}{4}$ of the remainder to the second child, and so on, until each of his four children were mentioned. The remainder of $20,250 went to his college. How much was the estate?

28 If 1, 4, and 9 are added to the first, second, and third terms, respectively, of an arithmetic progression with $d = 3$, a geometric progression is obtained. Find the *AP* and the common ratio of the *GP*.

21.3 INFINITE GEOMETRIC PROGRESSIONS

It is the purpose of this section to find the limit of the sum of a GP in which $|r| < 1$ and n increases indefinitely. We shall let $s(n)$ represent the sum of the first n terms of a GP, and then the symbol $\lim_{n \to \infty} s(n)$ stands for the limit of $s(n)$ as n increases indefinitely. Furthermore, the statement

$$\lim_{n \to \infty} s(n) = s$$

means that if n is sufficiently large, $s(n)$ will differ from s by an amount that is less than any number chosen in advance.

We shall use formula (21.5) in the form

$$s(n) = \frac{a}{1 - r}(1 - r^n) \tag{1}$$

Since $|r| < 1$, it follows that $|r| > |r^2| > |r^3| > \cdots > |r^n|$, and it can be proved† that r^n can be made arbitrarily small by choosing n sufficiently

---

† We shall prove this statement for $0 < r < 1$, and we shall first assume that r is a rational fraction. Any positive rational fraction can be expressed in the form $k/(k+t)$, where k and t are positive integers. Consequently, $r^n = [k/(k+t)]^n$. We shall let

$$\left(\frac{k}{k+t}\right)^n = x$$

Hence,
$$\log\left(\frac{k}{k+t}\right)^n = \log x$$

and it follows that
$$n[\log k - \log (k+t)] = \log x \tag{2}$$

Since k and $k+t$ are positive integers and $t > 0$, we have $\log k < \log (k+t)$. Consequently, $\log k - \log (k+t) = -d$, where d is a positive constant. Therefore, by substituting this value in the left member of (2), we have $-nd = \log x$. Therefore, by the definition of logarithms, we obtain $x = 10^{-nd} = 1/10^{nd}$. Hence we may make the value of x arbitrarily small by taking n sufficiently large, and since $x = r^n$, we have $\lim_{n \to \infty} r^n = 0$, provided that r is a positive rational fraction less than 1.

If r is a positive irrational number less than 1, then two rational fractions R_1 and R_2 exist such that $0 < R_1^n < r^n < R_2^n < 1$. Consequently, since $\lim R_1^n = \lim_{n \to \infty} R_2^n = 0$, it follows that $\lim_{n \to \infty} r^n = 0$.

large. Hence, in (1), r^n approaches zero as n increases definitely. Therefore, from (1) we have

Sum of an
infinite GP

$$\lim s(n) = \frac{a}{1-r}$$

This statement is usually written in the form

$$s = \frac{a}{1-r} \qquad |r| < 1 \tag{21.7}$$

EXAMPLE 1   Find the limit of the sum of the terms in a GP in which $a = 1$, $r = \frac{1}{2}$, and n increases indefinitely.

Solution   If we substitute the given values in (21.7), we obtain

$$s = \frac{1}{1 - \frac{1}{2}} = \frac{1}{\frac{1}{2}} = 2$$

EXAMPLE 2   If in a GP, $a = 3$, $r = \frac{2}{3}$, and n is infinite, find the limit of the sum of the terms.

Solution   If we substitute the given values in (21.7), we get

$$s = \frac{3}{1 - \frac{2}{3}} = 9$$

EXAMPLE 3   Find the fraction that is represented by $0.126126126 \cdots$.

Solution   The repeating decimal fraction $0.126126126 \cdots$ is equal to $0.126 + 0.000126 + 0.000000126 + \cdots$. This expression is an infinite GP with $a = 0.126$ and $r = 0.001$. Consequently, if we substitute these values in (21.7), we see that

$$s = \frac{0.126}{1 - 0.001} = \frac{126}{999} = \frac{14}{111}$$

is the desired fraction.

21.4 ARITHMETIC MEANS

Arithmetic means

In any AP the terms between a and l are called the *arithmetic means* of a and l. If an AP consists of three terms, the second term is *the arithmetic mean* of the first and third. For example, in the AP 3, 11, 19, the arithmetic mean of 3 and 19 is 11.

If three numbers a, m, and l form an AP, then $m = a + d$ and $l = a + 2d$. Furthermore, since $a + d = \frac{1}{2}(a + a + 2d)$, we have

$$m = \tfrac{1}{2}(a + l) \tag{21.8}$$

If in an AP we know the first and last terms and the number of terms, we can find d by the use of (21.1) and then can get the means by using the definition of an AP.

EXAMPLE 1 Find the arithmetic mean of $\frac{1}{4}$ and $\frac{5}{8}$.

Solution By use of (21.8) we have

$$m = \tfrac{1}{2}(\tfrac{1}{4} + \tfrac{5}{8}) = \tfrac{7}{16}$$

EXAMPLE 2 Obtain the five arithmetic means of 3 and 12.

Solution To get the five arithmetic means of 3 and 12, we first use (21.1) with $a = 3$, $l = 12$, and $n = 5 + 2 = 7$, and get

$$12 = 3 + (7 - 1)d$$
$$12 = 3 + 6d$$
$$6d = 9 \quad \blacktriangleleft \text{ solving for } 6d$$
$$d = \tfrac{3}{2} \quad \blacktriangleleft \text{ dividing by 6 and reducing to lowest terms}$$

Hence, by the definition of an AP, the means are $3 + \frac{3}{2}$, $3 + 3$, $3 + \frac{9}{2}$, $3 + 6$, and $3 + \frac{15}{2}$, that is, they are $\frac{9}{2}$, 6, $\frac{15}{2}$, 9, and $\frac{21}{2}$.

21.5 GEOMETRIC MEANS

Geometric means

The numbers between the first and last terms of a GP are called *geometric means*. To obtain a specified number of geometric means between two numbers, we use (21.4) to find r and then use the definition of a GP to get the required means. If a, m, and l is a GP in which the common ratio is r, then $m = ar$ and $l = ar^2$. Consequently, since $m^2 = a^2r^2 = a(ar^2) = al$, it follows that

$$m = \pm\sqrt{al} \tag{21.9}$$

Therefore, the geometric mean of two numbers is plus or minus the square root of their product.

EXAMPLE 1   Find five geometric means of 3 and 12,288.

Solution   Since $a = 3$, $l = 12,288$, and $n = 2 + 5 = 7$, we shall use (21.4) to obtain r.

$12,288 = 3r^{7-1}$ ◄ substituting $a = 3$, $l = 12,288$, and $n = 7$ in (21.4)
$r^6 = \frac{12288}{3} = 4,096$ ◄ simplifying and solving for r^6
$r = \pm 4$ ◄ since $4^6 = 4,096$ and $(-4)^6 = 4,096$

Hence the required means are 12, 48, 192, 768, and 3,072 for $r = 4$ and -12, 48, -192, 768, and $-3,072$ for $r = -4$.

EXAMPLE 2   Find the geometric mean of 4 and 625.

Solution   If we designate the geometric mean by m, we have, by use of (21.9),

$$m = \pm\sqrt{4(625)} = \pm\sqrt{2,500} = \pm 50$$

21.6 HARMONIC PROGRESSIONS

A *harmonic progression* is a sequence of numbers whose reciprocals form an arithmetic progression. For example, $\frac{1}{2}$, $\frac{1}{5}$, $\frac{1}{8}$, $\frac{1}{11}$, $\frac{1}{14}$ is a harmonic progression, since 2, 5, 8, 11, 14 is an arithmetic progression.

In order to obtain the harmonic means between two numbers, we get the arithmetic means of the reciprocals of the numbers and then take the reciprocals of the arithmetic means thus found.

EXAMPLE   Obtain the four harmonic means of 4 and 20.

Solution   To find the four harmonic means of 4 and 20, we first get the four arithmetic means of $\frac{1}{4}$ and $\frac{1}{20}$ by use of (21.1) and the following procedure.

$\frac{1}{20} = \frac{1}{4} + (6-1)d$ ◄ substituting $a = \frac{1}{4}$, $l = \frac{1}{20}$, and $d = 4 + 2 = 6$ in (21.1)
$5d = \frac{1}{20} - \frac{1}{4}$ ◄ solving for $(6-1)d = 5d$
$= -\frac{1}{5}$ ◄ combining fractions in right member
$d = -\frac{1}{25}$ ◄ dividing by 5

Consequently, the arithmetic means of $\frac{1}{4}$ and $\frac{1}{20}$ are $\frac{1}{4} - \frac{1}{25} = \frac{21}{100}$, $\frac{21}{100} - \frac{1}{25} =$ $\frac{17}{100}$, $\frac{17}{100} - \frac{1}{25} = \frac{13}{100}$, and $\frac{13}{100} - \frac{1}{25} = \frac{9}{100}$. Hence the four harmonic means between 4 and 20 are $\frac{100}{21}$, $\frac{100}{17}$, $\frac{100}{13}$, and $\frac{100}{9}$.

Exercise 21.3   Infinite Geometric Progressions, Means, Harmonic Progressions

Find the sum of the infinite geometric progression in each of problems 1 to 8.

1 $a = 6$, $r = \frac{1}{2}$ **2** $a = 5$, $r = \frac{3}{5}$

3 $a = 7$, $r = -\frac{1}{6}$ **4** $a = 7$, $r = -\frac{3}{4}$

5 $0.777 \cdots$ **6** $3.272727 \cdots$

7 $1.818181 \cdots$ **8** $0.135135135 \cdots$

9 Find the sum of the terms of the infinite GP $1, \frac{1}{3}, \frac{1}{9}, \cdots$.

10 Find the sum of the infinite GP with $a = 4$ and fourth term $\frac{1}{2}$.

11 Find the first term of the infinite GP with $s = 6$ and $r = \frac{2}{3}$.

12 Find the ratio in the infinite GP with $s = 8$ and $a = 4$.

If n is a positive integer, find the sum of all fractions of the form given in each of problems 13 to 16.

13 $\left(\frac{3}{4}\right)^n$ **14** $\left(\frac{2}{5}\right)^n$ **15** $\left(\frac{5}{7}\right)^n$ **16** $\left(\frac{7}{9}\right)^n$

17 If a ball rebounds $\frac{3}{5}$ as far as it falls, how far will it travel if dropped from 10 feet?

18 An alumna gave an oil field to her university. If the university received \$230,000 from the field the first year and $\frac{2}{3}$ as much each year thereafter as during the immediately preceding year, how much did the college realize from the field?

19 Assume that potatoes shrink one-half as much each week as during the immediately preceding one. If a dealer stores 1,000 pounds when the price is n cents per pound and if the weight decreases to 950 during the first week, for what range of n can he afford to hold the potatoes until the price rises to $(n + 1)$ cents per pound?

20 A subdivision is laid out to contain 86 lots. If the highest price of a lot is \$12,000 and each of the others is 0.95 the price of the next higher lot, about how much is received for the entire subdivision?

21 Insert three arithmetic means between 2 and 14.

22 Insert four arithmetic means between 3 and 28.

23 Find the arithmetic mean of 2 and 14.

24 Find the arithmetic mean of 3 and 28.

25 Find the geometric mean of $\frac{1}{2}$ and 2.

26 Find the geometric mean of 4 and 256.
27 Insert four geometric means between $\frac{1}{3}$ and 81.
28 Find two sets of three geometric means between 2 and 32.
29 Find three harmonic means between 1 and $\frac{1}{9}$.
30 Find four harmonic means between $-\frac{1}{3}$ and $\frac{1}{7}$.
31 Find two harmonic means between $\frac{1}{3}$ and $\frac{1}{27}$.
32 Find the harmonic mean of 1 and $\frac{1}{99}$.

Classify the progression given in each of problems 33 to 40 and give the next two terms.

33 $1, \frac{1}{3}, \frac{1}{5}, \frac{1}{7}$
34 $8, 4, 2, 1$
35 $13, 9, 5, 1$
36 $-2, 1, 4, 7$
37 $2, 6, 18, 54$
38 $25, 17, 9, 1$
39 $1, \frac{1}{3}, \frac{1}{9}, \frac{1}{27}$
40 $\frac{1}{2}, \frac{1}{5}, \frac{1}{8}, \frac{1}{11}$

Exercise 21.4 Review

1 Write out the seven terms of an AP with $a = 3$ and $d = -2$.
2 Write out the six terms of a GP with $a = 2$ and $r = 2$.
3 What are the other three terms of a harmonic progression with first term $\frac{1}{2}$ and fifth term $\frac{1}{18}$?
4 Find the sixth term and the sum of the six terms of a GP that begins 6, 3.
5 Find the sixth term and the sum of the six terms of an AP that begins 6, 3.
6 If in an AP, $a = 7$, $d = -3$, and $s = 5$, find l and n.
7 If in a GP, $a = 36$, $l = \frac{4}{9}$, and $n = 5$, find s and r.
8 Find the sum of all integers between 5 and 76 that are divisible by 6.
9 If n is a positive integer, find the sum of all fractions of the form $(\frac{4}{5})^n$.
10 Find the value of x if $2x - 1$, $3x + 1$, and $6x + 5$ form an AP.
11 Find the value of x if x, $x + 3$, and $3x + 3$ form a GP.
12 If a boy received \$300 from a certain source when he was 13 days old, and on each anniversary thereof, he receives $\frac{2}{3}$ as much as he got a year earlier, about how much will be to his credit from this source by the time the children in the neighborhood think he is an old man?
13 Find the replacement set for x so that the sum

$$\frac{2}{3x + 2} + \frac{4}{(3x + 2)^2} + \frac{8}{(3x + 2)^3} + \cdots$$

exists.

14 If 1, 0, and 1 are added to the first, second, and third terms, respectively, of an AP with $d = 3$, a GP is obtained. Find the AP and the common ratio of the GP.

15 If $a \neq b$ and both are positive numbers, show that their geometric mean is less than their arithmetic mean.

16 Show that x, y, z is a GP if $y - x$, $2y$, $y - z$ is a harmonic progression.

17 Find the replacement for a so that a, $2a^2$, $3a^3$ is an AP.

Mathematical Induction

Generalization is a method that is quite useful in mathematics as well as in other sciences. There are two commonly used ways of arriving at a generalization.

One way is to work first with the general case itself as we did in deriving the quadratic formula. If this is done, we need only identify any particular case in which we are interested as being a special case of the generalization or general form. We then work from the general to the particular case. If this is done, the logic used is called *deduction*.

Another method of procedure is to begin by examining a number of specific cases and try to find a pattern into which they all fit. If this can be done, we state the generalization as a law. If this procedure is used, the method of logic is known as *induction*.

The danger of induction is that the specific cases may be special cases regardless of how many we use, and consequently, the generalization based on them may be erroneous. Nonmathematical sciences must often be content to work with this limitation which is called *incomplete induction*. In mathematics, however, we are able to avoid the use of incomplete induction by use of a method of proof that is called *mathematical induction* or *complete induction*.

22.1 METHOD OF MATHEMATICAL INDUCTION

We shall use the method of mathematical induction in Chap. 23 to prove the binomial theorem; here we shall examine the method itself. For example, if we let $n = 0, 1, 2, 3$ in $q(n) = n^2 - n + 41$, we find that $q(n)$ becomes $41, 41, 43$, and 47, respectively. These numbers are primes; that is, no one of them is divisible by any integer other than itself and unity and each is greater than 1. If we calculate the value of $q(n)$ for each integral value of n up to and including 40, we see that $q(n)$ represents the same type of integer, and this suggests that $q(n)$ represents a prime number for every integral value assigned to n. However, if n is equal to 41, $q(n) = 41^2 - 41 + 41 = 41^2$, which is not a prime.

On the other hand, $1 + 3 = 4 = 2^2$, $1 + 3 + 5 = 9 = 3^2$, $1 + 3 + 5 + 7 = 16 = 4^2$. These results suggest that the sum of the first n odd integers is n^2; that is,

$$1 + 3 + 5 + \cdots + (2n - 1) = n^2 \qquad (22.1)$$

Since repeated verification of the truth of Eq. (22.1) for particular values of n does not constitute a proof, we must find some other means of demonstrating its general validity. The type of reasoning involved in mathematical induction is illustrated by the following hypothetical example: Suppose that a certain goal can be reached by a sequence of steps that

459

are successive but of unknown number. Suppose, further, that a person in the process of achieving this goal can be assured that it will always be possible for her to take the next step. Then; regardless of all other circumstances, she knows that she can ultimately attain the goal.

In order to apply this method of reasoning to the proof of the statement in Eq. (22.1), we assume that the statement is true for some definite but unknown integral value of n, that is, for $n = k$. Then we show that it *necessarily* follows that the statement is true for the *next* integer, $k + 1$. Hence, if we can show that the statement is true for some number, say $k = 3$, we know that it is true for the next integer, $k = 4$, and, by proceeding in the same manner, for all following integers. We shall show how this is done in Example 1.

EXAMPLE 1   Prove Eq. (22.1) by mathematical induction.

Solution   We shall first rewrite Eq. (22.1) and have

$$1 + 3 + 5 + \cdots + (2n - 1) = n^2 \tag{1}$$

We then assume that (1) is true for $n = k$ and obtain

$$1 + 3 + 5 + \cdots + (2k - 1) = k^2 \tag{2}$$

Next we write (1) with $n = k + 1$ and get

$$1 + 3 + 5 + \cdots + (2k + 1) = (k + 1)^2 \tag{3}$$

Now we shall prove that the truth of (3) necessarily follows from (2). The last term in the left member of (3) is the $(k + 1)$st term of (1); hence the next to the last is the kth term and, therefore, is $(2k - 1)$. Consequently, we can write (3) in the form

$$1 + 3 + 5 + \cdots + (2k - 1) + (2k + 1) = (k + 1)^2 \tag{4}$$

In order to prove that (4) is true, provided we assume the truth of (2), we notice that the left member of (4) is the corresponding member of (2) increased by $(2k + 1)$, that is, by the $(k + 1)$st term of (1). Hence, we add $2k + 1$ to each member of (2), thus obtaining

$$1 + 3 + 5 + \cdots + (2k + 1) = k^2 + 2k + 1 = (k + 1)^2$$

which is the same as (4). Therefore, if (2) is true, (4) is true. That is, (1) is true for $n = k + 1$ if it is true for $n = k$.

Evidently, (1) is true for $n = 1$. Hence, (1) holds for $n = 1, 2, 3, 4, 5$, and so on.

The formal process of a proof by mathematical induction consists of the following five steps:

Steps in proof
by mathematical
induction

1 Assume that the theorem or statement to be proved is true for n equal to a particular but unspecified integer k, and express this assumption in symbolic form.

2 Obtain a symbolic statement of the theorem for $n = k + 1$.

3 Prove that if the equation in the statement in step 1 is true, the equation in the statement in step 2 is true also.

4 Verify the theorem for the least integral value q of n for which it has a meaning.

5 Using the conclusion in step 3, show by successive steps that the theorem is true for $n = q + 1$ since it is true for the integer q of step 4; furthermore, that it is true for $n = q + 2$ since it is true for $n = q + 1; \ldots$; and finally, that it is true for $n = q + m$ since it is true for $n = q + (m - 1)$, regardless of the positive integral value of m.

No general directions can be given for carrying out the work of step 3. However, the following additional examples illustrate a procedure that can frequently be followed.

EXAMPLE 2   Prove that Demoivre's theorem,

$$[r(\cos\theta + i\sin\theta)]^n = r^n(\cos n\theta + i\sin n\theta) \tag{1}$$

is true if n is a positive integer.

Solution   According to step 1, we assume that (1) is true for $n = k$, thus obtaining

$$[r(\cos\theta + i\sin\theta)]^k = r^k(\cos k\theta + i\sin k\theta) \tag{2}$$

We write (1) with $n = k + 1$, and we get

$$[r(\cos\theta + i\sin\theta)]^{k+1} = r^{k+1}[\cos(k+1)\theta + i\sin(k+1)\theta] \tag{3}$$

In order to prove that the truth of (3) follows from (2), we multiply each member of the latter by $r(\cos\theta + i\sin\theta)$ since this will give us a new equation whose left member is the same as that of (3). We thus have

$$\begin{aligned}[r(\cos\theta + i\sin\theta)]^{k+1} &= [r(\cos\theta + i\sin\theta)][r^k(\cos k\theta + i\sin k\theta)]\\ &= r^{k+1}[\cos(k+1)\theta + i\sin(k+1)\theta]\end{aligned}$$

Hence, (1) is true for $n = k + 1$ if it is true for $n = k$.

Equation (2) is trivial but true for $k = 1$. Hence, (1) is true for $n = 1, 2, 3, \ldots$.

EXAMPLE 3   Prove that $x^n - y^n$ is divisible by $x - y$.

Solution   We first assume that $x^k - y^k$ is divisible by $x - y$, or that

$$\frac{x^k - y^k}{x - y} = q(x, y) \qquad (1)$$

where

$$q(x, y) = x^{k-1} + x^{k-2}y + \cdots + xy^{k-2} + y^{k-1} \qquad (2)$$

When $n = k + 1$, the quotient of $x^n - y^n$ and $x - y$ can be expressed in the form

$$\frac{x^{k+1} - y^{k+1}}{x - y} = \frac{x^{k+1} - xy^k + xy^k - y^{k+1}}{x - y} \qquad \blacktriangleleft \text{ since } xy^k - xy^k = 0$$

$$= \frac{x(x^k - y^k) + y^k(x - y)}{x - y}$$

$$= x\left(\frac{x^k - y^k}{x - y}\right) + y^k\left(\frac{x - y}{x - y}\right)$$

$$= x[q(x, y)] + y^k$$

$$= x^k + x^{k-1}y + \cdots + x^2y^{k-2} + xy^{k-1} + y^k$$

by substituting the value of $q(x, y)$ from (2).

When $k = 2$, (1) becomes

$$\frac{x^2 - y^2}{x - y} = x + y$$

Hence, $x^n - y^n$ is divisible by $x - y$ when $n = 1, 2, 3, 4, 5$, and so on.

In most of the problems of Exercise 22.1, the work of step 3 can be carried out by adding the $(k + 1)$st term of the formula under consideration to each member of the equation obtained in step 1.

Exercise 22.1   Mathematical Induction

Show that the equation in each of problems 1 to 24 is true for all integers $n \geq 1$ by use of mathematical induction.

1   $1 + 4 + 7 + \cdots + (3n - 2) = n(3n - 1)/2$
2   $1 + 2 + 3 + \cdots + n = n(n + 1)/2$
3   $5 + 9 + 13 + \cdots + (4n + 1) = n(2n + 3)$
4   $3 + 7 + 11 + \cdots + (4n - 1) = n(2n + 1)$
5   $3 + 5 + 7 + \cdots + (2n + 1) = n(n + 2)$

6 $\quad 2 + 7 + 12 + \cdots + (5n - 3) = n(5n - 1)/2$

7 $\quad 2 + 5 + 8 + \cdots + (3n - 1) = n(3n + 1)/2$

8 $\quad 6 + 11 + 16 + \cdots + (5n + 1) = n(5n + 7)/2$

9 $\quad 2 + 6 + 12 + \cdots + n(n + 1) = n(n + 1)(n + 2)/3$

10 $\quad 4 + 10 + 18 + \cdots + n(n + 3) = n(n + 1)(n + 5)/3$

11 $\quad 2 + 14 + 36 + \cdots + n(5n - 3) = n(n + 1)(5n - 2)/3$

12 $\quad -3 - 2 + 3 + \cdots + n(2n - 5) = n(n + 1)(4n - 13)/6$

13 $\quad 1^2 + 2^2 + 3^2 + \cdots + n^2 = n(n + 1)(2n + 1)/6$

14 $\quad 1^3 + 2^3 + 3^3 + \cdots + n^3 = n^2(n + 1)^2/4$

15 $\quad 1^2 + 3^2 + 5^2 + \cdots + (2n - 1)^2 = n(2n - 1)(2n + 1)/3$

16 $\quad 1^3 + 3^3 + 5^3 + \cdots + (2n - 1)^3 = n^2(2n^2 - 1)$

17 $\quad 2 + 2^2 + 2^3 + \cdots + 2^n = 2(2^n - 1)$

18 $\quad 3 + 3^2 + 3^3 + \cdots + 3^n = 3(3^n - 1)/2$

19 $\quad 6 + 6^2 + 6^3 + \cdots + 6^n = 6(6^n - 1)/5$

20 $\quad 7 + 7^2 + 7^3 + \cdots + 7^n = 7(7^n - 1)/6$

21 $\quad \dfrac{1}{1 \cdot 2} + \dfrac{1}{2 \cdot 3} + \dfrac{1}{3 \cdot 4} + \cdots + \dfrac{1}{n(n + 1)} = \dfrac{n}{n + 1}$

22 $\quad \dfrac{1}{2 \cdot 3} + \dfrac{1}{3 \cdot 4} + \dfrac{1}{4 \cdot 5} + \cdots + \dfrac{1}{(n + 1)(n + 2)} = \dfrac{n}{2(n + 2)}$

23 $\quad \dfrac{1}{3 \cdot 4} + \dfrac{1}{4 \cdot 5} + \dfrac{1}{5 \cdot 6} + \cdots + \dfrac{1}{(n + 2)(n + 3)} = \dfrac{n}{3(n + 3)}$

24 $\quad \dfrac{1}{1 \cdot 2} + \dfrac{5}{2 \cdot 3} + \dfrac{11}{3 \cdot 4} + \cdots + \dfrac{n^2 + n - 1}{n(n + 1)} = \dfrac{n^2}{n + 1}$

25 $\quad$ Show that $4^n - 1$ is divisible by 3.

26 $\quad$ Show that $5^n - 1$ is divisible by 4.

27 $\quad$ Show that $3^{2n+1} + 1$ is divisible by 4.

28 $\quad$ Show that $4^{2n} - 1$ is divisible by 5.

29 $\quad$ Show that $x^{2n+1} + y^{2n+1}$ is divisible by $x + y$.

30 $\quad$ Show that $x^{2n-1} - y^{2n-1}$ is divisible by $x - y$.

31 $\quad$ Show that $x^{2n} - y^{2n}$ is divisible by $x - y$.

32 $\quad$ Show that $x^{2n} - y^{2n}$ is divisible by $x + y$.

23

The Binomial Theorem

We examined the square of a binomial in Chap. 1 and later used it in deriving the quadratic formula. It is the purpose of this chapter to develop a formula for use in finding the expansion of a binomial to any integral power and for obtaining any desired number of terms in the expansion of a binomial to a fractional or negative power.

23.1 THE BINOMIAL FORMULA

As an aid in deciding on a possible expansion, we shall consider the powers given below that can be obtained by multiplication.

$$(x + y)^2 = x^2 + 2xy + y^2$$
$$(x + y)^3 = x^3 + 3x^2y + 3xy^2 + y^3$$
$$(x + y)^4 = x^4 + 4x^3y + 6x^2y^2 + 4xy^3 + y^4$$

We can readily verify the following properties of the expansion of $(x + y)^n$ for $n = 2$, 3, and 4 by examining the expansions given above.

Properties of the expansion of $(x + y)^n$

1 The first term in the expansion is x^n.

2 The second term is $nx^{n-1}y$.

3 The exponent of x decreases by one from term to term, and the exponent of y increases by 1.

4 If we multiply the coefficient of any term in the expansion by the exponent of x in that term and then divide by the number of the term, we obtain the coefficient of the next term.

5 The nth, next to last, term of the expansion is nxy^{n-1}.

6 The $(n + 1)$st, last, term is y^n.

7 The sum of the exponents of x and y in any term is n.

If we assume that these seven properties hold for all positive integral n, we can write the first five terms in the expansion of $(x + y)^n$ as follows:

First term $= x^n$ ◀ by Property 1

Second term $= nx^{n-1}y$ ◀ by Property 2

Third term $= \dfrac{n(n-1)}{2}x^{n-2}y^2$ ◀ by Properties 4 and 3

Fourth term $= \dfrac{n(n-1)(n-2)}{3 \times 2}x^{n-3}y^3$ ◀ by Properties 4 and 3

Fifth term $= \dfrac{n(n-1)(n-2)(n-3)}{4 \times 3 \times 2}x^{n-4}y^4$ ◀ by Properties 4 and 3

We can continue this process until we have

nth term $= nxy^{n-1}$ ◀ by Property 5

$(n-1)$st term $= y^n$ ◀ by Property 6

465

We are now in position to form the sum of the above terms and obtain the binomial formula. However, if we introduce a new notation at this point, we can write the expansion in a slightly more compact form. The product of any positive integer n and all positive integers less than n is called *factorial n,* and it is designated by the symbol $n!$. For example,

Factorial *n*, or *n*!

$$3! = 3 \times 2 \times 1 = 6 \qquad \text{and} \qquad 5! = 5 \times 4 \times 3 \times 2 \times 1 = 120$$

Now, we notice that $4 \times 3 \times 2 = 4 \times 3 \times 2 \times 1 = 4!, 3 \times 2 = 3 \times 2 \times 1 = 3!,$ and $2 = 2 \times 1 = 2!,$ and we write

Binomial formula $$(x + y)^n = x^n + nx^{n-1}y + \frac{n(n-1)}{2!}x^{n-2}y^2 + \frac{n(n-1)(n-2)}{3!}x^{n-3}y^3$$

$$+ \frac{n(n-1)(n-2)(n-3)}{4!}x^{n-4}y^4$$

$$+ \cdots + nxy^{n-1} + y^n \qquad \textbf{(23.1)}$$

Equation (23.1) is called the *binomial formula,* and the statement that it is true is called the *binomial theorem.*

EXAMPLE 1   Use the binomial formula to obtain the expansion of $(2a + b)^6$.

Solution   We first apply Eq. (23.1) with $x = 2a$, $y = b$, and $n = 6$. Thus,

$$(2a + b)^6 = (2a)^6 + 6(2a)^5b + \frac{6 \times 5}{2!}(2a)^4b^2 + \frac{6 \times 5 \times 4}{3!}(2a)^3b^3$$

$$+ \frac{6 \times 5 \times 4 \times 3}{4!}(2a)^2b^4 + \frac{6 \times 5 \times 4 \times 3 \times 2}{5!}(2a)b^5$$

$$+ \frac{6 \times 5 \times 4 \times 3 \times 2 \times 1}{6!}b^6$$

Now simplifying the coefficients, and raising $2a$ to the indicated powers, we obtain

$$(2a + b)^6 = 64a^6 + 6(32a^5)b + 15(16a^4)b^2 + 20(8a^3)b^3$$
$$+ 15(4a^2)b^4 + 6(2a)b^5 + b^6$$

Finally, we perform the indicated multiplication in each term and get

$$(2a + b)^6 = 64a^6 + 192a^5b + 240a^4b^2 + 160a^3b^3 + 60a^2b^4 + 12ab^5 + b^6$$

The computation of the coefficients can, in most cases, be performed mentally by use of Property 7, and thus we can avoid writing the first step in the expansion in the above example.

EXAMPLE 2   Expand $(a - 3b)^5$.

Solution   The first term in the expansion is a^5, and the second is $5a^4(-3b)$ by Properties 1 and 2. To get the coefficient of the third term, we use Property 4, multiply 5 by 4, and divide the product by 2, thus obtaining 10. Hence, the third term is $10a^3(-3b)^2$. Similarly, the fourth term is

$$\tfrac{30}{3} a^2(-3b)^3 = 10a^2(-3b)^3$$

By continuing this process, we obtain the following expansion:

$$(a - 3b)^5 = a^5 + 5a^4(-3b) + 10a^3(-3b)^2 + 10a^2(-3b)^3 + 5a(-3b)^4 + (-3b)^5$$
$$= a^5 - 15a^4b + 90a^3b^2 - 270a^2b^3 + 405ab^4 - 243b^5$$

It should be noted that we carry the second term of the binomial, $-3b$, through the first step of the expansion as a single term. Then we raise $-3b$ to the indicated power and simplify the result.

23.2   THE *r*th TERM OF THE BINOMIAL FORMULA

In Sec. 23.1 we explained and illustrated the method for obtaining any term of a binomial expansion. It was obtained from the immediately preceding term. That is a disadvantage since it requires that all preceding terms be formed before any desired term can be obtained. We shall now develop a formula that enables us to write out the general *r*th term without dependence on previous terms. We shall consider the fourth term in Eq. (23.1) and notice the following properties:

Properties of the fourth term

1 The exponent of y in the fourth term is 1 less than the number 4 of the term, i.e., 3.

2 The exponent of x is n minus the exponent of y, i.e., $n - 3$.

3 The denominator of the coefficient is the factorial of the exponent of y, i.e., 3!.

4 The first factor in the numerator of the coefficient is n, the second is $n - 1$, and the number of factors is the same as the exponent of y with each factor 1 less than the immediately preceding one.

We can readily verify the fact that the above four properties hold for other terms of the expansion. If we assume that they hold for each term, hence for the *r*th term, we can say that:

Properties of the *r*th term

1 The exponent of y in the *r*th term is $r - 1$.

2 The exponent of x is $n - (r - 1) = n - r + 1$.

3   The denominator of the coefficient is $(r-1)!$.

4   The numerator of the coefficient is $n(n-1)(n-2)\cdots(n-r+2)$.

Therefore, we say that *the rth term in the expansion of* $(x+y)^n$ *is*

Formula for
the *r*th term

$$\frac{n(n-1)(n-2)\cdots(n-r+2)}{(r-1)!}\,x^{n-r+1}y^{r-1} \qquad\qquad (23.2)$$

EXAMPLE   Find the sixth term in the expansion of $(3a-b^2)^8$.

Solution   In this problem, $x=3a$, $y=-b^2$, $n=8$, and $r=6$.   If we substitute these values in (23.2), we find that the sixth term in the expansion is

$$\frac{(8)(7)(6)(5)(4)}{5!}\,(3a)^3(-b^2)^5 = -1{,}512a^3b^{10}$$

Exercise 23.1   Expansion of Binomials

Find the expansion of the binomial in each of problems 1 to 12 by use of the binomial formula.

1 $(a+b)^8$	**2** $(a-x)^7$	**3** $(b-y)^5$
4 $(x+y)^6$	**5** $(a-2y)^5$	**6** $(x+3b)^6$
7 $(2b+x)^7$	**8** $(3a+b)^4$	**9** $(2a-3b^2)^3$
10 $(3x-2b^3)^5$	**11** $(2b^2+3x)^6$	**12** $(5x^2+2y)^4$

Find the first four terms of the expansion of the binomial in each of problems 13 to 16.

13 $(a+y)^{33}$ **14** $(x-y)^{51}$ **15** $(m-2y)^{101}$ **16** $(b+3c)^{42}$

Find the indicated power of the number in each of problems 17 to 20 and round off to four decimal places.

17 $(1+.04)^5$ **18** 1.05^4 **19** 1.03^6 **20** 1.06^3

Find the specified term of the expansion in each of problems 21 to 32.

21 Fifth term of $(x-2y)^7$	**22** Fourth term of $(2a-c)^6$
23 Sixth term of $(3x+y)^9$	**24** Third term of $(x+4y)^8$
25 Fourth term of $(a-a^{-1})^7$	**26** Sixth term of $(2x-x^{-2})^9$
27 Seventh term of $(x^2+2y)^{11}$	**28** Fifth term of $(3x+y^3)^8$
29 Middle term of $(x+2y^{1/2})^6$	**30** Middle term of $(x-3y^{1/4})^8$

31   The term in $(x+2y)^{10}$ that involves x^7.

32   The term in $(3x-y^{1/2})^{13}$ that involves y^4.

23.3 PROOF OF THE BINOMIAL FORMULA

We have seen that the binomial formula is true for $n = 1, 2, 3, 4,$ and $5,$ and we shall prove by mathematical induction that it is true for all integral values of n. In order to do this, we assume it is true for $n = k$, and we show that it follows that it is true for $n = k + 1$ and, hence, for all positive integers. Under the assumption that it is true for $n = k$, we have

$$(x + y)^k = x^k + kx^{k-1}y + \cdots + \frac{k(k-1) \cdots (k-r+3)x^{k-r+2}y^{r-2}}{(r-2)!}$$

$$+ \frac{k(k-1) \cdots (k-r+2)x^{k-r+1}y^{r-1}}{(r-1)!} + \cdots + y^k \quad (23.3)$$

This expansion shows the first, second, $(r-1)$st, rth, and last terms.

Multiplying each member of this assumed equation by $x + y$ and writing the first and last terms and those which contain y^{r-1} [these are obtained by multiplying the last term in the first line of Eq. (23.3) by y and the first term of the next line by x], we have

$$(x + y)^{k+1} = x^{k+1} + \cdots + \frac{k(k-1) \cdots (k-r+3)x^{k-r+2}y^{r-1}}{(r-2)!}$$

$$+ \frac{k(k-1) \cdots (k-r+2)x^{k-r+2}y^{r-1}}{(r-1)!} + \cdots + y^{k+1}$$

By collecting, we find that the coefficient of $x^{k-r+2}y^{r-1}$ is

$$\frac{k(k-1) \cdots (k-r+3)}{(r-2)!} + \frac{k(k-1) \cdots (k-r+2)}{(r-1)!}$$

$$= \frac{k(k-1) \cdots (k-r+3)(r-1)}{(r-2)!(r-1)}$$

$$+ \frac{k(k-1) \cdots (k-r+3)(k-r+2)}{(r-1)!}$$

If we factor this, we see that the coefficient is

$$\frac{k(k-1) \cdots (k-r+3)[(r-1)+(k-r+2)]}{(r-1)!}$$

$$= \frac{k(k-1) \cdots (k-r+3)(k+1)}{(r-1)!}$$

Hence the term in the product of $x + y$ and $(x + y)^k$ which involves y^{r-1} is

$$\frac{(k+1)k(k-1) \cdots (k-r+3)x^{k-r+2}y^{r-1}}{(r-1)!}$$

By using this as a formula for all terms after the first, we see that the product is

$$(x + y)^{k+1} = x^{k+1} + (k + 1)x^k y + \cdots$$

$$+ \frac{(k + 1)k(k - 1) \cdots (k - r + 3)x^{k-r+2}y^{r-1}}{(r - 1)!} + \cdots + y^{k+1} \quad \textbf{(23.4)}$$

Equation (23.4) is readily seen to be the one obtained by replacing k by $k + 1$ in Eq. (23.3). Hence we have shown that Eq. (23.3) is true for $n = k + 1$ if true for $n = k$. This fact, along with our knowledge that it is true for $n = 1, 2, 3, 4,$ and 5, enables us to say that it is true for $n = 5 + 1 = 6$, hence for $6 + 1 = 7; \ldots$; and consequently, for all positive integers.

23.4 BINOMIAL THEOREM FOR FRACTIONAL AND NEGATIVE EXPONENTS

The proof of the binomial formula for fractional and negative exponents is beyond the scope of this book; however, we shall point out some elementary applications of it. It should be noted that the expansion of $(x + y)^n$ for n not a positive integer has no last term since the coefficient never becomes zero; hence it is impossible to complete the series, and we must be content with any desired or indicated number of terms. The following fact can be established, although the proof will not be given.

The binomial expansion of $x + y$ for fractional and negative exponents is valid if and only if the value of y is between the values of x and $-x$.

EXAMPLE 1   What are the first four terms in the expansion of $(2 + x)^{1/2}$? In what interval is the expansion valid?

Solution   The expansion is

$$(2 + x)^{1/2} = 2^{1/2} + (\tfrac{1}{2} \times 2^{-1/2}x) + \frac{\tfrac{1}{2} \times (-\tfrac{1}{2}) \times 2^{-3/2}x^2}{2!}$$

$$+ \frac{\tfrac{1}{2} \times (-\tfrac{1}{2}) \times (-\tfrac{3}{2}) \times 2^{-5/2}x^3}{3!} + \cdots$$

$$= \sqrt{2} + \frac{x}{2\sqrt{2}} - \frac{x^2}{16\sqrt{2}} + \frac{x^3}{64\sqrt{2}} - \cdots$$

$$= \sqrt{2}\left(1 + \frac{x}{4} - \frac{x^2}{32} + \frac{x^3}{128} - \cdots\right)$$

It is valid if and only if $-2 < x < 2$.

EXAMPLE 2

Determine an approximation to the square root of 10.

Solution

$$\sqrt{10} = 10^{1/2} = (9+1)^{1/2} = (3^2+1)^{1/2}$$

$$= (3^2)^{1/2} + \tfrac{1}{2} \times (3^2)^{-1/2} \times 1 + \frac{\tfrac{1}{2} \times (-\tfrac{1}{2}) \times (3^2)^{-3/2} \times 1^2}{2}$$

$$+ \frac{\tfrac{1}{2} \times (-\tfrac{1}{2}) \times (-\tfrac{3}{2}) \times (3^2)^{-5/2} \times 1^3}{2 \times 3} + \cdots$$

$$= 3 + (\tfrac{1}{2} \times 3^{-1}) - (\tfrac{1}{8} \times 3^{-3}) + (\tfrac{1}{16} \times 3^{-5}) - \cdots$$

$$= 3 + \tfrac{1}{6} - \tfrac{1}{216} + \tfrac{1}{3,888}$$

$$= 3 + 0.16667 - 0.00463 + 0.00026$$

$$= 3.16230$$

By comparing the four terms in the above expansion, we see that their values decrease very rapidly. The rate of this decrease increases as the expansion is carried further. In fact, the fifth term is -0.0000178, and when this is combined with the other four terms, we obtain $\sqrt{10} = 3.1622822$, or, rounded to four decimal places, 3.1623. Hence we conclude that this is the correct value of $\sqrt{10}$ to five figures. Obviously, the expansion can be extended until we obtain any desired degree of accuracy.

Exercise 23.2 Binomials with Negative and Fractional Exponents

Find the first four terms of the expansion of the binomial in each of problems 1 to 20. Find the range of the variable for which the expansion is valid in 17 to 20.

1 $(x+y)^{-3}$	**2** $(a+b)^{-2}$	**3** $(a-y)^{-5}$
4 $(x-b)^{-4}$	**5** $(2a-y)^{-1}$	**6** $(x-2y)^{-3}$
7 $(3x+2y)^{-2}$	**8** $(5x+3y)^{-3}$	**9** $(x+x^{-1})^{-4}$
10 $(x^2-2x^{-1})^{-1}$	**11** $(x^{-2}-3x)^{-4}$	**12** $(x^{-3}-2x^2)^{-2}$
13 $(y+1)^{1/3}$	**14** $(1+y)^{1/3}$	**15** $(25-x)^{1/2}$
16 $(32-x)^{1/5}$	**17** $(16+x)^{-1/4}$	**18** $(x+9)^{-1/2}$
19 $(2+x)^{-1}$	**20** $(4+x^2)^{-2}$	

In problems 21 to 28, obtain the first four terms of the expansion and round off to four decimal places if more than four are obtained.

21 $\sqrt{101} = (10^2+1)^{1/2}$	**22** $\sqrt[3]{999}$	**23** $\sqrt[3]{28}$
24 $\sqrt[5]{31}$	**25** 1.04^{-5}	**26** 1.06^{-4}
27 1.07^{-3}	**28** 1.05^{-6}	

24

Trigonometric Equations

In this chapter we shall find angles that satisfy equations which involve at least one trigonometric function value of one or more angles. Such an equation is called a *trigonometric equation,* and each angle that satisfies it is called a root or solution of the equation.

Trigonometric equation

If functions of θ alone are involved, we shall find only those solutions between zero and 2π including zero but not 2π, although an equation may have solutions and none of them be in that interval. For example, the smallest nonnegative solution of $\tan(\theta/9) = 1$ is $\theta = 9\pi/4$.

If $k\theta$ is the only angle involved, we shall begin the solution by finding the values of $k\theta$ from zero to $k2\pi$ including zero but not $k2\pi$, so as to be able to get the values of θ from zero to 2π that satisfy the equation.

To solve a trigonometric equation for the angle, we first solve it for a function value of the angle and then find the angle by use of tables. Methods for doing this are discussed in this chapter.

24.1 EQUATIONS THAT CONTAIN ONLY ONE ANGLE AND ONLY ONE FUNCTION VALUE

If a trigonometric equation contains only one function value of only one angle, we solve for that function value by the methods of Chaps. 4, 11, and 17. Then the angles can be found by use of a table or a knowledge of the function values of $\pi/6$ and $\pi/4$ and their multiples.

EXAMPLE 1   Solve the equation $4 \tan\theta = \sqrt{3} + \tan\theta$.

Solution   If we add $-\tan\theta$ to each member and combine coefficients, we have

$$3 \tan\theta = \sqrt{3}$$

$$\tan\theta = \frac{\sqrt{3}}{3} \quad \blacktriangleleft \text{ dividing by 3}$$

Therefore, $\theta = \pi/6$ and $7\pi/6$ are the elements of the solution set such that $0 \le \theta < 2\pi$.

EXAMPLE 2   Solve the equation $3 \sec^2\theta - 5 \sec\theta - 2 = 0$.

Solution   The given equation is a quadratic and can be solved by any of the methods of Chap. 11. By factoring the left member, we obtain

$$(3 \sec\theta + 1)(\sec\theta - 2) = 0$$

$$\sec\theta = -\tfrac{1}{3}, 2$$

If $\sec \theta = 2$, then $\theta = \pi/3$ and $5\pi/3$; furthermore, these are the only solutions in the interval $0 \le \theta < 2\pi$ since the secant of an angle is never between 1 and -1. Hence, the solution set is $\{\pi/3, 5\pi/3\}$.

24.2 EQUATIONS WITH ONE MEMBER FACTORABLE AND THE OTHER ZERO

If one member of an equation is zero and the other is factorable so that each factor contains only one function value of only one angle, we can solve the equation by the factoring method.

EXAMPLE 1   Solve $\tan^2 \theta - \tan \theta = 0$.

Solution   By factoring the left member of the given equation, we get

$$\tan \theta (\tan \theta - 1) = 0$$

Now, by setting each factor equal to zero, we obtain $\tan \theta = 0$ and $\tan \theta - 1 = 0$. Therefore,

$$\tan \theta = 0, 1$$

Consequently, $\{0, \pi, \pi/4, 5\pi/4\}$ is the solution set.

EXAMPLE 2   Solve the equation $2 \sin \theta \cos 2\theta - 2 \cos 2\theta - \sin \theta + 1 = 0$.

Solution   The left member of the given equation in factored form is

$$(\sin \theta - 1)(2 \cos 2\theta - 1) = 0$$

There are two angles in the equation but each factor contains only one function value of only one angle. Setting the first factor equal to zero gives

$$\sin \theta - 1 = 0$$
$$\sin \theta = 1$$
$$\theta = \frac{\pi}{2}$$

The second factor contains the angle 2θ, and we want values of θ in $0 \le \theta < 2\pi$; hence, we must find the values of 2θ such that $0 \le 2\theta < 4\pi$. Equating the second factor to zero gives

$$2 \cos 2\theta - 1 = 0$$

Hence,
$$\cos 2\theta = \tfrac{1}{2}$$

$$2\theta = \frac{\pi}{3}, \frac{5\pi}{3}, \frac{7\pi}{3}, \frac{11\pi}{3}$$

and
$$\theta = \frac{\pi}{6}, \frac{5\pi}{6}, \frac{7\pi}{6}, \frac{11\pi}{6}$$

Therefore the solution set is $\{\pi/2, \pi/6, 5\pi/6, 7\pi/6, 11\pi/6\}$.

Exercise 24.1   Solution of Trigonometric Equations

Find all values of θ such that $0 \le \theta < 2\pi$ which satisfy the following equations.

1 $2 \cos \theta - 1 = 0$
2 $2 \sin \theta - \sqrt{3} = 0$
3 $3 \cot \theta - \sqrt{3} = 0$
4 $\csc \theta - 2 = 0$
5 $4 \sin^2 \theta - 1 = 0$
6 $\tan^2 \theta - 3 = 0$
7 $2 \csc^2 \theta - 3 \csc \theta - 2 = 0$
8 $\sqrt{3} \cot^2 \theta - 2 \cot \theta - \sqrt{3} = 0$
9 $2 \cos^2 \theta + 3 \cos \theta - 2 = 0$
10 $\sqrt{3} \sec^2 \theta - 2 \sec \theta = 0$
11 $2 \sin^2 \theta + \sin \theta - 1 = 0$
12 $2\sqrt{3} \cos^2 \theta + \cos \theta - 2\sqrt{3} = 0$
13 $4 \sin^3 \theta - \sin \theta = 0$
14 $4 \cos^3 \theta - 3 \cos \theta = 0$
15 $\sqrt{3} \tan^3 \theta - \tan^2 \theta + 2\sqrt{3} \tan \theta - 2 = 0$
16 $\sqrt{3} \sec^3 \theta - 2 \sec^2 \theta + 4\sqrt{3} \sec \theta - 8 = 0$
17 $\sqrt{3} \cos \theta \tan \theta - \cos \theta = 0$
18 $2 \sin \theta \cos \theta + \sin \theta = 0$
19 $2 \sin \theta \cos \theta - \sin \theta + 2 \cos \theta - 1 = 0$
20 $\sec \theta \tan \theta - 2 \tan \theta - \sec \theta + 2 = 0$
21 $2 \cos 2\theta - 1 = 0$
22 $\tan 2\theta - \sqrt{3} = 0$
23 $2 \cos 3\theta + 1 = 0$
24 $\sqrt{3} \csc 4\theta - 2 = 0$
25 $2 \sin 2\theta \sec \theta - \sec \theta - 2 \sin 2\theta + 1 = 0$

26 $\csc \theta \tan 2\theta + \tan 2\theta - \sqrt{3} \csc \theta - \sqrt{3} = 0$
27 $2 \sin \theta \tan 3\theta - \sqrt{3} \tan 3\theta + 2\sqrt{3} \sin \theta - 3 = 0$
28 $2 \cos 3\theta \tan 2\theta + 2 \cos 3\theta - \tan 2\theta - 1 = 0$

24.3 EQUATIONS REDUCIBLE TO A FORM SOLVABLE BY FACTORING

Many equations that cannot be solved by factoring can be reduced to a solvable form by use of the fundamental identities.

EXAMPLE 1 Solve the equation $\sin 3\theta + \sin \theta - \sin 2\theta = 0$.

Solution If we apply the identity $\sin x + \sin y = 2 \sin \frac{1}{2}(x + y) \cos \frac{1}{2}(x - y)$ to the first two terms, we get

$$2 \sin 2\theta \cos \theta - \sin 2\theta = 0$$
$$\sin 2\theta (2 \cos \theta - 1) = 0 \qquad \blacktriangleleft \text{ factoring}$$

Now, by equating each factor to zero and solving, we get

$$\sin 2\theta = 0$$
$$2\theta = 0, \pi, 2\pi, 3\pi$$
$$\theta = 0, \frac{\pi}{2}, \pi, \frac{3\pi}{2}$$

and
$$2 \cos \theta - 1 = 0$$
$$\cos \theta = \tfrac{1}{2}$$
$$\theta = \frac{\pi}{3}, \frac{5\pi}{3}$$

Hence, the solution set is $\{0, \pi/2, \pi, 3\pi/2, \pi/3, 5\pi/3\}$.

EXAMPLE 2 Solve the equation $\cos 2\theta - 2 \cos \theta = 0$.

Solution If we apply the formula for the cosine of twice an angle to the first term in the equation and rearrange terms, we get

$$2 \cos^2 \theta - 2 \cos \theta - 1 = 0$$

Now, on solving by the quadratic formula, we have

$$\cos \theta = \frac{2 \pm \sqrt{(-2)^2 - 4(2)(-1)}}{2(2)} = \frac{1 \pm \sqrt{3}}{2}$$
$$= 1.3660, -0.3660$$

Since the cosine of an angle is never larger than 1, there are no values of θ corresponding to $\cos \theta = 1.3660$. Therefore, the only solutions are those for which $\cos \theta = -0.3660$. We want only those for which $0 \leq \theta < 2\pi$; therefore, the angles are in the second and third quadrant and the cosine of the reference angle of each is 0.3660. By use of tables and interpolation, we find that the reference angle is 68°32′. The solution set is then {111°28′, 248°32′} since these are the second- and third-quadrant angles between 0° and 360° which have a reference angle of 68°32′.

24.4   SOLUTION OF $A \sin \theta + B \cos \theta = C$

In order to solve an equation of the form

$$A \sin \theta + B \cos \theta = C \tag{1}$$

we shall let

$$A = r \cos \alpha \tag{2}$$

and

$$B = r \sin \alpha \tag{3}$$

where r is a positive number. Then (1) becomes

$$r(\cos \alpha \sin \theta + \sin \alpha \cos \theta) = C$$

Now, by making use of the fact that the expression in parentheses is $\sin (\theta + \alpha)$ and dividing by r, we can write

$$\sin (\theta + \alpha) = \frac{C}{r} \tag{4}$$

If r is known, the value of $\sin (\theta + \alpha)$ is determined, and we can find $\theta + \alpha$ by use of a table of sines of angles. Consequently, we can find θ provided α is known. Now that we see it is desirable to know the value of r and of α, we shall see how to find them.

If we add the squares of the corresponding members of (2) and (3), we have

$$A^2 + B^2 = r^2(\cos^2 \alpha + \sin^2 \alpha)$$
$$= r^2$$

Therefore, since r is a positive number, we must have

$$r = \sqrt{A^2 + B^2} \tag{5}$$

To determine α, we divide each member of (3) by the corresponding member of (2), and get

$$\tan \alpha = \frac{B}{A} \tag{6}$$

Consequently, α is an angle whose tangent is B/A, but it must be so selected that both (2) and (3) are satisfied.

EXAMPLE

Solve the equation $\sqrt{3} \sin \theta + 1 \cos \theta = 1$.

Solution

If we compare the given equation and (1), we see that $A = \sqrt{3}$ and $B = 1$; hence, by use of (5), $r = 2$. Furthermore, $\tan \alpha = 1/\sqrt{3}$ by use of (6). Consequently, α can be 30° or any other angle whose tangent is $1/\sqrt{3}$ provided (2) and (3) are satisfied. They are satisfied for $\alpha = 30°$. Now, the given equation can be written in the form

$$\sin (\theta + 30°) = \tfrac{1}{2}$$

by use of (4) and the values just determined for α and r. We must find values of $\theta + 30°$ in $30° \leq \theta + 30° < 390°$ in order to obtain values of θ in the interval $0 \leq \theta < 360°$. Therefore, the desired values are $\theta + 30° = 30°$ and 150°, from which $\theta = 0°$ and 120° are the nonnegative solutions less than 360°. Hence, the solution set is $\{0, 120°\}$.

Exercise 24.2 Solution of Trigonometric Equations

Solve the following equations for values of θ such that $0 \leq \theta < 2\pi$.

1	$2 \sin^2 \theta + 3 \cos \theta - 3 = 0$	**2**	$5 \sin \theta - 1 - 2 \cos^2 \theta = 0$
3	$4 \tan^2 \theta - 3 \sec^2 \theta = 0$	**4**	$\sec^2 \theta + \tan \theta - 1 = 0$
5	$2 \sin \theta + \sqrt{3} - 2 = \sqrt{3} \csc \theta$	**6**	$2 \sin \theta - 5 + 2 \csc \theta = 0$
7	$\cot \theta + 1 - 2 \tan \theta = 0$	**8**	$2 \cos \theta + 1 - \sec \theta = 0$
9	$\cot 2\theta - \tan \theta = 0$	**10**	$\cos^2 \theta - \cos 2\theta = 2$
11	$\tan 2\theta + \cot \theta = 0$	**12**	$\sin 2\theta - \cos^2 \theta + \sin^2 \theta = 0$
13	$\cot \theta + \tan \dfrac{\theta}{2} = 0$	**14**	$\cot \dfrac{\theta}{2} + 1 + \cos \theta = 0$
15	$\tan \tfrac{1}{2}\theta + \sin 2\theta = \csc \theta$	**16**	$\cos \theta - 1 + \tan \tfrac{1}{2}\theta = 0$

17 $\cos 3\theta \cos 2\theta + \sin 3\theta \sin 2\theta = \sin 2\theta$
18 $\sin 3\theta \cos 2\theta - \cos 3\theta \sin 2\theta = \sin 2\theta$
19 $\sin 2\theta \cos 3\theta - \sin 3\theta \cos 2\theta = 0$
20 $\sin 3\theta \cos \theta - \cos 3\theta \sin \theta = 0$

21 $\sin 3\theta + \sin \theta = 0$ **22** $\cos \theta + \cos 3\theta = 0$

23 $\sin 3\theta - \sin \theta = 0$ **24** $\cos 3\theta - \cos \theta = 0$

25 $3 \sin \theta - 4 \cos \theta = 4.5$ **26** $\sin \theta + \cos \theta = \sqrt{2}/2$

27 $5 \sin \theta + 12 \cos \theta = 9.1026$ **28** $\sin \theta + \sqrt{3} \cos \theta = 2$

25

Inverse Trigonometric Functions

We studied functions and inverse functions in Chap. 6 and trigonometric functions in Chaps. 7, 10, and 15. In this chapter, we shall apply the concept of the inverse of a function to the trigonometric functions.

25.1 THE INVERSE TRIGONOMETRIC FUNCTIONS

In Sec. 6.9, we defined the inverse of a function $\{(x, y)|y = f(x)\}$ as the function, if one exists, that is obtained by interchanging the elements in each of the ordered pairs that constitute the function. We also stated that the defining equation of the inverse is obtained by solving the defining equation $y = f(x)$ for x in terms of y and then interchanging x and y.

We shall now consider the sine function

$$\sin = \{(x, y)|y = \sin x\}$$

and the inverse relation

$$\sin^{-1} = \{(x, y)|x = \sin y\} \tag{1}$$

Arcsine   The term *arcsine* is also used to indicate the inverse relation of sine. This relation is not a function unless the value of y is restricted since there is more than one value of y for some given values of x. For example, if $x = \sqrt{3}/2$, then $\sqrt{3}/2 = \sin y$ and y may be $\pi/3$ or any angle whose sine is positive and whose reference angle is $\pi/3$. Consequently y may be $\pi/3 + 2n\pi$ or $2\pi/3 + 2n\pi$ where n is an integer. These two sets of values can be combined in set notation as $y = \{\pi/3 + 2n\pi\} \cup \{2\pi/3 + 2n\pi\}$.

We indicate that the equation $x = \sin y$ in (1) has been solved for y by writing $y = \sin^{-1} x$ or $y = \arcsine x$ and that the value of y has been restricted so as to have a function by writing $y = \text{Sin}^{-1} x$ or $y = \text{Arcsine } x$. Consequently, we indicate the inverse function of

$$\{x, y)|y = \sin x\}$$

by writing

$$\{(x, y)|y = \text{Sin}^{-1} x\}$$

or, if we care to show the restriction on y, we write

$$\textbf{Arcsine} = \{(x, y)|y = \textbf{Sin}^{-1}x, -\pi/2 \leq y \leq \pi/2\} \tag{25.1}$$

The inverse functions   Similar notations are used for the other inverse functions. They are

$$\textbf{Arccosine} = \{(x, y)|y = \textbf{Cos}^{-1} x, 0 \leq y \leq \pi\} \tag{25.2}$$
$$\textbf{Arctangent} = \{(x, y)|y = \textbf{Tan}^{-1} x, -\pi/2 < y < \pi/2\} \tag{25.3}$$
$$\textbf{Arccotangent} = \{(x, y)|y = \textbf{Cot}^{-1} x, 0 < y < \pi\} \tag{25.4}$$
$$\textbf{Arcsecant} = \{(x, y)|y = \textbf{Sec}^{-1} x, 0 \leq y \leq \pi, y \neq \pi/2\} \tag{25.5}$$
$$\textbf{Arccosecant} = \{(x, y)|y = \textbf{Csc}^{-1} x, -\pi/2 \leq y \leq \pi/2, y \neq 0\} \tag{25.6}$$

Many older books and some current ones refer to the inverse trigono-

481

metric relations as "the inverse functions" and to the inverse functions as "the principal values of the inverse functions."

If we make use of (25.1), (25.3), and (25.5) and the function values of multiples of $\pi/6$, we see that

$$\text{Arcsine } \tfrac{1}{2} = \text{Sin}^{-1} \tfrac{1}{2} = \pi/6 \qquad \blacktriangleleft \text{ from (25.1)}$$
$$\text{Arctan } (-\sqrt{3}) = \text{Tan}^{-1}(-\sqrt{3}) = -\pi/3 \qquad \blacktriangleleft \text{ from (25.3)}$$
$$\text{Arcsec } 2 = \text{Sec}^{-1} = \pi/3 \qquad \blacktriangleleft \text{ from (25.5)}$$

25.2   GRAPHS OF THE INVERSE TRIGONOMETRIC FUNCTIONS

The two equations

$$y = \text{Arcsin } x \qquad\qquad (1)$$

and

$$x = \sin y \qquad\qquad (2)$$

express the same relation between x and y for $-\pi/2 \leq y \leq \pi/2$. Consequently, they have the same graph. We shall obtain that graph by use of (2) and shall require that $-\pi/2 \leq y = \text{Arcsin } x \leq \pi/2$, since $-\pi/2$ to $\pi/2$ is the range of $y = \text{Arcsin } x$. By assigning values to y and computing each corresponding value of $x = \sin y$, we obtain the following table.

y	$-\pi/2$	$-\pi/3$	$-\pi/4$	$-\pi/6$	0	$\pi/6$	$\pi/4$	$\pi/3$	$\pi/2$
$x = \sin y$	-1	$-\sqrt{3}/2$	$-\sqrt{2}/2$	$-1/2$	0	$1/2$	$\sqrt{2}/2$	$\sqrt{3}/2$	1

If we plot the points determined by the number pairs (x, y) of this table, we obtain the portion of the graph indicated by the darker line in Fig. 25.1. The lighter colored portions were obtained by assigning values from $\pi/2$ to 2π and from $-\pi/2$ to -2π to y, but they are not part of the graph of $y = \text{Arcsin } x$.

The graphs of $y = \text{Arccos } x$ and $y = \text{Arctan } x$ are shown in Figs. 25.2 and 25.3. In each case, the dark part of the curve was obtained by use of the values of y in the range and is called the *principal branch* by some writers.

25.3   TRANSFORMING INVERSE FUNCTIONS

It is often desirable to find a trigonometric function of an inverse trigonometric function or an inverse function of a trigonometric function or to express one inverse trigonometric function in terms of one or more others. The procedure in such problems will be illustrated by two examples.

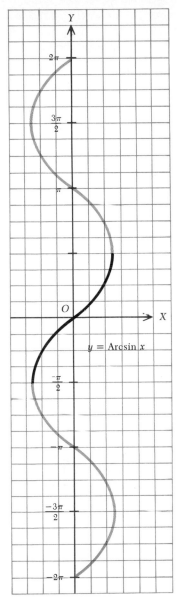

Figure 25.1

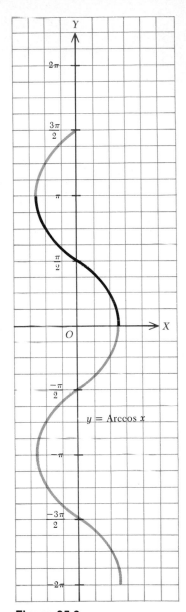

Figure 25.2

EXAMPLE 1   Evaluate $\cos (\text{Tan}^{-1} u)$ for $u < 0$.

Solution   If we let $\theta = \text{Arctan } u$, then $\tan \theta = u$ for $u < 0$. We know that θ must be such that $-\pi/2 < \theta \leq 0$. We can construct Fig. 25.4 by use of the definition of the tangent of an angle and the pythagorean theorem. The latter

483

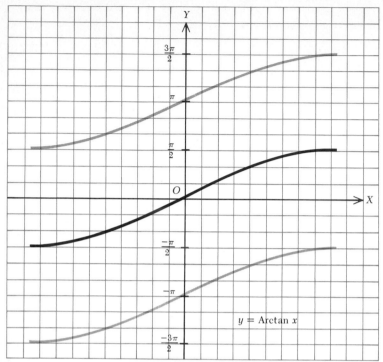

$y = \text{Arctan } x$

Figure 25.3

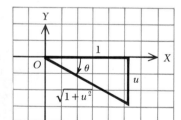

Figure 25.4

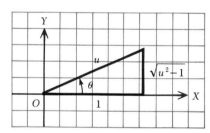

Figure 25.5

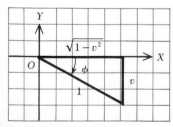

Figure 25.6

enables us to see that the hypotenuse is $\sqrt{1 + u^2}$. Now from the figure we find that $\cos \theta = \cos (\text{Tan}^{-1} u) = 1/\sqrt{1 + u^2}$.

EXAMPLE 2   Evaluate $\sin (\text{Arcsec } u - \text{Arcsin } v)$, for $u > 0$, and $v < 0$.

Solution   If we let $\theta = \text{Arcsec } u$ and $\phi = \text{Arcsin } v$, then $0 \leq \theta < \pi/2$ and $-\pi/2 \leq \phi \leq 0$; furthermore, $\sec \theta = u$ and $\sin \phi = v$. The two angles are shown in Figs. 25.5 and 25.6.

Now that we have figures from which the function of the angles can be obtained, we see that

$$
\begin{aligned}
\sin (\text{Arcsec } u - \text{Arcsin } v) &= \sin (\theta - \phi) \\
&= \sin \theta \cos \phi - \cos \theta \sin \phi \\
&= \frac{\sqrt{u^2 - 1}}{u} \frac{\sqrt{1 - v^2}}{1} - \frac{1}{u} \frac{v}{1}
\end{aligned}
$$

Exercise 25.1   Evaluation of Inverse Functions

Find the value of the inverse function in each of problems 1 to 8.

1 $\text{Arccos } \frac{1}{2}$
2 $\text{Arcsin } 1$
3 $\text{Arccot } \sqrt{3}$
4 $\text{Arccot } 0$
5 $\text{Arcsec } (-\sqrt{2})$
6 $\text{Arccsc } (-2)$
7 $\text{Sin}^{-1} (\sqrt{2}/2)$
8 $\text{Cos}^{-1} (\sqrt{3}/2)$

Evaluate the function of the angle in each of problems 9 to 20.

9 $\cos (\text{Arccos } 0.43)$
10 $\tan (\text{Arccot } 1.7)$
11 $\cot (\text{Tan}^{-1} 0.59)$
12 $\sec (\text{Cos}^{-1} 0.31)$
13 $\csc (\text{Cos}^{-1} \sqrt{3}/2)$
14 $\sin (\text{Tan}^{-1} \sqrt{3})$
15 $\cos (\text{Arcsin } \frac{1}{2})$
16 $\tan (\text{Arccsc } 2/\sqrt{3})$
17 $\cot (\text{Arccos } 0.7)$
18 $\sec (\text{Arcsin } 0.3)$
19 $\csc (\text{Cot}^{-1} 1.4)$
20 $\sin (\text{Sec}^{-1} 2.1)$

Assume that u and v are greater than 1 and find the function taken in each of problems 21 to 40.

21 $\sec (\text{Arcsec } u)$
22 $\cot (\text{Arccot } u)$
23 $\sin [\text{Arccsc } (-u)]$
24 $\tan [\text{Arccot } (-u)]$
25 $\cos 2 \left(\text{Arccos } \dfrac{1}{u} \right)$
26 $\csc 2 \left(\text{Arcsin } \dfrac{1}{u} \right)$
27 $\tan 2 (\text{Arcsec } u)$
28 $\sin 2 \left(\text{Arccos } \dfrac{-1}{u} \right)$

29 $\sin\left(\pi + \text{Arcsin}\dfrac{1}{u}\right)$ **30** $\cos\left(\pi + \text{Arcsec } u\right)$

31 $\tan\left(\pi - \text{Arctan } u\right)$ **32** $\tan\left(\pi - \text{Arccsc } u\right)$

33 $\sin\frac{1}{2}\left(\text{Arccos}\dfrac{-1}{u}\right)$ **34** $\cos\frac{1}{2}\left(\text{Arccos}\dfrac{-1}{u}\right)$

35 $\cos\frac{1}{2}\left[\text{Arccos}\left(\dfrac{\sqrt{u^2 - 1}}{u}\right)\right]$ **36** $\tan\frac{1}{2}(\text{Arcsec } u)$

37 $\cos\left(\text{Arccos}\dfrac{1}{u} + \text{Arcsin}\dfrac{1}{v}\right)\bigg]$ **38** $\sin\left(\text{Arcsin}\dfrac{1}{u} + \text{Arccos}\dfrac{1}{v}\right)$

39 $\tan\left(\text{Arctan } u + \text{Arcsec } v\right)$ **40** $\tan\left(\text{Arccot } u - \text{Arccsc } v\right)$

25.4 INVERSE TRIGONOMETRIC IDENTITIES

Identity   An equation that contains inverse trigonometric functions is called an *identity* if (1) its solution set consists of all replacements for the variable that are common to the domains of definition of the functions involved and if (2) the equation is satisfied by all replacements for the variable between two specified numbers. Thus,

$$\text{Arcsin } u = \text{Arccot } \frac{\sqrt{1 - u^2}}{u}$$

is an identity for $-1 \le u \le 1$, $u \ne 0$, since the two members are defined and equal for all such values of u. It is not an identity for $u > 1$, since Arcsin u is not defined for such values. The equation

$$\text{Arccos } u = \frac{\pi}{2} - \text{Arcsin } u$$

is an identity for $-1 \le u \le 1$ but is not defined for other values. Finally,

$$\text{Arccos } u = \text{Arctan } \frac{\sqrt{1 - u^2}}{u}$$

is an identity for $0 < u \le 1$ but is not an identity for $-1 < u < 0$, since for such values $\pi/2 < \text{Arccos } u < \pi$ and $-\pi/2 < \text{Arctan }(\sqrt{1 - u^2}/u) < 0$.

Proof that an equation involving inverse trigonometric functions is an identity consists of showing that for all values of the variable under consideration, there is a value of each inverse function such that the two members of the equation are equal.

EXAMPLE 1   Prove that

$$\text{Arccos } u = \frac{\pi}{2} - \text{Arcsin } u \qquad \text{for } -1 \le u \le 1 \tag{1}$$

is an identity.

Solution   If we let Arccos $u = \theta$, then $\cos \theta = u$ and

$$\text{Arcsin } u = \text{Arcsin } (\cos \theta) = \frac{\pi}{2} - \theta \qquad \blacktriangleleft \text{ since } \sin(\pi/2 - \theta) = \cos\theta$$

Consequently, the right member of (1) is $\pi/2 - (\pi/2 - \theta) = \theta$, as is the left member.   Therefore, (1) is an identity.

EXAMPLE 2   Prove that

$$2 \text{ Arccos } \frac{1}{u} = \text{Arccos } \frac{2 - u^2}{u^2} \qquad \text{for } u \geq 1 \qquad (2)$$

is an identity.

Solution   If we let Arccos $(1/u) = \theta$, then $\cos \theta = 1/u$ and the left member of (2) is 2θ. Furthermore,

$$\frac{2 - u^2}{u^2} = \frac{2}{u^2} - 1 = 2 \cos^2 \theta - 1 = \cos 2\theta$$

Consequently, the right member of (2) becomes Arccos $\cos 2\theta = 2\theta$, as is the left member.   Therefore (2) is an identity.

At times the sum or difference of two inverse trigonometric functions can be simplified by replacing it by a single function, as illustrated in the following example.

EXAMPLE 3   Prove that

$$\text{Sin}^{-1} \tfrac{31}{33} - \text{Cos}^{-1} \tfrac{7}{11} = \text{Sin}^{-1} \tfrac{1}{3} \qquad (3)$$

Solution   If we let $\text{Sin}^{-1} \tfrac{31}{33} = A$, $\text{Cos}^{-1} \tfrac{7}{11} = B$, and $\text{Sin}^{-1} \tfrac{1}{3} = C$, then $\sin A = \tfrac{31}{33}$, $\cos B = \tfrac{7}{11}$, and $\sin C = \tfrac{1}{3}$.   Furthermore, all three angles are between 0 and $\pi/2$, since they are principal values and their function values are positive. Consequently, $-\pi/2 < A - B < \pi/2$ and $0 < C < \pi/2$.   Therefore, $A - B = C$ if $\sin(A - B) = \sin C$.   We shall now find $\sin A$, $\cos A$, $\sin B$, and $\cos B$, since they are needed in obtaining $\sin(A - B)$.   They are

$$\sin A = \sin(\text{Sin}^{-1} \tfrac{31}{33}) = \tfrac{31}{33}$$

$$\cos A = \sqrt{1 - \sin^2 A} = \sqrt{1 - (\tfrac{31}{33})^2} = \sqrt{1 - \tfrac{961}{1089}}$$

$$= \frac{8\sqrt{2}}{33}$$

$$\cos B = \cos(\text{Cos}^{-1} \tfrac{7}{11}) = \tfrac{7}{11}$$

$$\sin B = \sqrt{1 - \left(\frac{7}{11}\right)^2} = \sqrt{1 - \frac{49}{121}} = \frac{6\sqrt{2}}{11}$$

Consequently,

$$\sin (A - B) = \sin A \cos B - \cos A \sin B$$

$$= \frac{31}{33}\left(\frac{7}{11}\right) - \left(\frac{8\sqrt{2}}{33}\right)\left(\frac{6\sqrt{2}}{11}\right) = \frac{217 - 96}{33(11)} = \frac{1}{3}$$

and (3) is true, since $\sin C$ is also $\frac{1}{3}$.

Exercise 25.2   Inverse Trigonometric Identities

Prove that the statement in each of problems 1 to 12 is true.

1 Arcsin $(2/\sqrt{5})$ + Arcsin $(1/\sqrt{5})$ = $\pi/2$
2 Arccos $(-1/\sqrt{5})$ − Arccos $(2/\sqrt{5})$ = $\pi/2$
3 Arccos $(-3/\sqrt{13})$ − Arcsin $(3/\sqrt{13})$ = $\pi/2$
4 Arccos $(\frac{4}{5})$ + Arctan $(\frac{1}{7})$ = $\pi/2$
5 Arctan $(\frac{5}{3})$ − Arctan $(\frac{1}{4})$ = $\pi/4$
6 Arctan 2 − Arctan $(\frac{1}{3})$ = $\pi/4$
7 Arctan $(\frac{3}{7})$ + Arcsin $(2/\sqrt{29})$ = $\pi/4$
8 Arctan $(\frac{1}{5})$ + Arctan $(\frac{2}{3})$ = $\pi/4$
9 Arctan $(\frac{3}{4})$ − Arctan $(\frac{5}{8})$ = Arctan $(\frac{4}{47})$
10 Arcsin $(\frac{3}{5})$ + Arccos $(\frac{12}{13})$ = Arccos $(\frac{33}{65})$
11 Arctan $(\frac{1}{2})$ − Arcsin $(2/\sqrt{5})$ = Arctan $(\frac{4}{-3})$
12 Arctan $(1/\sqrt{10})$ + Arccos $(\frac{7}{11})$ = Arcsin $(\frac{31}{33})$

Prove that the following statements are true for every value of u and v or for all values in the specified domain.

13 $\text{Arcsin } u + \text{Arccos } u = \dfrac{\pi}{2}, -1 \leq u \leq 1$

14 $\text{Arctan } u + \text{Arccsc } \sqrt{1 + u^2} = \dfrac{\pi}{2}$

15 $\text{Arcsin } \dfrac{u}{\sqrt{1 + u^2}} + \text{Arcsin } \dfrac{1}{\sqrt{1 + u^2}} = \dfrac{\pi}{2}, u > 0$

16 $\text{Arcsin } \dfrac{u}{\sqrt{1 + u^2}} + \text{Arccot } u = \dfrac{\pi}{2}, u > 0$

17 $\text{Arctan } u = \text{Arcsin } \dfrac{u}{\sqrt{1 + u^2}}$

18 $\text{Arcsec } \dfrac{\sqrt{u^2 + 1}}{u} = \text{Arccot } u, u > 0$

19 $\text{Arccos } \dfrac{u}{\sqrt{1 + u^2}} = \text{Arctan } \dfrac{1}{u}$

20 $\text{Arcsec } u = \text{Arcsin } \dfrac{\sqrt{u^2 - 1}}{u}, \; u > 0$

21 $2 \text{ Arctan} \dfrac{1}{u} = \text{Arctan} \dfrac{2u}{u^2 - 1}, \; u > 1$

22 $2 \text{ Arccos} \dfrac{u}{\sqrt{u^2 + 1}} = \text{Arcsin} \dfrac{2u}{u^2 + 1}, \; 0 \leq u \leq 1$

23 $\text{Arcsin} \dfrac{2u}{u^2 + 1} = 2 \text{ Arcsin} \dfrac{1}{\sqrt{u^2 + 1}}, \; u \geq 1$

24 $\text{Arccsc} \dfrac{u^2 + 1}{2u} = 2 \text{ Arccot } u, \; u > 1$

25 $\text{Arccot} \dfrac{1 + u}{1 - u} = -\tfrac{1}{2} \text{ Arctan} \dfrac{u^2 - 1}{2u}, \; u > 1$

26 $\text{Arctan} \dfrac{1}{u} = \tfrac{1}{2} \text{ Arccsc} \dfrac{u^2 + 1}{2u}, \; u \geq 1$

27 $\text{Arctan} \dfrac{1}{u} = \tfrac{1}{2} \text{ Arctan} \dfrac{2u}{1 - u^2}, \; u > 1$

28 $\text{Arctan } u = \tfrac{1}{2} \text{ Arccos} \dfrac{1 - u^2}{1 + u^2}, \; u > 0$

29 $2 \text{ Arctan } u = \text{Arcsin} \dfrac{u}{\sqrt{1 + u^2}} + \text{Arccos} \dfrac{1}{\sqrt{1 + u^2}}, \; u > 0$

30 $\text{Arctan } u - \text{Arccot } u = \text{Arcsin} \dfrac{u^2 - 1}{u^2 + 1}, \; u > 0$

31 $\text{Arccos} \dfrac{1}{\sqrt{u^2 + 1}} - \text{Arccos} \dfrac{u}{\sqrt{u^2 + 1}} = \text{Arccos} \dfrac{2u}{u^2 + 1}, \; u \geq 1$

32 $\text{Arcsin} \dfrac{1}{\sqrt{1 + u^2}} - \text{Arctan } u = \text{Arccos} \dfrac{2u}{1 - u^2}, \; u > 1$

33 $\text{Arctan } u + \text{Arctan } v = \text{Arctan} \dfrac{u + v}{1 - uv}, \; uv \neq 1$

34 $\text{Arcsin} \dfrac{u}{\sqrt{1 - u^2}} + \text{Arccos} \dfrac{\sqrt{v^2 - 1}}{v} = \text{Arcsec} \dfrac{v\sqrt{u^2 + 1}}{u + \sqrt{v^2 - 1}},$
$u \geq 1, \; v \geq \sqrt{2}$

35 $\text{Arcsin } u + \text{Arcsin } v = \text{Arctan} \dfrac{u\sqrt{1 - v^2} + v\sqrt{1 - u^2}}{\sqrt{(1 - u^2)(1 - v^2)} - uv}, \; u^2 + v^2 < 1$

36 $\text{Arctan } u - \text{Arcsin} \dfrac{v}{\sqrt{1 + v^2}} = \text{Arctan} \dfrac{u - v}{1 + uv}, \; u > 0, \; v > 0$

26

Permutations and Combinations

It is frequently necessary or desirable to be able to calculate the number of ways that the elements in a set can be arranged and to determine the number of ways in which they can be combined into subsets. For example, the government must furnish a unique social security number for each citizen and future citizens, and each state must furnish a different combination of letters and numbers for each license plate issued any year.

26.1 TERMINOLOGY

Element
Combination
Permutation

Our work in this chapter will deal with collections and arrangements of symbols, objects, or events. Each symbol, object, or event will be called an *element*; furthermore, each set of elements will be called a *combination,* and each unique arrangement of the elements will be called a *permutation.* Consequently, a permutation is determined by the elements that form it and the order in which they appear, whereas a combination is a set of elements without regard to order. All the elements may be used in a permutation or combination, but it is not necessary to use all of them.

26.2 THE FUNDAMENTAL PRINCIPLE

Fundamental
principle

Our work on combinations and permutations will be based on the following principle: *If an event can occur in h_1 ways and if, after it has occurred, a second event can occur in h_2 ways, then the two events can occur in h_1h_2 ways in the indicated order.*

EXAMPLE 1

In how many ways can a boy and a girl be selected from a group of five boys and six girls?

Solution

The boy can be selected in five ways, and after that is done, the girl can be selected in six ways; hence a boy and a girl can be selected in $5 \times 6 = 30$ ways.

We can extend the fundamental principle by thinking of the first two events as a single one that can happen in (h_1h_2) ways; hence, if after the first two events have occurred, a third can happen in h_3 ways, then the three can happen in $(h_1h_2)h_3 = h_1h_2h_3$ ways. If then a fourth can occur in h_4 ways, the four can happen in $(h_1h_2h_3)h_4 = h_1h_2h_3h_4$ ways. This procedure can be continued event by event until we find that n events can happen in $h_1h_2h_3 \cdots h_n$ ways in the order indicated by the subscripts.

EXAMPLE 2   How many automobile license plates can be made if the inscription on each contains two different letters followed by three digits?

Solution   There are 26 letters; therefore, the first of the two letters can be chosen in 26 ways. Since the two letters must be different, there are only 25 ways in which the second letter can be chosen. The first digit can be selected in 10 ways; and since the three digits may be repeated, the second can be chosen in 10 ways and the third in 10. Consequently, from the two letters and three digits, license plates can be made up in $26 \times 25 \times 10 \times 10 \times 10 = 650,000$ ways.

26.3 PERMUTATIONS OF n DIFFERENT ELEMENTS TAKEN r AT A TIME

The definition of a permutation was given in Sec. 26.1 but is repeated here for ready reference. Each arrangement of a set of n elements is called a *permutation* of the set.

The arrangements *abc, acb, bac, bca, cab,* and *cba* constitute the six permutations of the letters *a, b,* and *c.* Furthermore, *ab, ba, ac, ca, bc,* and *cb* are the six permutations of the same three letters taken two at a time.

We shall let $P(n, r)$ represent the number of permutations of n elements taken r at a time and shall develop a formula for evaluating the symbol. The symbols $_nP_r$ and $P_r{}^n$ are also used. We can fill the position 1 in the arrangement in n ways. After position 1 has been filled, we have $n - 1$ choices for position 2, then $n - 2$ choices for position 3, and finally, $n - (r - 1)$ choices for position r. Hence, since $n - (r - 1) = n - r + 1$, we have

$$P(n, r) = n(n - 1)(n - 2) \cdots (n - r + 1)$$

If we multiply the right member of this equation by $\dfrac{(n - r)!}{(n - r)!}$, where, as defined in Sec. 23.2, $(n - r)!$ is the product of the integer $n - r$ and all the positive integers less than $n - r$, we get

Permutation of n elements taken r at a time

$$\frac{n(n - 1)(n - 2) \cdots (n - r + 1)(n - r)!}{(n - r)!} = \frac{n!}{(n - r)!}$$

Therefore, $$P(n, r) = \frac{n!}{(n - r)!} \qquad (26.1)$$

In order to obtain the number of permutations of n elements taken n at a time, we let $r = n$ in Eq. (26.1). Thus, we obtain

Permutation of
n elements taken
n at a time

$$P(n, n) = n! \qquad (26.2)$$

since, by definition, $0! = 1$.

EXAMPLE 1

Find the number of four-digit numbers that can be formed from the digits 1, 2, 3, and 4 if no digit is repeated.

Solution

Here we have four elements to be taken four at a time; by Eq. (26.2), the number is 4!, or $4 \times 3 \times 2 \times 1 = 24$.

EXAMPLE 2

Six people enter a room that contains 10 chairs. In how many ways can they be seated?

Solution

Since only six of the chairs are to be occupied, the number of different seating arrangements is equal to the number of permutations of 10 elements taken 6 at a time and, by Eq. (26.1), is

$$P(10, 6) = \frac{10!}{(10 - 6)!} = \frac{10 \times 9 \times 8 \times 7 \times 6 \times 5 \times (4!)}{(4!)}$$

$$= 10 \times 9 \times 8 \times 7 \times 6 \times 5 = 151{,}200$$

26.4 PERMUTATIONS OF n ELEMENTS NOT ALL DIFFERENT

If the n elements of a set are not all different, the problem of determining the number of permutations of the set presents a new aspect. For example, there are two permutations of the letters a and b, namely, ab and ba, but there is only one permutation of the letters a and a, since neither letter can be distinguished from the other and the two can therefore be put in only one *unique* arrangement. We shall suppose that s members of the set are alike and then designate the $n - s$ different elements by $t_1, t_2, \ldots, t_{n-s}$. The permutations, that is, the distinguishable arrangements of the n elements, will then depend only on the arrangement of the elements $t_1, t_2, \ldots, t_{n-s}$. We can therefore obtain all the permutations of the n elements by distributing $t_1, t_2, \ldots, t_{n-s}$ in the n positions in all possible ways and then filling in the vacant places with the s identical elements. This amounts to distributing $n - s$ elements in the positions $1, 2, 3, \ldots, n$ in as many ways as possible, that is, to finding the number of permutations of n elements taken $n - s$ at a time. By Eq. (26.1), we have

$$P(n, n - s) = \frac{n!}{[n - (n - s)]!} = \frac{n!}{s!}$$

Hence, if s members of a set of n elements are alike, the number of permutations of the n elements taken n at a time is equal to the number of permutations of n things taken n at a time divided by the number of permutations of s things taken s at a time. By a repeated application of this principle, we can derive the following theorem: *If, in a set of n elements, there are g groups, the first containing n_1 members all of which are alike; the second containing n_2 which are alike; the third, n_3 which are alike; and so on to the gth group, which has n_g members alike; then, the number of permutations of the n elements taken n at a time is given by*

Theorem on permutations of like elements

$$\frac{n!}{n_1!n_2! \cdots n_g!}$$

EXAMPLE 1   In how many ways can the letters of the word "Connecticut" be arranged?

Solution   The solution of this problem involves two factors: (1) the number of permutations of 11 letters taken 11 at a time and (2) the letters n and t two times and the letter c three times. Hence, the number of unique arrangements is given by

$$\frac{11!}{2!2!3!} = \frac{11 \times 10 \times 9 \times 8 \times 7 \times 6 \times 5 \times 4 \times 3 \times 2 \times 1}{2 \times 1 \times 2 \times 1 \times 3 \times 2 \times 1} = 1,663,200$$

26.5 CYCLIC PERMUTATIONS

Cyclic permutation   If a permutation of n elements is such that each element is adjacent to two others, we say that we have a *cyclic,* or circular, permutation. For example, a set of n elements arranged around a round table is a cyclic permutation. In a cyclic permutation, we can get all the arrangements by leaving one element in a fixed position and permuting the other $(n - 1)$ elements. Therefore, the number of cyclic permutations of n elements is $(n - 1)!$

Exercise 26.1 Fundamental Principle, Permutations

1 How many seven-digit numbers can be made if digits can be repeated and the first digit cannot be zero?

2 How many seven-digit numbers can be made if no digit can be repeated and the first digit cannot be zero?

3 How many nine-digit social security numbers can be made?

4 How many auto license plates can be made if the inscription on each consists of five digits, none repeated, and a letter other than L and O?

5 How many bouquets consisting of an amaryllis, a Dutch iris, and a bearded iris can be chosen from 15 amaryllises, 18 Dutch irises, and 6 bearded irises.

6 In how many ways can a horse, a cow, and two goats be selected from a pen of eight of each variety?

7 In how many ways can a frog, a tadpole, a snake, and a piece of string be given to four children so that each child receives a gift?

8 In how many ways can an aardvark, a whooping crane, and a hyena be distributed among six zoos?

Give the meaning and value of the symbol in each of problems 9 to 12.

9 $P(8, 3)$ 10 $P(8, 5)$ 11 $P(10, 4)$ 12 $P(7, 2)$

Show that the equation in each of problems 13 to 16 is an identity.

13 $P(n, n-1) = P(n, n)$ 14 $P(n, n) = 2P(n, n-2)$
15 $P(n, n) = r!P(n, n-r)$
16 $P(n, n-r+1) = P(r, 1)P(n, n-r)$

If all letters are used in each permutation, how many permutations can be made from the letters in the word in each of problems 17 to 20?

17 Democrat   18 Republican
19 Louisiana   20 Tennessee

21 After he has selected his nine players and named the person who is to hit for the pitcher, how many batting orders can a coach have?

22 In how many ways can 10 children be assigned 10 seats that are in a row?

23 How many football teams can be selected from 11 students if each can play each position?

24 In how many orders can a student visit all of them if there are 12 buildings on a campus?

25 Ten cows enter a barn with six available stalls. In how many ways can the stalls be used?

26 In how many ways can four girls and three boys be put alternately in a row of seven chairs?

27 In how many ways can four boys and four girls be seated on a bench if one of the girls refuses to be seated next to a boy?

28 In how many ways can a boy make his selection if he has three pairs of shoes, four pairs of trousers, and five shirts?

29 In how many ways can six children, including a pair of twins and a set of triplets, be placed in 10 homes if all members of a multiple birth must go to the same home and no home can take more than one child unless the children are members of a multiple birth?

30 In how many ways can four huskies, three hounds, and five terriers be placed in a row if dogs of a breed must be together?

31 Fifteen people enter a plane and find there are five single seats on one side of the aisle and five double ones on the other. In how many ways can they be seated?

32 The president, two vice presidents, and six directors of a firm are to be seated in the nine seats at the head table at a banquet. If the president is to have the center seat and is to have a vice president immediately on his right and his left, in how many ways can the nine persons be seated?

26.6 COMBINATIONS

The definition of a combination is given in Sec. 26.1, but it will be repeated here for ready reference: A set of elements that is taken without regard to the order in which the elements are arranged is called a *combination*. According to this definition, the six permutations *xyz, xzy, yzx, yxz, zxy,* and *zyx* are only one combination of the letters *x, y,* and *z*.

There is only one combination of *n* elements taken *n* at a time, but the problem of determining the number of combinations of *n* elements taken *r* at a time is both interesting and important. We shall designate the combination of *n* elements taken *r* at a time by $C(n, r)$, and it is the number of subsets of *r* elements that may be formed from a set of *n* elements. Since the number of permutations of a combination of *r* elements taken *r* at a time is *r*!, we have

$$P(n, r) = r!C(n, r)$$

Hence $$C(n, r) = \frac{P(n, r)}{r!}$$

Combinations of *n* elements taken *r* at a time

Therefore, the number of combinations of n elements taken r at a time is

$$C(n, r) = \frac{n!}{(n - r)!r!} \tag{26.3}$$

as obtained by replacing $P(n, r)$ by its value as given in Eq. (26.1).

EXAMPLE 1   How many committees of 7 members can be formed from a group of 25 individuals?

Solution   The number of committees is equal to the number of combinations of 25 elements taken 7 at a time. Hence it is

$$C(25, 7) = \frac{25!}{18!7!} = \frac{25 \times 24 \times 23 \times 22 \times 21 \times 20 \times 19 \times (18!)}{(18!) \times 7 \times 6 \times 5 \times 4 \times 3 \times 2 \times 1} = 480{,}700$$

EXAMPLE 2

A business firm wishes to employ six men and three boys. In how many ways can the selection be made if nine men and five boys are available?

Solution

The six men can be selected from the nine in $C(9, 6)$ ways, and the three boys from the five in $C(5, 3)$. Hence the number of ways in which the selection of the employees can be made is

$$C(9, 6)C(5, 3) = \frac{9!}{(9 - 6)!6!} \frac{5!}{(5 - 3)!3!}$$

$$= \frac{9 \times 8 \times 7 \times (6!)}{3 \times 2 \times (6!)} \frac{5 \times 4 \times (3!)}{2 \times (3!)}$$

$$= 840$$

26.7 THE SUM OF CERTAIN COMBINATIONS

If we multiply and divide the expression in Eq. (26.2) by $(n - r + 1)!$, we obtain

$$\frac{n(n - 1) \cdots (n - r + 2)[(n - r + 1)!]}{(r - 1)!(n - r + 1)!} x^{n-r+1}y^{r-1}$$

$$= \frac{n!}{(r - 1)![n - (r - 1)]!} x^{n-r+1}y^{r-1} = C(n, r - 1)x^{n-r+1}y^{r-1}$$

Therefore, we can express the expansion of $(x + y)^n$ in the form

$$(x + y)^n = x^n + C(n, 1)x^{n-1}y + C(n, 2)x^{n-2}y^2 + \cdots$$
$$+ C(n, n - 1)\,xy^{n-1} + C(n, n)y^n \quad \textbf{(26.4)}$$

Equation (26.4) can be used to obtain an expression for the total number of combinations of n elements taken one at a time, two at a time, and so on, to n at a time. If we let $x = y = 1$ in Eq. (26.4), we get

$$(1 + 1)^n = 1 + C(n, 1) + C(n, 2) + C(n, 3) + \cdots + C(n, n - 1) + C(n, n)$$

Now, by replacing $(1 + 1)^n$ by 2^n and adding -1 to each member, we see that *the total number of combinations obtained by taking n elements one at a time, two at a time, and so on, to n at a time is* $2^n - 1$. For example, the total number of combinations of 8 elements taken one at a time, two at a time, ..., eight at a time is $2^8 - 1 = 255$.

Exercise 26.2   Combinations

1   Show that $C(17, 6) = C(17, 11)$.
2   Show that $8C(15, 5) = 3C(16, 10)$.
3   Show that $2C(9, 5) = 3C(9, 3)$.
4   Show that $5C(14, 5) = 14C(13, 4)$.
5   How many combinations of 5 different letters can be selected from the 20 consonants?
6   Ten buildings are arranged in a circle and a straight walk connects each of them with all the others.   How many walks are used?
7   How many triangles are determined by 20 points if no three of them are in a line?
8   In how many ways can three books be selected from 1,000?
9   How many different bridge hands can be selected from a deck of cards?
10   How many combinations are possible if four cards are drawn from a deck of 52?
11   In how many ways can three dimes be chosen from $1.40 worth of dimes?
12   In how many ways can Tom pay Jack $25 if he has $150 in $5 bills?
13   Alice has $100 in $10 bills and $100 in $5 bills.   In how many ways can she pay a motel bill of $15?
14   In how many ways could Alice in problem 13 pay a bill of $20?
15   In how many ways can two cows and two horses be selected from eight cows and nine horses?
16   A hotel has one room that can accommodate four adults and another that can take care of two.   In how many ways can six adults occupy the rooms?
17   In how many ways can four mules be put in three stalls if no stall can be occupied by more than two mules and each stall can accommodate two?
18   In how many ways can 2 pilots and 6 parachute jumpers be chosen from 5 pilots and 20 jumpers?
19   In how many ways can two Texans, two Kansans, and two Alaskans be chosen from 25 people from one state, 30 from another, and 35 from the third?
20   In how many ways can 2 Armenians, 2 Frenchmen, 2 Englishmen, and 3 others be chosen from 7 Armenians, 5 Frenchmen, 6 Englishmen, and 13 other people?
21   How many sums of money can be formed from a halfpenny, a 2-cent piece, a 3-cent piece, a 20-cent piece, a $3 gold piece, and a $4 gold piece?   NOTE: Each of these coins has been minted as a U.S. coin.
22   How many sets of more than 7 coins can be formed from 10 coins?
23   How many sums of more than 20 cents can be formed from the coins of problem 21?

24 In how many ways can $2 be withdrawn from a cash register that contains 6 silver dollars, 8 half-dollars, and 10 quarters.

25 How many clubs of more than five girls can be formed by eight girls?

26 How many groups of 2 or more people can be formed from 20 people?

27 How many groups of less than four chairs can be formed from nine chairs?

28 How many groups of seven dresses or less can be formed from nine dresses?

Probability

The word "probability" is used loosely by the layman to indicate a vague likelihood that something will happen. It is often used synonymously with "chance." The likelihood that something will or will not happen under specified circumstances has application in a wide variety of human pursuits. The gambler, the statistician, the economist, and the engineer are among those who make use of probability. It is the purpose of this chapter to attempt to bring order out of the chaos caused by the loose use of the word "probability."

27.1 A SAMPLE SPACE AND PROBABILITY

Sample space
Outcome, or
sample point
Event

If an experiment is undertaken, there is a set of possible results associated with it. The set of all possible results of an experiment is called a *sample space* for the experiment. Each element of a sample space is called an *outcome* or *sample point*. Any subset of a sample space is called an *event*. If a die is cast, it may stop with 1, 2, 3, 4, 5, or 6 on top; hence, {1, 2, 3, 4, 5, 6} is the sample space, and any one of these elements is an outcome or sample point. Furthermore, any combination of one or more of them is an event.

Random

If a die is made accurately and rolled honestly, it is as likely to stop with one number up as with another. Thus, each of the outcomes is equally likely, and we say the outcome is *random*. We shall assume that all outcomes in experiments discussed in this chapter are random in the sense that they are equally likely. Under those conditions, we say that the experiment is a *random experiment*.

Random
experiment
Permissible

A set of outcomes of a random experiment is called *permissible* if and only if we assign exactly one element of the set to each performance of the experiment. For example, {1, 2, 3, 4, 5, 6} is a permissible set of outcomes of one toss of a die. We shall deal exclusively with permissible sets of outcomes in this chapter, and we shall use the following notation:

Symbol	Meaning
S	A sample space of permissible random outcomes
$n(S)$	The number of elements in S
E	A set of specified outcomes in S. Hence $E \subseteq S$
$n(E)$	The number of elements in E
$p(E)$	The probability that E will happen, or more briefly, the probability of E

Mathematical
probability

If a probability is predicted in advance of an experiment on the basis of an analysis of all possibilities of occurrence, we say that we have a *mathematical* or *a priori probability*.

The probability $p(E)$ of an event $E \subseteq S$ is defined as follows:

$$p(E) = \frac{n(E)}{n(S)}$$

Since $n(E)$ and $n(S)$ are numbers and $E \subseteq S$, then $n(E) \leq n(S)$, and it follows that $p(E)$ is a number less than or equal to 1.

EXAMPLE 1

Find the probability of tossing a 4 in one toss of a die.

Solution

The sample space is $S = \{1, 2, 3, 4, 5, 6\}$, and the specified outcome is that the die falls with 4 up. Hence $E = \{4\}$. Therefore,

$$p(4) = \frac{n(4)}{n(S)} = \frac{1}{6}$$

EXAMPLE 2

If one card is drawn from a standard deck of 52 cards, find the probability that the card will be a jack.

Solution

Here $S = \{x | x \text{ is a card in a deck of 52 cards}\}$. Hence $n(S) = 52$. Furthermore, $E = \{\text{club jack, diamond jack, heart jack, spade jack}\}$, so $n(E) = 4$.
   Therefore,

$$p(E) = \frac{n(E)}{n(S)} = \frac{4}{52} = \frac{1}{13}$$

EXAMPLE 3

Each of the three-digit numbers that can be formed using the integers from 1 to 9, with no digit repeated, is written on a card. The cards are then stacked and shuffled. If one card is drawn from the stack, find the probability that the sum of the digits in the number on it will be 10.

Solution

Here $S = \{x | x \text{ is a card bearing a three-digit number formed with the digits from 1 to 9 with no digit repeated}\}$. The number of cards in S is $P(9, 3) = 9 \cdot 8 \cdot 7 = 504$. Therefore, $n(S) = 504$. Furthermore,

$$E = \{x | x \text{ is a card in } S \text{ whose digit sum is 10}\}$$

The sets of three different digits whose sum is 10 are $\{1, 2, 7\}$, $\{1, 3, 6\}$, $\{1, 4, 5\}$, and $\{2, 3, 5\}$. Since the digits in each of these four sets can be arranged in $P(3, 3) = 3! = 6$ ways, there are $4 \cdot 6 = 24$ cards in E. Hence $n(E) = 24$. Therefore,

$$p(E) = \frac{n(E)}{n(S)} = \frac{24}{504} = \frac{1}{21}$$

27.2 MATHEMATICAL EXPECTATION

Mathematical expectation

Value of an expectation

If the probability that an event will occur in one trial is p, then the *expected* number of occurrences in n trials is np. The latter product is called the *mathematical expectation* of the event.

If a person is to receive D dollars if an event occurs and the probability of its occurrence is p, then the *value of the expectation* is Dp dollars.

EXAMPLE

If the probability of a man's living to age sixty-five is $\frac{3}{5}$, find the value of his expectation if he is to receive \$1,000 if alive at age sixty-five.

Solution

The value of the man's expectation of \$1,000 at age sixty-five is the product of the probability of his living to age sixty-five and the amount he is to receive if alive. Therefore, $\$1,000 \times \frac{3}{5} = \$600$ is the expectation.

27.3 MUTUALLY EXCLUSIVE EVENTS

In this section, we shall derive and explain the use of the first two of several theorems on probability. If E_1 and E_2 represent two events and if $n(E_1)$ and $n(E_2)$ denote the number of elements in sets E_1 and E_2, respectively, then

$$n(E_1) \geq 0 \quad \text{and} \quad n(E_2) \geq 0 \tag{1}$$
$$n(\varnothing) = 0 \tag{2}$$
$$n(E_1 \cup E_2) = n(E_1) + n(E_2) - n(E_1 \cap E_2) \tag{3}$$

The situation for (3) is shown in Fig. 27.1 with 7 elements in E_1, 4 in E_2, and 2 elements in common, and in Fig. 27.2 for the case in which $E_1 \cap E_2 = \varnothing$. Quite often, we say that two events are *mutually exclusive* or *disjoint* to indicate that $E_1 \cap E_2 = \varnothing$. If two events are mutually exclusive, then $n(E_1 \cap E_2) = 0$ and (3) becomes

$$n(E_1 \cup E_2) = n(E_1) + n(E_2) \tag{4}$$

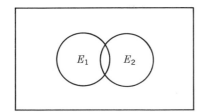

Figure 27.1

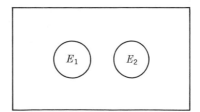

Figure 27.2

We shall now state and prove a theorem concerning the probability of two events.

If E_1 and E_2 are any two events, mutually exclusive or not, with probabilities $p(E_1)$ and $p(E_2)$, then the probability of E_1 or E_2 is

$$p(E_1 \cup E_2) = p(E_1) + p(E_2) - p(E_1 \cap E_2) \qquad (27.1)$$

In proof of this theorem, we begin by noting that

$$p(E_1 \cup E_2) = \frac{n(E_1 \cup E_2)}{n(S)} \qquad \blacktriangleleft \text{ definition of probability}$$

$$= \frac{n(E_1) + n(E_2) - n(E_1 \cap E_2)}{n(S)} \qquad \blacktriangleleft \text{ by (3)}$$

$$= \frac{n(E_1)}{n(S)} + \frac{n(E_2)}{n(S)} - \frac{n(E_1 \cap E_2)}{n(S)}$$

by dividing each term of the numerator separately by the denominator. Consequently, by use of the definition of probability, we have

$$p(E_1 \cup E_2) = p(E_1) + p(E_2) - p(E_1 \cap E_2)$$

as stated in the theorem.

EXAMPLE 1

If 1 card is drawn from a deck of 52 playing cards, find the probability that it will be red or a king.

Solution

If E_1 represents the set of red cards and E_2 represents the set of kings, then $n(E_1) = n(\text{reds}) = 26$, $n(E_2) = n(\text{kings}) = 4$, and $n(E_1 \cap E_2) = n(\text{red} \cap \text{king}) = 2$. Therefore, $p(E_1) = \frac{26}{52} = \frac{1}{2}$, $p(E_2) = \frac{4}{52} = \frac{1}{13}$, and $p(E_1 \cap E_2) = \frac{2}{52} = \frac{1}{26}$. Consequently, the probability that the card that is drawn will be red or a king is

$$p(E_1 \cup E_2) = p(E_1) + p(E_2) - p(E_1 \cap E_2)$$
$$= \tfrac{1}{2} + \tfrac{1}{13} - \tfrac{1}{26}$$
$$= \tfrac{14}{26} = \tfrac{7}{13}$$

If the events considered in Eq. (27.1) are mutually exclusive, then $n(E_1 \cap E_2) = 0$ and

$$p(E_1 \cup E_2) = \frac{n(E_1 \cup E_2)}{n(S)}$$

$$= \frac{n(E_1) + n(E_2) - n(E_1 \cap E_2)}{n(S)}$$

$$= \frac{n(E_1) + n(E_2)}{n(S)} \qquad \blacktriangleleft \text{ since } n(E_1 \cap E_2) = 0$$

$$= p(E_1) + p(E_2)$$

Consequently, we have shown:

Mutually exclusive, or disjoint

If E_1 and E_2 are mutually exclusive events, then the probability that one of them will occur in a single trial is

$$p(E_1 \cup E_2) = p(E_1) + p(E_2) \qquad (27.2a)$$

EXAMPLE 2

If the probability of team A winning its conference championship in football is $\frac{1}{8}$ and the probability of team B of the same conference winning the championship is $\frac{1}{3}$, what is the probability that team A or team B will be the champion?

Solution

Since the events are mutually exclusive, the probability of A or B becoming the champion is the sum of the separate probabilities, as given by Eq. (27.2a). Consequently,

$$p(A \cup B) = p(A) + p(B)$$
$$= \tfrac{1}{8} + \tfrac{1}{3} = \tfrac{11}{24}$$

The theorem on mutually exclusive events can be extended to any finite number of events by considering $(E_1 \cup E_2)$ as an event and adding a third event E_3, then considering $(E_1 \cup E_2 \cup E_3)$ as an event and adding another. If this is continued, we reach the conclusion that *the probability that some one of a set of mutually exclusive events will occur in a single trial is the sum of the probabilities of the separate events.* Symbolically,

$$p(E_1 \cup E_2 \cup E_3 \cup \cdots \cup E_{n-1} \cup E_n) = p(E_1) + p(E_2) + \cdots + p(E_n) \qquad (27.2)$$

Exercise 27.1 Probability, Expectation, Mutually Exclusive Events

1 If one card is drawn at random from a standard deck, find the probability that it will be (*a*) a heart, (*b*) a black card, (*c*) a red queen.
2 Find the probability that a card drawn from a standard deck will be (*a*) a 2, 3, 4, or 5, (*b*) a number between 3 and 9, inclusive.
3 Find the probability that the card in problem 2 will be (*a*) a spade, (*b*) a club or a diamond, (*c*) a black 7.
4 Find the probability that the card in problem 2 will be (*a*) a 2, 5, 8, or jack, (*b*) a 7 or higher.
5 Each of the two-digit numbers with no digit repeated is written on a card. The cards are shuffled and one is drawn at random. Find the probability that the first digit will be 8.
6 What is the probability that the number on the card drawn in problem 5 will be an odd number?
7 What is the probability that the number on the card drawn in problem 5 will be divisible by 9?

8 What is the probability that the sum of the digits in the number on the card drawn in problem 5 will be divisible by 3?

9 Find the probability of throwing a prime number in one toss of a die.

10 In one toss of a die, find the probability of throwing (*a*) a 6, (*b*) an even number.

11 If one die is tossed, find the probability of throwing (*a*) a 2, (*b*) an odd number, (*c*) a 3 or a 5.

12 If one die is tossed, find the probability of throwing (*a*) a 4, (*b*) a 1, a 2, or a 4, (*c*) a 1, 2, 3, 4, 5, or 6.

13 If a pair of dice are tossed, find the probability of throwing (*a*) a 2, (*b*) a 3, (*c*) a 4.

14 If two dice are tossed, find the probability of throwing (*a*) a 7, (*b*) a 6.

15 Find the probability that each will be a 9 if three cards are drawn from a deck without any being replaced.

16 Find the probability that all will be clubs if five cards are drawn from a deck and none replaced.

17 A box contains seven $10 bills, six $50 bills, and eight $100 bills. If two are drawn at random simultaneously, find the probability that the sum drawn is (*a*) $20, (*b*) $60.

18 A bag contains six $10 bills, seven $50 bills, and five $100 bills. If two are drawn simultaneously at random, find the probability that the sum will be (*a*) $20, (*b*) $150.

19 If three bills are drawn simultaneously at random from the bag in problem 18, find the probability that their sum will be (*a*) $30, (*b*) $200.

20 If three bills are drawn simultaneously at random from a bag that contains four $10 bills, five $50 bills, and six $100 bills, what is the probability that no two of them will be of the same denominations?

21 If John is to receive $1,378 provided he draws a red card higher than a nine from a deck, what is the value of his expectation?

22 If Tom is to receive $125 provided he is alive at the end of nine years, find his expectation provided the probability of his living that long is .56.

23 The probability of Jack killing a deer is $\frac{2}{3}$. Find his expectation if he is to receive $75 provided he kills a deer.

24 Find her expectation if Dorothy is to receive $600 for performing a task and the likelihood of her doing it is $\frac{2}{5}$.

25 If the probability of marrying James is $\frac{1}{2}$ for Jane and $\frac{1}{5}$ for Martha, what is the probability that one of them will marry him?

26 If the probability that Abner will go to California for a certain summer is $\frac{1}{6}$ and the probability that he will go to Maine for the same summer is $\frac{1}{4}$, what is the probability that he will go to one of them?

27 If the probabilities that teams A, B, and C will win a round robin tournament are $\frac{1}{3}$, $\frac{1}{4}$, and $\frac{1}{7}$, what is the probability that one of them will win it? That none of them will win it?

28   A class has a membership of 40. Five of them are math majors and three are physics majors. If one person is chosen at random to speak to the department head, what is the probability that the student will be neither a math or physics major?

29   If the probability of being nominated for state senator is $\frac{1}{3}$ for Long, $\frac{1}{5}$ for Short, and $\frac{1}{10}$ for Tall, find the probability that one of them will be nominated.

30   Find the probability that either a 5 or a 6 will be obtained in one toss of a pair of dice.

31   Find the probability that neither a 7 nor a 12 will be thrown on one toss of a pair of dice.

32   A bag contains four green balls, six black ones, and five white ones. If one ball is drawn from the bag, find the probability that it will be either green or white.

27.4   INDEPENDENT AND DEPENDENT EVENTS

In this section, we shall consider the probability of the occurrence of both of two successive events E_1 and E_2. We shall use $E_1 \cap E_2$ to indicate that both E_1 and E_2 occur and $p(E_2|E_1)$ to denote the probability that E_2 occurs after E_1 has occurred. If the occurrence of E_2 is not affected by the occurrence or nonoccurrence of E_1, then the events are *independent* and $p(E_2|E_1) = p(E_2)$; furthermore, if $p(E_2|E_1) \neq p(E_2)$, then E_2 and E_1 are *dependent* events.

Independent
events
Dependent
events

We shall now prove that if E_1 *and* E_2 *are two events in a sample space, then*

$$p(E_1 \text{ and } E_2) = p(E_1 \cap E_2) = p(E_1)p(E_2|E_1) \qquad (27.3a)$$

In proving this theorem, we begin by noting that by the definition of probability, we have

$$p(E_1 \cap E_2) = \frac{n(E_1 \cap E_2)}{n(S)}$$
$$= \frac{n(E_1 \cap E_2)}{n(E_1)} \frac{n(E_1)}{n(S)}$$
$$= p(E_2|E_1)\, p(E_1)$$

as stated in the theorem.

EXAMPLE 1   If one ball is drawn from a bag that contains three white and two black balls and is not replaced before a second ball is drawn, what is the probability that both will be white?

Solution   The probability of getting a white ball on the first draw is $p(E_1) = \frac{3}{5}$ since the bag contains five balls and three of them are white. After a white ball is drawn and not replaced, the bag contains four balls including two white ones. Hence, the probability of getting a white ball on the second draw is $p(E_2|E_1) = \frac{2}{4} = \frac{1}{2}$. Consequently the desired probability is

$$
\begin{aligned}
p(E_1 \cap E_2) &= p(E_1)p(E_2|E_1) \\
&= \left(\tfrac{3}{5}\right)\left(\tfrac{1}{2}\right) \\
&= \tfrac{3}{10}
\end{aligned}
$$

The theorem of this section can be extended to any finite number of events by adding one at a time. If we consider $E_1 \cap E_2$ as an event and add the event E_3, we get

$$
\begin{aligned}
p(E_1 \cap E_2 \cap E_3) &= \frac{n(E_1 \cap E_2 \cap E_3)}{n(S)} \\
&= \frac{n(E_1)}{n(S)} \frac{n(E_1 \cap E_2)}{n(E_1)} \frac{n(E_1 \cap E_2 \cap E_3)}{n(E_1 \cap E_2)}
\end{aligned}
$$

Hence

$$p(E_1 \cap E_2 \cap E_3) = p(E_1)\, p(E_2|E_1)\, p(E_3|E_2 \cap E_1) \qquad (27.3b)$$

This procedure can be continued until we have

$$
\begin{aligned}
p(E_1 \cap E_2 \cap E_3 \cap \cdots \cap E_n) = \\
p(E_1)p(E_2|E_1)p(E_3|E_2 \cap E_1) \cdots p(E_n|E_{n-1} \cap \cdots \cap E_2 \cap E_1) \qquad (27.3)
\end{aligned}
$$

EXAMPLE 2   If the probability that Gene and Jean will become engaged is $\frac{3}{7}$, if the probability that they will be married if engaged is $\frac{4}{5}$, and if the probability that they will have children if married is $\frac{7}{12}$, find the probability that they will have children.

Solution   If the events are taken in the order stated in the problem, we have $p(E_1) = \frac{3}{7}$, $p(E_2|E_1) = \frac{4}{5}$, and $p(E_3|E_2 \cap E_1) = \frac{7}{12}$. Consequently, by use of (27.3) with $n = 3$ or (27.3b), we have

$$
\begin{aligned}
p(E_1 \cap E_2 \cap E_3) &= \left(\tfrac{3}{7}\right)\left(\tfrac{4}{5}\right)\left(\tfrac{7}{12}\right) \\
&= \tfrac{1}{5}
\end{aligned}
$$

If the events E_1 and E_2 used in the theorem that includes (27.3a) are independent, then $p(E_2|E_1) = p(E_2)$ and the theorem becomes: *if E_1 and E_2 are two independent events, then*

$$p(E_1 \cap E_2) = p(E_1)p(E_2) \qquad (27.4a)$$

EXAMPLE 3   A bag contains three white and two black balls. Find the probability of drawing a white ball and then a black one if the first is replaced before the second is drawn.

Solution   The events are independent since the first is replaced before the second is drawn; furthermore, $p(E_1) = \frac{3}{5}$ and $p(E_2) = \frac{2}{5}$. Therefore, $p(E_1 \cap E_2) = (\frac{3}{5})(\frac{2}{5}) = \frac{6}{25}$.

We can get an extension of the above theorem on independent events from (27.3) by making use of the fact that, for independent events, $p(E_n|E_{n-1} \cap \cdots \cap E_2 \cap E_1)$ becomes $p(E_n)$. Therefore, we know that *if $E_1, E_2, \cdots, E_n$ are independent events, then*

$$p(E_1 \cap E_2 \cap \cdots \cap E_n) = p(E_1)p(E_2) \cdots p(E_n) \qquad (27.4)$$

EXAMPLE 4   Find the probability that three graduates of the class of 1950 will meet for the fiftieth anniversary if the separate probabilities of their attending are $\frac{4}{9}$, $\frac{1}{4}$, and $\frac{3}{17}$.

Solution   The desired probability is the product of the separate probabilities since the events are independent. Therefore,

$$p(E_1 \cap E_2 \cap E_3) = (\tfrac{4}{9})(\tfrac{1}{4})(\tfrac{3}{17}) = \tfrac{1}{51}$$

Exercise 27.2   Independent and Dependent Events

1   The probabilities that A and B will win separate contests are 0.3 and 0.5. Find the probability that (*a*) both will win, (*b*) neither will win.

2   Tom, Dick, and Harry are in separate contests and their probabilities of winning are $\frac{5}{8}$, $\frac{3}{4}$, and $\frac{2}{5}$. Find the probability that all three will win.

3   In problem 2, find the probabilities of each being the only one to win.

4   If the probabilities of Jones and Draboroski making B or better in this course are $\frac{2}{3}$ and $\frac{3}{4}$, find the probability that (*a*) both will make B or A, (*b*) exactly one will make C or worse.

5   A bag contains five white balls and seven black ones. If three balls are drawn successively and none replaced, what is the probability that the first two are white and the third black?

6   The probability that a student-faculty committee will recommend Dean Burnside for president of Tobias University is $\frac{1}{4}$, and the probability that she will be selected if recommended is $\frac{1}{3}$. Find the probability of her being (*a*) selected, (*b*) recommended and not selected.

7   If three balls are withdrawn from a bag that contains four white and eight black balls and none replaced, find the probability that all three will be white.

8   A farmer estimates that if the equivalent in moisture of 1.5 inches of

rain falls in April, the probabilities of his raising 100 bushels of oats, 55 bushels of wheat, and 40 bushels of barley per acre are $\frac{2}{3}$, $\frac{3}{4}$, and $\frac{5}{6}$. If the probability of receiving 1.5 inches of rain in April is 0.3, find the probability that he will harvest (*a*) all three of the specified amounts, (*b*) the specified amount of oats but not of the other two.

9 A coin is drawn from and replaced in a bag that contains four Indian head and six Lincoln pennies, and then another is drawn. Find the probability that both will be of the Lincoln variety.

10 Find the probability of obtaining a head and a ten if a coin and a die are tossed.

11 The probability that a college will win the first game of the season is $\frac{2}{3}$ and the probability that it will win the last game is $\frac{3}{5}$. Find the probability that the team will win exactly one of them.

12 A couple decides to take a world cruise in four years if both are alive. What is the probability of their taking the trip if the probabilities of their being alive are 0.91 and 0.88?

13 The probabilities that Ms. Simoneaux will win the first primary, the second primary, and general election, if in them, are $\frac{3}{8}$, $\frac{1}{6}$, and $\frac{1}{12}$, respectively. Find the probability that she will (*a*) reach the general election, (*b*) win the general election. Assume that failure in any contest prohibits participation in those that follow.

14 Most universities require that a candidate for the Ph.D. degree must pass a qualifying and a final examination in order to receive the degree. If experience shows that $\frac{7}{9}$ of those who take the qualifying exam pass it and $\frac{9}{10}$ of those taking the final pass it, what is the probability of a student passing both examinations?

15 If two balls are drawn from a bag that contains three red, five black, and two white balls and the first is not replaced before the second is drawn, find the probability that the first one drawn is red and the second one white.

16 A detail of 10 men for a raid is to be selected from a group of 80. What is the probability that a specified man will make and survive the raid if the probability of being lost in the raid is 0.1?

17 One bag contains seven black and five white balls and another contains five black and three white ones. If one ball is drawn from each bag, find the probability that both will be (*a*) black, (*b*) white, (*c*) the same color.

18 The probabilities that Jones, Brown, and Abramowitz will hit the bull's eye in one shot are $\frac{3}{5}$, $\frac{1}{3}$, and $\frac{3}{4}$. What is the probability that at least one of them will hit the bull's eye if each has one shot?

19 What is the probability of throwing a 12 on one toss of a pair of dice and following it by a 2 on the next toss?

20 If, from 12 girls and 10 boys, five contestants for a quiz game are chosen by lot, find the probability that all will be (*a*) boys, (*b*) of the same sex.

21 The probability that John will be in a certain section of algebra is $\frac{1}{4}$ and the probability of his making an A if in that section is $\frac{1}{3}$. Find the probability of (*a*) his being in that section and making an A, (*b*) being in that section and not making an A.

22 If three beans are drawn one at a time and not replaced from a bag that contains four white and eight black beans, find the probability that all three will be white.

23 Find the probability that none of the beans drawn in problem 22 will be white.

24 Find the probability that the three beans drawn in problem 22 will be of the same color.

25 Find the probability of getting a 5 and two tails if two coins and a pair of dice are tossed.

26 Find the probability of obtaining at least one tail and a number greater than 4 if three coins and a die are cast.

27 Find the probability that neither a 7 or a 12 will be thrown on the first toss of a pair of dice and that a 2 will be thrown on the second toss.

28 A will provides that a trust fund of \$2,000 will be payable in 10 years to the oldest then living of three brothers. The probability that the oldest will live 10 years is $\frac{3}{4}$, that the next will live 10 years is $\frac{7}{9}$, and that the youngest will live 10 years is $\frac{4}{5}$. Find the value of the expectation of each.

27.5 REPEATED TRIALS OF AN EVENT

If we know the probability of an event occurring in one trial, then the probability of its happening a given number of times in n trials is given by the following theorem:

If p is the probability that an event will occur in one trial, then the probability that it will occur exactly r times in n trials is equal to

$$C(n, r)p^r(1-p)^{n-r}$$

We prove the above theorem as follows: The r trials can be selected from the n trials in $C(n, r)$ ways, by Sec. 18.5. The probability that the event will occur r times and fail the remaining $n-r$ times is $p^r(1-p)^{n-r}$, by Sec. 27.3, since the trials are independent and $1-p$ is the probability of the event failing in any trial. By the theorem of Sec. 27.3, the desired probability is therefore $C(n, r)p^r(1-p)^{n-r}$. The theorem has the following corollary:

If p is the probability that an event will occur in one trial, then the probability that it will occur at least r times in n trials is equal to

$$p^n + C(n, n-1)p^{n-1}(1-p) + C(n, n-2)p^{n-2}(1-p)^2 + \cdots$$
$$+ C(n, r+1)p^{r+1}(1-p)^{n-r-1} + C(n, r)p^r(1-p)^{n-r} \quad (27.5)$$

28

Introduction to Polar Coordinates

Heretofore we have employed the cartesian coordinate system and have located a point in the plane by means of the directed distances from the coordinate axes. Occasionally it is more convenient to employ other coordinate systems for dealing with certain situations, and the one most frequently employed is the polar coordinate system. In that system, a point is located in a plane by means of a distance and an angle. The polar forms of the equations of many curves are less complicated than the cartesian forms, and the polar coordinate system is especially appropriate for dealing with problems involving rotations. We shall discuss the polar coordinate system in this chapter.

28.1 THE POLAR COORDINATE SYSTEM

Polar axis
Pole
Modulus
Radius vector
Amplitude
Vectorial angle

The frame of reference for the polar coordinate system is a ray or half line and a starting point. The ray is called *the polar axis* and its starting point is the *pole*. A point P is located a distance r from the pole and the angle θ is measured from the polar axis to the ray from the pole through the point as shown in Fig. 28.1. The distance r is called *modulus* or *radius vector* and the angle θ is known as the *amplitude* or *vectorial angle*. The amplitude is positive or negative according to whether the direction of rotation from the polar axis to OP is counterclockwise or clockwise. It is positive in Fig. 28.1. Distances measured from the pole along the radius vector are positive, whereas distances measured along the extension of the radius vector through the pole are negative. A point is determined if a value r of the radius vector and a value θ of the vectorial angle are given, and it is designated by (r, θ) as in Fig. 28.1. The perpendicular to the polar axis through the pole is called the *normal axis*.

In rectangular coordinates there is a one-to-one correspondence between points in the plane and ordered number pairs (x, y). In polar coordinates there is exactly one point for each ordered number pair (r, θ), but if (r, θ) is a pair of coordinates for a point, then $(r, \theta + k360°)$ is another pair of coordinates for any integral value of k since θ and $\theta + k360°$ determine the same direction. Since in $(-r, 180 + \theta)$, r is preceded by a negative

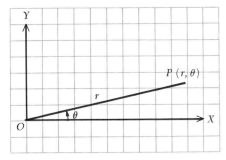

Figure 28.1

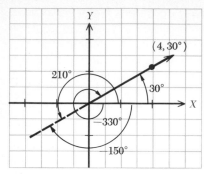

Figure 28.2

sign, it is understood that the distance r is to be measured along the extension through the pole of the terminal side of the vectorial angle; hence, $(-r, \theta + 180°)$ represents the same point as (r, θ). Furthermore, $(-r, \theta + 180° + k360°)$ represents the same point for integral values of k.

EXAMPLE

Locate the point $(4, 30°)$ and give three other pairs of coordinates for it.

Solution

The point $(4, 30°)$ is four units out on a ray that makes an angle of $30°$ with the polar axis and is shown in Fig. 28.2. The angle $-330°$ has the same terminal side as $30°$; hence $(4, -330°)$ is another pair of coordinates for $(4, 30°)$. Furthermore, the terminal side of $-150°$ and of $210°$ is the extension of $30°$ through the pole; consequently, $(-4, -150°)$ and $(-4, 210°)$ are two more pairs of coordinates for $(4, 30°)$.

28.2 INTERCEPTS AND SYMMETRY

The amount of labor required for sketching a curve whose equation is known is often metrically decreased if we make use of intercepts and symmetry. We also find it helpful to extend the polar axis through the pole and to draw a perpendicular to the polar axis through the pole. This perpendicular is called the *normal axis*. A desirable first step in sketching is to determine the intercepts. The intercepts of $r = f(k\theta)$ are the values of r for which θ is an integral multiple of $90°$. We can get the intercepts by solving the equations in r that are obtained by putting θ equal to $0°, 90°, 180°, 270°, -$ in $f(r, \theta) = 0$. The points $(r, n90°)$ are called *intercept points*.

Normal axis
Intercepts

EXAMPLE 1

Find the intercepts of the graph defined by $r = 1 + \sin \theta$.

Solution

In this problem it is not necessary to assign values greater than $360°$ to θ since $\sin \theta$ is periodic and the period is $360°$. Consequently, we obtain the required intercepts by assigning $0°, 180°, 90°$, and $270°$ to θ. Thus

514

$$r = 1 + \sin 0 = 1 + 0 = 1$$
$$r = 1 + \sin 180° = 1 + 0 = 1$$
$$r = 1 + \sin 90° = 1 + 1 = 2$$
$$r = 1 + \sin 270° = 1 + (-1) = 0$$

Consequently, the intercepts on the polar axis and on its extension are 1. The normal intercepts are 2 and 0.

The next step in the procedure for sketching a polar curve is to determine the axes of symmetry. It should be recalled that a curve is symmetrical with respect to a line if the line bisects every chord of the curve that is perpendicular to the line. Furthermore, a curve is symmetrical with respect to a point if the point bisects every chord of the curve that passes through the point. There are many tests that enable us to determine the axes or points of symmetry of a curve. The following three will suffice for our purposes:

1   If the equation $r = f(\theta)$ is unchanged when θ is replaced by $-\theta$, the curve is symmetrical with respect to the polar axis or its extension.

2   If $r = f(\theta)$ is not changed when θ is replaced by $180° - \theta$, the curve is symmetrical with respect to the normal axis.

3   If $r = f(\theta)$ is not changed when θ is replaced by $180° + \theta$, the curve is symmetrical with respect to the pole.

We shall prove the above statements by use of Fig. 28.3.

Proof ▶   1   Since $r = f(\theta) = f(-\theta)$, the points $P_1(r, -\theta)$ and $P_2(r, \theta)$ are the extremities of a chord of the curve. Furthermore, the triangle

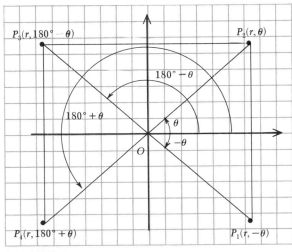

Figure 28.3

P_1OP_2 is isosceles; hence the polar axis bisects the angle at O. Therefore the polar axis is the perpendicular bisector of the chord P_1P_2, and the curve is symmetrical with respect to the polar axis.

2   Here we have $r = f(\theta) = f(180° - \theta)$ with P_2 and P_3 on the curve. Furthermore, the triangle P_2OP_3 is isosceles with the angle at O bisected by the normal axis. Consequently, the normal axis is the perpendicular bisector of the chord P_3P_2, and the curve is symmetrical with respect to the normal axis.

3   Since $r = f(\theta) = f(180° + \theta)$, the points $P_4(r, 180° + \theta)$, O, and $P_2(r, \theta)$ are on a straight line segment with O as the midpoint and P_2 and P_4 on the curve. Hence the pole O bisects the chord P_4P_2, and the curve is symmetrical with respect to the pole.

As previously inferred, these tests do not exhaust the possibilities. For example, a curve is symmetrical with respect to the polar axis if its polar equation is unchanged when θ is replaced by $\pi - \theta$ and r by $-r$ simultaneously even though symmetry with respect to the polar axis is not revealed when test 1 is used.

EXAMPLE 2   Apply the tests for symmetry to the equation $r = 1 + \sin \theta$.

Solution   If we replace θ by $-\theta$, we obtain

$$r = 1 + \sin(-\theta)$$
$$= 1 - \sin \theta \qquad \blacktriangleleft \text{ since } \sin(-\theta) = -\sin \theta$$

Hence, since the equation is changed by this replacement, this test does not reveal symmetry with respect to the polar axis or to its extension.
If θ is replaced by $180° - \theta$, we have

$$r = 1 + \sin(180° - \theta)$$
$$= 1 + \sin \theta \qquad \blacktriangleleft \text{ since } \sin(180° - \theta) = \sin \theta$$

Consequently, since the equation is not changed by this replacement, the graph is symmetrical with respect to the normal axis.
The replacement of θ by $180° + \theta$ yields

$$r = 1 + \sin(180° + \theta)$$
$$= 1 - \sin \theta \qquad \blacktriangleleft \text{ since } \sin(180° + \theta) = -\sin \theta$$

Therefore this test fails to reveal symmetry with respect to the pole.

We shall conclude this discussion by stating the four steps in a procedure for sketching the graph of a polar equation:

1   Determine the intercepts of the graph and plot the points corresponding to each.

2 Apply the three tests for symmetry.

3 If necessary, determine the coordinates of a few additional points on the graph and plot the points.

4 Sketch the curve.

EXAMPLE 3

Sketch the graph of $r = 1 + \sin \theta$.

Solution

In Example 1 we found that the intercepts of the graph of this equation on the polar axis and its extension are each 1 and that the normal intercepts are 0 and 2. We plot these points as indicated in Fig. 28.4.

In Example 2 we found that the graph is symmetrical with respect to the normal axis.

We now assign $-60°$, $-30°$, $30°$, and $60°$ to θ, compute each corresponding value of r, and obtain the following table:

θ	$-60°$	$-30°$	$30°$	$60°$
r	0.1	0.5	1.5	1.9

After plotting these points and drawing a smooth curve through them and through those points determined by the polar and normal intercepts, we obtain the curve at the right of the normal axis in Fig. 28.4.

Since the normal axis is an axis of symmetry, the left half of the curve will be a reflection of the right half with respect to the normal axis and can be sketched as indicated in Fig. 28.4.

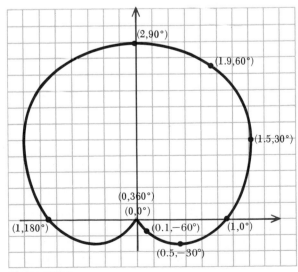

Figure 28.4

Frequently we can obtain an approximate sketch of a curve by considering the variation of r as θ increases through certain intervals. We shall illustrate the procedure in Example 4.

EXAMPLE 4   Sketch the graph of the equation $r = \cos 2\theta$.

Solution   The tests for symmetry reveal the fact that the curve is symmetrical with respect to both axes and the pole. We shall discover the intercepts as we proceed with the following argument.

As θ varies from 0 to 45°, 2θ varies from 0 to 90°, and $r = \cos 2\theta$ decreases from 1 to 0. Hence the portion of the curve yielded by corresponding values of r and θ in the interval $0 \leq \theta \leq 45°$ is the curve numbered 1 in Fig. 28.5. Similarly, the information tabulated below enables us to sketch the portions of the curve indicated.

Vectorial angle	Radius vector	Corresponding portion of curve in Fig. 28.5
$45° \leq \theta \leq \ \ 90°$	r decreases from 0 to -1	2
$90° \leq \theta \leq 135°$	r increases from -1 to 0	3
$135° \leq \theta \leq 180°$	r increases from 0 to 1	4

Similarly, by considering the variation of r as θ passes through the intervals $180° \leq \theta \leq 225°$, $225° \leq \theta \leq 270°$, $270° \leq \theta \leq 315°$, and $315° \leq 0 \leq 360°$, we obtain the portions of the graph numbered 5, 6, 7, and 8 in the figure.

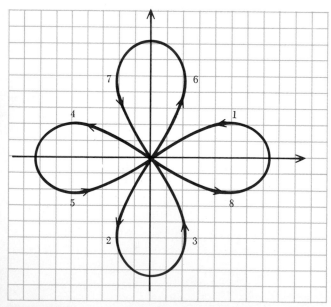

Figure 28.5

In the above discussion we made no use of the knowledge that we have about the symmetry of the curve. If we use this information, we need only consider the intervals $0 \leq \theta \leq 45°$ and $45° \leq \theta \leq 90°$, and we obtain the curves numbered 1 and 2. Then since the graph is symmetrical with respect to the polar axis, we can sketch curves 7 and 8, and finally the curves numbered 3, 4, 5, and 6 by use of symmetry with respect to the normal axis.

Exercise 28.1   Graphs

Plot the points whose coordinates are given in problems 1 to 4.

1 $(3, 60°)$, $(-3, 60°)$, $(3, -60°)$, $(-3, -60°)$
2 $(2, 45°)$, $(-2, 45°)$, $(2, -45°)$, $(-2, -45°)$
3 $(6, 30°)$, $(-6, 30°)$, $(6, -30°)$, $(-6, -30°)$
4 $(4, 210°)$, $(-4, 210°)$, $(4, -210°)$, $(-4, -210°)$

Find three other pairs of polar coordinates with $|\theta| < 360°$ for each point in problems 5 to 8.

5 $(-7, -60°)$ | **6** $(5, -135°)$
7 $(-4, 30°)$ | **8** $(6, 120°)$

Find the intercept points, test for symmetry, and sketch the graph of the curve whose equation is given in each of problems 9 to 48.

9 $r = 3$ | **10** $r = 5$
11 $r + 4 = 0$ | **12** $r = -2$
13 $\theta = 60°$ | **14** $\theta = -45°$
15 $\theta = 150°$ | **16** $\theta = 240°$
17 $r \sin \theta = 2$ | **18** $r = -3 \csc \theta$
19 $r = 4 \sec \theta$ | **20** $r \cos \theta = -1$
21 $r = 3 \cos \theta$ | **22** $r = -4 \cos \theta$
23 $r = -2 \sin \theta$ | **24** $r = 8 \sin \theta$
25 $r = 3 \cos 2\theta$ | **26** $r = \sin 4\theta$
27 $r = 2 \sin 3\theta$ | **28** $r = 5 \cos 5\theta$
29 $r = 2(\cos \theta + 1)$ | **30** $r = 3(\cos \theta - 1)$
31 $r = 4(\sin \theta - 1)$ | **32** $r = 2(\sin \theta + 1)$
33 $r = 2 \sin \theta + 1$ | **34** $r = 3 + 2 \sin \theta$
35 $r = 2 \cos \theta + 3$ | **36** $r = 3 \cos \theta - 2$

37 $r = \dfrac{2}{1 - \cos \theta}$ | **38** $r = \dfrac{3}{1 + \sin \theta}$

39 $r = \dfrac{4}{2 + \cos \theta}$ | **40** $r = \dfrac{6}{3 - \sin \theta}$

41 $r = \dfrac{4}{1 + 2 \sin \theta}$ **42** $r = \dfrac{5}{1 - 3 \cos \theta}$

43 $r = \dfrac{2}{4 - \cos \theta}$ **44** $r = \dfrac{2}{1 - 3 \sin \theta}$

45 $r^2 = 4 \sin \theta$ **46** $r^2 = 9 \cos \theta$

47 $r = 4 \cos^2 \theta$ **48** $r = 9 \sin^2 \theta$

28.3 TRANSFORMATION OF COORDINATES

Frequently it is desirable to transform an equation from one system of coordinates to an equation in the other system. We accomplish this by use of the relations that we shall next develop.

As indicated in Fig. 28.6 we can superimpose the polar plane on the cartesian plane so that the pole in the former coincides with the origin in the latter and the polar axis coincides with the positive half of the X axis. Next we choose a point P in the plane with polar coordinates (r, θ) and cartesian coordinates (x, y). Now we see from the figure that

Rectangular to polar

$$x = r \cos \theta \quad \text{and} \quad y = r \sin \theta \qquad (28.1)$$

We employ Eqs. (28.1) to transform an equation in cartesian coordinates into one in polar coordinates, as illustrated in the following examples.

EXAMPLE 1   Transform the equation $x^2 + y^2 - x + 3y = 3$ to a polar equation.

Solution   We replace x by $r \cos \theta$ and y by $r \sin \theta$ and get

$$r^2 \cos^2 \theta + r^2 \sin^2 \theta - r \cos \theta + 3r \sin \theta = 3$$

We now simplify this equation by the following steps:

$r^2 (\cos^2 \theta + \sin^2 \theta) - r(\cos \theta - 3 \sin \theta) = 3$ ◀ by the distributive axiom

$r^2 - r(\cos \theta - 3 \sin \theta) = 3$ ◀ since $\sin^2 \theta + \cos^2 \theta = 1$

EXAMPLE 2   Prove that a polar form of $y^2 - 3x^2 + 12x - 9 = 0$ is

$$r = \frac{3}{1 + 2 \cos \theta}$$

Solution   We replace x and y by $r \cos \theta$ and $r \sin \theta$, respectively, and get

$r^2 \sin^2 \theta - 3r^2 \cos^2 \theta + 12r \cos \theta - 9 = 0$

$r^2 - r^2 \cos^2 \theta - 3r^2 \cos^2 \theta + 12r \cos \theta - 9 = 0$ ◀ replacing $\sin^2 \theta$ by $1 - \cos^2 \theta$

$r^2 - 4r^2 \cos^2 \theta + 12r \cos \theta - 9 = 0$ ◀ combining terms

$r^2 - (4r^2 \cos^2 \theta - 12 \cos \theta + 9) = 0$ ◀ by the distributive axiom

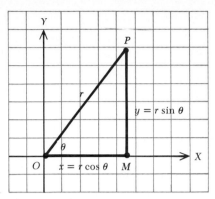

Figure 28.6

We shall now factor the left member of the equation.

$$r^2 - (2r \cos \theta - 3)^2 = 0$$
$$(r + 2r \cos \theta - 3)(r - 2r \cos \theta + 3) = 0 \tag{1}$$

Consequently, any pair of numbers that satisfies either of the following equations also satisfies (1):

$$r + 2r \cos \theta - 3 = 0 \tag{2}$$
$$r - 2r \cos \theta + 3 = 0 \tag{3}$$

Therefore the graph of (1) is the curve composed of the graphs of (2) and (3). We shall next show that the graph of (3) is the same as the graph of (2).

We first note that (r, θ) and $(-r, 180° + \theta)$ are coordinates of the same point since the angles differ by 180° and the radii differ only in sign. Now if $P(r', \theta')$ is on the graph of (2), we have

$$r' + 2r' \cos \theta' - 3 = 0 \tag{4}$$

Furthermore, P is also a point on the graph of (3) since $(-r', 180° + \theta')$ satisfies (3), as we shall show next. If we replace (r, θ) by $(-r', 180° + \theta')$ in the left member of (3), we have

$$-r' + 2r' \cos (180° + \theta') + 3 = -r' - 2r' \cos \theta' + 3 \qquad \blacktriangleleft \text{ since } \cos (180° + \theta')$$
$$= -\cos \theta'$$
$$= -(r' + 2r' \cos \theta' - 3) \qquad \blacktriangleleft \text{ by (4)}$$
$$= -0$$
$$= 0$$

Consequently, any point on the graph of (2) is a point on the graph of (3). By a similar argument we can prove that any point on the graph of (3) is a point on the graph of (2). Therefore the graphs of (2) and (3) coincide. Furthermore, this curve is the graph of (1). Hence in the remainder of this discussion we shall consider only Eq. (2).

521

If $\theta \neq 2\pi/3$, we can solve (2) for r and get

$$r = \frac{3}{1 + 2 \cos \theta} \tag{5}$$

This is the polar form required.

We next note that if $\theta = 2\pi/3$, no value of r exists. If, however, θ approaches $2\pi/3$, then $|1 + 2 \cos \theta|$ becomes less than any preassigned positive number. Hence $|r|$ increases indefinitely, and it follows that the graph of $\theta = 2\pi/3$ is an asymptote of the graph of (5).

In order to transform an equation from polar coordinates to cartesian coordinates, we employ one or more of the following relations. Each is evident from Fig. 28.6.

Polar to
rectangular

$$\sin \theta = \frac{y}{r} \qquad \cos \theta = \frac{x}{r} \qquad r = \sqrt{x^2 + y^2} \tag{28.2}$$

$$\arctan \frac{y}{x} = \theta \qquad \textbf{provided } x \neq 0 \qquad r = \sqrt{x^2 + y^2} \tag{28.3}$$

If the equation involves sines or cosines, we use (28.2), and it is usually advisable first to replace $\sin \theta$ and $\cos \theta$ by y/r and x/r, respectively, to simplify the result obtained, and then to replace r by $\sqrt{x^2 + y^2}$. If the resulting equation involves radicals of the second order, it should be rationalized.

EXAMPLE 3   Transform the equation $r = 4/(2 - 3 \sin \theta)$ to an equation in cartesian coordinates.

Solution   We proceed as follows:

$$r = \frac{4}{2 - 3y/r} \qquad \blacktriangleleft \text{ replacing } \sin \theta \text{ by } y/r$$

$$= \frac{4r}{2r - 3y} \qquad \blacktriangleleft \text{ simplifying the complex fraction}$$

$$2r - 3y = 4 \qquad \blacktriangleleft \text{ multiplying each member by } (2r - 3y)/r$$
$$2\sqrt{x^2 + y^2} = 4 + 3y \qquad \blacktriangleleft \text{ replacing } r \text{ by } \sqrt{x^2 + y^2} \text{ and adding } 3y \text{ to each member}$$

$$4x^2 + 4y^2 = 9y^2 + 24y + 16 \qquad \blacktriangleleft \text{ equating the squares of each member}$$
$$4x^2 - 5y^2 - 24y = 16 \qquad \blacktriangleleft \text{ adding } -9y^2 - 24y \text{ to each member}$$

If the polar equation involves tangents and cotangents, it is advisable to use Eqs. (28.3).

EXAMPLE 4   Convert the equation $r = \tan \theta + \cot \theta$ to an equation in cartesian coordinates.

Solution   Since $\tan \theta \neq -\cot \theta$, then $r \neq 0$, and the graph of

$$r = \tan \theta + \cot \theta$$

does not pass through the pole. Therefore in the transformation of this equation to cartesian coordinates, we must remember that $(x, y) \neq (0, 0)$. Then by (28.3), $\arctan (y/x) = \theta$, and it follows that $\tan \theta = y/x$ and $\cot \theta = x/y$. Now since $r = \sqrt{x^2 + y^2}$, we proceed as follows:

$$\sqrt{x^2 + y^2} = \frac{y}{x} + \frac{x}{y} \qquad \blacktriangleleft \text{ replacing } r \text{ by } \sqrt{x^2 + y^2}, \tan \theta \text{ by } y/x,$$
$$\text{and } \cot \theta \text{ by } x/y$$

$$xy\sqrt{x^2 + y^2} = x^2 + y^2 \qquad \blacktriangleleft \text{ multiplying each member by } xy, \, xy \neq 0$$

$$xy = \sqrt{x^2 + y^2} \qquad \blacktriangleleft \text{ dividing each member by}$$
$$\sqrt{x^2 + y^2}, \, \sqrt{x^2 + y^2} \neq 0$$

$$x^2 y^2 = x^2 + y^2 \qquad \blacktriangleleft \text{ equating the squares of the members}$$

where $x \neq 0$ and $y \neq 0$.

Exercise 28.2   Transformations

Express the equation in each of problems 1 to 20 in terms of polar coordinates.

1 $x = 3$	**2** $y = 5$
3 $x + 2y = 1$	**4** $3x - 4y = 7$
5 $x^2 + y^2 = 9$	**6** $x^2 + y^2 = 4$
7 $x^2 - y^2 = 36$	**8** $y^2 - x^2 = 16$
9 $y^2 = 6x$	**10** $x^2 = 4y$
11 $x^2 + 4y^2 = 8y$	**12** $x^2 - 4y^2 = 4y$
13 $xy = 4$	**14** $y^2 = x^3$
15 $x(x^2 + y^2) = y^2$	**16** $(2a - x)y^2 = x(x - a)^2$
17 $(x^2 + y^2)^2 = 2a^2 xy$	**18** $x^4 - y^4 = 2xy$
19 $x(x^2 + y^2) = a(3x^2 - y^2)$	**20** $(x^2 + y^2)^{3/2} = x^2 - y^2 - 2xy$

Express the equation in each of problems 21 to 40 in terms of rectangular coordinates.

21 $r = 3 \sec \theta$	**22** $r = 5 \csc \theta$
23 $r(\cos \theta + 2 \sin \theta) = 1$	**24** $r = 7/(3 \cos \theta - 4 \sin \theta)$
25 $r^2 = 9$	**26** $r = 2$

27 $r^2 = 36 \sec 2\theta$ **28** $r^2 \cos 2\theta = -16$

29 $r = 6 \cot \theta \csc \theta$ **30** $r = 4 \tan \theta \sec \theta$

31 $r = 8 \sin \theta$ **32** $r = 4/(\cos \theta \cot \theta - 4 \sin \theta)$

33 $r^2 = 8 \csc 2\theta$ **34** $r = \sec \theta \tan^2 \theta$

35 $r = \sin \theta \tan \theta$

36 $r(2a - r \cos \theta) = \cot \theta \csc \theta (r \cos \theta - a)^2$

37 $r^2 = a^2 \sin 2\theta$ **38** $r^2 = \tan 2\theta$

39 $r = a(3 \cos \theta - \sin \theta \tan \theta)$ **40** $r = \cos 2\theta - \sin 2\theta$

Appendix

TABLES

TABLE I COMMON LOGARITHMS

N	0	1	2	3	4	5	6	7	8	9
10	0000	0043	0086	0128	0170	0212	0253	0294	0334	0374
11	0414	0453	0492	0531	0569	0607	0645	0682	0719	0755
12	0792	0828	0864	0899	0934	0969	1004	1038	1072	1106
13	1139	1173	1206	1239	1271	1303	1335	1367	1399	1430
14	1461	1492	1523	1553	1584	1614	1644	1673	1703	1732
15	1761	1790	1818	1847	1875	1903	1931	1959	1987	2014
16	2041	2068	2095	2122	2148	2175	2201	2227	2253	2279
17	2304	2330	2355	2380	2405	2430	2455	2480	2504	2529
18	2553	2577	2601	2625	2648	2672	2695	2718	2742	2765
19	2788	2810	2833	2856	2878	2900	2923	2945	2967	2989
20	3010	3032	3054	3075	3096	3118	3139	3160	3181	3201
21	3222	3243	3263	3284	3304	3324	3345	3365	3385	3404
22	3424	3444	3464	3483	3502	3522	3541	3560	3579	3598
23	3617	3636	3655	3674	3692	3711	3729	3747	3766	3784
24	3802	3820	3838	3856	3874	3892	3909	3927	3945	3962
25	3979	3997	4014	4031	4048	4065	4082	4099	4116	4133
26	4150	4166	4183	4200	4216	4232	4249	4265	4281	4298
27	4314	4330	4346	4362	4378	4393	4409	4425	4440	4456
28	4472	4487	4502	4518	4533	4548	4564	4579	4594	4609
29	4624	4639	4654	4669	4683	4698	4713	4728	4742	4757
30	4771	4786	4800	4814	4829	4843	4857	4871	4886	4900
31	4914	4928	4942	4955	4969	4983	4997	5011	5024	5038
32	5051	5065	5079	5092	5105	5119	5132	5145	5159	5172
33	5185	5198	5211	5224	5237	5250	5263	5276	5289	5302
34	5315	5328	5340	5353	5366	5378	5391	5403	5416	5428
35	5441	5453	5465	5478	5490	5502	5514	5527	5539	5551
36	5563	5575	5587	5599	5611	5623	5635	5647	5658	5670
37	5682	5694	5705	5717	5729	5740	5752	5763	5775	5786
38	5798	5809	5821	5832	5843	5855	5866	5877	5888	5899
39	5911	5922	5933	5944	5955	5966	5977	5988	5999	6010
40	6021	6031	6042	6053	6064	6075	6085	6096	6107	6117
41	6128	6138	6149	6160	6170	6180	6191	6201	6212	6222
42	6232	6243	6253	6263	6274	6284	6294	6304	6314	6325
43	6335	6345	6355	6365	6375	6385	6395	6405	6415	6425
44	6435	6444	6454	6464	6474	6484	6493	6503	6513	6522
45	6532	6542	6551	6561	6571	6580	6590	6599	6609	6618
46	6628	6637	6646	6656	6665	6675	6684	6693	6702	6712
47	6721	6730	6739	6749	6758	6767	6776	6785	6794	6803
48	6812	6821	6830	6839	6848	6857	6866	6875	6884	6893
49	6902	6911	6920	6928	6937	6946	6955	6964	6972	6981
50	6990	6998	7007	7016	7024	7033	7042	7050	7059	7067
51	7076	7084	7093	7101	7110	7118	7126	7135	7143	7152
52	7160	7168	7177	7185	7193	7202	7210	7218	7226	7235
53	7243	7251	7259	7267	7275	7284	7292	7300	7308	7316
54	7324	7332	7340	7348	7356	7364	7372	7380	7388	7396
N	0	1	2	3	4	5	6	7	8	9

TABLE I COMMON LOGARITHMS (*continued*)

N	0	1	2	3	4	5	6	7	8	9
55	7404	7412	7419	7427	7435	7443	7451	7459	7466	7474
56	7482	7490	7497	7505	7513	7520	7528	7536	7543	7551
57	7559	7566	7574	7582	7589	7597	7604	7612	7619	7627
58	7634	7642	7649	7657	7664	7672	7679	7686	7694	7701
59	7709	7716	7723	7731	7738	7745	7752	7760	7767	7774
60	7782	7789	7796	7803	7810	7818	7825	7832	7839	7846
61	7853	7860	7868	7875	7882	7889	7896	7903	7910	7917
62	7924	7931	7938	7945	7952	7959	7966	7973	7980	7987
63	7993	8000	8007	8014	8021	8028	8035	8041	8048	8055
64	8062	8069	8075	8082	8089	8096	8102	8109	8116	8122
65	8129	8136	8142	8149	8156	8162	8169	8176	8182	8189
66	8195	8202	8209	8215	8222	8228	8235	8241	8248	8254
67	8261	8267	8274	8280	8287	8293	8299	8306	8312	8319
68	8325	8331	8338	8344	8351	8357	8363	8370	8376	8382
69	8388	8395	8401	8407	8414	8420	8426	8432	8439	8445
70	8451	8457	8463	8470	8476	8482	8488	8494	8500	8506
71	8513	8519	8525	8531	8537	8543	8549	8555	8561	8567
72	8573	8579	8585	8591	8597	8603	8609	8615	8621	8627
73	8633	8639	8645	8651	8657	8663	8669	8675	8681	8686
74	8692	8698	8704	8710	8716	8722	8727	8733	8739	8745
75	8751	8756	8762	8768	8774	8779	8785	8791	8797	8802
76	8808	8814	8820	8825	8831	8837	8842	8848	8854	8859
77	8865	8871	8876	8882	8887	8893	8899	8904	8910	8915
78	8921	8927	8932	8938	8943	8949	8954	8960	8965	8971
79	8976	8982	8087	8993	8998	9004	9009	9015	9020	9025
80	9031	9036	9042	9047	9053	9058	9063	9069	9074	9079
81	9085	9090	9096	9101	9106	9112	9117	9122	9128	9133
82	9138	9143	9149	9154	9159	9165	9170	9175	9180	9186
83	9191	9196	9201	9206	9212	9217	9222	9227	9232	9238
84	9243	9248	9253	9258	9263	9269	9274	9279	9284	9289
85	9294	9299	9304	9309	9315	9320	9325	9330	9335	9340
86	9345	9350	9355	9360	9365	9370	9375	9380	9385	9390
87	9395	9400	9405	9410	9415	9420	9425	9430	9435	9440
88	9445	9450	9455	9460	9465	9469	9474	9479	9484	9489
89	9494	9499	9504	9509	9513	9518	9523	9528	9533	9538
90	9542	9547	9552	9557	9562	9566	9571	9576	9581	9586
91	9590	9595	9600	9605	9609	9614	9619	9624	9628	9633
92	9638	9643	9647	9652	9657	9661	9666	9671	9675	9680
93	9685	9689	9694	9699	9703	9708	9713	9717	9722	9727
94	9731	9736	9741	9745	9750	9754	9759	9763	9768	9773
95	9777	9782	9786	9791	9795	9800	9805	9809	9814	9818
96	9823	9827	9832	9836	9841	9845	9850	9854	9859	9863
97	9868	9872	9877	9881	9886	9890	9894	9899	9903	9908
98	9912	9917	9921	9926	9930	9934	9939	9943	9948	9952
99	9956	9961	9965	9969	9974	9978	9983	9987	9991	9996
N	0	1	2	3	4	5	6	7	8	9

TABLE II   TRIGONOMETRIC FUNCTIONS

Angles	Sines		Cosines		Tangents		Cotangents		Angles
	Nat.	Log.	Nat.	Log.	Nat.	Log.	Nat.	Log.	
0° 00′	.0000	∞	1.0000	0.0000	.0000	∞	∞	∞	90° 00′
10	.0029	7.4637	1.0000	0000	.0029	7.4637	343.77	2.5363	50
20	.0058	7648	1.0000	0000	.0058	7648	171.89	2352	40
30	.0087	9408	1.0000	0000	.0087	9409	114.59	0591	30
40	.0116	8.0658	.9999	0000	.0116	8.0658	85.940	1.9342	20
50	.0145	1627	.9999	0000	.0145	1627	68.750	8373	10
1° 00′	.0175	8.2419	.9998	9.9999	.0175	8.2419	57.290	1.7581	89° 00′
10	.0204	3088	.9998	9999	.0204	3089	49.104	6911	50
20	.0233	3668	.9997	9999	.0233	3669	42.964	6331	40
30	.0262	4179	.9997	9999	.0262	4181	38.188	5819	30
40	.0291	4637	.9996	9998	.0291	4638	34.368	5362	20
50	.0320	5050	.9995	9998	.0320	5053	31.242	4947	10
2° 00′	.0349	8.5428	.9994	9.9997	.0349	8.5431	28.636	1.4569	88° 00′
10	.0378	5776	.9993	9997	.0378	5779	26.432	4221	50
20	.0407	6097	.9992	9996	.0407	6101	24.542	3899	40
30	.0436	6397	.9990	9996	.0437	6401	22.904	3599	30
40	.0465	6677	.9989	9995	.0466	6682	21.470	3318	20
50	.0494	6940	.9988	9995	.0495	6945	20.206	3055	10
3° 00′	.0523	8.7188	.9986	9.9994	.0524	8.7194	19.081	1.2806	87° 00′
10	.0552	7423	.9985	9993	.0553	7429	18.075	2571	50
20	.0581	7645	.9983	9993	.0582	7652	17.169	2348	40
30	.0610	7857	.9981	9992	.0612	7865	16.350	2135	30
40	.0640	8059	.9980	9991	.0641	8067	15.605	1933	20
50	.0669	8251	.9978	9990	.0670	8261	14.924	1739	10
4° 00′	.0698	8.8436	.9976	9.9989	.0669	8.8446	14.301	1.1554	86° 00′
10	.0727	8613	.9974	9989	.0729	8624	13.727	1376	50
20	.0756	8783	.9971	9988	.0758	8795	13.197	1205	40
30	.0785	8946	.9969	9987	.0787	8960	12.706	1040	30
40	.0814	9104	.9967	9986	.0816	9118	12.251	0882	20
50	.0843	9256	.9964	9985	.0846	9272	11.826	0728	10
5° 00′	.0872	8.9403	.9962	9.9983	.0875	8.9420	11.430	1.0580	85° 00′
10	.0901	9545	.9959	9982	.0904	9563	11.059	0437	50
20	.0929	9682	.9957	9981	.0934	9701	10.712	0299	40
30	.0958	9816	.9954	9980	.0963	9836	10.385	0164	30
40	.0987	9945	.9951	9979	.0992	9966	10.078	0034	20
50	.1016	9.0070	.9948	9977	.1022	9.0093	9.7882	0.9907	10
6° 00′	.1045	9.0192	.9945	9.9976	.1051	9.0216	9.5144	0.9784	84° 00′
10	.1074	0311	.9942	9975	.1080	0336	9.2553	9664	50
20	.1103	0426	.9939	9973	.1110	0453	9.0098	9547	40
30	.1132	0539	.9936	9972	.1139	0567	8.7769	9433	30
40	.1161	0648	.9932	9971	.1169	0678	8.5555	9322	20
50	.1190	0755	.9929	9969	.1198	0786	8.3450	9214	10
7° 00′	.1219	9.0859	.9925	9.9968	.1228	9.0891	8.1443	0.9109	83° 00′
10	.1248	0961	.9922	9966	.1257	0995	7.9530	9005	50
20	.1276	1060	.9918	9964	.1287	1096	7.7704	8904	40
30	.1305	1157	.9914	9963	.1317	1194	7.5958	8806	30
40	.1334	1252	.9911	9961	.1346	1291	7.4287	8709	20
50	.1363	1345	.9907	9959	.1376	1385	7.2687	8615	10
8° 00′	.1392	9.1436	.9903	9.9958	.1405	9.1478	7.1154	0.8522	82° 00′
10	.1421	1525	.9899	9956	.1435	1569	6.9682	8431	50
20	.1449	1612	.9894	9954	.1465	1658	6.8269	8342	40
30	.1478	1697	.9890	9952	.1495	1745	6.6912	8255	30
40	.1507	1781	.9886	9950	.1524	1831	6.5606	8169	20
50	.1536	1863	.9881	9948	.1554	1915	6.4348	8085	10
9° 00′	.1564	9.1943	.9877	9.9946	.1584	9.1997	6.3138	0.8003	81° 00′
	Nat.	Log.	Nat.	Log.	Nat.	Log.	Nat.	Log.	

Angles	Cosines		Sines		Cotangents		Tangents		Angles

TABLE II TRIGONOMETRIC FUNCTIONS (*continued*)

Angles	Sines		Cosines		Tangents		Cotangents		Angles
	Nat.	Log.	Nat.	Log.	Nat.	Log.	Nat.	Log.	
9° 00′	.1564	9.1943	.9877	9.9946	.1584	9.1997	6.3138	0.8003	81° 00′
10	.1593	2022	.9872	9944	.1614	2078	6.1970	7922	50
20	.1622	2100	.9868	9942	.1644	2158	6.0844	7842	40
30	.1650	2176	.9863	9940	.1673	2236	5.9758	7764	30
40	.1679	2251	.9858	9938	.1703	2313	5.8708	7687	20
50	.1708	2324	.9853	9936	.1733	2389	5.7694	7611	10
10° 00′	.1736	9.2397	.9848	9.9934	.1763	9.2463	5.6713	0.7537	80° 00′
10	.1765	2468	.9843	9931	.1793	2536	5.5764	7464	50
20	.1794	2538	.9838	9929	.1823	2609	5.4845	7391	40
30	.1822	2606	.9833	9927	.1853	2680	5.3955	7320	30
40	.1851	2674	.9827	9924	.1883	2750	5.3093	7250	20
50	.1880	2740	.9822	9922	.1914	2819	5.2257	7181	10
11° 00′	.1908	9.2806	.9816	9.9919	.1944	9.2887	5.1446	0.7113	79° 00′
10	.1937	2870	.9811	9917	.1974	2953	5.0658	7047	50
20	.1965	2934	.9805	9914	.2004	3020	4.9894	6980	40
30	.1994	2997	.9799	9912	.2035	3085	4.9152	6915	30
40	.2022	3058	.9793	9909	.2065	3149	4.8430	6851	20
50	.2051	3119	.9787	9907	.2095	3212	4.7729	6788	10
12° 00′	.2079	9.3179	.9781	9.9904	.2126	9.3275	4.7046	0.6725	78° 00′
10	.2108	3238	.9775	9901	.2156	3336	4.6382	6664	50
20	.2136	3296	.9769	9899	.2186	3397	4.5736	6603	40
30	.2164	3353	.9763	9896	.2217	3458	4.5107	6542	30
40	.2193	3410	.9757	9893	.2247	3517	4.4494	6483	20
50	.2221	3466	.9750	9890	.2278	3576	4.3897	6424	10
13° 00′	.2250	9.3521	.9744	9.9887	.2309	9.3634	4.3315	0.6366	77° 00′
10	.2278	3575	.9737	9884	.2339	3691	4.2747	6309	50
20	.2306	3629	.9730	9881	.2370	3748	4.2193	6252	40
30	.2334	3682	.9724	9878	.2401	3804	4.1653	6196	30
40	.2363	3734	.9717	9875	.2432	3859	4.1126	6141	20
50	.2391	3786	.9710	9872	.2462	3914	4.0611	6086	10
14° 00′	.2419	9.3837	.9703	9.9869	.2493	9.3968	4.0108	0.6032	76° 00′
10	.2447	3887	.9696	9866	.2524	4021	3.9617	5979	50
20	.2476	3937	.9689	9863	.2555	4074	3.9136	5926	40
30	.2504	3986	.9681	9859	.2586	4127	3.8667	5873	30
40	.2532	4035	.9674	9856	.2617	4178	3.8208	5822	20
50	.2560	4083	.9667	9853	.2648	4230	3.7760	5770	10
15° 00′	.2588	9.4130	.9659	9.9849	.2679	9.4281	3.7321	0.5719	75° 00′
10	.2616	4177	.9652	9846	.2711	4331	3.6891	5669	50
20	.2644	4223	.9644	9843	.2742	4381	3.6470	5619	40
30	.2672	4269	.9636	9839	.2773	4430	3.6059	5570	30
40	.2700	4314	.9628	9836	.2805	4479	3.5656	5521	20
50	.2728	4359	.9621	9832	.2836	4527	3.5261	5473	10
16° 00′	.2756	9.4403	.9613	9.9828	.2867	9.4575	3.4874	0.5425	74° 00′
10	.2784	4447	.9605	9825	.2899	4622	3.4495	5378	50
20	.2812	4491	.9596	9821	.2931	4669	3.4124	5331	40
30	.2840	4533	.9588	9817	.2962	4716	3.3759	5284	30
40	.2868	4576	.9580	9814	.2994	4762	3.3402	5238	20
50	.2896	4618	.9572	9810	.3026	4808	3.3052	5192	10
17° 00′	.2924	9.4659	.9563	9.9806	.3057	9.4853	3.2709	0.5147	73° 00′
10	.2952	4700	.9555	9802	.3089	4898	3.2371	5102	50
20	.2979	4741	.9546	9798	.3121	4943	3.2041	5057	40
30	.3007	4781	.9537	9794	.3153	4987	3.1716	5013	30
40	.3035	4821	.9528	9790	.3185	5031	3.1397	4969	20
50	.3062	4861	.9520	9786	.3217	5075	3.1084	4925	10
18° 00′	.3090	9.4900	.9511	9.9782	.3249	9.5118	3.0777	0.4882	72° 00′
	Nat.	Log.	Nat.	Log.	Nat.	Log.	Nat.	Log.	
Angles	Cosines		Sines		Cotangents		Tangents		Angles

TABLE II TRIGONOMETRIC FUNCTIONS (*continued*)

Angles	Sines		Cosines		Tangents		Cotangents		Angles
	Nat.	Log.	Nat.	Log.	Nat.	Log.	Nat.	Log.	
18° 00′	.3090	**9.4900**	.9511	**9.9782**	.3249	**9.5118**	3.0777	**0.4882**	72° 00′
10	.3118	4939	.9502	9778	.3281	5161	3.0475	4839	50
20	.3145	4977	.9492	9774	.3314	5203	3.0178	4797	40
30	.3173	5015	.9483	9770	.3346	5245	2.9887	4755	30
40	.3201	5052	.9474	9765	.3378	5287	2.9600	4713	20
50	.3228	5090	.9465	9761	.3411	5329	2.9319	4671	10
19° 00′	.3256	**9.5126**	.9455	**9.9757**	.3443	**9.5370**	2.9042	**0.4630**	71° 00′
10	.3283	5163	.9446	9752	.3476	5411	2.8770	4589	50
20	.3311	5199	.9436	9748	.3508	5451	2.8502	4549	40
30	.3338	5235	.9426	9743	.3541	5491	2.8239	4509	30
40	.3365	5270	.9417	9739	.3574	5531	2.7980	4469	20
50	.3393	5306	.9407	9734	.3607	5571	2.7725	4429	10
20° 00′	.3420	**9.5341**	.9397	**9.9730**	.3640	**9.5611**	2.7475	**0.4389**	70° 00′
10	.3448	5375	.9387	9725	.3673	5650	2.7228	4350	50
20	.3475	5409	.9377	9721	.3706	5689	2.6985	4311	40
30	.3502	5443	.9367	9716	.3739	5727	2.6746	4273	30
40	.3529	5477	.9356	9711	.3772	5766	2.6511	4234	20
50	.3557	5510	.9346	9706	.3805	5804	2.6279	4196	10
21° 00′	.3584	**9.5543**	.9336	**9.9702**	.3839	**9.5842**	2.6051	**0.4158**	69° 00′
10	.3611	5576	.9325	9697	.3872	5879	2.5826	4121	50
20	.3638	5609	.9315	9692	.3906	5917	2.5605	4083	40
30	.3665	5641	.9304	9687	.3939	5954	2.5386	4046	30
40	.3692	5673	.9293	9682	.3973	5991	2.5172	4009	20
50	.3719	5704	.9283	9677	.4006	6028	2.4960	3972	10
22° 00′	.3746	**9.5736**	.9272	**9.9672**	.4040	**9.6064**	2.4751	**0.3936**	68° 00′
10	.3773	5767	.9261	9667	.4074	6100	2.4545	3900	50
20	.3800	5798	.9250	9661	.4108	6136	2.4342	3864	40
30	.3827	5828	.9239	9656	.4142	6172	2.4142	3828	30
40	.3854	5859	.9228	9651	.4176	6208	2.3945	3792	20
50	.3881	5889	.9216	9646	.4210	6243	2.3750	3757	10
23° 00′	.3907	**9.5919**	.9205	**9.9640**	.4245	**9.6279**	2.3559	**0.3721**	67° 00′
10	.3934	5948	.9194	9635	.4279	6314	2.3369	3686	50
20	.3961	5978	.9182	9629	.4314	6348	2.3183	3652	40
30	.3987	6007	.9171	9624	.4348	6383	2.2998	3617	30
40	.4014	6036	.9159	9618	.4383	6417	2.2817	3583	20
50	.4041	6065	.9147	9613	.4417	6452	2.2637	3548	10
24° 00′	.4067	**9.6093**	.9135	**9.9607**	.4452	**9.6486**	2.2460	**0.3514**	66° 00′
10	.4094	6121	.9124	9602	.4487	6520	2.2286	3480	50
20	.4120	6149	.9112	9596	.4522	6553	2.2113	3447	40
30	.4147	6177	.9100	9590	.4557	6587	2.1943	3413	30
40	.4173	6205	.9088	9584	.4592	6620	2.1775	3380	20
50	.4200	6232	.9075	9579	.4628	6654	2.1609	3346	10
25° 00′	.4226	**9.6259**	.9063	**9.9573**	.4663	**9.6687**	2.1445	**0.3313**	65° 00′
10	.4253	6286	.9051	9567	.4699	6720	2.1283	3280	50
20	.4279	6313	.9038	9561	.4734	6752	2.1123	3248	40
30	.4305	6340	.9026	9555	.4770	6785	2.0965	3215	30
40	.4331	6366	.9013	9549	.4806	6817	2.0809	3183	20
50	.4358	6392	.9001	9543	.4841	6850	2.0655	3150	10
26° 00′	.4384	**9.6418**	.8988	**9.9537**	.4877	**9.6882**	2.0503	**0.3118**	64° 00′
10	.4410	6444	.8975	9530	.4913	6914	2.0353	3086	50
20	.4436	6470	.8962	9524	.4950	6946	2.0204	3054	40
30	.4462	6495	.8949	9518	.4986	6977	2.0057	3023	30
40	.4488	6521	.8936	9512	.5022	7009	1.9912	2991	20
50	.4514	6546	.8923	9505	.5059	7040	1.9768	2960	10
27° 00′	.4540	**9.6570**	.8910	**9.9499**	.5095	**9.7072**	1.9626	**0.2928**	63° 00′
	Nat.	Log.	Nat.	Log.	Nat.	Log.	Nat.	Log.	

Angles	Cosines		Sines		Cotangents		Tangents		Angles

TABLE II TRIGONOMETRIC FUNCTIONS (*continued*)

Angles	Sines		Cosines		Tangents		Cotangents		Angles
	Nat.	Log.	Nat.	Log.	Nat.	Log.	Nat.	Log.	
27° 00′	.4540	9.6570	.8910	9.9499	.5095	9.7072	1.9626	0.2928	63° 00′
10	.4566	6595	.8897	9492	.5132	7103	1.9486	2897	50
20	.4592	6620	.8884	9486	.5169	7134	1.9347	2866	40
30	.4617	6644	.8870	9479	.5206	7165	1.9210	2835	30
40	.4643	6668	.8857	9473	.5243	7196	1.9074	2804	20
50	.4669	6692	.8843	9466	.5280	7226	1.8940	2774	10
28° 00′	.4695	9.6716	.8829	9.9459	.5317	9.7257	1.8807	0.2743	62° 00′
10	.4720	6740	.8816	9453	.5354	7287	1.8676	2713	50
20	.4746	6763	.8802	9446	.5392	7317	1.8546	2683	40
30	.4772	6787	.8788	9439	.5430	7348	1.8418	2652	30
40	.4797	6810	.8774	9432	.5467	7378	1.8291	2622	20
50	.4823	6833	.8760	9425	.5505	7408	1.8165	2592	10
29° 00′	.4848	9.6856	.8746	9.9418	.5543	9.7438	1.8040	0.2562	61° 00′
10	.4874	6878	.8732	9411	.5581	7467	1.7917	2533	50
20	.4899	6901	.8718	9404	.5619	7497	1.7796	2503	40
30	.4924	6923	.8704	9397	.5658	7526	1.7675	2474	30
40	.4950	6946	.8689	9390	.5696	7556	1.7556	2444	20
50	.4975	6968	.8675	9383	.5735	7585	1.7437	2415	10
30° 00′	.5000	9.6990	.8660	9.9375	.5774	9.7614	1.7321	0.2386	60° 00′
10	.5025	7012	.8646	9368	.5812	7644	1.7205	2356	50
20	.5050	7033	.8631	9361	.5851	7673	1.7090	2327	40
30	.5075	7055	.8616	9353	.5890	7701	1.6977	2299	30
40	.5100	7076	.8601	9346	.5930	7730	1.6864	2270	20
50	.5125	7097	.8587	9338	.5969	7759	1.6753	2241	10
31° 00′	.5150	9.7118	.8572	9.9331	.6009	9.7788	1.6643	0.2212	59° 00′
10	.5175	7139	.8557	9323	.6048	7816	1.6534	2184	50
20	.5200	7160	.8542	9315	.6088	7845	1.6426	2155	40
30	.5225	7181	.8526	9308	.6128	7873	1.6319	2127	30
40	.5250	7201	.8511	9300	.6168	7902	1.6212	2098	20
50	.5275	7222	.8496	9292	.6208	7930	1.6107	2070	10
32° 00′	.5299	9.7242	.8480	9.9284	.6249	9.7958	1.6003	0.2042	58° 00′
10	.5324	7262	.8465	9276	.6289	7986	1.5900	2014	50
20	.5348	7282	.8450	9268	.6330	8014	1.5798	1986	40
30	.5373	7302	.8434	9260	.6371	8042	1.5697	1958	30
40	.5398	7322	.8418	9252	.6412	8070	1.5597	1930	20
50	.5422	7342	.8403	9244	.6453	8097	1.5497	1903	10
33° 00′	.5446	9.7361	.8387	9.9236	.6494	9.8125	1.5399	0.1875	57° 00′
10	.5471	7380	.8371	9228	.6536	8153	1.5301	1847	50
20	.5495	7400	.8355	9219	.6577	8180	1.5204	1820	40
30	.5519	7419	.8339	9211	.6619	8208	1.5108	1792	30
40	.5544	7438	.8323	9203	.6661	8235	1.5013	1765	20
50	.5568	7457	.8307	9194	.6703	8263	1.4919	1737	10
34° 00′	.5592	9.7476	.8290	9.9186	.6745	9.8290	1.4826	0.1710	56° 00′
10	.5616	7494	.8274	9177	.6787	8317	1.4733	1683	50
20	.5640	7513	.8258	9169	.6830	8344	1.4641	1656	40
30	.5664	7531	.8241	9160	.6873	8371	1.4550	1629	30
40	.5688	7550	.8225	9151	.6916	8398	1.4460	1602	20
50	.5712	7568	.8208	9142	.6959	8425	1.4370	1575	10
35° 00′	.5736	9.7586	.8192	9.9134	.7002	9.8452	1.4281	0.1548	55° 00′
10	.5760	7604	.8175	9125	.7046	8479	1.4193	1521	50
20	.5783	7622	.8158	9116	.7089	8506	1.4106	1494	40
30	.5807	7640	.8141	9107	.7133	8533	1.4019	1467	30
40	.5831	7657	.8124	9098	.7177	8559	1.3934	1441	20
50	.5854	7675	.8107	9089	.7221	8586	1.3848	1414	10
36° 00′	.5878	9.7692	.8090	9.9080	.7265	9.8613	1.3764	0.1387	54° 00′
	Nat.	Log.	Nat.	Log.	Nat.	Log.	Nat.	Log.	
Angles	Cosines		Sines		Cotangents		Tangents		Angles

TABLE II TRIGONOMETRIC FUNCTIONS (*continued*)

Angles	Sines		Cosines		Tangents		Cotangents		Angles
	Nat.	Log.	Nat.	Log.	Nat.	Log.	Nat.	Log.	
36° 00′	.5878	9.7692	.8090	9.9080	.7265	9.8613	1.3764	0.1387	54° 00′
10	.5901	7710	.8073	9070	.7310	8639	1.3680	1361	50
20	.5925	7727	.8056	9061	.7355	8666	1.3597	1334	40
30	.5948	7744	.8039	9052	.7400	8692	1.3514	1308	30
40	.5972	7761	.8021	9042	.7445	8718	1.3432	1282	20
50	.5995	7778	.8004	9033	.7490	8745	1.3351	1255	10
37° 00′	.6018	9.7795	.7986	9.9023	.7536	9.8771	1.3270	0.1229	53° 00′
10	.6041	7811	.7969	9014	.7581	8797	1.3190	1203	50
20	.6065	7828	.7951	9004	.7627	8824	1.3111	1176	40
30	.6088	7844	.7934	8995	.7673	8850	1.3032	1150	30
40	.6111	7861	.7916	8985	.7720	8876	1.2954	1124	20
50	.6134	7877	.7898	8975	.7766	8902	1.2876	1098	10
38° 00′	.6157	9.7893	.7880	9.8965	.7813	9.8928	1.2790	0.1072	52° 00′
10	.6180	7910	.7862	8955	.7860	8954	1.2723	1046	50
20	.6202	7926	.7844	8945	.7907	8980	1.2647	1020	40
30	.6225	7941	.7826	8935	.7954	9006	1.2572	0994	30
40	.6248	7957	.7808	8925	.8002	9032	1.2497	0968	20
50	.6271	7973	.7790	8915	.8050	9058	1.2423	0942	10
39° 00′	.6293	9.7989	.7771	9.8905	.8098	9.9084	1.2349	0.0916	51° 00′
10	.6316	8004	.7753	8895	.8146	9110	1.2276	0890	50
20	.6338	8020	.7735	8884	.8195	9135	1.2203	0865	40
30	.6361	8035	.7716	8874	.8243	9161	1.2131	0839	30
40	.6383	8050	.7698	8864	.8292	9187	1.2059	0813	20
50	.6406	8066	.7679	8853	.8342	9212	1.1988	0788	10
40° 00′	.6428	9.8081	.7660	9.8843	.8391	9.9238	1.1918	0.0762	50° 00′
10	.6450	8096	.7642	8832	.8441	9264	1.1847	0736	50
20	.6472	8111	.7623	8821	.8491	9289	1.1778	0711	40
30	.6494	8125	.7604	8810	.8541	9315	1.1708	0685	30
40	.6517	8140	.7585	8800	.8591	9341	1.1640	0659	20
50	.6539	8155	.7566	8789	.8642	9366	1.1571	0634	10
41° 00′	.6561	9.8169	.7547	9.8778	.8693	9.9392	1.1504	0.0608	49° 00′
10	.6583	8184	.7528	8767	.8744	9417	1.1436	0583	50
20	.6604	8198	.7509	8756	.8796	9443	1.1369	0557	40
30	.6626	8213	.7490	8745	.8847	9468	1.1303	0532	30
40	.6648	8227	.7470	8733	.8899	9494	1.1237	0506	20
50	.6670	8241	.7451	8722	.8952	9519	1.1171	0481	10
42° 00′	.6691	9.8255	.7431	9.8711	.9004	9.9544	1.1106	0.0456	48° 00′
10	.6713	8269	.7412	8699	.9057	9570	1.1041	0430	50
20	.6734	8283	.7392	8688	.9110	9595	1.0977	0405	40
30	.6756	8297	.7373	8676	.9163	9621	1.0913	0379	30
40	.6777	8311	.7353	8665	.9217	9646	1.0850	0354	20
50	.6799	8324	.7333	8653	.9271	9671	1.0786	0329	10
43° 00′	.6820	9.8338	.7314	9.8641	.9325	9.9697	1.0724	0.0303	47° 00′
10	.6841	8351	.7294	8629	.9380	9722	1.0661	0278	50
20	.6862	8365	.7274	8618	.9435	9747	1.0599	0253	40
30	.6884	8378	.7254	8606	.9490	9772	1.0538	0228	30
40	.6905	8391	.7234	8594	.9545	9798	1.0477	0202	20
50	.6926	8405	.7214	8582	.9601	9823	1.0416	0177	10
44° 00′	.6947	9.8418	.7193	9.8569	.9657	9.9848	1.0355	0.0152	46° 00′
10	.6967	8431	.7173	8557	.9713	9874	1.0295	0126	50
20	.6988	8444	.7153	8545	.9770	9899	1.0235	0101	40
30	.7009	8457	.7133	8532	.9827	9924	1.0176	0076	30
40	.7030	8469	.7112	8520	.9884	9949	1.0117	0051	20
50	.7050	8482	.7092	8507	.9942	9975	1.0058	0025	10
45° 00′	.7071	9.8495	.7071	9.8495	1.0000	0.0000	1.0000	0.0000	45° 00′
	Nat.	Log.	Nat.	Log.	Nat.	Log.	Nat.	Log.	
Angles	Cosines		Sines		Cotangents		Tangents		Angles

TABLE III SQUARES, SQUARE ROOTS, CUBES, CUBE ROOTS

No.	Sq.	Sq. Root	Cube	Cube Root	No.	Sq.	Sq. Root	Cube	Cube Root
1	1	1.000	1	1.000	51	2,601	7.141	132,651	3.708
2	4	1.414	8	1.260	52	2,704	7.211	140,608	3.733
3	9	1.732	27	1.442	53	2,809	7.280	148,877	3.756
4	16	2.000	64	1.587	54	2,916	7.348	157,464	3.780
5	25	2.236	125	1.710	55	3,025	7.416	166,375	3.803
6	36	2.449	216	1.817	56	3,136	7.483	175,616	3.826
7	49	2.646	343	1.913	57	3,249	7.550	185,193	3.849
8	64	2.828	512	2.000	58	3,364	7.616	195,112	3.871
9	81	3.000	729	2.080	59	3,481	7.681	205,379	3.893
10	100	3.162	1,000	2.154	60	3,600	7.746	216,000	3.915
11	121	3.317	1,331	2.224	61	3,721	7.810	226,981	3.936
12	144	3.464	1,728	2.289	62	3,844	7.874	238,328	3.958
13	169	3.606	2,197	2.351	63	3,969	7.937	250,047	3.979
14	196	3.742	2,744	2.410	64	4,096	8.000	262,144	4.000
15	225	3.873	3,375	2.466	65	4,225	8.062	274,625	4.021
16	256	4.000	4,096	2.520	66	4,356	8.124	287,496	4.041
17	289	4.123	4,913	2.571	67	4,489	8.185	300,763	4.062
18	324	4.243	5,832	2.621	68	4,624	8.246	314,432	4.082
19	361	4.359	6,859	2.668	69	4,761	8.307	328,509	4.102
20	400	4.472	8,000	2.714	70	4,900	8.367	343,000	4.121
21	441	4.583	9,261	2.759	71	5,041	8.426	357,911	4.141
22	484	4.690	10,648	2.802	72	5,184	8.485	373,248	4.160
23	529	4.796	12,167	2.844	73	5,329	8.544	389,017	4.179
24	576	4.899	13,824	2.884	74	5,476	8.602	405,224	4.198
25	625	5.000	15,625	2.924	75	5,625	8.660	421,875	4.217
26	676	5.099	17,576	2.962	76	5,776	8.718	438,976	4.236
27	729	5.196	19,683	3.000	77	5,929	8.775	456,533	4.254
28	784	5.291	21,952	3.037	78	6,084	8.832	474,552	4.273
29	841	5.385	24,389	3.072	79	6,241	8.888	493,039	4.291
30	900	5.477	27,000	3.107	80	6,400	8.944	512,000	4.309
31	961	5.568	29,791	3.141	81	6,561	9.000	531,441	4.327
32	1,024	5.657	32,768	3.175	82	6,724	9.055	551,368	4.344
33	1,089	5.745	35,937	3.208	83	6,889	9.110	571,787	4.362
34	1,156	5.831	39,304	3.240	84	7,056	9.165	592,704	4.380
35	1,225	5.916	42,875	3.271	85	7,225	9.220	614,125	4.397
36	1,296	6.000	46,656	3.302	86	7,396	9.274	636,056	4.414
37	1,369	6.083	50,653	3.332	87	7,569	9.327	658,503	4.431
38	1,444	6.164	54,872	3.362	88	7,744	9.381	681,472	4.448
39	1,521	6.245	59,319	3.391	89	7,921	9.434	704,969	4.465
40	1,600	6.325	64,000	3.420	90	8,100	9.487	729,000	4.481
41	1,681	6.403	68,921	3.448	91	8,281	9.539	753,571	4.498
42	1,764	6.481	74,088	3.476	92	8,464	9.592	778,688	4.514
43	1,849	6.557	79,507	3.503	93	8,649	9.644	804,357	4.531
44	1,936	6.633	85,184	3.530	94	8,836	9.695	830,584	4.547
45	2,025	6.708	91,125	3.557	95	9,025	9.747	857,375	4.563
46	2,116	6.782	97,336	3.583	96	9,216	9.798	884,736	4.579
47	2,209	6.856	103,823	3.609	97	9,409	9.849	912,673	4.595
48	2,304	6.928	110,592	3.634	98	9,604	9.899	941,192	4.610
49	2,401	7.000	117,649	3.659	99	9,801	9.950	970,299	4.626
50	2,500	7.071	125,000	3.684	100	10,000	10.000	1,000,000	4.642

TABLE IV   AMERICAN EXPERIENCE TABLE OF MORTALITY

Age	Number living	Number dying	Yearly probability of dying	Yearly probability of living	Age	Number living	Number dying	Yearly probability of dying	Yearly probability of living
10	100 000	749	0.007 490	0.992 510	53	66 797	1 091	0.016 333	0.983 667
11	99 251	746	0.007 516	0.992 484	54	65 706	1 143	0.017 396	0.982 604
12	98 505	743	0.007 543	0.992 457	55	64 563	1 199	0.018 571	0.981 429
13	97 762	740	0.007 569	0.992 431	56	63 364	1 260	0.019 885	0.980 115
14	97 022	737	0.007 596	0.992 404	57	62 104	1 325	0.021 335	0.978 665
15	96 285	735	0.007 634	0.992 366	58	60 779	1 394	0.022 936	0.977 064
16	95 550	732	0.007 661	0.992 339	59	59 385	1 468	0.024 720	0.975 280
17	94 818	729	0.007 688	0.922 312	60	57 917	1 546	0.026 693	0.973 307
18	94 089	727	0.007 727	0.992 273	61	56 371	1 628	0.028 880	0.971 120
19	93 362	725	0.007 765	0.992 235	62	54 743	1 713	0.031 292	0.968 708
20	92 637	723	0.007 805	0.992 195	63	53 030	1 800	0.033 943	0.966 057
21	91 914	722	0.007 855	0.992 145	64	51 230	1 889	0.036 873	0.963 127
22	91 192	721	0.007 906	0.992 094	65	49 341	1 980	0.040 129	0.959 871
23	90 471	720	0.007 958	0.992 042	66	47 361	2 070	0.043 707	0.956 293
24	89 751	719	0.008 011	0.991 989	67	45 291	2 158	0.047 647	0.952 353
25	89 032	718	0.008 065	0.991 935	68	43 133	2 243	0.052 002	0.947 998
26	88 314	718	0.008 130	0.991 870	69	40 890	2 321	0.056 762	0.943 238
27	87 596	718	0.008 197	0.991 803	70	38 569	2 391	0.061 993	0.938 007
28	86 878	718	0.008 264	0.991 736	71	36 178	2 448	0.067 665	0.932 335
29	86 160	719	0.008 345	0.991 655	72	33 730	2 487	0.073 733	0.926 267
30	85 441	720	0.008 427	0.991 573	73	31 243	2 505	0.080 178	0.919 822
31	84 721	721	0.008 510	0.991 490	74	28 738	2 501	0.087 028	0.912 972
32	84 000	723	0.008 607	0.991 393	75	26 237	2 476	0.094 371	0.905 629
33	83 277	726	0.008 718	0.991 282	76	23 761	2 431	0.102 311	0.897 689
34	82 551	729	0.008 831	0.991 169	77	21 330	2 369	0.111 064	0.888 936
35	81 822	732	0.008 946	0.991 054	78	18 961	2 291	0.120 827	0.879 173
36	81 090	737	0.009 089	0.990 911	79	16 670	2 196	0.131 734	0.868 266
37	80 353	742	0.009 234	0.990 766	80	14 474	2 091	0.144 466	0.855 534
38	79 611	749	0.009 408	0.990 592	81	12 383	1 964	0.158 605	0.841 395
39	78 862	756	0.009 586	0.990 414	82	10 419	1 816	0.174 297	0.825 703
40	78 106	765	0.009 794	0.990 206	83	8 603	1 648	0.191 561	0.808 439
41	77 341	774	0.010 008	0.989 992	84	6 955	1 470	0.211 359	0.788 641
42	76 567	785	0.010 252	0.989 748	85	5 485	1 292	0.235 552	0.764 448
43	75 782	797	0.010 517	0.989 483	86	4 193	1 114	0.265 681	0.734 319
44	74 985	812	0.010 829	0.989 171	87	3 079	933	0.303 020	0.696 980
45	74 173	828	0.011 163	0.988 837	88	2 146	744	0.346 692	0.653 308
46	73 345	848	0.011 562	0.988 438	89	1 402	555	0.395 863	0.604 137
47	72 497	870	0.012 000	0.988 000	90	847	385	0.454 545	0.545 455
48	71 627	896	0.012 509	0.987 491	91	462	246	0.532 468	0.467 532
49	70 731	927	0.013 106	0.986 894	92	216	137	0.634 259	0.365 741
50	69 804	962	0.013 781	0.986 219	93	79	58	0.734 177	0.265 823
51	68 842	1 001	0.014 541	0.985 459	94	21	18	0.857 143	0.142 857
52	67 841	1 044	0.015 389	0.984 611	95	3	3	1,000 000	0.000 000

ANSWERS TO SELECTED PROBLEMS

Exercise 1.1, page 6

1 $\{7, 14, 21\}$ **2** {New York, New Mexico, New Hampshire, New Jersey}
3 $\{o, a\}$ **5** $\{x|x$ is less than 13 and divisible by 3} **6** $\{x|x$ is an odd
integer less than 10} **7** $\{x|x$ is a non y vowel} **9** $A \cup B =$
$\{2, 3, 5, 7, 8, 12\}, A \cap B = \{2, 5\}, A - B = \{3, 8, 12\}$ **10** $A \cup B = A, A \cap B$
$= B, A - B = A$ **11** $A \cup B = \{2, 3, 4, 5, 6, 8, 10\}, A \cap B = \{4, 6\}, A - B =$
$\{2, 8, 10\}$ **13** $A \cup B = \{a, b, c, d, e, i, o, u\}, A \cap B = \{a, e\}, A - B =$
$\{b, c, d\}$ **14** $A \cup B = \{15, 5, 3, 1, 10, 2\}, A \cap B = \{5, 1\}, A - B = \{15, 3\}$
15 $A \cup B = \{x|x$ lettered in basketball or track at ABC College}, $A \cap B =$
$\{x|x$ lettered in basketball and track at ABC College}, $A - B = \{x|x$ lettered in
basketball but not in track at ABC College} **17** $A \cap B \cap C \cap D = \varnothing,$
$A \cup B \cup C = A, (A \cup C) - B = \{1, 3, 5, 7, 9\}$ **18** $(A - B) \cup (C - D) =$
$\{1, 3, 5, 7, 9, 10\}, (A \cup B) - (C \cup D) = \{2, 4, 6, 8\}, A - (B \cup C) - D = \varnothing$
21 $A \times B = \{(a, c), (a, d), (a, e), (b, c), (b, d), (b, e)\}$ **22** $A \times B =$
$\{(a, a), (a, b), (a, c), (b, a), (b, b), (b, \iota), (c, a), (c, b), (c, c)\}$
23 $\{(2, 5), (3, 5), (5, 5)\}$

Exercise 1.2, page 15

1 Positive integers **2** Integers **3** Integers **5** Rational numbers
6 Rational numbers **7** Rational numbers **9** Rational numbers
10 Rational numbers **11** Rational numbers **13** Integers
14 Rational numbers **15** Real numbers **17** $15 > 11, 7 < 13, 42 > 8\frac{1}{2}$
18 $3 > -5, -8 < -2, -1 > -3$ **19** $\frac{2}{5} > \frac{1}{3}, \frac{3}{7} < \frac{5}{9}, \frac{8}{17} > -\frac{9}{2}$ **21** 5
22 2 **23** -5 **25** 8 **26** 9 **27** 4

Exercise 1.3, page 16

1 {April, June, September, November} **2** $\{3, 6, 9, 12, 15\}$ **3** $\{x|x$ is a
one-digit odd number} **5** $\{3, 5, 8, 12, 17\}, \{3, 8, 17\}, \{5, 12\}$
6 $\{a, b, c, d, e, f, i, o, u, y\}, \{a, e\}, \{b, c, d, f\}$ **7** $\{6\}, \{3, 6\}, \{1, 2, 4, 5\}$
9 $(3, 2), (3, 5), (7, 2), (7, 5), (11, 2), (11, 5)$ **10** $(3, 8), (3, 9), (6, 8),$
$(6, 9), (9, 8), (9, 9)$ **11** Integer, negative integer, rational, rational
13 $3 > 2, \frac{2}{9} < \frac{1}{4}, 1.42 > \sqrt{2}$ **14** $-2, 2, 3$

Exercise 2.2, page 34

1 6 **2** 5 **3** -7 **5** 5 **6** 7 **7** zero **9** $3x - z$ **10** $8b$
11 $2p - r$ **13** $6ab + 2cd$ **14** $-qr$ **15** $8a^2b - ab^2$ **17** $6a - 7b$
18 $2x - z$ **19** $2d - q$ **21** $-5a^2b - 5ab^2 + 2ab$ **22** $-xy$
23 $6p^3d + pd^2 - pd$ **25** 15 **26** 62 **27** -51 **29** $a + 5b + c$
30 $-5a - b + 5c$ **31** $2x + y - z$ **33** 5 **34** 11 **35** 8

Exercise 2.3, page 41

9 $6x^5$ **10** $30x^7$ **11** $20x^9$ **13** $6x^3y^4$ **14** $10x^3y^4$ **15** $-12x^5y^7$
17 $4x^4y^6$ **18** $-27x^3y^9$ **19** $64x^9y^{12}$ **21** $6x^3 - 4x^2y$ **22** $6x^4y^2 - 9x^3y^3$
23 $6x^4 - 15x^3y$ **25** $6a^3b^3c^3 - 4ab^3c^5 - 10a^3b^2c^4$ **26** $15a^3b^2c^3 - 9a^3b^3c^3 -$
$6a^5b^3c^3$ **27** $6a^2b^4c^5 - 9a^3b^4c^7 + 12a^2b^5c^6$ **29** $6x^2 - 19xy + 15y^2$
30 $8x^2 + 14xy - 15y^2$ **31** $6a^2 + 7ab - 3b^2$ **33** $6a^3 - 13a^2b + 14ab^2 - 12b^3$
34 $6a^3 - 5a^2b - 12ab^2 - 4b^3$ **35** $10a^3 + 3a^2b - 16ab^2 + 3b^3$
37 $6x^4 - 5x^3y - 7x^2y^2 + 8xy^3 - 2y^4$ **38** $6x^4 + x^3y - 6x^2y^2 + 5xy^3 - 6y^4$
39 $15x^4 + 4x^3y + 4x^2y^2 + 7xy^3 - 2y^4$

Exercise 2.4, page 49

1 a^3 **2** b^4 **3** c^2 **5** ab^2 **6** a^3b^4 **7** $4a^4b^2$ **9** $3ac^2 - 2a^3b +$
$4b^2c$ **10** $5a^6b^2c^3 + 3a^3b^5 - 4ab^3c$ **11** $6a^2b^2c^2 - 4a^3b^3c^3 + 3a^4b^2$ **13** $x + 3y$
14 $2x - y$ **15** $2x + 5y$ **17** $2x + y$ **18** $x^2 + 2xy - y^2$
19 $x^2 - 2xy + y^2$ **21** $a + 2, 3$ **22** $a^2 + a + 2, 2$ **23** $2a^2 - a + 1, -4$
25 $13x - 5y$ **26** $-14x + 17y$ **27** $2a^3 - 5a^2b + 3ab^2 + 3b^3$ **29** $-x + 6$
30 $-9a + 20$ **31** $-3a^2 + ab + 10ac + 3a$ **33** $4a - 4ab - 2bc - 2c$
34 $6a^3 + 15a^2$ **35** $-22a + 90$

Exercise 2.5, page 53

1 $x^2 + 5xy + 6y^2$ **2** $x^2 + 2xy - 8y^2$ **3** $10x^2 - 3xy - y^2$ **5** $6b^2 + 19bc +$
$10c^2$ **6** $6b^2 + 23bc + 20c^2$ **7** $15b^2 + 38bc + 24c^2$ **9** $6x^2 - 13xy - 6y^2$
10 $12x^2 - 17xy + 6y^2$ **11** $18x^2 - 39xy + 20y^2$ **13** $12x^2 + 7xy - 12y^2$
14 $12x^2 - 32xy - 35y^2$ **15** $25x^2 - 5xy - 12y^2$ **17** $4x^2 + 4xy + y^2$
18 $9x^2 + 6xy + y^2$ **19** $x^2 - 6xy + 9y^2$ **21** $9a^2 + 30ab + 25b^2$
22 $4a^2 + 12ab + 9b^2$ **23** $4a^2 + 20ab + 25b^2$ **25** $16a^2 - 24ab + 9b^2$
26 $25a^2 - 20ab + 4b^2$ **27** $9a^2 - 30ab + 25b^2$ **29** $25x^2 - y^2$ **30** $9x^2 - y^2$
31 $x^2 - 16y^2$ **33** $4x^4 - 9y^2$ **34** $25x^2 - 4y^6$ **35** $16x^6 - 25y^4$
37 1,599 **38** 2,491 **39** 375 **41** 4,891 **42** 1,591 **43** 8,064
45 $a^2 + b^2 + 4 + 2ab + 4a + 4b$ **46** $a^2 + 4b^2 + 1 + 4ab + 2a + 4b$
47 $4a^2 + b^2 + 1 - 4ab + 4a - 2b$ **49** $9x^2 + y^2 + 4 + 6xy + 12x + 4y$
50 $4x^2 + 9y^2 + 1 - 12xy + 4x - 6y$ **51** $x^2 + 9y^2 + 16 - 6xy - 8x + 24y$
53 $x^2 - 6xy + 9y^2 - z^2$ **54** $4x^2 - y^2 + 2yz - z^2$ **55** $9x^2 - 4y^2 + 4yz - z^2$
57 $16x^2 - 24xy + 9y^2 - 4z^2$ **58** $16x^2 - 9y^2 - 12yz - 4z^2$ **59** $4x^2 + 12xz +$
$9z^2 - 25y^2$

Exercise 2.6, page 58

1 $2(x^2 - 3x + 1)$ **2** $3(2x - 3y + 1)$ **3** $3x^2(y - xy - 2)$
5 $3xy(2xy - 3xy^2 - 4)$ **6** $5a^2b(2ab^2 - 3a + 4)$ **7** $2a^2b^2(3a - 4 - 5b)$
9 $(a + 2)(a - 1 + b)$ **10** $(x + 2)(x - y)$ **11** $(x - y)(x - w)$
13 $(x + y)(x + z)$ **14** $(x - y)(y - z)$ **15** $(2x - 3y)(x - w)$
17 $(3x + y)(3x - y)$ **18** $(2x + 3y)(2x - 3y)$ **19** $(3x - 4y)(3x + 4y)$
21 $(2x - y^2)(2x + y^2)$ **22** $(5x^2 - y)(5x^2 + y)$ **23** $(3x^2 - 2y^3)(3x^2 + 2y^3)$
25 $(2a + 5b)^2$ **26** $(3a + 2b)^2$ **27** $(5x - 6y)^2$ **29** $(4x - 3y)^2$
30 $(9x - 4y)^2$ **31** $(5x + 7y)^2$ **33** $(2a - b - 4c^2)(2a - b + 4c^2)$
34 $(3a^2 + b + 5c^3)(3a^2 + b - 5c^3)$ **35** $(3a^2 + 2b^3 + 5c)(3a^2 - 2b^3 - 5c)$
37 $(2a + b + 1)^2$ **38** $(3a - 2b + 1)^2$ **39** $(2 - 4a - 5b)^2$
41 $(2a + b)(a + 2b)$ **42** $(3a + 2b)(2a + b)$ **43** $(5a + 2b)(2a + 3b)$
45 $(9x - 4y)(7x - 2y)$ **46** $(6x - 5y)(2x + 3y)$ **47** $(5x + 4y)(4x - 3y)$
49 $(4x - 7y)(7x - 4y)$ **50** $(9x - 4y)(7x - 5y)$ **51** $(8x + 3y)(3x - 4y)$
53 $(a + 2b)(a^2 - 2ab + 4b^2)$ **54** $(3a - b)(9a^2 + 3ab + b^2)$
55 $(2a - 3b)(4a^2 + 6ab + 9b^2)$ **57** $(a^3 - b^2)(a^6 + a^3b^2 + b^4)$
58 $(a^2 + b^3)(a^4 - a^2b^3 + b^6)$ **59** $(a^2 - 3b^2)(a^4 + 3a^2b^2 + 9b^4)$
61 $(a + 2)(a - 2)(a^2 + 4)$ **62** $(a^2 + 9)(a + 3)(a - 3)$
63 $(x - 3)(x^2 + 3x + 9)(x + 3)(x^2 - 3x + 9)$ **65** $(a^2 + 2a + 1)(a^2 - 2a + 1)$
66 $(x^2 + 3y^2 + 2xy)(x^2 + 3y^2 - 2xy)$ **67** $(x^2 + 2y^2 - 5xy)(x^2 + 2y^2 + 5xy)$

Exercise 2.8, page 64

7   8   **9** $x - 4y$ **10** $x - y + z$ **11** $4x - 5y + 2z$ **14** $-12x^5y^6z^5$
15 $8x^6y^9z^3$ **17** $6x^2 - 7xy - 20y^2$ **18** $6x^4 + 5x^3y + 4x^2y^2 - 11xy^3 - 4y^4$
19 $5x^3yz^4$ **21** $a^2 - a + 2, 10$ **22** $8a^2 - 5ab + 2a$ **23** $6a^2 + 7ab + 2b^2$
25 $16a^2 - 9b^2$ **26**   1,591   **27** $4a^2 + 12ab - 4a + 9b^2 - 6b + 1$
29 $(3x + 1)(x - 2)$ **30** $3ab^2(2a - b)(a + 3b)$ **31** $(5x - 4y)(5x + 4y)$
33 $(2a - b + 3)^2$ **34** $(a + b)(a^2 - ab + b^2)(a - b)(a^2 + ab + b^2)$

Exercise 3.1, page 71

1 $\dfrac{-8}{2 - x}$ **2** $\dfrac{-7}{x - 4}$ **3** $\dfrac{2b - 3a}{b^2 - a^2}$ **5** $\dfrac{9xy}{15y^2}$ **6** $\dfrac{10xy}{16x^2}$ **7** $\dfrac{6a^2b^4}{21a^4b^3}$

9 $\dfrac{2x^2 - xy - 3y^2}{x^2 - y^2}$ **10** $\dfrac{-6x^2 - 7xy + 5y^2}{-4x^2 + y^2}$ **11** $-\dfrac{(x - y)^2}{x^3 - y^3}$ **13** $\dfrac{2x}{3y}$

14 $\dfrac{y^2}{2x}$ **15** $\dfrac{3ab^3}{4c}$ **17** $\dfrac{a - 2}{a + 2}$ **18** $\dfrac{a + 1}{2a - 3}$ **19** $\dfrac{2a - 3}{3a - 2}$ **21** $\dfrac{3a + 5}{5a - 3}$

22 $\dfrac{2a - 1}{a}$ **23** $\dfrac{3a + 2}{2a - 3}$ **25** $\left\{\dfrac{2y^2}{xy^2}, \dfrac{3y}{xy^2}, \dfrac{5x}{xy^2}\right\}$ **26** $\left\{\dfrac{4xy^2}{x^3y^3}, \dfrac{3x^2}{x^3y^3}, \dfrac{2y}{x^3y^3}\right\}$

27 $\left\{\dfrac{-2x^2y}{x^5y^5}, \dfrac{3x^4}{x^5y^5}, \dfrac{2y^4}{x^5y^5}\right\}$ **29** $\left\{\dfrac{x^2 + 3xy + 2y^2}{x^2 - y^2}, \dfrac{2x^2 - 3xy + y^2}{x^2 - y^2}, \dfrac{x^2 - 3y^2}{x^2 - y^2}\right\}$

30 $\left\{\dfrac{x^2 + 5xy + 6y^2}{x^2 - xy - 6y^2}, \dfrac{x^2 - 5xy + 6y^2}{x^2 - xy - 6y^2}, \dfrac{x^2 - 6y^2}{x^2 - xy - 6y^2}\right\}$

31 $\left\{\dfrac{2x^2 - 5xy + 2y^2}{(x+y)(x-3y)(2x-y)}, \dfrac{x^2 - y^2}{(x+y)(x-3y)(2x-y)}, \dfrac{x^2 - 9y^2}{(x+y)(x-3y)(2x-y)}\right\}$

Exercise 3.2, page 73

1 $(19x + 1)/18$ **2** $(113x - 21)/30$ **3** $(-4x + 19)/7$

5 $(4x^2 - 27y^2 + 6z^2)/18xyz$ **6** $(-9x^2 - 12y^2 + 10z^2)/6xyz$

7 $(-7x^2 + 15y^2 + 4z^2)/10xyz$ **9** $(15x^2 + 9xy - 2y^2)/2y(3x + y)$

10 $(3x^2 - 2xy - 3y^2)/x(2x + y)$ **11** $(-2x^2 + 16xy + 9y^2)/2x(2x + 3y)$

13 $(3x^2 + 5xy + y^2)/(2x + y)(x + y)$ **14** $(5x^2 - 9xy + 3y^2)/(2x - y)(3x - y)$

15 $-7y^2/(x + 2y)(x + 3y)$ **17** $xy/(3x + 5y)(x + y)$ **18** $9x^2/(2x + 3y)(3x - y)$

19 $-3y^2/(2x + y)(3x - y)$ **21** 0 **22** 0 **23** $y^2/(3x - 2y)(2x + 3y)$

25 $(3x^2 - 6xy + 6y^2)/(x + 2y)(2x - y)(x - y)$

26 $(28x^2 - 49xy + 5y^2)/(3x + y)(2x - 3y)(x - y)$

27 $(-66xy + 12y^2)/(3x + 2y)(3x - y)(x - 2y)$ **29** $y^2/(x^2 - y^2)$

30 $6xy/(4x^2 - y^2)$ **31** $-3x^2/(9x^2 - 4y^2)$

33 $(x^2 + xy + 4y^2)/(2x - y)(2x + 3y)(x - y)$

34 $2(4x^2 + xy + y^2)/(x - 3y)(x + y)(3x + y)$

35 $2(x^2 + xy - 4y^2)/(x - y)(x + y)(2x + 5y)$

37 $(15x^3 + 20x^2y + 40xy^2 - 7y^3)/(x + 3y)(2x + y)(3x - y)$

38 $(-5x^3 - 18x^2y - 18xy^2 + 13y^3)/(2x + 5y)(x + 3y)(x - 2y)$

39 $(21x^3 + 31x^2y - 26xy^2 + 11y^3)/(2x + y)(13x + 4y)(3x + y)$

Exercise 3.3, page 76

1 $\dfrac{x^2}{yz}$ **2** $\dfrac{x^2z}{yw^2}$ **3** $\dfrac{2xy^2}{z^2w^2}$ **5** $\dfrac{3x^3z}{2y^2w^2}$ **6** $\dfrac{8x^4y^6}{21w^6z^4}$ **7** $\dfrac{6x^3w}{yz^3}$ **9** $\dfrac{5y}{6x}$

10 $\dfrac{7y}{x^2z^3}$ **11** $\dfrac{3xy^3w}{z^2}$ **13** $\dfrac{x^2}{3}$ **14** $\dfrac{xz}{6}$ **15** $\dfrac{2(x - 2y)}{y(2x - y)}$ **17** $\dfrac{x(y + 2z)}{y(y + z)}$

18 $\dfrac{x + y}{x}$ **19** $\dfrac{x^2}{y(2x - 3y)}$ **21** $\dfrac{y^2}{x}$ **22** $\dfrac{x(2z - y)}{2y - z}$ **23** $\dfrac{y(x - y)}{x(y - z)}$

25 $\dfrac{x^5}{y^4(z - 2y)}$ **26** $\dfrac{3}{5y^2}$ **27** $\dfrac{3y^2(x + 2y)}{25x(x - 2y)}$ **29** $\dfrac{w - 3}{3xw^2}$ **30** $\dfrac{y^2(x - 2y)}{2x(2x + y)}$

31 $\dfrac{3x^2}{y^2(x + y)}$ **33** $\dfrac{x + 3}{x - 4}$ **34** $\dfrac{(x - 1)(x - 5)}{(x - 2)(x + 2)}$ **35** $\dfrac{x - 2}{x + 2}$ **37** $\dfrac{x - 1}{x + 6}$

38 $\dfrac{x(x + 1)}{(2x + 1)(x + 3)}$ **39** 1

Exercise 3.4, page 79

1 3 **2** $\frac{18}{11}$ **3** 5 **5** $\frac{7}{6}$ **6** $\frac{15}{16}$ **7** -2 **9** $a/(a + 1)$

10 $(a + 1)/a$ **11** $(x + 2)/x$ **13** $(x^2 - 9)/5$ **14** $x/3$ **15** $(a - 8)/5$

17 $\dfrac{x - y}{x + 2y}$ **18** $\dfrac{x - y}{x + y}$ **19** $\dfrac{x + 2y}{x - y}$ **21** $\dfrac{3x - 2}{x - 1}$ **22** $\dfrac{x + 3}{2x + 1}$

23 $\dfrac{2x+3}{x-2}$ **25** $\dfrac{1}{a-1}$ **26** $\dfrac{2}{a-2}$ **27** $\dfrac{1}{(x+2)}$ **29** $\dfrac{-1}{x+1}$

30 $\dfrac{x+3}{-2(x+1)}$ **31** $\dfrac{-1}{x+1}$ **33** $\dfrac{-1}{2a+3}$ **34** $\dfrac{x-4}{x+6}$ **35** $\dfrac{1}{1-a}$

37 $\dfrac{x}{2y}$ **38** $\dfrac{x-3y}{-3xy}$ **39** $\dfrac{2x+y}{2x^2}$

Exercise 3.5, page 80

1 $\dfrac{-7}{x-3}$ **2** $\dfrac{20y^4z}{15xy^2z}$ **3** $\dfrac{3x^2+xy-4y^2}{x^2-y^2}$ **5** $4x^2y^2z$ **6** $\dfrac{a-2}{a+3}$ **7** $\dfrac{2a+1}{3a+4}$

9 $\left\{\dfrac{3xy}{x^3y^4}, \dfrac{-2x^2}{x^3y^4}, \dfrac{5y^2}{x^3y^4}\right\}$ **10** $\left\{\dfrac{4x^2-9y^2}{2x^2+3xy-2y^2}, \dfrac{x^2+xy-2y^2}{2x^2+3xy-2y^2}, \dfrac{3x+2y}{2x^2+3xy-2y^2}\right\}$

11 $\dfrac{19x+3}{12}$ **13** $\dfrac{9x^2+16xy+15y^2}{5y(3x+2y)}$ **14** $\dfrac{10(x^2-1)}{(x-3)(3x+1)}$ **15** $\dfrac{y}{x^2-9y^2}$

17 $\dfrac{2(x\bullet-5y)(x+5y)}{(2x+3y)(x-2y)(3x-y)}$ **18** $\dfrac{6x^5}{y^2zw^6}$ **19** $\dfrac{2xy^2w^4}{z^6}$ **21** $\dfrac{x}{z}$ **22** 1

23 $\dfrac{3}{y}$ **25** $\dfrac{5}{2}$ **26** 4 **27** $x-2$ **29** $\dfrac{-2}{3(x^2-1)}$ **30** $\dfrac{-2}{x+2}$ **31** $\dfrac{y}{z}$

Exercise 4.1, page 90

1 $\{3\}$ **2** $\{2\}$ **3** $\{0\}$ **5** $\{4\}$ **6** $\{-5\}$ **7** $\{-2\}$ **9** $\{12\}$
10 $\{10\}$ **11** $\{15\}$ **13** $\{18\}$ **14** $\{36\}$ **15** $\{\frac{3}{2}\}$ **17** $\{48\}$
18 $\{\varnothing\}$ **19** $\{24\}$ **21** $\{5\}$ **22** $\{4\}$ **23** $\{7\}$ **25** $\{2\}$
26 $\{5\}$ **27** $\{5\}$ **29** $\left\{\dfrac{a+b}{a-b}\right\}$ **30** $\left\{\dfrac{ab}{a-3b}\right\}$ **31** $\{a+2b\}$
33 $\{(a+b)/b\}$ **34** $\{1\}$ **35** $\{(a+b)/b\}$

Exercise 4.2, page 93

1 $\{5\}$ **2** $\{2\}$ **3** $\{2\}$ **5** $\{-4\}$ **6** $\{-2\}$ **7** $\{-6\}$ **9** $\{7\}$
10 $\{12\}$ **11** $\{8\}$ **13** $\{5\}$ **14** $\{3\}$ **15** $\{2\}$ **17** $\{\varnothing\}$
18 $\{\varnothing\}$ **19** $\{\varnothing\}$ **21** $\{(md-m+c)/c\}$ **22** $\{S(1-r)/(1-r^n)\}$
23 $\{pq/(p+q)\}$

Exercise 4.3, page 98

1 $\{x|x>2\}$ **2** $\{x|x>2\}$ **3** $\{x|x<4\}$ **5** $\{x|x>2\}$ **6** $\{x|x>2\}$
7 $\{x|x<-3\}$ **9** $\{x|x<-17/7\}$ **10** $\{x|x<6\}$ **11** $\{x|x>-6\}$

13 $\{x|x > 1\}$ **14** $\{x|x > 1\}$ **15** $\{x|x < 9\}$ **17** $\{x|x > -7\}$
18 $\{x|x > -4\}$ **19** $\{x|x < -2\}$ **21** $\{x|-1 < x < 5\}$
22 $\{x|-\frac{14}{3} < x < 2\}$ **23** $\{x|-2 < x < 4.5\}$ **25** $\{x|-4 > x\} \cup \{x|x > -\frac{4}{3}\}$
26 $\{x|x > -\frac{8}{5}\} \cup \{x|x < -2\}$ **27** $\{x|x > -1\} \cup \{x|x < -2\}$
29 $\{x|-2 < x < 6\}$ **30** {all x different from 9}
31 $\{x|x > -8\} \cup \{x|x < -16\}$

Exercise 4.4, page 101

1 $\{x|x < -2\} \cup \{x|x > 3\}$ **2** $\{x|x > 1\} \cup \{x|x < -4\}$
3 $\{x|x > 5\} \cup \{x|x < 3\}$ **5** $\{x|-4 < x < 2\}$ **6** $\{x|1 < x < 4\}$
7 $\{x|-4 < x < -2\}$ **9** $\{x|x > 2\} \cup \{x|x < -5\}$
10 $\{x|x > 4\} \cup \{x|x < -3\}$ **11** $\{x|3 < x < 4.25\}$
13 $\{x|x > -\frac{2}{3}\} \cup \{x|x < -9\}$ **14** $\{x|-0.75 < x < 7\}$
15 $\{x|x > 1.7\} \cup \{x|x < \frac{5}{3}\}$ **17** $\{x|x > 2\} \cup \{x|-\frac{3}{2} < x < \frac{5}{3}\}$
18 $\{x|x > 3.5\} \cup \{x|-3 < x < \frac{5}{3}\}$ **19** $\{x|x < -4.5\} \cup \{x|-\frac{7}{3} < x < \frac{3}{2}\}$
21 $\{x|x < -2.5\} \cup \{x|\frac{1}{3} < x < 4\}$ **22** $\{x|x < -3.5\} \cup \{x|\frac{7}{3} < x < 3\}$
23 $\{x|x > \frac{1}{2}\} \cup \{x|-3 < x < -\frac{3}{2}\}$

Exercise 4.5, page 102

1 $\{2\}$ **2** $\{3\}$ **3** $\{12\}$ **5** $\{6\}$ **6** $\{12\}$ **7** $\{3\}$ **9** $\{5\}$
10 $\{2\}$ **13** $IR/(l - IR)$ **14** $\{x|x > 3\}$ **15** $\{x|x > -2\}$
17 $\{x|x < 10\}$ **18** $\{x|-3 < x < 1\}$ **19** $\{x|x > -2\} \cup \{x|x < -3\}$
21 $\{x|x < -2\} \cup \{x|x > 1\}$ **22** $\{x|-1 < x < 3\}$ **23** $\{x|-4 < x < -2\}$
25 $\{x|-1 < x < 4\}$ **26** $\{x|x < -4\} \cup \{x|x > 6\}$
27 $\{x|x > 1.5\} \cup \{x|-3.5 < x < -2\}$

Exercise 5.1, page 107

1 128   **2** 243   **3** 243   **5** 16   **6** 243   **7** 81   **9** 729
10 729   **11** 4,096   **13** $\frac{9}{16}$ **14** $\frac{8}{125}$ **15** 104,976   **17** a^6b^9/c^3
18 $a^{12}b^8$ **19** $a^4b^8c^2/4d^{10}$ **21** $6x^5y^5$ **22** $-6x^5y^3z^2$ **23** $10xy^3z^4$
25 $2xy$ **26** $2x^4/3y^3$ **27** $12x^2y^5z^3/7$ **29** $108a^{10}b^5c^{11}$ **30** $2{,}500a^{14}b^8c^8$
31 $72a^{13}b^{11}c^{12}$ **33** $2a^4b^7$ **34** c^4/b^{10} **35** $a^4 3b^8c^5$ **37** $24b/cd^2$
38 $16a^6b^4/c^2$ **39** $8c^{12}/3a^7b^2$ **41** $x^{2a-1}y^4$ **42** x^{a-5}/y^7 **43** xy^{2b-8}
45 a^{2c} **46** a^{bc} **47** $a^{2(p-n)}b^{3(p-n)}$

Exercise 5.2, page 112

1 $\frac{1}{25}$ **2** $\frac{1}{81}$ **3** $\frac{1}{32}$ **5** $\frac{1}{4}$ **6** $\frac{1}{27}$ **7** 6   **9** $\frac{1}{3{,}125}$ **10** 27

11 $\frac{1}{16}$ **13** $\frac{3}{2}$ **14** $\frac{25}{9}$ **15** $\frac{27}{64}$ **17** $2a^2b$ **18** $3a^{-1}b^2$

19 $5a^3b^{-2}$ **21** pqr^{-2} **22** c^3d^{-3} **23** $6p^4q^{-2}r^{-2}$ **25** $\dfrac{3x^2z^4}{4y^5}$

26 $\dfrac{16yz^3}{27x^5}$ **27** $\dfrac{4xz}{5y}$ **29** $\dfrac{a^6}{b^4}$ **30** $\dfrac{b^6}{a^2}$ **31** $\dfrac{1}{a^8b^4}$ **33** a^4b^{14} **34** $\dfrac{b^9}{8a^9}$

35 $\dfrac{b^{14}}{a^{10}c^{18}}$ **37** $3a^2$ **38** $\dfrac{5}{a}$ **39** $\dfrac{-1}{x^3}$ **41** $\dfrac{y^2+yx+x^2}{xy(y+x)}$ **42** $\dfrac{xy^2-1}{1-x^3}$

43 $\dfrac{1}{y+x}$ **45** $\dfrac{-x-3}{(x-1)^3}$ **46** $\dfrac{-2x^2+3x-1}{(x-2)^4}$ **47** $\dfrac{16x+2}{(2x+1)^2(4x-1)^2}$

Exercise 5.3, page 116

1 2 **2** 3 **3** 2 **5** 0.2 **6** 0.1 **7** 0.5 **9** 4 **10** 8
11 27 **13** $\frac{1}{16}$ **14** $\frac{1}{4}$ **15** $\frac{1}{8}$ **17** 16 **18** 512 **19** 16
21 $5ab^2$ **22** $4ab^3$ **23** $3a^2b^4$ **25** $\sqrt[4]{xy^3}$ **26** $\sqrt[3]{xy^2}$ **27** $\sqrt[5]{x^2/y^3}$
29 $\sqrt[4]{x^3y}$ **30** $\sqrt[3]{1/x^2y}$ **31** $\sqrt[6]{1/xy^5}$ **33** $10x^{5/6}$ **34** $6x^{11/12}$
35 $15x^{8/15}$ **37** $\dfrac{3y^{1/5}}{x^{1/4}}$ **38** $3x^{1/9}y$ **39** $\dfrac{4x^{1/3}}{y^{3/2}}$ **41** $\dfrac{8p^{3/5}q^{2/3}r^{1/5}}{9}$

42 $\dfrac{9p^2q^{1/6}s^{7/6}}{5}$ **43** $\dfrac{3}{2p^3q^{1/8}}$ **45** x^4 **46** x^{4ab} **47** x

49 $\dfrac{14x+5}{(2x-1)^{1/3}(x+1)^{1/2}}$ **50** $\dfrac{13x+10}{(3x+2)^{1/3}(x+1)^{1/2}}$ **51** $\dfrac{9x+8}{(2x+3)^{3/4}(3x-1)^{1/2}}$

Exercise 5.4, page 120

1 $16\sqrt{2}$ **2** $9\sqrt{2}$ **3** $5\sqrt{6}$ **5** $2\sqrt[3]{6}$ **6** $5\sqrt[3]{3}$ **7** $2\sqrt[4]{14}$
9 $2ab^2\sqrt{3ab}$ **10** $3a^3b\sqrt{7ab}$ **11** $6ab^3\sqrt[3]{3a^2}$ **13** $3ab^2\sqrt[4]{3a^3b}$
14 $5ab\sqrt[4]{4a^2b^3}$ **15** $2ab^2\sqrt[3]{6ab^2}$ **17** $\sqrt{63x^3y}$ **18** $\sqrt[3]{56x^2y^4}$
19 $\sqrt[4]{48a^9b^7}$ **21** 30 **22** $4\sqrt[3]{36}$ **23** $\sqrt[3]{20/3}$ **25** $21x^2y^4$
26 $20x^3y^2$ **27** $10x^2y\sqrt[3]{25}$ **29** $3x^2y\sqrt{3/7}$ **30** $1.5x^3y^2$ **31** $9r^2y\sqrt{ry/7}$
33 $\sqrt{2xy}/x^2y$ **34** $y^4\sqrt{2x}/2x$ **35** $\sqrt{y/3y^3}$ **37** $x\sqrt{14xy}/2y$
38 $\sqrt{15xy/5y^3}$ **39** $2x^2\sqrt{xy}/7y^2$ **41** $\sqrt[3]{18x^2y^2}/3y$ **42** $\sqrt{50xy^2}/5y^3$
43 $\sqrt[4]{24x^3y^2}/2y$ **45** $\sqrt[6]{2a}$ **46** $\sqrt[12]{3b}$ **47** $\sqrt[4]{11c}$ **49** $2y\sqrt{xy}$
50 $x\sqrt{2y}$ **51** $2\sqrt[3]{x^2y}$ **53** $y\sqrt{2xy}$ **54** $\sqrt[3]{3xy^2}$ **55** $\sqrt[4]{4x^3y^2}$
57 $\sqrt[6]{a^3}, \sqrt[6]{a^2}, \sqrt[6]{a}$ **58** $\sqrt[18]{a^6}, \sqrt[18]{a^3}, \sqrt[18]{a^2}$ **59** $\sqrt[12]{16a^4b^4}, \sqrt[12]{27a^3b^6}, \sqrt[12]{25a^4b^2}$

Exercise 5.5, page 121

1 243 **2** 25 **3** 512 **5** 5,184 **6** a^4c^2/d^6 **7** $6x^5y^8$ **9** $4x^3y^6$
10 $72a^{13}b^8c^3$ **11** $16a^2b^{14}$ **13** y^6/x^6 **14** a^{4c} **15** $\frac{1}{27}$ **17** 8
18 $\frac{4}{81}$ **19** $3a^2b^2$ **21** x^2z/y^3 **22** a^2/b^4 **23** $a^6/16b^4$
25 $(b^2-ab+a^2)/ab(b-a)$ **26** $-3/(2x-1)^2$ **27** 3 **29** 36 **30** $\frac{1}{8}$
31 25 **33** $\sqrt[4]{ab^3}$ **34** $\sqrt[5]{b^2/a^3}$ **35** $6x^{7/12}$ **37** x^2 **38** $3\sqrt[3]{2}$

39 $2ab\sqrt[4]{5b}$ **41** $\sqrt[3]{24a}$ **42** $10\sqrt[3]{9}$ **43** $2xy\sqrt[4]{343y^3}$ **45** $2x\sqrt[3]{xy^2}/y$
46 $\sqrt[8]{3x}$ **47** $xy\sqrt{3y}$

Exercise 6.1, page 131

1 $(3, 4), (3, 6), (3, 13), (5, 4), (5, 6), (5, 13), (7, 4), (7, 6), (7, 13)$
2 $(c, g), (c, o), (a, g), (a, o), (t, g), (t, o)$ **3** $(-1, -3), (-1, -2),$
$(-1, -1), (0, -3), (0, -2), (0, -1), (1, -3), (1, -2), (1, -1)$
5 $(1, 8), (3, 6), (5, 4), (7, 2)$ **6** $(m, i), (a, h), (t, w)$
7 $(1, 7), (2, 8), (3, 9), (4, 10), (5, 11), (6, 12)$ **9** It is a function; there
is exactly one value of $\sqrt{x}$ for each x **10** It is a function; there is exactly
one value of x^2 for each value of x **11** A relation; there are two or more
values of y for some values of x **13** $(-1, -3), (0, -1), (1, 1), (2, 3), (3, 5)$
is a function   **14** $(3, 3), (6, 4), (9, 5), (12, 6)$ is a function   **15** $(-5, 0),$
$(0, 3), (5, 6), (10, 9), (15, 12)$ is a function   **17** $(-2, 8), (-1, 2), (0, -2),$
$(1, -4)$ is a function   **18** $(-4, 39), (-3, 22), (-2, 9), (-1, 0), (0, -5)$ is a
function   **19** $(-2, 2), (-2, -2), (-1, \sqrt{13}), (-1, -\sqrt{13}), (0, 4), (0, -4),$
$(1, \sqrt{13}), (1, -\sqrt{13}), (2, 2), (2, -2)$ is not a function   **21** $-7, -3, 5$
22 $15, 1, -7$ **23** $2, 22$ **25** $4, 9$ **26** $4x + 2, 3$ **27** $2x + h + 2$
29 $(2t + 1)/(4t - 1)$ **30** $6t$ **31** $(t - 1)/(t + 1), 0$ **33** $\{10, 13, 16, 19\}$
34 $\{-1, 1, 5, 13\}$ **35** $\{2\sqrt{3}, 2, \sqrt{3}, 2\}$ **37** $(1, \frac{3}{2}), (2, 2), (3, \frac{11}{4}),$
$(4, \frac{18}{5})$ **38** $(2, \frac{1}{2}), (3, \frac{1}{6}), (4, \frac{1}{2}), (5, \frac{1}{20})$ **39** $(2, 4), (3, 10), (4, 16),$
$(5, 22)$ **41** $\{(2, 3)\}$ **42** $\{(-3, -3)\}$ **43** $\{(\frac{2}{3}, 4)\}, \{(1, 5)\}$

Exercise 6.2, page 137

2 **(a)** The ray which bisects the first quadrant; **(b)** the Y axis; **(c)** the X axis;
(d) the bisector of the second and fourth quadrants   **3** **(a)** The line parallel
to the X axis and 6 units above it; **(b)** the line parallel to the Y axis and 5 units to
the left of it; **(c)** the line parallel to the Y axis and 9 units to the right of it;
(d) the line parallel to the X axis and 4 units below it   **5** -3 **6** 2 **7** $\frac{1}{2}$
9 3.5 **10** $\frac{4}{3}$ **11** -5 **13** zero   **14** zero   **15** $0,1$
17 none   **18** none   **19** $-1.6, 2.6$ **25** 3 **26** 2.3 **27** -0.8

Exercise 6.3, page 143

1 The graph is made up of the nine points $(1, 2), (1, 4), (1, 5), (2, 2), (2, 4),$
$(2, 5), (4, 2), (4, 4),$ and $(4, 5)$.   **2** The graph is made up of the three
points $(0, 0), (1, 1),$ and $(2, 2)$.   **3** The graph is made up of the nine points
$(-2, 0), (-2, 1), (-2, 2), (-1, 0), (-1, 1), (-1, 2), (0, 0), (0, 1),$ and $(0, 2)$.
17 $f^{-1} = \{(x, y) | y = \pm \sqrt{9 - x^2}\}$ is a relation with $-3 \le x \le 3$ as domain
18 $f^{-1} = \{(x, y) | y = \pm\frac{1}{2} \sqrt{8 - x^2}\}$ is a relation with $-2\sqrt{2} \le x \le 2\sqrt{2}$ as domain
19 $f^{-1} = \{(x, y) | y = \frac{1}{3}(x^2 + 2)\}$ is a function with all nonnegative numbers as
domain   **21** $f^{-1} = \{(x, y) | y = x - 1, x \ge 2\}$ **22** $f^{-1} = \{(x, y) | y = \frac{1}{2}(x + 1),$

$x \geq -5\}$ **23** $f^{-1} = \{(x, y)|y = \frac{1}{3}(x + 2), x \geq -8\}$ **25** $f^{-1} = \{(x, y)|y = \dfrac{4}{x - 1},$

$x > 1\}$ **26** $f^{-1} = \{(x, y)|y = \dfrac{3}{1 - x}, 0 < x < 1\}$ **27** $f^{-1} = \{(x, y)|y = \dfrac{x}{x - 1},$

$1 < x\}$

Exercise 6.4, page 143

1 $\{(a, t), (a, n), (e, t), (e, n), (i, t), (i, n)\}$ **2** $\{(s, t), (a, a), (m, r), (e, b)\}$
3 yes **5** $\{(-2, -4), (-1, -1), (0, 2), (1, 5), (2, 8)\}$, yes
6 no, more than one y for each x **7** 3 **9** $\{3, 5, 9, 11\}$
10 $\{(0, -2), (2, 0)\}$ **11** **(a)** parallel to the X axis and 3 units above it;
(b) the ray that bisects the second quadrant **15** $(-3.5, 0)$
21 $y = (3 + x)/(2x - 1), x \leq 4$ **22** $y = \pm\sqrt{16 - x^2}$ is a relation

Exercise 7.1, page 149

13 $(0, 8)$ **14** $(-7, 9)$ **15** $(1, 3)$ **17** $(-5, 3)$ **18** $(0, 9)$
19 $(0, 6)$ **21** 5 **22** 5 **23** 13 **25** $5, \sqrt{10}, 3\sqrt{5}$ **26** $5, \sqrt{106},$
$\sqrt{37}$ **27** $13, \sqrt{82}, 5$

Exercise 7.2, page 152

1 $\pi/5$ **2** $\pi/6$ **3** $\pi/20$ **5** $7\pi/4$ **6** $3\pi/20$ **7** $\pi/3$ **9** $49\pi/720$
10 $31\pi/180$ **11** $\pi/16$ **13** $\pi/54$ **14** $5\pi/48$ **15** $9\pi/50$
17 $751\pi/64,800$ **18** $269\pi/3,000$ **19** $219\pi/36,000$ **21** $31\pi/20$
22 $47\pi/20$ **23** 3.4π **25** $45°$ **26** $15°$ **27** $12°$ **28** $5°$
29 $72°$ **30** $135°$ **31** $105°$ **33** $28°7'30''$ **34** $26°40'$ **35** $35°$
37 $148°58'9''$ **38** $211°59'41''$ **39** $74°29'4''$

Exercise 7.3, page 154

1 1.49 centimeters **2** 8.92 centimeters **3** 32.0 centimeters
5 6.039 feet **6** 38.10 feet **7** 198.5 feet **9** 0.744 **10** 0.360
11 1.89 **13** 0.5715 **14** 1.687 **15** 0.3567 **17** $31\pi/30$ **18** $3\pi/5$
19 8π feet per hour **21** 2.91 inches **22** 15.3 seconds **23** 35.2

Exercise 7.4, page 159

1 5 **2** 13 **3** 17 **5** ±12 **6** ±8 **7** ±24 **9** ±15
10 ±7 **11** ±3 **13** 15 **14** -24 **15** $-\sqrt{33}$ **17** 24
18 7.5 **19** 72 **21** -36 **22** $\frac{8}{3}$ **23** -12; The ratios are given in

the same order as in Sec. 7.8. **25** $-\frac{4}{5}, \frac{3}{5}, -\frac{4}{3}, -\frac{3}{4}, \frac{5}{3}, -\frac{5}{4}$ **26** $\frac{15}{17}, -\frac{8}{17}$, $-\frac{15}{8}, -\frac{8}{15}, -\frac{17}{8}, \frac{17}{15}$ **27** $\frac{4}{5}, \frac{3}{5}, \frac{4}{3}, \frac{3}{4}, \frac{5}{3}, \frac{5}{4}$ **29** $-\frac{4}{5}, \frac{3}{5}, -\frac{4}{3}, -\frac{3}{4}, \frac{5}{3}, -\frac{5}{4}$, **30** $-\frac{15}{17}, \frac{8}{17}, -\frac{15}{8}, -\frac{8}{15}, \frac{17}{8}, -\frac{17}{15}$ **31** $\frac{24}{25}, \frac{7}{25}, \frac{24}{7}, \frac{7}{24}, \frac{25}{7}, \frac{25}{24}$

Exercise 7.5, page 167

The function values in problems 1 to 16 are given in the usual order, and the given one is included.

1 $\frac{4}{5}, \frac{3}{5}, \frac{4}{3}, \frac{3}{4}, \frac{5}{3}, \frac{5}{4}$ **2** $-\frac{12}{13}, \frac{5}{13}, -\frac{12}{5}, -\frac{5}{12}, \frac{13}{5}, -\frac{13}{12}$ **3** $-\frac{8}{17}, -\frac{15}{17}, \frac{8}{15}, \frac{15}{8}$, $-\frac{17}{15}, -\frac{17}{8}$ **5** $-\frac{12}{13}, \frac{5}{13}, -\frac{12}{5}, -\frac{5}{12}, \frac{13}{5}, -\frac{13}{12}$ **6** $\frac{5}{13}, -\frac{12}{13}, -\frac{5}{12}, -\frac{12}{5}, -\frac{13}{12}, \frac{13}{5}$

7 $-\frac{4}{5}, -\frac{3}{5}, \frac{4}{3}, \frac{3}{4}, -\frac{5}{3}, -\frac{5}{4}$ **9** $\pm\frac{24}{25}, \pm\frac{7}{25}, \frac{24}{7}, \frac{7}{24}, \pm\frac{25}{7}, \pm\frac{25}{24}$ **10** $\pm\frac{8}{17}, \frac{15}{17}$, $\pm\frac{8}{15}, \pm\frac{15}{8}, \frac{17}{15}, \pm\frac{17}{8}$ **11** $-\frac{12}{13}, \pm\frac{5}{13}, \mp\frac{12}{5}, \mp\frac{5}{12}, \pm\frac{13}{5}, -\frac{13}{12}$ **13** $\sqrt{3}/2, \frac{1}{2}, \sqrt{3}$, $1/\sqrt{3}, 2, 2/\sqrt{3}$; $1/\sqrt{2}, -1/\sqrt{2}, -1, -1, -\sqrt{2}, \sqrt{2}$; $-\frac{1}{2}, -\sqrt{3}/2, 1/\sqrt{3}, \sqrt{3}, -2/\sqrt{3}$, -2 **14** $-1/\sqrt{2}, 1/\sqrt{2}, -1, -1, \sqrt{2}, -\sqrt{2}$; $-\frac{1}{2}, \sqrt{3}/2, -1/\sqrt{3}, -\sqrt{3}, 2/\sqrt{3}$, -2; $-\sqrt{3}/2, -\frac{1}{2}, \sqrt{3}, 1/\sqrt{3}, -2, -2/\sqrt{3}$ **15** $0, -1, 0,$ no value, $-1,$ no value; $1/\sqrt{2}, -1/\sqrt{2}, -1, -1, -\sqrt{2}, \sqrt{2}$; $\sqrt{3}/2, -\frac{1}{2}, -\sqrt{3}, -1/\sqrt{3}, -2, 2/\sqrt{3}$
29 False **30** False **31** True **33** True **34** False **35** False
37 True **38** False **39** True **41** True **42** True **43** False

Exercise 7.6, page 168

1 $(0, -5)$ **2** $(6, 2)$ **3** 5 **5** $5, 13, 2\sqrt{65}$ **6** 0.3π **7** $31\pi/540$
9 $12°$ **10** $19°41'15''$ **11** $131°46'49''$ **13** 1.36 radians **14** 1.38
centimeters per minute **15** $24°36'14''$N latitude, $10°37'38''$W longitude
17 $-2.5, 6.5$ **18** $30, 34$ **19** $\sin\theta = \frac{3}{5}, \cos\theta = -\frac{4}{5}, \tan\theta = -\frac{3}{4}, \cot\theta =$
$-\frac{4}{3}, \sec\theta = -\frac{5}{4}, \csc\theta = \frac{5}{3}$ **21** $\sin\theta = -\frac{12}{13}, \cos\theta = -\frac{5}{13}, \tan\theta = \frac{12}{5}, \cot\theta = \frac{5}{12}$,
$\sec\theta = -\frac{13}{5}, \csc\theta = -\frac{13}{12}$ **22** $\cos\theta = -\frac{12}{13}, \tan\theta = -\frac{5}{12}, \cot\theta = -\frac{12}{5}, \sec\theta =$
$-\frac{13}{12}, \csc\theta = \frac{13}{5}$ **23** $\sin\theta = -\frac{3}{5}, \cos\theta = \frac{4}{5}, \cot\theta = -\frac{4}{3}, \sec\theta = \frac{5}{4}, \csc\theta = -\frac{5}{3}$
25 $\sin 60° = \sqrt{3}/2, \cos 60° = \frac{1}{2}, \tan 60° = \sqrt{3}, \cot 60° = 1/\sqrt{3}, \sec 60° = 2,$
$\csc 60° = 2/\sqrt{3}$; $\sin 150° = \frac{1}{2}, \cos 150° = -\sqrt{3}/2, \tan 150° = -1/\sqrt{3},$
$\cot 150° = -\sqrt{3}, \sec 150° = -2/\sqrt{3}, \csc 150° = 2$; $\sin 315° = -1/\sqrt{2},$
$\cos 315° = 1/\sqrt{2}, \tan 315° = -1, \cot 315° = -1, \sec 315° = \sqrt{2}, \csc 315° = -\sqrt{2}$

Exercise 9.1, page 182

9 $\sin 74°$ **10** $\cos 46°$ **11** $\tan 42°$ **13** $-\sec 58°$ **14** $\csc 18°$
15 $-\sin 79°$ **17** $\tan 83°$ **18** $\cot 62°$ **19** $-\sec 27°$ **21** $\cos 100°$,
$-\cos 80°$ **22** $-\tan 193°$, $-\tan 13°$ **23** $-\cot 286°$, $\cot 84°$
25 $-\csc 227°$, $\csc 47°$ **26** $-\sin 341°$, $-\sin 19°$ **27** $\cos 403°$, $\cos 43°$
29 $-\cot 298°$, $\cot 62°$ **30** $\sec 486°$, $-\sec 54°$ **31** $-\csc 196°$, $\csc 16°$

Exercise 9.2, page 186

1 .7400 **2** .3607 **3** .0987 **5** 1.000 **6** 2.9887 **7** .9621
9 −.9545 **10** .7046 **11** −.2979 **13** 5°10′ **14** 13°30′
15 73°30′ **17** 63°40′ **18** 56°30′ **19** 31°10′ **21** 37°30′
22 45°00′ **23** 39°40′ **25** .9467 **26** .8454 **27** .8915 **29** .8011
30 −.3426 **31** −.4975 **33** 76°39′ **34** 25°31′ **35** 20°39′
37 70°59′ **38** 82°36′ **39** 50°24′

Exercise 9.3, page 187

1 $-\cos 27°$ **2** $-\tan 12°$ **3** $\sec 14°$ **5** $-\cos 21°$ **6** $\sin 31°$
7 $-\tan 12°$ **9** .8158 **10** −1.9210 **11** −.7509 **13** −.9968
14 −.3830 **15** 85°50′ **17** 55°20′ **18** 22°10′ **19** 39°30′
21 73°56′ **22** 12°9′ **23** 79°54′

Exercise 10.1, page 194

1 $\sqrt{2}(1 + \sqrt{3})/4$ **2** $\sqrt{2}(1 - \sqrt{3})/4$ **3** $\frac{1}{2}$ **5** $-\frac{24}{25}, 0$ **6** $\frac{56}{65}, -\frac{16}{65}$
7 $\frac{220}{221}, \frac{140}{221}$ **9** $-\frac{119}{160}, 2\sqrt{13}/13$ **10** $-\frac{7}{25}, -2\sqrt{5}/5$ **11** $-\frac{161}{289}, -3\sqrt{34}/34$

Exercise 10.2, page 197

1 $\sqrt{2}(\sqrt{3} - 1)/4$ **2** $\sqrt{2}(\sqrt{3} + 1)/4$ **3** $\sqrt{3}/2$ **5** $-1, -\frac{7}{25}$
6 $-\frac{220}{221}, \frac{140}{221}$ **7** $\frac{4}{5}, \frac{44}{125}$ **9** $\frac{336}{625}, \frac{3}{5}$ **10** $-\frac{240}{289}, 5\sqrt{34}/34$
11 $\frac{24}{25}, 3\sqrt{10}/10$

Exercise 10.3, page 202

1 $(\sqrt{3} - 1)/(1 + \sqrt{3})$ **2** $-\sqrt{3}$ **3** $\sqrt{3}/3$ **5** $\frac{16}{63}, \frac{56}{33}$
6 $\frac{77}{36}, -\frac{13}{84}$ **7** $-\frac{240}{161}, 0$ **9** $-\frac{240}{161}, \frac{3}{5}$ **10** $\frac{336}{527}, \frac{4}{3}$ **11** $\frac{24}{7}, -\frac{1}{2}$
33 $1/\sqrt{10}, 3/\sqrt{10}$ **34** $2/\sqrt{13}, 3/\sqrt{13}$ **35** $1/\sqrt{17}, 4/\sqrt{17}$
37 $5/\sqrt{34}, 3/\sqrt{34}$ **38** $3/\sqrt{10}, 1/\sqrt{10}$ **39** $5/\sqrt{26}, 1/\sqrt{26}$

Exercise 10.5, page 205

1 $\frac{1}{2}\left(\sqrt{2 - \sqrt{3}}\right), \frac{1}{2}\left(\sqrt{2 + \sqrt{3}}\right), 2 - \sqrt{3}$ **2** $\sqrt{3}/2, -\frac{1}{2}, -\sqrt{3}$
3 $(\sqrt{6} + \sqrt{2})/4, (\sqrt{6} - \sqrt{2})/4, -(2 + \sqrt{3})$ **5** $-\frac{77}{85}, -\frac{13}{85}, -\frac{240}{289}$
6 $-\frac{171}{140}, \frac{21}{220}, -\frac{120}{119}$

Exercise 11.1, page 211

1 $\{2, -2\}$ **2** $\{-6, 6\}$ **3** $\{-\frac{1}{3}, \frac{1}{3}\}$ **5** $\{-\frac{9}{2}, \frac{9}{2}\}$ **6** $\{-\frac{11}{6}, \frac{11}{6}\}$
7 $\{-\frac{13}{3}, \frac{13}{3}\}$ **9** $\{1, 2\}$ **10** $\{3, 4\}$ **11** $\{-2, 3\}$ **13** $\{0, 5\}$
14 $\{-2, 1\}$ **15** $\{-4, 3\}$ **17** $\{-3, \frac{1}{2}\}$ **18** $\{-\frac{2}{3}, 1\}$ **19** $\{-2, -\frac{3}{2}\}$
21 $\{-\frac{4}{3}, \frac{3}{4}\}$ **22** $\{-\frac{3}{2}, \frac{5}{3}\}$ **23** $\{-\frac{3}{2}, \frac{1}{7}\}$ **25** $\{-\frac{5}{3}, \frac{7}{2}\}$ **26** $\{\frac{2}{5}, \frac{3}{2}\}$
27 $\{-\frac{4}{3}, -\frac{2}{7}\}$ **29** $\{-\frac{3}{7}, \frac{4}{5}\}$ **30** $\{-\frac{1}{6}, \frac{2}{5}\}$ **31** $\{-\frac{2}{7}, \frac{5}{6}\}$
33 $\{-\frac{5}{6}, -\frac{3}{4}\}$ **34** $\{-\frac{7}{2}, \frac{1}{15}\}$ **35** $\{-\frac{15}{7}, \frac{5}{6}\}$ **37** $\{-3c, 2c\}$
38 $\{-3d, 4d\}$ **39** $\{-2/c, 1/c\}$ **41** $\{-b/a, 2b/a\}$ **42** $\{-3a/b, 2a/b\}$
43 $\{a/b, -b/a\}$

Exercise 11.2, page 216

1 $\{1, -3\}$ **2** $\{-3, 5\}$ **3** $\{-7, 3\}$ **5** $\{2, 3\}$ **6** $\{3, 4\}$
7 $\{3, 5\}$ **9** $\{2 + i, 2 - i\}$ **10** $\{-1 + i, -1 - i\}$
11 $\{-3 + 2i, -3 - 2i\}$ **13** $\{\frac{1}{2}, 3\}$ **14** $\{-4, \frac{2}{3}\}$ **15** $\{-3, \frac{2}{5}\}$
17 $\{-\frac{3}{2}, \frac{1}{3}\}$ **18** $\{-\frac{2}{3}, \frac{3}{2}\}$ **19** $\{-\frac{2}{3}, \frac{3}{4}\}$ **21** $\{1 + \sqrt{7}, 1 - \sqrt{7}\}$
22 $\{2 + \sqrt{3}, 2 - \sqrt{3}\}$ **23** $\{-2 + \sqrt{2}, -2 - \sqrt{2}\}$
25 $\left\{\dfrac{-1 + \sqrt{2}}{2}, \dfrac{-1 - \sqrt{2}}{2}\right\}$ **26** $\left\{\dfrac{2 + \sqrt{3}}{3}, \dfrac{2 - \sqrt{3}}{3}\right\}$
27 $\{1 + \sqrt{5}/3, 1 - \sqrt{5}/3\}$ **29** $\{-b, 2c\}$ **30** $\{b, 3c\}$ **31** $\{-a/2, 2b\}$
33 $\{-a + ib, -a - ib\}$ **34** $\{-2a + 3ib, -2a - 3ib\}$ **35** $\left\{\dfrac{a + ib}{2}, \dfrac{a - ib}{2}\right\}$
37 $\{-b/a, 2b/a\}$ **38** $\{2a/b, -3a/b\}$ **39** $\{a/b, -b/a\}$

Exercise 11.3, page 220

1 $\{4 + \sqrt{21}, 4 - \sqrt{21}\}$ **2** $\{-2 + \sqrt{10}, -2 - \sqrt{10}\}$
3 $\left\{\dfrac{3 + \sqrt{23}}{2}, \dfrac{3 - \sqrt{23}}{2}\right\}$ **5** $\left\{\dfrac{4 + \sqrt{38}}{11}, \dfrac{4 - \sqrt{38}}{11}\right\}$ **6** $\left\{\dfrac{3 + \sqrt{35}}{2}, \dfrac{3 - \sqrt{35}}{2}\right\}$
7 $\{-2 + \sqrt{19}, -2 - \sqrt{19}\}$ **9** $\left\{\dfrac{-5 + 3i}{2}, \dfrac{-5 - 3i}{2}\right\}$ **10** $\left\{\dfrac{-2 + i}{2}, \dfrac{-2 - i}{2}\right\}$
11 $\left\{\dfrac{1 + i}{2}, \dfrac{1 - i}{2}\right\}$ **13** $\left\{\dfrac{-9 + i\sqrt{199}}{35}, \dfrac{-9 - i\sqrt{199}}{35}\right\}$ **14** $\{\frac{7}{5}, -\frac{5}{6}\}$
15 $\left\{\dfrac{5 + \sqrt{53}}{2}, \dfrac{5 - \sqrt{53}}{2}\right\}$ **17** $\{2a/b, -a/b\}$ **18** $\{c/3d, -2d/c\}$
19 $\{-\frac{1}{2}, -a/3\}$ **21** $\left\{\dfrac{-3 + i\sqrt{6}}{cd}, \dfrac{-3 - i\sqrt{6}}{cd}\right\}$ **22** $\left\{\dfrac{3 + i}{2p}, \dfrac{3 - i}{2p}\right\}$
23 $\left\{\dfrac{-3 + i}{10b}, \dfrac{-3 - i}{10b}\right\}$ **25** $\{-2, -1, 1, 2\}$ **26** $\{-3, -1, 1, 3\}$
27 $\{-1, -1, 1, 1\}$ **29** $\{-1, 1, -3i, 3i\}$ **30** $\{-2i, -i, i, 2i\}$
31 $\{-3, 3, -2i, 2i\}$ **33** $\{3, 2\}$ **34** $\{\frac{5}{2}, 4\}$ **35** $\{-\frac{1}{5}, \frac{1}{2}\}$
37 $\{-2, -1, 1, 2, -2i, -i, i, 2i\}$ **38** $\{-3, -1, 1, 3, -3i, -i, i, 3i\}$

39 $\{-1, -\frac{1}{2}, \frac{1}{2}, 1, -i, -i/2, i/2, i\}$ **41** $\{-2\sqrt{2}, -2, 2, 2\sqrt{2}\}$
42 $\{-3, -1, 1, 3\}$ **43** $\{1, 2, 4, -1\}$ **45** $\{4, -10\}$ **46** $\{-1, 0\}$
47 $\{-18, 8\}$ **49** $\{4, -\frac{1}{2}\}$ **50** $\{\frac{1}{2}, -\frac{5}{2}\}$ **51** $\{12, -6\}$

Exercise 11.4, page 224

1 $\{3\}$ **2** $\{5\}$ **3** $\{-1\}$ **5** $\{6\}$ **6** $\{7\}$ **7** $\{0, 3\}$ **9** $\{0, -4\}$
10 $\{-2, -3\}$ **11** $\{-1, 2\}$ **13** $\{0, 18\}$ **14** $\{-2\}$ **15** $\{3\}$
17 $\{\frac{1}{2}, 1\}$ **18** $\{\frac{1}{3}\}$ **19** $\{\frac{3}{2}\}$ **21** $\{1\}$ **22** $\{2, -\frac{123}{46}\}$
23 $\{-1, \frac{2}{3}\}$ **25** $\{3, -2\}$ **26** $\{0\}$ **27** $\{2\}$ **29** $\{-1\}$ **30** $\{-2\}$
31 $\{4\}$ **33** $\{a, -2a\}$ **34** $\{b, 2b\}$ **35** $\{2a, 2a(a^2 + b^2)/(a - b)^2\}$

Exercise 11.5, page 230

1 $D = 1$, rational and unequal, 5, 6   **2** $D = 1$, rational and unequal, -3, 2
3 $D = 121$, rational and unequal, $\frac{5}{12}, -\frac{1}{6}$ **5** $D = 0$, rational and equal, 6, 9
6 $D = 0$, rational and equal, -8, 16   **7** $D = 0$, rational and equal, $-\frac{4}{3}, \frac{4}{9}$
9 $D = 72$, irrational and unequal, 6, -9 **10** $D = 128$, irrational and
unequal, $-8, -16$ **11** $D = 84$, irrational and unequal, $-\frac{1}{2}, -\frac{5}{4}$
13 $D = -76$, imaginary, $-\frac{1}{2}, \frac{5}{4}$ **14** $D = -47$, imaginary, $\frac{3}{7}, \frac{2}{7}$
15 $D = -32$, imaginary, 2, 9   **17** $D = 2$, real and unequal, $-\sqrt{6}$, 1
18 $D = 13$, real and unequal, $\sqrt{5}/\sqrt{2}, -1$ **19** $D = 16$, real and unequal,
$2\sqrt{2}, -2$ **21** True   **22** True   **23** True   **25** False, wrong product
26 False, wrong product   **27** False, wrong product   **29** False, wrong
sum   **30** False, wrong sum   **31** False, wrong sum   **33** $x^2 - 3x + 2 = 0$
34 $x^2 - 2x - 15 = 0$ **35** $15x^2 - 2x - 8 = 0$ **37** $x^2 - 4x - 1 = 0$
38 $x^2 - 6x + 2 = 0$ **39** $x^2 - 4x + 40 = 0$ **41** $3x - 2, 4x - 3$
42 $5x - 2, 2x + 1$ **43** $D = -31$ is not a perfect square   **45** $D = 20$ is not
a perfect square   **46** $x - 3 + i, x - 3 - i$ **47** $2x - 1 - 2i, 2x - 1 + 2i$

Exercise 11.6, page 235

1 $V(\frac{3}{2}, -\frac{13}{4})$, up   **2** $V(-1, 1)$, up   **3** $V(-\frac{3}{2}, -\frac{21}{4})$, up   **5** $V(\frac{1}{4}, \frac{17}{8})$,
down   **6** $V(\frac{3}{4}, \frac{17}{8})$, down   **7** $V(-\frac{1}{3}, \frac{10}{3})$, down   **9** $V(1, -4)$, up
10 $V(-1, -2)$, up   **11** $V(2, 7)$, down   **13** $\{x|x < -1\} \cup \{x|x > 6\}$
14 $\{x|x < 2\} \cup \{x|x > 3\}$ **15** $\{x|x < -3\} \cup \{x|x > 2\}$
17 $\{x|1 < x < 2\}$ **18** $\{x|3 < x < 5\}$ **19** $\{x|-2 < x < 5\}$
21 $\{x|x < 1\} \cup \{x|x > 3\}$ **22** $\{x|x < 2\} \cup \{x|x > 5\}$ **23** $\{x|-3 < x < 2\}$
25 $\varnothing$ **26** all except 4   **27** all x **29** all x **30** $\{x|-4 < x < -\frac{3}{2}\}$
31 $\{x|-\frac{2}{3} < x < \frac{3}{2}\}$

Exercise 11.7, page 236

1 $\{-3, 1\}$ **2** $\{-\frac{1}{3}, \frac{3}{2}\}$ **3** $\{-\frac{2}{3}, -\frac{1}{4}\}$ **5** $\{-c, 3c/2\}$ **6** $\{-2a/3, -a\}$

7 $\left\{\dfrac{-2 + \sqrt{3}}{3}, \dfrac{-2 - \sqrt{3}}{3}\right\}$ **9** $\{3b/2, -b\}$ **10** $\{2c, 3c/2\}$

11 $\left\{\dfrac{-3 + i\sqrt{3}}{2}, \dfrac{-3 - i\sqrt{3}}{2}\right\}$ **13** $\{-3, -1, 1, 3\}$ **14** $\{-1, 1, -3i, 3i\}$

15 $\{0, -\frac{7}{6}\}$ **17** $\{3\}$ **18** $\{5, \frac{11}{3}\}$ **19** $\{7\}$ **21** $D = 1$, rational and unequal, $-7, 12$ **22** $D = 0$, rational and equal, 8, 16   **23** $D = -23$, imaginary, $-\frac{3}{2}, 2$ **25** $4x^2 - 13x - 12 = 0$ **26** $x^2 - 6x + 34 = 0$

27 **(a)** $D = 49$, factorable; **(b)** $D = 45$, not factorable into rational factors
29 $(\frac{5}{4}, \frac{49}{8})$

Exercise 12.2, page 249

1 $\{(3, 2)\}$ **2** $\{(-2, 3)\}$ **3** $\{(3, \frac{3}{2})\}$ **5** $\{(1.5, -2.5)\}$
6 $\{(2, -1)\}$ **7** $\{(0, 1.2)\}$ **9** Dependent   **10** Dependent
11 Dependent   **13** Inconsistent   **14** Inconsistent   **15** Inconsistent
17 $\{(2.4, 6.2), (-0.9, -3.7)\}$ **18** $\{(-2, 4, 2, 6), (0.4, 5.4)\}$
19 $\{(1.6, 1.2), (-3.6, -0.5)\}$ **21** $\{(4, 2), (2.9, -1.7)\}$
22 $\{(2.4, 2.7), (1.1, -1.8)\}$ **23** $\{(2, 2), (0.8, -1.3)\}$
25 $\{(3, 4), (3, -4)\}$ **26** $\{(12, 5), (12, -5)\}$
27 $\{(1, 2), (1, -2), (4, 4), (4, -4)\}$ **29** $\{(2, 5), (2, -5), (-2, 5), (-2, -5)\}$
30 $\{(4, 6), (4, -6), (-4, 6), (-4, -6)\}$
31 $\{(1, 3), (1, -3), (-1, 3), (-1, -3)\}$

Exercise 12.3, page 255

1 $\{(1, 2)\}$ **2** $\{(2, 3)\}$ **3** $\{(3, -1)\}$ **5** Dependent   **6** $\{(-1, 0)\}$
7 $\{(2, 3)\}$ **9** $\{(1, 5)\}$ **10** $\{(-2, -1)\}$ **11** Inconsistent
13 $\{(2, 4)\}$ **14** $\{(3, -6)\}$ **15** $\{(\frac{1}{2}, \frac{1}{3})\}$ **17** $\{(3, \frac{1}{2}), (3, -\frac{1}{2}),$
$(-3, \frac{1}{2}), (-3, -\frac{1}{2})\}$ **18** $\{(1, 1), (1, -1), (-1, 1), (-1, -1)\}$ **19** $\{(\frac{1}{2}, 1),$
$(\frac{1}{2}, -1), (-\frac{1}{2}, 1), (-\frac{1}{2}, -1)\}$ **21** $\{(\sqrt{2}, 1), (\sqrt{2}, -1), (-\sqrt{2}, 1),$
$(-\sqrt{2}, -1)\}$ **22** $\{(2, \frac{1}{2}), (2, -\frac{1}{2}), (-2, \frac{1}{2}), (-2, -\frac{1}{2})\}$ **23** $\{(2, 2),$
$(2, -2), (-2, 2), (-2, -2)\}$ **25** $\{(2, \sqrt{3}), (2, -\sqrt{3}), (-2.6, 2i\sqrt{2.01}),$
$(-2.6, -2i\sqrt{2.01})\}$ **26** $\{(1, i), (1, -i), (\frac{19}{7}, i\sqrt{457}/7), (\frac{19}{7}, -i\sqrt{457}/7)\}$
27 $\{(0, \frac{1}{2}), (0, -\frac{1}{2}), (2, \frac{1}{2}), (2, -\frac{1}{2})\}$ **29** $\{(3, 1), (-3, 1), (4, 2), (-4, 2)\}$
30 $\{(\frac{2}{5}, \frac{1}{5}), (-\frac{2}{5}, \frac{1}{5}), (1, -\frac{1}{2}), (-1, -\frac{1}{2})\}$ **31** $\{(i, 2), (-i, 2),$
$(\sqrt{62}/17, -\frac{5}{17}), (-\sqrt{62}/17, -\frac{5}{17})\}$ **33** $\{(-2, \frac{1}{2}), (1, 2)\}$ **34** $\{(1 + i, 1),$
$(1 - i, 1)\}$ **35** $\{(1 + i, 1), (1 - i, 1)\}$ **37** $\{(1, 2), (-\frac{1}{6}, -\frac{3}{2})\}$
38 $\{(1, 1 + i), (1, 1 - i)\}$ **39** $\{(3, 1), (-\frac{47}{21}, -\frac{7}{3})\}$

Exercise 12.4, page 259

1 $\{(2, 3)\}$ **2** $\{(3, 1)\}$ **3** $\{(2, -1)\}$ **5** $\{(3, 2)\}$ **6** $\{(-2, 1)\}$
7 $\{(2, 1)\}$ **9** $\{(1, \frac{1}{2})\}$ **10** $\{(2, \frac{1}{3})\}$ **11** $\{(\frac{1}{2}, \frac{1}{3})\}$ **13** $\{(5, 7),$
$(3, 1)\}$ **14** $\{(3, 2), (7, 8)\}$ **15** $\{(2, 5), (1, 3)\}$ **17** $\{(b, a),$
$(-b, -a)\}$ **18** $\{(3b, -b/a), (-b, 3b/a)\}$ **19** $\{(-b/m, 0), (0, b)\}$

21 $\{(2, 10), (-2, -2), (-\frac{3}{2}, -4), (\frac{3}{2}, 5)\}$ **22** $\{(4, 1), (0, -1), (0, \frac{1}{3}),$
$(-\frac{4}{3}, \frac{1}{3})\}$ **23** $\{(3, 16), (-3, -2), (2, 8), (-2, -4)\}$ **25** $\{(2, 14),$
$(-2, 10), (2i, -12 + 2i), (-2i, -12 - 2i)\}$ **26** $\{(0, 1), (-2, -1),$
$(1 + \sqrt{3}, \sqrt{3}), (1 - \sqrt{3}, -\sqrt{3})\}$ **27** $\{(-2, 1), (-4, -1), (3 + \sqrt{7}, \sqrt{7}),$
$(3 - \sqrt{7}, -\sqrt{7})\}$ **29** $\left\{ (10, 2), (6, -2), \left(\dfrac{-17 + \sqrt{17}i}{2}, \dfrac{\sqrt{17}i}{2}\right), \right.$
$\left. \left(\dfrac{-17 - \sqrt{17}i}{2}, \dfrac{-\sqrt{17}i}{2}\right) \right\}$ **30** $\{(2, \frac{1}{2}), (0, -\frac{1}{2}), (-3 + \sqrt{3}i, \sqrt{3}i/2),$
$(-3 - \sqrt{3}i, -\sqrt{3}i/2)\}$ **31** $\{(6, 1), (-3, -1), (14 + 4\sqrt{14}, \sqrt{14}),$
$(14 - 4\sqrt{14}, -\sqrt{14})\}$ **33** $\{(2\sqrt{2}, \sqrt{2}/2), (-2\sqrt{2}, -\sqrt{2}/2), (1, 2), (-1, -2)\}$
34 $\{(5, -4), (-5, 4), (2i\sqrt{6}, 5i\sqrt{6}/3), (-2i\sqrt{6}, -5i\sqrt{6}/3)\}$ **35** $\{(3, 1),$
$(-3, -1), (\sqrt{6}/2, \sqrt{6}), (-\sqrt{6}/2, -\sqrt{6})\}$ **37** $\{(1, 2), (-1, -2), (-2i, i),$
$(2i, -i)\}$ **38** $\{(3\sqrt{2}/4, \sqrt{2}/2), (-3\sqrt{2}/4, -\sqrt{2}/2), (\frac{1}{2}, \frac{3}{2}), (-\frac{1}{2}, -\frac{3}{2})\}$
39 $\{(1, 2), (-1, -2), (-2i, i), (2i, -i)\}$

Exercise 12.5, page 265

1 $\{(3, 3), (-3, -3), (2, -1), (-2, 1)\}$ **2** $\{(3, -1), (-3, 1), (4, 2),$
$(-4, -2)\}$ **3** $\{(3, 2), (-3, -2), (1, -1), (-1, 1)\}$ **5** $\{(1, -1), (-1, 1),$
$(\sqrt{2}/2, -\sqrt{2}), (-\sqrt{2}/2, \sqrt{2})\}$ **6** $\{(1, 1), (1, -2), (-1, -1), (-1, 2)\}$
7 $\{(2, -1), (-2, 1), (25i\sqrt{69}/69, 14i\sqrt{69}/69), (-25i\sqrt{69}/69, -14i\sqrt{69}/69)\}$
9 $\{(3, -1), (-3, 1), (2i/3, 5i/3), (-2i/3, -5i/3)\}$ **10** $\{(0, i), (0, -i),$
$(2i, i), (-2i, -i)\}$ **11** $\{(2, -2), (-2, 2), (8/\sqrt{41}, 2/\sqrt{41}), (-8/\sqrt{41}, -2/\sqrt{41})\}$
13 $\{(1, -\frac{2}{3}), (-1, \frac{2}{3}), (0, i\sqrt{3}/3), (0, -i\sqrt{3}/3)\}$ **14** $\{(3, \frac{1}{2}), (-3, -\frac{1}{2}),$
$(2i, 2.5i), (-2i, -2.5i)\}$ **15** $\{(2\sqrt{3}, 5\sqrt{3}), (-2\sqrt{3}, -5\sqrt{3}), (\sqrt{3}, -2\sqrt{3}),$
$(-\sqrt{3}, 2\sqrt{3})\}$ **17** $\{(3, -1), (\frac{1}{2}, \frac{3}{2})\}$ **18** $\{(3, 2), (-1, -1)\}$
19 $\{(3, 5), (2, 6)\}$ **21** $\{(2, 1), (\frac{17}{5}, -\frac{9}{5})\}$ **22** $\{(3, 2), (-1, 2)\}$
23 $\{(1, 1)\}$ **25** $\{(-2, -3), (-\frac{7}{2}, -\frac{3}{2})\}$ **26** $\{(3, 2), (4, 3)\}$
27 $\{(4, 1), (3, 2)\}$ **29** $\{(4, 1), (1, 4), (2, 5), (5, 2)\}$
30 $\left\{ (1, 1), (1, 1), \left(\dfrac{-1 + i\sqrt{3}}{2}, \dfrac{-1 - i\sqrt{3}}{2}\right), \left(\dfrac{-1 - i\sqrt{3}}{2}, \dfrac{-1 + i\sqrt{3}}{2}\right) \right\}$
31 $\{(4, -2), (-2, 4), (3, 1), (1, 3)\}$ **33** $\{(1, 2), (2, 1), (3, 2), (2, 3)\}$
34 $\{(4, 5), (5, 4), (-3, 6), (6, -3)\}$ **35** $\{(-1, -1)\}$
37 $\{(2, 3), (3, 2)\}$ **38** $\{(2, -1), (-1, 2)\}$ **39** $\{(3, 3)\}$

Exercise 12.6, page 271

1 Lloyd, \$37.50; Brad, \$40   **2** Brian, \$110; Bruce, \$95   **3** 6 days
5 32 girl scouts, 48 campfire girls   **6** 2 summer months, 7 nonsummer
months   **7** 5 miles   **9** \$1.76, 50 cents   **10** \$120, \$360   **11** \$2,520,
\$3,170   **13** 60 miles going, 50 miles returning   **14** 95 by 60 feet
15 260 miles per hour, 20 miles per hour   **17** New York to Boston, 229
miles; New York to Philadelphia, 91 miles   **18** $\frac{1}{4}$ hour by car, 2 hours by
plane, $\frac{1}{2}$ hour by limousine   **19** 15 sedans, 20 sports cars, 10 station wagons
21 5, 9   **22** 15 by 36 feet   **23** 100 by 150 feet   **25** 5 by 12 feet
26 15 by 15 feet, 10 by 15 feet   **27** 4 by 9 inches or 3 by 12 inches
29 120 shares at \$65 apiece   **30** 2 inches, 5 inches   **31** Side: 8 inches,
radius: 7 inches; or side: $\frac{332}{25}$ inches, radius, $\frac{91}{25}$ inches

Exercise 12.7, page 276

25 $(-3, 2), (1, -3), (2, 1)$ **26** $(2, 3), (-1, -2), (3, -4), (4, 0)$
27 $(0, 0), (1, -3), (2, -4), (5, -1), (4, 0)$

Exercise 12.8, page 283

1 $x = 2 + 3t, y = 3 + 5t$ **2** $x = 3 - 4t, y = 5 - 3t$ **3** $x = -4 + 6t, y = -5t$
5 $5, 13$ **6** $-8, 7$ **7** $-16, -1$ **9** $3 + 28t, 3, 31$ **10** $-3 + 7t, -3, 4$
11 $-14 + 33t, -14, 19$ **13** $13, 20$ **14** $-12, 10$ **15** $-14, 18$
17 $(-3, 0), (0, 0), (0, 2); -7, 1$ **18** $(-3, 0), (0, -3), (0, 0); -5, 10$
19 $(-1, 0), (2, 0), (0, 2); -7, 1$ **21** $(-2, 0), (0, -1), (3, 0), (0, 3); -10, -5$
22 $(-2, 0), (0, -3), (4, 0), (0, 2); -5, 13$ **23** $(-1, 0), (0, -1), (1, 0),$
$(0, 1); 4, 12$ **25** $(-2, 2), (-1, -1), (3, -2), (1, 3); 0, 16$ **26** $(-4, 0),$
$(1, -1), (3, 1), (-1, 2); -2, 21$ **27** $(-4, 0), (0, -4), (2, 0), (0, 3), (-3, 2);$
$-22, 29$ **29** No pokers, 5 tongs   **30** Pepper, 10 acres; rhubarb, 3 acres;
tomatoes, 7 acres   **31** A, 3 pounds; B, 2 pounds

Exercise 12.9, page 286

7 $\{(2, 1)\}$ **9** $\{(1, -1), (-\frac{2}{3}, \frac{3}{2})\}$ **10** $\{(\frac{1}{2}, \frac{1}{3})\}$ **11** $\{(2, 1), (2, -1),$
$(-2, 1), (-2, -1)\}$ **13** $\{(1, 2), (-1, -2), (5/\sqrt{14}, -1/\sqrt{14}),$
$(-5/\sqrt{14}, 1/\sqrt{14})\}$ **14** $\{(3, -2), (-3, 2), (2, -3), (-2, 3)\}$
15 $\{(1, -2), (-2, 1)\}$ **17** Vertices are $(-3, 1), (4, -2), (1, 2)$;
extrema are -4 and 23.   **18** Vertices are $(-3, 4), (-1, -1), (6, 0), (5, 2)$;
extrema are -10 and 25.   **19** 400 A of cotton, 300 A of maize, and 1,100 A
of wheat

Exercise 13.1, page 291

1 $a = 2, b = 3, c = -1, d = 0$ **2** $a = 1, b = 5, c = -2, d = 3$ **3** $a = 3,$
$b = 2, c = 3, d = 2$ **5** $\begin{bmatrix} -2 & -3 & -1 \\ 0 & 4 & -5 \end{bmatrix}$ **6** $\begin{bmatrix} 4 & 6 & 2 \\ 0 & -8 & 10 \end{bmatrix}$ **7** $\begin{bmatrix} 0 & 0 & 0 \\ 0 & 0 & 0 \end{bmatrix}$

9 $\begin{bmatrix} -3 & 5 & 5 \\ 3 & -4 & 4 \end{bmatrix}$ **10** $\begin{bmatrix} 7 & 1 & -3 \\ -3 & -4 & 6 \end{bmatrix}$ **11** $\begin{bmatrix} 19 & 0 & -10 \\ -9 & -8 & 13 \end{bmatrix}$

13 $\begin{bmatrix} 2 & 0 \\ 3 & -4 \\ 1 & 5 \end{bmatrix}$ **14** $\begin{bmatrix} -5 & 3 \\ 2 & 0 \\ 4 & -1 \end{bmatrix}$ **15** Not defined   **25** $\begin{bmatrix} 7 & 11 \\ 1 & 5 \end{bmatrix}$

26 $\begin{bmatrix} -5 & 4 \\ 4 & 7 \end{bmatrix}$ **27** $\begin{bmatrix} -4 & -2 \\ 3 & 3 \end{bmatrix}$ **29** $\begin{bmatrix} -8 & -13 & 14 \\ 4 & 11 & -11 \\ -7 & -19 & 12 \end{bmatrix}$

30 $\begin{bmatrix} 2 & 2 & 9 \\ -11 & -2 & -22 \\ 7 & -3 & 15 \end{bmatrix}$ **31** $\begin{bmatrix} -2 & -11 & 7 \\ 2 & -2 & -3 \\ 9 & -22 & 15 \end{bmatrix}$ **37** $\begin{bmatrix} 11 & 1 & 9 \\ 19 & 2 & 14 \end{bmatrix}$

38 $\begin{bmatrix} 0 & 2 \\ 5 & -1 \\ -2 & 6 \end{bmatrix}$ **39** $\begin{bmatrix} -1 & 1 & 8 \\ -8 & -1 & 6 \\ 8 & 3 & 9 \end{bmatrix}$

Exercise 13.2, page 296

1 $\{2, 1\}$ **2** $\{3, -2\}$ **3** $\{1, -1\}$ **5** $\{1, 2, 3\}$ **6** $\{2, 1, -1\}$
7 $\{3, 1, \frac{1}{2}\}$ **9** $\{1, 2, 0, -1\}$ **10** $\{-1, -1, 1, 0\}$ **11** $\{2, -3, 1, -1\}$

13 $-\frac{1}{11}\begin{bmatrix} -4 & -3 \\ -1 & 2 \end{bmatrix}$ **14** $\frac{1}{10}\begin{bmatrix} 1 & 3 \\ -3 & 1 \end{bmatrix}$ **15** $-\frac{1}{25}\begin{bmatrix} -4 & -3 \\ -3 & 4 \end{bmatrix}$ **17** $\frac{1}{2}\begin{bmatrix} 0 & 1 & 1 \\ 1 & -1 & 0 \\ 1 & 0 & -1 \end{bmatrix}$

18 $-\frac{1}{28}\begin{bmatrix} -7 & -5 & 1 \\ -7 & 7 & -7 \\ 7 & -3 & -5 \end{bmatrix}$ **19** $-\frac{1}{16}\begin{bmatrix} -22 & -2 & 10 \\ -10 & 2 & 6 \\ -1 & 5 & -1 \end{bmatrix}$

Exercise 13.3, page 301

1 3   **2** 11   **3** -10 **5** -1 **6** -19 **7** 14   **9** 13   **10** 15
11 23   **13** $as - ce$ **14** $ca - rt$ **15** $-3ma - th$ **17** -12
18 -58 **19** -28 **21** 21   **22** 9   **23** -9 **25** 0   **26** 0
27 51   **29** -46 **30** -17 **31** 85

Exercise 13.4, page 306

17 $\{28\}$ **18** $\{42\}$ **19** $\{-6\}$ **21** $\{3\}$ **22** $\{8\}$ **23** $\{3, -1\}$
25 $\{1, \frac{3}{2}\}$ **26** $\{3, -3\}$ **27** $\{2\}$

Exercise 13.5, page 313

1 $\{(2, 1)\}$ **2** $\{(1, -1)\}$ **3** $\{(4, 3)\}$ **5** $\{(3, -5)\}$ **6** $\{(-2, 3)\}$
7 $\{(4, -5)\}$ **9** $\{(\frac{1}{2}, \frac{1}{4})\}$ **10** $\{(\frac{1}{6}, \frac{1}{3})\}$ **11** $\{(\frac{2}{5}, \frac{3}{5})\}$ **13** $\{(1, 1, 3)\}$
14 $\{(2, 1, 1)\}$ **15** $\{(1, 2, 2)\}$ **17** $\{(4, -3, -2)\}$ **18** $\{(-2, -2, 3)\}$
19 $\{(5, -4, -3)\}$ **21** $\{(\frac{1}{2}, \frac{1}{4}, -\frac{1}{4})\}$ **22** $\{(\frac{1}{3}, -\frac{1}{6}, -\frac{1}{6})\}$
23 $\{(\frac{1}{4}, -\frac{1}{8}, \frac{1}{8})\}$ **25** $\{(a, b, a - b)\}$ **26** $\{(2a, a + b, a - b)\}$
27 $\{(a, a - b, a + b)\}$ **29** $\{(1, 0, 3)\}$ **30** $\{(2, 1, -2)\}$
31 $\{(2, -1, -3)\}$

Exercise 13.6, page 314

1 $x = 3, y = -2$ **2** $x = 2, y = 1, z = -1$ **3** $\begin{bmatrix} 1 & x & y \\ 0 & 6 & x \end{bmatrix}$

5 $\begin{bmatrix} 11 & 0 & 33 & 20 \\ 0 & -11 & 22 & 12 \\ 4 & -8 & 28 & 16 \end{bmatrix}$ **6** $\begin{bmatrix} -17 & 12 & 12 \\ 32 & -16 & 2 \end{bmatrix}$ **7** $\begin{bmatrix} 3 & 2 & 0 \\ 3 & 2 & 1 \\ 5 & 3 & 2 \end{bmatrix}$

9 $x = 1, y = 3$ **10** $x = 2, y = 1, z = -3$ **11** $\frac{1}{10}\begin{bmatrix} 4 & -2 \\ -1 & 3 \end{bmatrix}$

13 $\frac{1}{4}\begin{bmatrix} 2 & 2 & 1 \\ 6 & 2 & -1 \\ 2 & 2 & -1 \end{bmatrix}$ **15** 17   **17** 1   **18** 18   **19** 12   **25** 9

26 $3, -\frac{1}{2}$ **27** $\{4, -5\}$ **29** $\{-1, 1, 2\}$ **30** $\{2, 0, 3\}$

Exercise 14.1, page 320

1 $\frac{14}{1}$ **2** $\frac{44}{7}$ **3** $\frac{1}{16}$ **5** $\frac{10}{3}$ **6** $\frac{12}{5}$ **7** $\frac{16}{3}$ **9** 18 eggs per hen
10 13 miles per gallon   **11** 56 miles per hour   **13** $115 per acre
14 $1.89 per pound   **15** $1,300 per month   **17** 0.84   **18** 2.2
19 80   **21** 3.15   **22** 4   **23** 10   **25** 2   **26** 3   **27** 5, -11
29 ± 4 **30** ± 9 **31** ± 10 **33** 18   **34** 25   **35** 48   **37** 4
38 9   **39** 7   **41** 3, 4   **42** 5, 10   **43** 9, 6

Exercise 14.2, page 324

1 $w = kt, k = 3$ **2** $s = k/v, k = 33$ **3** $m = kpq, k = 9$ **5** $\frac{3}{5}$ **6** 20
7 7   **9** 161 per second   **10** 128   **11** 1,875 pounds
13 6,000 pounds   **14** 100 pounds   **15** 0.054 inch   **17** 48 dynes
18 1.04 pounds per square inch   **19** 300 ergs   **21** 15 feet per second
22 7.2 ohms   **23** 144 tons   **25** 14,000 pounds   **26** $\frac{27}{128}$ inches
27 3,200 pounds

Exercise 14.3, page 326

1 $\frac{8}{15}$ **2** 5 cents per apple   **3** 0.4   **5** 2, -3 **6** 12   **7** 12.5
9 $x = 3, y = 6$ **10** 8 seconds after the shot   **11** 15.5 pounds

Exercise 15.2, page 339

1 $2\pi/3, 2$ **2** $\pi, 4$ **3** $\pi/5, \infty$ **5** $2\pi, \infty$ **6** $8\pi, \infty$ **7** $6\pi, 7$

9 2   **10** 4   **11** 3   **13** $\frac{1}{2}$ **14** $\frac{1}{2}$ **15** 1   **17** $\pi/2$, $\pi/2$ units to the left   **18** $2\pi/3$, $2\pi/3$ units to the left   **19** $\pi/2$, $3\pi/4$ units to the left
21 4π, 2π units to the left   **22** $\pi/2$, 8π units to the left
23 4π, 2π units to the right

Exercise 15.3, page 340

7 $\pi, 5$ **9** $\pi/3, \infty$ **10** $a = 3$, $b = 2$ **11** $a = 0.5$, $b = \frac{2}{3}$
13 $\pi/3$, ∞, $2\pi/3$ units to the left   **14** π, 3, $\pi/2$ units to the right
15 4π, 2, $\pi/2$ units to the left

Exercise 16.1, page 346

1 5, 2   **2** 3, -1 **3** 7, 5   **5** 2, -7 **6** $-2, -2$ **7** 3, 6
9 $(6, -2)$ **10** $(-1, 5)$ **11** $(0, 0)$ **13** $4 - 2i$ **14** $-i$
15 $7 + 2i$ **17** $(1, 4)$ **18** $(-5, 1)$ **19** $(0, 0)$ **21** $1 + i$
22 $1 - 7i$ **23** -10 **25** $(2, 14)$ **26** $(11, -10)$ **27** $(-3, 5)$
29 $-4 + 19i$ **30** $16 + 11i$ **31** $1 + 18i$ **33** $(\frac{3}{5}, \frac{14}{5})$ **34** $(\frac{7}{26}, -\frac{17}{26})$
35 $(\frac{15}{17}, \frac{8}{17})$ **37** $(\frac{11}{19} - \frac{13i}{29})$ **38** $(\frac{4}{29} + \frac{19i}{29})$ **39** $(\frac{-56}{65} + \frac{33i}{65})$
41 $(17, 6)$ **42** $(5, 10)$ **43** $11 - 16i$ **45** $\frac{2}{5} - 11i/5$ **46** $\frac{19}{29} - 4i/29$
47 $(\frac{8}{13}, \frac{1}{13})$

Exercise 16.2, page 350

13 $(7, 1)$ **14** $(-4, 5)$ **15** $(1, 2)$ **17** $(1, 2)$ **18** $(-1, 3)$
19 $(4, -2)$ **21** $(-8, 2)$, $2\sqrt{17}$ arctan $1/(-4)$ **22** $(-6, 9)$, $3\sqrt{13}$, arctan $-\frac{3}{2}$ **23** $(16, -9)$, $\sqrt{337}$, arctan $-\frac{9}{16}$ **25** $(-36, 24)$, $12\sqrt{13}$, arctan $2/(-3)$ **26** $(30, -20)$, $10\sqrt{13}$, arctan $(-2)/3$ **27** $(-10, 28)$, $2\sqrt{221}$, arctan $14/(-5)$ **29** 2 cis 60°   **30** $\sqrt{2}$ cis 315°   **31** 2 cis 150°
33 2 cis 120°   **34** $2\sqrt{3}$ cis 300°   **35** 1 cis 270°   **37** 5 cis 53°10′
38 25 cis 343°40′   **39** 13 cis 112°40′   **41** $\sqrt{34}$ cis 59°
42 $\sqrt{85}$ cis 319°20′   **43** $\sqrt{13}$ cis 123°40′   **45** $\sqrt{41}$ cis 38°40′
46 $5\sqrt{2}$ cis 351°50′   **47** $\sqrt{29}$ cis 111°50′

Exercise 16.3, page 354

1 10 cis 45° = $5\sqrt{2}(1 + i)$ **2** 12 cis 30° = $6(\sqrt{3} + i)$ **3** 21 cis 60° = $10.5(1 + \sqrt{3}i)$ **5** 6 cis 120° = $3(-1 + \sqrt{3}i)$ **6** 15 cis 135° = $7.5\sqrt{2}(-1 + i)$
7 35 cis 150° = $17.5(-\sqrt{3} + i)$ **9** 6 cis 210° = $-3(\sqrt{3} + i)$
10 10 cis 225° = $-5\sqrt{2}(1 + i)$ **11** 35 cis 240° = $-17.5(1 + \sqrt{3}i)$
13 4 cis 30° = $2(\sqrt{3} + i)$ **14** 5 cis 45° = $2.5\sqrt{2}(1 + i)$ **15** 11 cis 60° = $5.5(1 + \sqrt{3}i)$ **17** 5 cis 120° = $2.5(-1 + \sqrt{3}i)$ **18** 5 cis 135° = $2.5\sqrt{2}(-1 + i)$ **19** 13 cis 150° = $6.5(-\sqrt{3} + i)$ **21** 4 cis 225° = $-2\sqrt{2}(1 + i)$ **22** 2 cis 240° = $-(1 + \sqrt{3}i)$ **23** 2 cis 300° = $1 - \sqrt{3}i$

25 $2 \text{ cis } 30° = \sqrt{3} + i$ **26** $2 \text{ cis } 45° = \sqrt{2}(1 + i)$ **27** $3 \text{ cis } 90° = 3i$
29 $(\sqrt{2} \text{ cis } 45°)(2 \text{ cis } 300°) = 2\sqrt{2} \text{ cis } 345°$ **30** $(2 \text{ cis } 150°)(2 \text{ cis } 60°) =$
$4 \text{ cis } 210°$ **31** $(2 \text{ cis } 30°)(2 \text{ cis } 120°) = 4 \text{ cis } 150°$ **33** $(\sqrt{2} \text{ cis } 315°)^3 =$
$2\sqrt{2} \text{ cis } 225°$ **34** $(1 \text{ cis } 90°)^5 = 1 \text{ cis } 90°$ **35** $(\sqrt{14} \text{ cis } 135°)^4 = 196 \text{ cis } 180°$

37 $\dfrac{2\sqrt{3} \text{ cis } 150°}{2 \text{ cis } 300°} = \sqrt{3} \text{ cis } (-150°)$ **38** $\dfrac{2\sqrt{3} \text{ cis } 330°}{\sqrt{2} \text{ cis } 135°} = \sqrt{6} \text{ cis } 195°$

39 $\dfrac{5\sqrt{2} \text{ cis } 225°}{2 \text{ cis } 330°} = 2.5\sqrt{2} \text{ cis } (-105°)$ **41** $\dfrac{(2 \text{ cis } 330°)(1 \text{ cis } 90°)}{2 \text{ cis } 180°} = 1 \text{ cis } 240°$

42 $\dfrac{(4 \text{ cis } 0°)(2 \text{ cis } 270°)}{2 \text{ cis } 150°} = 4 \text{ cis } 120°$

43 $\dfrac{(5\sqrt{2} \text{ cis } 135°)(2 \text{ cis } 300°)}{2 \text{ cis } 210°} = 5\sqrt{2} \text{ cis } 225°$

Exercise 16.4, page 358

1 $32 \text{ cis } 90°$ **2** $81 \text{ cis } 0°$ **3** $125 \text{ cis } 180°$ **5** $128 \text{ cis } 60°$
6 $32 \text{ cis } 150°$ **7** $4 \text{ cis } 180°$ **9** $32 \text{ cis } 120°$ **10** $8 \text{ cis } 90°$
11 $8\sqrt{2} \text{ cis } 315°$ **13** $13^4 \text{ cis } 90°40'$ **14** $17^3 \text{ cis } 174°$ **15** $5^5 \text{ cis } 85°50'$
17 $34\sqrt{34} \text{ cis } 183°$ **18** $4{,}225 \text{ cis } 118°40'$ **19** $74^3 \text{ cis } 327°$
21 $2^{1/3} \text{ cis } (20° + k120°), k = 0, 1, 2$ **22** $1 \text{ cis } (60° + k120°), k = 0, 1, 2$
23 $2 \text{ cis } (45° + k180°), k = 0, 1$ **25** $3 \text{ cis } k90°, k = 0, 1, 2, 3$ **26** $2^{1/4} \text{ cis }$
$(82°30' + k90°), k = 0, 1, 2, 3$ **27** $2^{1/10} \text{ cis } (45° + k72°), k = 0, 1, 2, 3, 4$
29 $2 \text{ cis } k60°, k = 0, 1, \dots, 5$ **30** $1 \text{ cis } (45° + k60°), k = 0, 1, \dots, 5$
31 $2^{1/4} \text{ cis } (45° + k360°/7), k = 0, 1, \dots, 6$ **33** $2^{1/8} \text{ cis } (30° + k45°),$
$k = 0, 1, \dots, 7$ **34** $1 \text{ cis } (22°30' + k45°), k = 0, 1, \dots, 7$ **35** $2^{1/18} \text{ cis }$
$(35° + k40°), k = 0, 1, \dots, 8$ **37** $3 \text{ cis } (90° + k180°), k = 0, 1$
38 $2 \text{ cis } (135° + k180°), k = 0, 1$ **39** $3 \text{ cis } (90° + k120°), k = 0, 1, 2$
41 $2 \text{ cis } (45° + k90°), k = 0, 1, 2, 3$ **42** $1 \text{ cis } (67°30' + k90°), k = 0, 1, 2, 3$
43 $2 \text{ cis } (18° + k72°), k = 0, 1, 2, 3, 4$

Exercise 16.5, page 359

1 $4, -1$ **2** $(5, 1)$ **3** $(2, -2)$ **5** $2 - 3i$ **6** $2 + 23i$ **7** $48 + 7i$
9 $\sqrt{13} \text{ cis } 33°40'$ **10** $2\sqrt{2} \text{ cis } 195°$ **11** $4 \text{ cis } 30°$ **13** $(1/\sqrt{2}) \text{ cis } 75°$
14 $4 \text{ cis } 180°$ **15** $32 \text{ cis } 60°$ **17** $2^{1/4} \text{ cis } (60° + k90°), k = 0, 1, 2, 3$
18 $1 \text{ cis } (90° + k180°), k = 0, 1$ **19** $2 \text{ cis } (54° + k72°), k = 0, 1, 2, 3, 4$

Exercise 17.1, page 365

1 -3 **2** -26 **3** -6 **5** 3 **6** 5 **7** 22 **21** $\{1, 3, -4\}$
22 $\{6, -5, 2\}$ **23** $\{2, \frac{1}{2}, -\frac{2}{3}\}$ **25** $\{\frac{3}{2}, 1, -1\}$ **26** $\{\frac{5}{3}, -1, \frac{3}{2}\}$
27 $\{a/3, -2a, 2, -3\}$ **29** $x^2 + 4x - 1, 1$ **30** $x^2 - x + 2, 2$
31 $2x^2 + x + 1, 5$ **33** $5x^3 + 2x^2 - 11x + 9, -6$ **34** $7x^3 + x^2 + x - 4, 17$

35 $2x^3 + 2x^2 + 3x + 5, 12$ **37** $x^4 - 4x^3 - 3x^2 - x - 6, -5$
38 $x^4 - 2x^2 - 2x - 5, -24$ **39** $2x^5 + 7x^3 + x^2 - 3x + 4, -1$

Exercise 17.2, page 372

1 $\{-3, 1, 2\}$ **2** $\{-3, -2, 1\}$ **3** $\{-1, 2, 4\}$ **5** $\{1.3, -0.8, 2\}$
6 $\{-2.2, 0.7, -2\}$ **7** $\{1.5, 1.2, -0.5\}$ **9** $\{0.6, -1.6, 2.4, 0.4\}$
10 $\{0.4, -2.4, 0.6, -3.6\}$ **11** $\{1, -1.5, 0.6, -1.6\}$ **13** $5; 2, 2; -\frac{1}{3}, 3$
14 $6; -\frac{3}{2}, 4; -\frac{4}{3}, 2$ **15** $7; -3, 3; \frac{1}{2}, 2; -\frac{3}{4}, 2$ **17** $13; \frac{3}{2}, 6; -\frac{1}{3}, 4; -\frac{5}{2}, 3$
18 $11; -\frac{8}{7}, 1; -\frac{7}{8}, 7; \frac{1}{3}, 3$ **19** $12; -\frac{11}{3}, 3; \frac{4}{5}, 2; \frac{9}{8}, 7$ **21** $-2, 4; -2$ and $-1, 0$ and $1, 3$ and 4 **22** $-2, 4; -2$ and $-1, -1$ and $0, 3$ and 4
23 $-1, 6; -1$ and $0, 2$ and $3, 3$ and 4 **25** $-3, 3; -3$ and $-2, 0$ and $1, 1$ and 2
26 $-4, 1; -4$ and $-3, -1$ and $0, 0$ and 1 **27** $-5, 1; -5$ and $-4, -1$ and $0, 0$ and 1 **29** $-5, 3; -5$ and $-4, -1$ and $0, 0$ and $1, 2$ and 3
30 $-3, 5; -3$ and $-2, -1$ and $0, 0$ and $1, 4$ and 5 **31** $-8, 1; -5$ and $-4, -4$ and $-3, -1$ and $0, 0$ and 1

Exercise 17.3, page 376

1 $\{1, 2, -3\}$ **2** $\{-1, 2, -3\}$ **3** $\{-1, -2, 4\}$ **5** $\{-3, 2, \frac{1}{2}\}$
6 $\{-5, -2, \frac{2}{3}\}$ **7** $\{4, 3, -\frac{1}{2}\}$ **9** $\{-1, \frac{1}{2}, -\frac{2}{3}\}$ **10** $\{-2, -\frac{1}{3}, \frac{5}{2}\}$
11 $\{3, \frac{1}{2}, -\frac{3}{2}\}$ **13** $\{2, 1 - \sqrt{2}, 1 + \sqrt{2}\}$ **14** $\{-3, 1 - \sqrt{3}, 1 + \sqrt{3}\}$
15 $\{-1, (1 - \sqrt{5})/2, (1 + \sqrt{5})/2\}$ **17** $\{-\frac{3}{2}, i, -i\}$ **18** $\{-\frac{1}{2}, 2i, -2i\}$
19 $\{\frac{2}{7}, (-1 + \sqrt{3}i)/2, (-1 - \sqrt{3}i)/2\}$ **21** $\{-1, 2, -\frac{3}{2}, \frac{1}{3}\}$
22 $\{-3, 1, \frac{2}{3}, -\frac{3}{2}\}$ **23** $\{-2, -3, \frac{1}{4}, -\frac{1}{2}\}$ **25** $\{-2, 3, \sqrt{5}, -\sqrt{5}\}$
26 $\{-1, 2, (-1 + \sqrt{5})/2, (-1 - \sqrt{5})/2\}$
27 $\{\frac{3}{2}, 2, (-3 + \sqrt{21})/2, (-3 - \sqrt{21})/2\}$ **29** $\{\frac{1}{2}, -\frac{3}{2}, i, -i\}$
30 $\{-\frac{2}{3}, \frac{5}{2}, 2i, -2i\}$ **31** $\{2, -\frac{1}{2}, (1 + \sqrt{7}i)/4, (1 - \sqrt{7}i)/4\}$

Exercise 17.4, page 380

1 0.26 **2** 0.21 **3** 0.39 **5** 0.35 **6** 0.37 **7** 0.43 **9** 2.41
10 0.13 **11** 1.28 **13** -1.839 **14** -1.604 **15** -1.937
17 -1.414 **18** -0.866 **19** -1.236 **21** $-3.26, -1.34, 1.60$
22 $-4.77, -3.12, 1.88$ **23** $-0.26, 1.66, 4.60$
25 $2.41, -0.41, 1.37, -0.37$ **26** $3.41, -0.59, 0.79, 2.21$
27 $-2.73, 0.73, 1.37, -0.37$

Exercise 17.5, page 381

1 $x^2 - 4x + 3, -3$ **2** $2x^3 - 3x^2 - 9x - 1, 1$ **3** $4x^4 + x^3 + x^2 - 4x - 1, -2$
5 $1, 3; -\frac{3}{2}, 4; \frac{2}{3}, 1$ **9** $-2, 3; -2$ and $-1, 0$ and $1, 2$ and 3
10 $-4, 4; -3$ and $-2, -2$ and $-1, 0$ and $1, 3$ and 4 **11** $\{\frac{1}{2}, 1, \frac{3}{2}\}$
13 $\{-\frac{3}{2}, \frac{2}{3}, 1, 2i, -2i\}$ **14** 3.65 **15** 3.41

Exercise 18.1, page 387

1 372.4   **2** 83.29   **3** 0.7842   **5** 0.003123   **6** 0.0844   **7** 485.8
9 $1.235(10^4)$ **10** $3.721(10^5)$ **11** $8.754(10^4)$ **13** 2,359   **14** 5,794
15 $3.824(10^5)$ **17** $3.782(10^3)$ **18** $3.706(10)$ **19** $8.059(10^4)$
21 $3.52(10^{-2})$ **22** $5.968(10^{-1})$ **23** $4.78(10^{-4})$ **25** $3.780(10^3)$
26 $2.8500(10^4)$ **27** $6.000(10^3)$ **29** $3.78(10^3)$ **30** $2.85(10^4)$
31 $6.00(10^3)$ **33** $2.0(10^2)$ **34** $1.1(10)$ **35** $4.53(10^2)$
37 $7.90(10^3)$ **38** 3.04   **39** $5.7(10)$ **41** 3.6   **42** $1.6(10)$
43 $7.2(10)$ **45** $3.82(10)$ **46** 6.99   **47** $2.96(10)$ **49** $3.048(10)$
50 8.6   **51** $6.8(10^{-1})$

Exercise 18.2, page 391

1 $\log_5 25 = 2$ **2** $\log_2 32 = 5$ **3** $\log_3 81 = 4$ **5** $\log_3 \frac{1}{9} = -2$
6 $\log_4 \frac{1}{64} = -3$ **7** $\log_5 \frac{1}{625} = -4$ **9** $\log_{1/2} 8 = -3$ **10** $\log_{1/3} 3 = -1$
11 $\log_{1/5} 3,125 = -5$ **13** $\log_{49} 7 = \frac{1}{2}$ **14** $\log_{64} 4 = \frac{1}{3}$ **15** $\log_{64} 32 = \frac{5}{6}$
17 $5^2 = 25$ **18** $2^5 = 32$ **19** $7^3 = 343$ **21** $3^{-2} = \frac{1}{9}$ **22** $2^{-3} = \frac{1}{8}$
23 $5^{-4} = \frac{1}{625}$ **25** $4^{3/2} = 8$ **26** $27^{2/3} = 9$ **27** $16^{5/4} = 32$ **29** 4
30 2   **31** 3   **33** $\frac{3}{4}$ **34** $\frac{3}{4}$ **35** $\frac{2}{3}$ **37** 32   **38** 27   **39** 25
41 9   **42** 32   **43** 27   **45** 3   **46** 5   **47** 2   **49** 9   **50** 8
51 81   **53** 14   **54** 17   **55** 20   **57** 8   **58** 6   **59** 6
61 30   **62** 28   **63** 4

Exercise 18.3, page 396

1 2   **2** 3   **3** 1   **5** 1   **6** 0   **7** 1   **9** -1 **10** -3
11 -2 **13** -6 **14** -2 **15** -1 **17** 0.5752   **18** 0.9926
19 0.3118   **21** 1.5899   **22** 1.7007   **23** 1.8280   **25** 2.8500
26 2.7316   **27** 2.9330   **29** $9.5092 - 10$ **30** $9.8519 - 10$
31 $9.8351 - 10$ **33** $8.4871 - 10$ **34** $8.4409 - 10$ **35** $8.7752 - 10$
37 $7.7160 - 10$ **38** $6.3385 - 10$ **39** $7.9652 - 10$ **41** 3.4412
42 3.7663   **43** 3.8846   **45** 1.8911   **46** 2.2735   **47** 0.8198
49 $9.7576 - 10$ **50** $9.9396 - 10$ **51** $8.6995 - 10$ **53** 1.8935
54 1.8936   **55** 2.9070   **57** $8.0959 - 10$ **58** $9.5793 - 10$
59 $7.3247 - 10$ **61** 1.6974   **62** 3.9922   **63** 0.4496

Exercise 18.4, page 398

1 2.13   **2** 30.7   **3** 579   **5** 0.925   **6** 0.0178   **7** 0.00294
9 $(4.62)(10^3)$ **10** $(3.71)(10^4)$ **11** $(8.14)(10^6)$ **13** $(9.07)(10^{-3})$
14 $(1.18)(10^{-4})$ **15** $(6.63)(10^{-1})$ **17** 22.0   **18** 314   **19** 7.37
21 $6.00(10^3)$ **22** $2.06(10^5)$ **23** $1.55(10^4)$ **25** $6.39(10^{-1})$
26 $5.23(10^{-2})$ **27** $7.94(10^{-3})$ **29** 196.5   **30** 30.84   **31** 15.46

33 $5.976(10^4)$ **34** $6.321(10^6)$ **35** $4.774(10^5)$ **37** $5.027(10^{-1})$
38 $5.984(10^{-3})$ **39** $5.826(10^{-2})$ **41** 0.8010 **42** $4.511(10^{-3})$
43 $2.878(10^{-2})$

Exercise 18.5, page 401

1 $\log y - \log (b - y)$ **2** $\log x - \log (x + a)$ **3** $\log a + \log (x - a)$
5 $\log c + 2 \log x$ **6** $\frac{1}{2} \log (x + a) - 3 \log x$
7 $2 \log a + 3 \log x - \log b - \frac{1}{2} \log (x + b)$ **9** $\log a/b$ **10** $\log x/y$
11 $\log a (x + a)$ **13** $\log ab^2$ **14** $\log \sqrt{a}\, b^3$ **15** $\log \dfrac{a^2 b^3}{c\sqrt{x + a}}$
17 228   **18** 101   **19** 223   **21** 963   **22** 754   **23** 393
25 4.00   **26** 1.16   **27** 0.0666   **29** 1.70   **30** 1.47   **31** 0.386
33 7.81   **34** 0.684   **35** 2.75   **37** 52.6   **38** 19.2   **39** 2.14
41 4.48   **42** 1.45   **43** 3.38   **45** 1.970   **46** 0.9904   **47** 0.07255
49 4.314   **50** 3.785   **51** 4.059   **53** 0.03076   **54** 0.5783
55 71.70   **57** 1.184   **58** 1.703   **59** 2.176

Exercise 18.6, page 403

1 2.03   **2** 2.10   **3** 7.21   **5** 9.75   **6** 3.78   **7** 2.44   **9** 2.99
10 1.60   **11** 6.30   **13** 4.54   **14** 4.01   **15** 4.33   **17** 2.32
18 3.30   **19** 3.84   **21** 0.819   **22** 3.15   **23** 1.99

Exercise 18.7, page 405

1 $x = \log y$ **2** $x = \log (1/y)$ **3** $x = \log \sqrt{5/y}$ **5** $x = \log_e (y \pm \sqrt{y^2 - 1})$
6 $x = \log_e (y + \sqrt{y^2 + 1})$ **7** $x = (e^{-y} - e^y)/2$ **9** $\{4\}$ **10** $\{3\}$
11 $\{6\}$ **13** $\{2, -2\}$ **14** $\{4, -1\}$ **15** $\{1, 3\}$ **17** $\{5, 36\}$
18 $\{-0.31\}$ **19** $\{0.19\}$ **21** $\{2\}$ **22** $\{2\}$ **23** $\{6\}$ **25** $\{3\}$
26 $\{5\}$ **27** $\{-2\}$ **29** $\{0.31, 0.88\}$ **30** $\{0.42, 1.58\}$
31 $\{0.23, 0.035\}$

Exercise 18.9, page 408

1 $7.824(10^2)$ **2** $2.399(10^3)$ **3** $3.988(10)$ **5** $5.986(10^{-2})$
6 $8.768(10^{-3})$ **7** $2.2(10^2)$ **9** 6.57   **10** 72.12   **11** $\log_3 243 = 5$
13 5   **14** 3   **15** 216   **17** 2.358   **18** 2.7688   **19** 200
21 2.402   **22** 36.8   **23** 562   **25** 2.15   **26** 5.59   **27** -0.818
29 $\frac{1}{2} \log (1 + y)/(1 - y)$ **30** $\{3.29\}$ **31** $\{3.5\}$ **33** $(500, 20)$

Exercise 19.1, page 416

1 $A = 60°$, $a = 6$, $b = 6\sqrt{3}$ **2** $B = 30°$, $b = 18$, $a = 18\sqrt{3}$
3 $B = 30°$, $a = 16\sqrt{3}$, $c = 32$ **5** $A = 45°$, $b = 20$, $c = 20\sqrt{2}$
6 $A = 45°$, $a = 7\sqrt{2}$, $c = 98$ **7** $A = B = 45°$, $a = 17$
9 $B = 48°50'$, $a = 606$, $b = 693$ **10** $B = 57°30'$, $a = 4.37$, $b = 6.86$
11 $B = 16°20'$, $a = 250$, $c = 261$ **13** $A = 52°30'$, $b = 494$, $c = 812$
14 $A = 64°40'$, $b = 2.69$, $c = 6.28$ **15** $A = 28°30'$, $B = 61°30'$, $b = 5.88$
17 $A = 46°$, $B = 44°$, $a = 0.429$ **18** $A = 62°20'$, $B = 27°40'$, $a = 3.39$
19 $A = 39°50'$, $B = 50°10'$, $c = 82.3$ **21** $A = 52°48'$, $a = 5.689$, $b = 4.318$
22 $A = 32°26'$, $a = 3.740$, $b = 5.886$ **23** $B = 8°46'$, $a = 0.7799$, $b = 0.1203$
25 $A = 73°23'$, $b = 1,410$, $c = 4,930$ **26** $A = 18°14'$, $b = 40.85$, $c = 43.01$
27 $B = 37°7'$, $a = 7.537$, $c = 9.452$ **29** $A = 43°54'$, $B = 46°6'$, $a = 6.848$
30 $A = 57°31'$, $B = 32°29'$, $a = 3.162$ **31** $A = 39°55'$, $B = 50°5'$, $c = 88.48$
33 $B = 50°50'$, $b = 3.98(10^3)$, $c = 5.13(10^3)$
34 $B = 26°40'$, $b = 1.37(10^2)$, $c = 3.06(10^2)$
35 $A = 62°20'$, $b = 3.80(10^3)$, $a = 7.25(10^3)$
37 $A = 54°42'$, $B = 35°18'$, $c = 4.562(10^4)$
38 $A = 41°25'$, $B = 48°35'$, $c = 6.836(10^5)$
39 $A = 46°40'$, $B = 43°20'$, $b = 8.468(10^4)$

Exercise 19.2, page 420

1 164 meters   **2** 154 miles   **3** 13.6 feet   **5** 142 feet   **6** 24°
7 11.7 feet   **9** 3,101 feet   **10** 7.458 feet   **11** 2.805 feet
13 14.4 feet   **14** 19.9 feet   **15** 25.1 feet   **17** 810 yards S66°W
18 243 feet   **19** Train, 5.2 minutes   **21** 11.3 feet   **22** 453 square feet
23 21.6 miles   **25** 85°30', 25.5 miles per hour   **26** 627 miles, 299°30'
27 595 yards   **29** N38°E   **30** 188°50', 198 miles per hour
31 2,149 feet

Exercise 19.3, page 422

1 $a = 18\sqrt{3}$, $c = 36$, $B = 30°$ **2** $b = 14.4$, $c = 18.3$, $B = 51°50'$
3 $a = 46.2$, $b = 18.0$, $B = 21°20'$ **5** $b = 43.4$, $A = 42°20'$, $B = 47°40'$
6 $b = 1.99(10^3)$, $c = 3.07(10^3)$, $A = 49°40'$
7 $a = 3.92(10^4)$, $b = 1.71(10^4)$, $A = 66°30'$
9 207 meters, $A = 51°40'$ **10** 302 miles west, 364 miles north

Exercise 20.1, page 428

1 $C = 80°10'$, $b = 293$, $c = 329$, $K = 2.99(10^4)$ **2** $C = 85°10'$, $b = 115$, $c = 452$, $K = 2.45(10^4)$ **3** $C = 77°50'$, $a = 0.150$, $c = 0.165$, $K = 7.92(10^{-3})$

5 $B = 39°20'$, $a = 49.2$, $c = 42.5$, $K = 663$ **6** $B = 78°0'$, $a = 4.54(10^{-2})$, $c = 5.02(10^{-2})$, $K = 1.12(10^{-3})$ **7** $B = 87°10'$, $b = 4.22$, $c = 2.09$, $K = 3.94$
9 $A = 99°10'$, $a = 0.450$, $b = 0.284$, $K = 4.30(10^{-2})$ **10** $A = 95°10'$, $a = 214$, $b = 174$, $K = 9.47(10^3)$ **11** $A = 50°10'$, $a = 17.9$, $b = 22.9$, $K = 98.5$
13 $C = 82°16'$, $b = 3,137$, $c = 3,596$, $K = 3.468(10^6)$ **14** $C = 67°44'$, $a = 58.91$, $c = 57.13$, $K = 1,074$ **15** $C = 101°43'$, $a = 0.4500$, $b = 0.6036$, $K = 0.1330$
17 $B = 64°59'$, $a = 2.482(10^{-2})$, $c = 1.620(10^{-2})$, $K = 1.822(10^{-4})$
18 $A = 78°38'$, $a = 856.2$, $b = 617.9$, $K = 2.229(10^5)$ **19** $A = 103°57'$, $b = 4.579$, $c = 4.896$, $K = 10.88$ **21** 229 feet
22 21.7 centimeters, 29.3 centimeters   **23** 1.41 kilometers, 1.64 kilometers

Exercise 20.2, page 432

1 One   **2** Two   **3** Two   **5** None   **6** None   **7** One
9 $A = 77°50'$, $C = 52°$, $a = 0.989$, $K = 0.303$, $A' = 1°50'$, $C' = 128°$, $a' = 0.0324$, $K' = 0.00992$ **10** None   **11** $A = 137°20'$, $B = 14°20'$, $a = 33.7$, $K = 98.4$
13 9.0   **14** 25   **15** 1.2   **17** 7.3   **18** 91   **19** 1.09
21 $A = 68°$, $K = 818$ **22** $B = 79°10'$, $K = 159$ **23** $C = 89°$, $K = 635$,
25 $A = 42°30'$, $K = 103$ **26** $A = 41°$, $K = 219$ **27** $A = 49°30'$, $K = 436$
29 47 feet   **30** 561 pounds   **31** 273 miles per hour, 229°

Exercise 20.3, page 435

1 $A = 48°$, $B = 84°$, $C = 17$, $K = 1.5(10^3)$ **2** $A = 32°$, $B = 110°$, $c = 2.4(10^2)$, $K = 2.4(10^2)$ **3** $A = 42°$, $C = 62°$, $b = 62$, $K = 1.2(10^3)$ **5** $B = 59°30'$, $C = 81°50'$, $a = 147$, $K = 1.97(10^4)$ **6** $B = 51°10'$, $C = 64°30'$, $a = 606$, $K = 1.43(10^5)$ **7** $B = 54°50'$, $C = 37°50'$, $a = 871$, $K = 1.91(10^5)$
9 $A = 53°10'$, $B = 78°10'$, $c = 311$, $K = 5.03(10^4)$ **10** $B = 74°20'$, $A = 46°20'$, $c = 324$, $K = 4.28(10^4)$ **11** $A = 28°10'$, $B = 37°30'$, $c = 355$, $K = 1.99(10^4)$
13 $A = 41°$, $C = 66°24'$, $b = 3431$, $K = 3.708(10^6)$ **14** $A = 44°19'$, $C = 34°27'$, $b = 6,314$, $K = 9.086(10^6)$ **15** $A = 33°57'$, $C = 62°51'$, $b = 1.062(10^4)$, $K = 2.828(10^7)$ **17** 10 miles per hour, 111°   **18** 311 miles per hour, 213°40'   **19** 1.9(10^3)

Exercise 20.4, page 438

1 $A = 76°$, $B = 54°$, $C = 50°$, $K = 1.7(10^2)$ **2** $A = 48°$, $B = 72°$, $C = 60°$, $K = 0.17$ **3** $A = 132°$, $B = 26°$, $C = 20°$, $K = 8.23$ **5** $A = 78°40'$, $B = 36°40'$, $C = 64°40'$, $K = 3.81(10^4)$ **6** $A = 77°20'$, $B = 57°$, $C = 45°40'$, $K = 2.57(10^3)$ **7** $A = 57°20'$, $B = 75°20'$, $C = 47°$, $K = 3.71$ **9** $A = 23°20'$, $B = 52°20'$, $C = 104°20'$, $K = 1.47(10^{-4})$ **10** $A = 76°$, $B = 50°20'$, $C = 54°$, $K = 383$ **11** $A = 103°20'$, $B = 47°$, $C = 29°40'$, $K = 1.45(10^5)$
13 $A = 60°40'$, $B = 77°20'$, $C = 42°$, $K = 0.1255$ **14** $A = 78°6'$, $B = 50°8'$, $C = 51°44'$, $K = 1,790$ **15** $A = 64°12'$, $B = 61°26'$, $C = 54°20'$, $K = 3.209(10^7)$ **17** 8.6 feet   **18** 45 cubic feet   **19** 82°

Exercise 20.5, page 439

1 $a = 293$, $c = 567$, $C = 74°40'$, $K = 8.07(10^4)$ **2** $a = 240$, $c = 316$, $B = 102°50'$, $K = 3.69(10^4)$ **3** $B = 57°$, $C = 86°$, $K = 157$; $B' = 123°$, $C' = 20°$, $K' = 53.9$, $c = 25$, $c' = 8.5$ **5** $A = 46°20'$, $C = 86°10'$, $c = 341$ **6** $A = 39°$ **7** $C = 101°$ **9** $a = 6.0$ **10** $C = 74°$ **11** $B = 141°40'$ **13** $32°30'$ **14** $17°$

Exercise 21.1, page 445

1 $3, 5, 7, 9, 11, 13$ **2** $11, 8, 5, 2, -1, -4, -7$ **3** $-4, -1, 2, 5, 8$ **5** $l = 14$, $s = 40$ **6** $l = -1$, $s = 56$ **7** $s = 21$, $d = -1$ **9** $a = 17$, $d = -2$ **10** $a = 1$, $d = 4$ **11** $d = 3$, $n = 6$ **13** $n = 6$, $s = 72$ **14** $n = 6$, $s = 12$ **15** $d = 2$, $l = 15$ **17** $a = 12$, $s = -21$ **18** $a = 11$, $s = 14$ **19** $a = 1$, $d = 1$ **21** $l = -13$, $n = 7$ **22** $l = -7$, $n = 7$ **23** $a = 22$, $n = 7$ **25** $a = 19$, $n = 7$ **26** $a = 2$, $n = 6$ **27** $a = 37$, $n = 8$ **29** 432 **30** 196 **31** 272 feet, $1,296$ feet **33** $98, 80.5$ **34** 159 **35** 17 **37** 2 **38** $x = 9$, $y = 27$ **39** 3

Exercise 21.2, page 450

1 $2, 6, 18, 54, 162, 486$ **2** $1, -3, 9, -27, 81, -243, 729$ **3** $2, \pm 4, 8, 16, 32$ **5** $l = 81$, $s = 121$ **6** $l = 64$, $s = 43$ **7** $r = \frac{1}{2}$, $s = \frac{31}{2}$ **9** $a = 5$, $r = -1$ **10** $a = 343$, $r = \frac{1}{7}$ **11** $n = 5$, $r = \frac{1}{4}$ **13** $a = 2$, $s = 242$ **14** $l = 64$, $s = 127$ **15** $l = -\frac{1}{16}$, $a = 1$ **17** $n = 5$, $l = 1$ **18** $l = 3,125$, $n = 6$ **19** $a = 128$, $n = 8$ **21** 26th **22** 2, $\$5, 120$ **23** $1,715.06$ **25** $\$672.28$ **26** 126 **27** $\$64,000$

Exercise 21.3, page 455

1 12 **2** 12.5 **3** 6 **5** $\frac{7}{9}$ **6** $\frac{36}{11}$ **7** $\frac{20}{11}$ **9** 1.5 **10** 8 **11** 2 **13** 3 **14** $\frac{2}{3}$ **15** 2.5 **17** 35 **18** $\$690,000$ **19** $n < 9$ **21** $5, 8, 11$ **22** $8, 13, 18, 23$ **23** 8 **25** ± 1 **26** ± 32 **27** $1, 3, 9, 27$ **29** $\frac{1}{3}, \frac{1}{5}, \frac{1}{7}$ **30** $-1, 1, \frac{1}{3}, \frac{1}{5}$ **31** $\frac{1}{11}, \frac{1}{19}$ **33** Harmonic, $\frac{1}{9}, \frac{1}{11}$ **34** Geometric, $\frac{1}{2}, \frac{1}{4}$ **35** Arithmetic, $-3, -7$ **37** Geometric, $162, 486$ **38** Arithmetic, $-7, -15$ **39** Geometric, $\frac{1}{81}, \frac{1}{243}$

Exercise 21.4, page 456

1 $3, 1, -1, -3, -5, -7, -9$ **2** $2, 4, 8, 16, 32, 64$ **3** $\frac{1}{6}, \frac{1}{10}, \frac{1}{14}$ **5** $-9, -9$ **6** $n = 5$, $l = -5$ **7** $r = \pm \frac{1}{3}$, $s = \frac{484}{9}, \frac{244}{9}$ **9** 4 **10** -1 **11** $3, -\frac{3}{2}$ **13** $\{x | x > 0\} \cup \{x | x < -\frac{4}{3}\}$ **14** $1, 4, 7; 2$ **15** $0, \frac{1}{3}, 1$

Exercise 23.1, page 468

1 $a^8 + 8a^7b + 28a^6b^2 + 56a^5b^3 + 70a^4b^4 + 56a^3b^5 + 28a^2b^6 + 8ab^7 + b^8$

2 $a^7 - 7a^6x + 21a^5x^2 - 35a^4x^3 + 35a^3x^4 - 21a^2x^5 + 7ax^6 - a^7$

3 $b^5 - 5b^4y + 10b^3y^2 - 10b^2y^3 + 5by^4 - y^5$

5 $a^5 - 10a^4y + 40a^3y^2 - 80a^2y^3 + 80ay^4 - 32y^5$

6 $x^6 + 18x^5b + 135x^4b^2 + 540x^3b^3 + 1{,}215x^2b^4 + 1{,}458xb^5 + 729b^6$

7 $128b^7 + 448b^6x + 672b^5x^2 + 560b^4x^3 + 280b^3x^4 + 84b^2x^5 + 14bx^6 + x^7$

9 $8a^3 - 36a^2b^2 + 54ab^4 - 27b^6$

10 $243x^5 - 810x^4b^3 + 1{,}080x^3b^6 - 720x^2b^9 + 240xb^{12} - 32b^{15}$

11 $64b^{12} + 576b^{10}x + 2{,}160b^8x^2 + 4{,}320b^6x^3 + 4{,}860b^4x^4 + 2{,}916b^2x^5 + 729x^6$

13 $a^{33} + 33a^{32}y + 528a^{31}y^2 + 5{,}456a^{30}y^3$

14 $x^{51} - 51x^{50}y + 1{,}275x^{49}y^2 - 20{,}825x^{48}y^3$

15 $m^{101} - 202m^{100}y + 20{,}200m^{99}y^2 - 1{,}333{,}200m^{98}y^3$

17 1.2167 **18** 1.2155 **19** 1.1941 **21** $560x^3y^4$ **22** $-160a^3c^3$

23 $10{,}206x^4y^5$ **25** $-35a$ **26** $-2{,}016x^{-6}$ **27** $29{,}568x^{10}y^6$

29 $160x^3y^{3/2}$ **30** $5{,}670x^4y$ **31** $960x^7y^3$

Exercise 23.2, page 471

1 $x^{-3} - 3x^{-4}y + 6x^{-5}y^2 - 10x^{-6}y^3$ **2** $a^{-2} - 2a^{-3}b + 3a^{-4}b^2 - 4a^{-5}b^3$

3 $a^{-5} + 5a^{-6}y + 15a^{-7}y^2 + 35a^{-8}y^3$ **5** $a^{-1}/2 + a^{-2}y/4 + a^{-3}y^2/8 + a^{-4}y^3/16$

6 $x^{-3} + 6x^{-4}y + 24x^{-5}y^2 + 80x^{-6}y^3$ **7** $x^{-2}/9 - 4x^{-3}y/27 + 4x^{-4}y^2/27 - 32x^{-5}y^3/243$

9 $x^{-4} - 4x^{-6} + 10x^{-8} - 20x^{-10}$ **10** $x^{-2} + 2x^{-5} + 4x^{-8} + 8x^{-11}$

11 $x^8 + 12x^{11} + 90x^{14} + 540x^{17}$ **13** $y^{1/3} + y^{-2/3}/3 - y^{-5/3}/9 - 5y^{-8/3}/81$

14 $1 + y/3 - y^2/9 + 5y^3/81$ **15** $5 + x/10 - x^2/1{,}000 - x^3/50{,}000$

17 $\frac{1}{2} - x/128 + 5x^2/16{,}384 - 15x^3/1{,}048{,}576, \; -16 < x < 16$

18 $x^{-1/2} - 9x^{-3/2}/2 + 243x^{-5/2}/8 - 3645x^{-7/2}/16, \; x > 9$

19 $\frac{1}{2} - x/4 + x^2/8 - x^3/16, \; -2 < x < 2$ **21** 10.0499 **22** 9.9967

23 3.0366 **25** 0.8219 **26** 0.7921 **27** 0.8163

Exercise 24.1, page 475

1 $\pi/3, 5\pi/3$ **2** $\pi/3, 2\pi/3$ **3** $\pi/3, 4\pi/3$ **5** $\pi/6, 5\pi/6, 7\pi/6, 11\pi/6$

6 $\pi/3, 2\pi/3, 4\pi/3, 5\pi/3$ **7** $\pi/6, 5\pi/6$ **9** $\pi/3, 5\pi/3$ **10** $\pi/6, 11\pi/6$

11 $3\pi/2, \pi/6, 5\pi/6$ **13** $0, \pi, \pi/6, 5\pi/6, 7\pi/6, 11\pi/6$

14 $\pi/2, 3\pi/2, \pi/6, 5\pi/6, 7\pi/6, 11\pi/6$ **15** $\pi/6, 7\pi/6$

17 $\pi/2, 3\pi/2, \pi/6, 7\pi/6$ **18** $0, \pi, 2\pi/3, 4\pi/3$ **19** $3\pi/2, \pi/3, 5\pi/3$

21 $\pi/6, 5\pi/6, 7\pi/6, 11\pi/6$ **22** $\pi/6, \pi/3, 7\pi/6, 5\pi/3$

23 $2\pi/9, 4\pi/9, 8\pi/9, 10\pi/9, 14\pi/9, 16\pi/9$ **25** $0, \pi/12, 5\pi/12, 13\pi/12, 17\pi/12$

26 $3\pi/2, \pi/6, 2\pi/3, 7\pi/6, 5\pi/3$

27 $\pi/3, 2\pi/3, 2\pi/9, 5\pi/9, 8\pi/9, 11\pi/9, 14\pi/9, 17\pi/9$

Exercise 24.2, page 478

1 $0, \pi/3, 5\pi/3$ **2** $\pi/6, 5\pi/6$ **3** $\pi/3, 2\pi/3, 4\pi/3, 5\pi/3$
5 $\pi/2, 4\pi/3, 5\pi/3$ **6** $\pi/6, 5\pi/6$ **7** $\pi/4, 5\pi/4, 153°26', 333°26'$
9 $\pi/6, 5\pi/6, 7\pi/6, 11\pi/6$ **10** None **11** $\pi/2, 3\pi/2$ **13** None
14 $\pi, 3\pi/2$ **15** $\pi/4, \pi/2, 3\pi/4, 5\pi/4, 3\pi/2, 7\pi/4$ **17** $\pi/2, 3\pi/2, \pi/6, 5\pi/6$
18 $0, \pi, \pi/3, 5\pi/3$ **19** $0, \pi$ **21** $0, \pi/2, \pi, 3\pi/2$
22 $\pi/4, \pi/2, 5\pi/4, 7\pi/4, 3\pi/2, 3\pi/4$ **23** $0, \pi, \pi/4, 3\pi/4, 5\pi/4, 7\pi/4$
25 $117°20', 169°0'$ **26** $7\pi/12, 23\pi/12$ **27** $68°10', 337°10'$

Exercise 25.1, page 485

1 $\pi/3$ **2** $\pi/2$ **3** $\pi/6$ **5** $3\pi/4$ **6** $-\pi/6$ **7** $\pi/4$ **9** 0.43
10 $1/1.7$ **11** $1/0.59$ **13** 2 **14** $\sqrt{3}/2$ **15** $\sqrt{3}/2$ **17** $7/\sqrt{51}$
18 $1/\sqrt{0.91}$ **19** $\sqrt{74}/5$ **21** u **22** u **23** $-1/u$ **25** $(2 - u^2)/u^2$
26 $u^2/2\sqrt{u^2 - 1}$ **27** $2\sqrt{u^2 - 1}/(2 - u^2)$ **29** $-1/u$ **30** $-1/u$
31 $-u$ **33** $\sqrt{2u(u + 1)}/2u$ **34** $\sqrt{2u(u - 1)}/2u$
35 $\sqrt{2u(u + \sqrt{u^2 - 1})}/2u$ **37** $(\sqrt{v^2 - 1} - \sqrt{u^2 - 1})/uv$
38 $(1 + \sqrt{u^2 - 1}\,\sqrt{v^2 - 1})/uv$ **39** $(u + \sqrt{v^2 - 1})/(1 - u\sqrt{v^2 - 1})$

Exercise 26.1, page 494

1 $9{,}000{,}000$ **2** $9(9!)$ **3** 10^9 **5** $1{,}620$ **6** $3{,}584$ **7** 24
9 $6{,}720$ **10** 336 **11** $151{,}200$ **17** $43{,}120$ **18** $3{,}668{,}800$
19 $90{,}720$ **21** $362{,}880$ **22** $3{,}628{,}800$ **23** $39{,}916{,}800$ **25** $151{,}200$
26 144 **27** $8{,}640$ **29** 720 **30** $103{,}686$ **31** $15!$

Exercise 26.2, page 498

5 $15{,}504$ **6** 45 **7** 188 **9** $52!/39!13!$ **10** $270{,}725$ **11** 364
13 $1{,}140$ **14** $6{,}790$ **15** $4{,}032$ **17** 36 **18** $387{,}600$
19 $77{,}647{,}500$ **21** 63 **22** 56 **23** 55 **25** 37 **26** $1{,}048{,}555$
27 121

Exercise 27.1, page 505

1 $\frac{1}{4}, \frac{1}{2}, \frac{1}{26}$ **2** $\frac{4}{13}, \frac{5}{13}$ **3** $\frac{1}{4}, \frac{1}{2}, \frac{1}{26}$ **5** $\frac{8}{81}$ **6** $\frac{41}{81}$ **7** $\frac{1}{9}$ **9** $\frac{1}{2}$
10 $\frac{1}{6}, \frac{1}{2}$ **11** $\frac{1}{6}, \frac{1}{2}, \frac{1}{3}$ **13** $\frac{1}{36}, \frac{1}{18}, \frac{1}{12}$ **14** $\frac{1}{6}, \frac{5}{36}$ **15** $\frac{1}{5,525}$
17 $\frac{1}{10}, \frac{1}{5}$ **18** $\frac{5}{51}, \frac{35}{153}$ **19** $\frac{1}{204}, \frac{125}{816}$ **21** $\$265$ **22** $\$70$ **23** $\$50$
25 $\frac{7}{10}$ **26** $\frac{5}{12}$ **27** $\frac{61}{84}, \frac{23}{84}$ **29** $\frac{19}{30}$ **30** $\frac{11}{36}$ **31** $\frac{29}{36}$

Exercise 27.2, page 509

1 0.15, 0.05 **2** $\frac{3}{16}$ **3** Tom, $\frac{3}{32}$; Dick, $\frac{27}{160}$, Harry, $\frac{3}{80}$ **5** $\frac{7}{66}$
6 $\frac{1}{2}, \frac{1}{6}$ **7** $\frac{1}{55}$ **9** $\frac{9}{25}$ **10** $\frac{1}{12}$ **11** $\frac{7}{15}$ **13** $\frac{1}{16}, \frac{1}{192}$ **14** 0.7
15 $\frac{1}{15}$ **17** $\frac{35}{96}, \frac{5}{32}, \frac{25}{48}$ **18** $\frac{14}{15}$ **19** $\frac{1}{1,296}$ **21** $\frac{1}{2}, \frac{1}{6}$ **22** $\frac{1}{55}$
23 $\frac{14}{55}$ **25** $\frac{1}{36}$ **26** $\frac{7}{24}$ **27** $\frac{29}{1,296}$

Exercise 28.1, page 519

5 $(7, -240°), (7, 120°), (-7, 300°)$ **6** $(-5, 225°), (5, 45°), (5, -315°)$
7 $(4, 210°), (4, -150°), (-4, -330°)$

In problems 9 to 47, the intercept points are given and are followed by a semicolon, then the axes and point of symmetry, if any, are given.

9 $(3, 0), (3, 90°), (3, 180°), (3, 270°)$; both axes, the pole
10 $(5, 0), (5, 90°), (5, 180°), (5, 270°)$; both axes, the pole
11 $(4, 0), (4, 90°), (4, 180°), (4, 270°)$; both axes, the pole
13 Pole; pole **14** Pole; pole **15** Pole; pole **17** $(2, 90°)$; normal axis **18** $(0, 90°), (-3, 0)$; polar axis **19** $(4, 0°)$; polar axis
21 $(3, 0), (0, 90°)$; polar axis **22** $(-4, 0), (0, 90°)$; polar axis **23** $(0, 0)$, $(-2, 90°)$; normal axis **25** $(3, 0), (-3, 90°), (3, 180°), (-3, 270°)$; polar axis, normal axis, pole **26** $(0, 0), (0, 90°), (0, 180°), (0, 270°)$; pole
27 $(0, 0), (-2, 90°), (0, 180°), (2, 270°)$ **29** $(0, 4), (2, 90°), (0, 180°)$, $(2, 270°)$; polar axis **30** $(0, 0), (3, 90°), (-6, 180°), (-3, 270°)$; polar axis
31 $(-4, 0), (0, 90°), (-4, 180°), (-8, 270°)$; normal axis
33 $(1, 0), (3, 90°), (1, 180°), (-1, 270°)$; normal axis
34 $(3, 0), (5, 90°), (3, 180°), (1, 270°)$; normal axis
35 $(5, 0), (3, 90°), (1, 180°), (3, 270°)$; polar axis
37 $(2, 90°), (1, 180°), (2, 270°)$; polar axis **38** $(3, 0), (\frac{3}{2}, 90°), (3, 180°)$; normal axis **39** $(\frac{4}{3}, 0), (2, 90°), (4, 180°), (2, 270°)$; polar axis
41 $(4, 0), (\frac{4}{3}, 90°), (4, 180°), (-4, 270°)$; normal axis
42 $(-\frac{5}{2}, 0), (5, 90°), (\frac{5}{4}, 180°), (5, 270°)$; polar axis
43 $(\frac{2}{3}, 0), (\frac{1}{2}, 90°), (\frac{2}{5}, 180°), (\frac{1}{2}, 270°)$; polar axis
45 $(0, 0), (2, 90°), (-2, 90°), (0, 180°)$; normal axis
46 $(3, 0), (-3, 0), (0, 90°), (0, 270°)$; polar axis
47 $(4, 0), (0, 90°), (4, 180°), (0, 270°)$; polar axis, pole, normal axis

Exercise 28.2, page 523

Problems 21 to 39 are the polar forms of the equations given in rectangular form in problems 1 to 19.

Index

$$\cos\left(\frac{\pi}{2} - B\right) = \sin B \tag{10.2}$$

$$\sin\left(\frac{\pi}{2} - B\right) = \cos B \tag{10.3}$$

$$\cos(A + B) = \cos A \cos B - \sin A \sin B \tag{10.4}$$

$$\cos 2A = \cos^2 A - \sin^2 A \tag{10.5a}$$

$$= 2\cos^2 A - 1 \tag{10.5b}$$

$$= 1 - 2\sin^2 A \tag{10.5c}$$

$$\cos\tfrac{1}{2}\theta = \pm\sqrt{\frac{1 + \cos\theta}{2}} \tag{10.6}$$

$$\sin(A + B) = \sin A \cos B + \cos A \sin B \tag{10.7}$$

$$\sin 2A = 2\sin A \cos A \tag{10.8}$$

$$\sin\tfrac{1}{2}\theta = \pm\sqrt{\frac{1 - \cos\theta}{2}} \tag{10.9}$$

$$\sin(A - C) = \sin A \cos C - \cos A \sin C \tag{10.10}$$

$$\tan(A + B) = \frac{\tan A + \tan B}{1 - \tan A \tan B} \tag{10.11}$$

$$\tan 2A = \frac{2\tan A}{1 - \tan^2 A} \tag{10.12}$$

$$\tan\tfrac{1}{2}\theta = \frac{1 - \cos\theta}{\sin\theta} \tag{10.13a}$$

$$\tan\tfrac{1}{2}\theta = \frac{\sin\theta}{1 + \cos\theta} \tag{10.13b}$$

$$\tan(A - C) = \frac{\tan A - \tan C}{1 + \tan A \tan C} \tag{10.14}$$

If $(x + d)^2 = k$, then $x + d = \pm\sqrt{k}$ (11.1)

$$\pm\sqrt{-n} = \pm\sqrt{n}\,i,\ n > 0 \tag{11.2}$$

If $ax^2 + bx + c = 0$, then

$$x = \frac{-b \pm \sqrt{b^2 - 4ac}}{2a} \tag{11.5}$$

$$r = \frac{-b + \sqrt{D}}{2a},\ s = \frac{-b - \sqrt{D}}{2a}$$

$$\text{if } D = b^2 - 4ac \tag{11.7}$$

$$r + s = \frac{-b}{a} \tag{11.8}$$

$$rs = \frac{c}{a} \tag{11.9}$$

If $a = b$ and $c = d$, then $a \pm c = b \pm d$ (12.1)

If $\dfrac{a}{b} = \dfrac{c}{d}$, then $ad = bc$ (14.2)

If $\dfrac{a}{b} = \dfrac{c}{d}$, then $\dfrac{b}{a} = \dfrac{d}{c}$ and $\dfrac{a}{c} = \dfrac{b}{d}$ (14.3)

If $\dfrac{a}{b} = \dfrac{c}{d}$, then $\dfrac{a \pm b}{b} = \dfrac{c \pm d}{d}$ (14.4)

If $\dfrac{a}{b} = \dfrac{c}{d}$, then $\dfrac{a + b}{a - b} = \dfrac{c + d}{c - d}$ (14.5)

The mean proportionals to a and b are $\pm\sqrt{ab}$ (14.6)

If $a{:}b{:}c = x{:}y{:}z$, then

$$\frac{a + b + c}{x + y + z} = \frac{a}{x} = \frac{b}{y} = \frac{c}{z} \tag{14.7}$$

If $f(\theta + p) = f(\theta)$ for all θ, then $f(\theta)$ is periodic with period p (15.1)

$(a, b) = (c, d)$ if and only if $a = c$ and $b = d$ (16.1)

$(a, b) + (c, d) = (a + c, b + d)$ (16.2)

$(a, b)(c, d) = (ac - bd, ad + bc)$ (16.3)

(a, b) and $a + bi$ are two forms of the same complex number (16.5)

If c is real, then $(a, b)c = (ac, bc)$ (16.3a)

$(a, b) - (c, d) = (a - c, b - d)$ (16.6)

$$\frac{(a, b)}{(c, d)} = \left(\frac{ac + bd}{c^2 + d^2}, \frac{bc - ad}{c^2 + d^2}\right) \tag{16.7}$$

If $\mathbf{v}$ is a vector or complex number, then $|\mathbf{v}| = r = \sqrt{x^2 + y^2}$ (16.8)

$$\theta = \arctan\frac{y}{x} \tag{16.9}$$

$$z = r(\cos\theta + i\sin\theta) \tag{16.10}$$

$$(r\operatorname{cis}\theta)(R\operatorname{cis}\phi) = rR\operatorname{cis}(\theta + \phi) \tag{16.11}$$

$$\frac{r\operatorname{cis}\theta}{R\operatorname{cis}\phi} = \frac{r}{R}\operatorname{cis}(\theta - \phi) \tag{16.12}$$

$$(r\operatorname{cis}\theta)^n = r^n\operatorname{cis}n\theta \tag{16.13}$$

$\log_b N = L$ if and only if $b^L = N$ (18.1)

$\log_b MN = \log_b M + \log_b N$ (18.2)